— PRAISE FOR —

Herbal Formularies for Health Professionals

"Jill Stansbury's *Herbal Formularies for Health Professionals* reflect not only the maturity of contemporary biomedical herbal medicine but also the mastery gained from years of clinical practice. Throughout the text, a thoughtful integration of traditional context is blended with science-principled herbalism. If you practice herbal medicine, or want to dive in deeply with a master in the field, these books will be in a prized place on your shelves."

—BEVIN CLARE, professor, Maryland University of Integrative Health; president, American Herbalists Guild

"Over the five volumes of *Herbal Formularies for Health Professionals*, Jill Stansbury has shown us what a remarkable practitioner she is, presenting nuanced therapeutic protocols for a wide array of conditions that exemplify cutting edge medical knowledge and are firmly rooted in a deep understanding of plants and traditional herbalism. Amazingly, all of this is done in a highly accessible manner that conveys the personal knowledge that Jill has gained over decades of practice. There is, quite frankly, nothing else like it, and no one else like Jill."

—JIM MCDONALD, herbalist and founder, herbcraft.org

"*Herbal Formularies for Health Professionals* is an essential reference set for all practitioners who use, or aspire to use, botanical medicine. The coverage is comprehensive, from the first volume on digestion and elimination through to this ultimate volume on immune, EENT, and musculoskeletal conditions. I appreciate as well the book's opening chapter on the art of formulation, which provides an excellent description of how to best address the precise presentation of a patient, rather than a general diagnosis or condition."

—LINNEA WARDWELL, cofounder, Herbal Educational Services, BotanicalMedicine.org

"Jill Stansbury is a quintessential medicine woman in today's world. *Herbal Formularies for Health Professionals* are the texts my clinic students have been waiting for."

—KAT MAIER, author of *Energetic Herbalism*, director, Sacred Plant Traditions

"Dr. Stansbury is a prescient, loving steward of the Standing Green Nation. Deeply loved by herbalists globally, her mentoring and clinical mastery powerfully manifest in this set of formularies. The volumes themselves are beautiful objects whose rich content and elegant design are exemplary and respectful of this hard-earned, cumulative wisdom."

—DAVID J. SCHLEICH, PhD, president, National University of Natural Medicine

"There are many herb books, but few address the fine points of clinical herbal prescribing. To remedy this deficiency, Dr. Jill Stansbury, ND, has written perhaps the best herbal formulary since the days of the renowned Eclectic physicians. . . . Each section and page clearly shows the skill, experience, and expertise of the author and will help even the seasoned naturopathic physician or herbalist improve their clinical practice and success in treating their patients."

—DAVID WINSTON, RH (AHG), dean, David Winston's Center for Herbal Studies; founder, Herbal Therapeutics Research Library

HERBAL FORMULARIES FOR HEALTH PROFESSIONALS

VOLUME 5

HERBAL FORMULARIES FOR HEALTH PROFESSIONALS

VOLUME 5

IMMUNOLOGY, ORTHOPEDICS, AND OTOLARYNGOLOGY

INCLUDING ALLERGIES, THE IMMUNE SYSTEM, THE MUSCULOSKELETAL SYSTEM, AND THE EYES, EARS, NOSE, MOUTH, AND THROAT

DR. JILL STANSBURY, ND

Chelsea Green Publishing
White River Junction, Vermont
London, UK

Project Manager: Patricia Stone
Editor: Fern Marshall Bradley
Assistant Editor: Natalie Wallace
Editorial Assistance: Alexandra Loch-Mally
Copy Editor: Deborah Heimann
Proofreader: Diane Durrett
Indexer: Shana Milkie
Designer: Melissa Jacobson

Printed in the United States of America.
First printing September 2021.
10 9 8 7 6 5 4 3   24 25 26 27 28

ISBN: 978-1-60358-857-7

Library of Congress Cataloging-in-Publication Data
Names: Stansbury, Jill, author.
Title: Herbal formularies for health professionals. Volume I, Digestion and elimination, including the gastrointestinal system, liver and gallbladder, urinary system, and the skin / Dr. Jill Stansbury.
Other titles: Digestion and elimination, including the gastrointestinal system, liver and gallbladder, urinary system, and the skin
Description: White River Junction, Vermont : Chelsea Green Publishing, [2017] | Includes bibliographical references and index.
Identifiers: LCCN 2017044410 | ISBN 9781603587075 (hardcover) | ISBN 9781603587082 (ebook)
Subjects: | MESH: Formularies as Topic | Phytotherapy—methods | Digestive System Diseases—drug therapy | Urologic Diseases—drug therapy | Skin Diseases—drug therapy
Classification: LCC RM666.H33 | NLM QV 740.1 | DDC 615.3/21—dc23
LC record available at https://lccn.loc.gov/2017044410

Chelsea Green Publishing
White River Junction, Vermont USA
London, UK

www.chelseagreen.com

To the Standing Green Nation—
all plants everywhere as living, breathing, sentient beings.
May we learn to treat them with respect and care.

And to all the traditional peoples, cultures, and nations
who have lovingly entered into sacred relationship with
the earth and the plants, and contributed to the art
and craft of herbal wisdom shared in these pages.

— CONTENTS —

Suppression of Natural Medicine

The five volumes of *Herbal Formularies for Health Professionals* explore a wide range of folkloric traditions regarding herbal medicines and cite thousands of modern molecular research studies, cell culture studies, and clinical trials on the effects of herbs on neurotransmitters, hormone receptors, gene regulation, and inflammatory enzymes. Research studies can provide valuable information, but they can also be problematic. In the introduction to Volume 3, I focused on issues that impact research investigations into natural products, including bias, barriers to publishing, and "spin"—all of which occur in even the most respected mainstream scientific medical journals. I feel compelled to return to this subject because of the recent alarming increase in the suppression of natural medicine. In 2009, Marcia Angell, a US physician and the first female editor-in-chief of *The New England Journal of Medicine*, reported on corruption in medical research, writing, "It is simply no longer possible to believe much of the clinical research that is published, or to rely on the judgment of trusted physicians or authoritative medical guidelines. I take no pleasure in this conclusion, which I reached slowly and reluctantly over my two decades as an editor of *The New England Journal of Medicine*."[1] Richard Smith, a 25-year staffer at the *British Medical Journal*, concurs with this view, writing that "most of what appears in peer-reviewed journals is scientifically weak."[2] Although authoritative voices are speaking out, I contend that, more than 10 years after Angell's dramatic statement, medical research is still subject to corruption. In addition, there appear to be structured efforts underway to suppress, if not throttle, the practice of natural medicine both within and outside of the research arena.

Although I have long known that suppression of natural medicine is a real issue both historic and current, never have I experienced the tacit suppression of my profession so personally as in recent years. In the late fall of 2019, I attempted to run a Google ad to announce holiday specials and events in my apothecary. To my astonishment, the ad was denied for violating Google's "community standards." When I questioned this, I was referred to rules against content involving terrorism, sexual exploitation, profanity, the sale of counterfeit goods, and . . . "unapproved substances." I experienced a moment of outrage upon learning that the beloved selection of herbs and medicines listed on my website had triggered a process that grouped me with terrorists and other nefarious bedfellows. Grocers and department stores sell herbal and nutritional supplements, too, and yet their ads can be seen on Google. I learned that colleagues had had experiences similar to mine; it appeared that Google was blocking only natural medicine businesses from running announcements. I advertised elsewhere. A month later, I alerted my property insurer that I had made some updates and improvements to my business property. Long story short, I ended up being dropped by the insurer because they "didn't have an appetite for alternative medicine." Upon seeking out another insurer, I quickly learned that all of them had a line of questioning that eventually led to: "Do you practice homeopathy?" I seriously doubted whether the inquirers even knew what homeopathy was, but I learned that responding "yes" would lead to my dismissal. Every single insurer appeared to have been instructed to reject people in my line of work; it seemed I might have had an easier time insuring my business if I sold pesticides, chainsaws, or firearms rather than natural medicines.

In all the chaos of the 2020 pandemic year, the censorship deepened. Clinicians' merchant accounts were cancelled for offering "immune support," which was suddenly purported to be a violation of Federal Trade Commission (FTC) rules. The American Association of Naturopathic Physicians alerted its members in 2020 that the FTC was presently issuing "warning letters" to alternative medicine practitioners and informed us that offering information on "supporting the immune system" could "trigger a cascade of negative consequences." Some of the negative outcomes include negative press when one's professional business is shut down by Facebook for offering information purported to be false or making claims about health deemed unscientific. Indeed, throughout 2020 Google and Facebook removed posts offering encouragement for keeping one's immune system healthy that were from small independent alternative practitioners but allowed such posts to stand when put forth from hospitals or large medical establishments. A private physician could cite the same studies on vitamin C or zinc as public health agencies and be flagged for posting potentially false information. The public was instructed to listen only to the Centers for Disease Control and Prevention (CDC), but that institution's weekly reversals and gyrations, distribution of faulty test kits, miscounting of test results, and other confusing statements and actions hardly instilled confidence. Those who advised sensible food choices and nutrition to support good health were censored for spreading fake news, and health practitioners who espoused basic tenets of healthy living were called out as being anti-science; some were outright punished. Eventually, a variety of entities sued Facebook for government-sponsored censorship.

A Historical Record of Suppression

Censorship of valid health information—and on the flip side, the advancement of research favorable to the pharmaceutical industry—is hardly new, although the veil obscuring the money trail behind the suppressive efforts is becoming so thin that it is nearly nonexistent. According to a *New York Times* article about Dr. Angell, during her time as editor of *The New England Journal of Medicine*, "she vetted manuscripts that omitted any mention of a drug's side effects, and studies that were weighted to make a drug look good; she repeatedly heard about studies never submitted for publication because they made a drug look bad."[3] Furthermore, researchers are allowed to conduct and publish studies even when they have financial stakes in the drug under study. The National Institutes of Health (NIH) is allowed to recommend therapies for which it and its individual employees hold patents. The CDC is able to use taxpayer funding to develop vaccines and therapies, which it then charges taxpayers fees—sometimes very high fees—to access. When governmental agencies are stakeholders who stand to profit from the very policies they implement, flagrant conflicts of interest necessitate that we question

the agencies' priorities: Are they focused on policies to protect public health or to promote their own profit?

Excessive regulation of historically safe and effective natural medicines similarly calls into question the motives of public health agencies. Natural medicine professionals in the United States have struggled to mature their various disciplines for well over a century despite such regulations and interventions. In his book *Divided Legacy*, medical historian Harris Coulter explores how the American Medical Association (AMA) was first organized to oppose herbal medicine, hydrotherapy, homeopathy, and other alternative medical disciplines, which were so popular among the general public that some allopathic physicians saw clinicians working in these disciplines as a threat to their livelihood.[4] The American Institute for Homeopathy was the first professional medical association that arose in the United States. Established in 1844, it was a robust organization with a large membership and many thriving, well-established hospitals. Not to be outdone, the AMA, which organized in 1847, almost immediately positioned themselves in a moral high tower and began to question the ethics of other medical practitioners. In an era characterized by a glut of medical practitioners, with no one profession being particularly more scientific than another, the AMA claimed that it was the most scientific and that those practicing homeopathy, traditional medicine, midwifery, or physical medicine were less scientific and, therefore, charlatans.

Some writers point to the efforts of John D. Rockefeller, the United States' first billionaire oil baron, as furthering the AMA's anti-traditional-medicine stance by promoting and funding the use of petrochemical-derived medicines by member physicians. Perhaps, though, the intent was less about crippling other types of medicine and more about promoting the new pharmaceuticals as a promising cutting edge therapy. Rockefeller may have viewed it as both a business opportunity and a philanthropic endeavor. Rockefeller invested heavily in pharmaceutical manufacturing and had a vested interest in physicians' use of such medicines over natural medicines. He effectively enticed, pressured, and swayed the health care profession to embrace pharmaceuticals, despite the distrust with which they were greeted when they first became available. These petrochemical medicines, such as the first synthetic version of beeswax (acquired when Standard Oil purchased the first company to make petroleum jelly), were often intended to mimic the action of traditional herbal medicines. Scientists identified the active compounds in the herbal medicines and revealed their chemical structures through the emerging discipline of organic chemistry. The naturally occurring plant compounds could not be patented, but synthetic petrochemicals could be, ushering in the era of petrochemical-based pharmaceuticals. The drug industry has since grown into the most powerful industry in the world, earning its moniker of Big Pharma.

One arm of the AMA, the Council on Medical Education, at its very first meeting (1904) defined educational criteria that favored white males and made it difficult for women or people of color to be admitted to medical school. The criteria also sought to mandate that only graduates of the Council's favored educational programs would be entitled to legally practice. In 1905 the Council (funded by Rockefeller and another oil and steel magnate, Andrew Carnegie) financed the Flexner Report, a project often referred to as leading to the demise of alternative medicine in the United States.

The Carnegie Foundation hired Abraham Flexner, assisted by the secretary of the AMA, to evaluate the state of medical education in the United States. Education of medical students was poorly regulated and organized at the time; there were no clear standards, and curricula differed greatly among institutions. Medical training programs varied from extended apprenticeships to two-year and four-year programs. Flexner issued his report to the Council in 1910, which then mandated that all medical schools use Johns Hopkins' medical curricula as the gold standard and required that schools remove courses in traditional healing methods under threat of losing their accreditation. When a few successful alternative medicine hospitals and training programs resisted, the Council established ever more rigorous obstacles, rules, and regulations. Meeting the requirements was difficult and expensive, and over time the practices of herbalism, homeopathy, naturopathy, midwifery, physiomedicalism, and electromagnetic therapies were effectively suppressed as the smaller institutions lacked the resources to oppose the mandates. Rockefeller also founded a General Education Board (GEB), which dispersed millions of dollars in support, ostensibly to help with the transition of medical schools' curricula, but which effectively bought the schools' allegiance to the pharmaceutical model of medicine. Those willing to accept the pharmaceutical model and teach a pharmaceutical-based curriculum received funding for faculty salaries, shaping the landscape of medical education up to the present time.

In this manner, the once-broad diversity of medical disciplines collapsed, and the survivors were primarily institutions that taught pharmaceutical and surgical therapies exclusively. The number of medical schools and accredited doctors in the United States were both drastically reduced, and the income of the average physician increased dramatically. Older practitioners who used natural therapies were regarded as "quacks," a term derived from the Dutch "kwakzalver," originally used to refer to physicians who used mercury compounds as therapeutic agents.

In the early 1900s, there were more than a dozen naturopathic medical schools in the United States. By 1950, there was just one left, hanging by a financial thread. By the 1970s, its students resorted to building their own clinic in the evenings and on weekends! During this transition, historically Black medical schools were closed as well. Only two survived, and the expectation was that their graduates would treat only Black patients and that their programs would be held in less esteem. The Flexner Report also advised that Black physicians should treat only Black patient populations and raised the benefit that health promotion in Black communities would better protect white people from common diseases such as hookworm and tuberculosis. Rising up from near extinction, many traditional medical disciplines saw a resurgence in both professional applicants and in public demand for their services in the 1980s. Interest remained so robust that allopathic schools embraced the demand and added complementary and alternative medicine (CAM) programs and functional medicine branches to their school coursework and to their clinical offerings. Naturopathic and chiropractic schools and midwifery programs are also once again accredited, but the effort to limit the growth and maturation of such professions is ever present.

Pharmaceutical Giants in Control

In the 1960s, there was a resurgence of interest in health food, organic gardening, and the health benefits of herbs and supplements. Since then, many of the mom-and-pop health food and supplement companies founded in the garages or kitchens of hippies have grown into highly successful and lucrative businesses that have been bought by corporate conglomerates such as Proctor & Gamble, Nestlé, and Amazon. Scientific research into natural products exploded in the aftermath of this resurgence. And as popular nutritional or herbal supplements have become multimillion-dollar commodities backed by molecular research and clinical trials, natural products have attracted the interest of governmental regulatory entities, which are swooping in to restrict use and control versions of the products or isolated molecules for themselves. Big Pharma has also moved to own natural products and to shift control of them to the government entities it prefers—those that have financial ties to the pharmaceutical industry. In a few short decades, alternative medicine practitioners have witnessed an astounding shift. Where once their medicines were criticized as unscientific and worthless quackery, those same medicines are now considered too powerful to be administered freely by practitioners.

At the time of this writing, the FDA is removing numerous natural substances from the hands of private practitioners and restricting the compounding of such medicines by individual physicians and compounding pharmacists alike via a series of hearings to evaluate the safety of individual natural substances. These hearings arose out of new FDA efforts aimed at regulating over-the-counter (OTC, or nonprescription) medications, initiated in 1980.[5] Via the OTC Drug Review, the FDA began to methodically evaluate the safety and efficacy of thousands of OTC drug products—from toothpaste, to skin products for acne, to herbal and nutritional supplements—and produced formal documents to categorize and legislate such products; now one of the largest and most complex regulatory undertakings ever at the FDA. Included in this quagmire of rulemaking processes was the formation of a federal committee that solicited "nominations" of natural substances to review in establishing a list of bulk drug substances for compounding. This bulk substances list is intended to comprise substances that are neither equipped with a US Pharmacopeia or National Formulary monograph, nor "components of FDA-approved drug products."[6] In other words, those substances that are not already approved for use by the FDA are now up for review.

The FDA's "Interim Policy on Compounding Using Bulk Drug Substances Under Section 503A of the Federal Food, Drug, and Cosmetic Act" outlines the three categories into which nominated substances are placed. Category 1 comprises substances that are eligible for evaluation, category 2 are those that "raise significant safety concerns" and therefore will immediately become illegal to compound, and category 3 are those that the FDA has determined were "nominated without adequate support," and therefore will not receive a hearing unless they are resubmitted[7] and will fall immediately

under the control of the FDA. All of these substances presently under review are still legal to compound until the FDA completes its hearings. However, if the FDA adopts the decisions made during the hearings into a legal ruling, all the substances that were not approved will become illegal to compound by compounding pharmacists and physicians.

Of the 310 substances that were nominated as part of this hearing process, the FDA decreed that 242 did not warrant a hearing or require any expert testimony.[8] These substances were immediately relegated to category 3. These 242 substances include acidophilus, alfalfa, asparagus, peppermint oil, papaya enzymes, *Ginkgo*, and numerous nutrients (vitamins, enzymes, and minerals). The remaining category 1 substances under review are hardly in a better position: The hearings typically lead to the herbs in question being deemed unsafe to compound or sell without restriction. Already declared unsafe from the category 1 list are numerous plants and substances named throughout the five volumes of *Herbal Formularies*, including glutamine, *Boswellia*, artemisinin, curcumin, and glycyrrhizin.[9]

This does not necessarily mean that the substances won't be available at all; it does portend a new layer of restriction regarding in-office compounding of such agents, such that even the simple process of blending herbs into a tea or tincture could violate compounding laws. Unless a substance under review can be verified as the *only* therapy available for a certain medical condition, the FDA representatives advise the committee that the substance should not be approved as safe or suitable for herbalists to handle in their offices.[10] And who will be able to compound and sell these substances? Only those entities, such as the pharmaceutical giants, licensed to compound drugs would escape FDA prosecution; *not* naturopathic physicians, herbalists, or compounding pharmacists, even though these are the entities who have the most clinical experience with these substances.

The surprising fact that the regulatory agencies are reviewing substances such as nettles and *Aloe vera* for public safety provides a deeper and darker context to the censorship of my Google ad and cancellation of my property insurance. There are pharmaceutical companies producing patentable synthetic versions of some of the same natural compounds the FDA committee is presently restricting. In my mind, this raises the question of whether natural compounds such as curcumin are being reviewed and possibly restricted precisely so they won't compete with synthetic commercial products. The process is shrouded in the illusion of public safety, but reeks of the profit motive.

Naturopathic physician Dr. Paul Anderson of the Anderson Medical Specialty Associates in Seattle, Washington, was one of the subject matter experts appointed to testify at the US FDA review hearings. According to Dr. Anderson, those who testify against the safety profile and approval are allotted the lion's share of the time in the hearings. Those who provide authentic clinical experience with an herb and offer testimony sharing current published research are given just a few minutes to rebut. The individuals who advise the committee to restrict access to natural substances are typically pharmaceutical company allies or others who have clear conflicts of interest, "but the FDA waives these conflicts," says Dr. Anderson—an admission that is recorded in the public records.[11]

Dr. Anderson's testimony makes it clear that no amount of research, experience, or cogent arguments can assuage the committee's bias against the safety of the substances under review in the hands of skilled herbalists. According to Dr. Anderson, the "experts" who argue to restrict natural substances cite outdated research or claim that there is no research to support the safety of a substance. When shown evidence to the contrary, the committee members blatantly and willfully choose to ignore it. For example, Dr. Anderson recounts, "There was [a] natural substance where the FDA material stated that there is no modern evidence supporting its use, but we had 15 citations in peer-reviewed journals just in the last 5 years. I showed these to them, and the response was, 'That might be true, but we don't have to consider any of these references.' It's possible to rebut what the FDA expert says with up-to-date evidence, but if the FDA doesn't want to take it into account, they don't have to."[12]

In this manner, the hearings are theater. They may provide the illusion of a platform of serious review, but in fact they are a mechanism created to gain control of our most important herbal medicines. Based on Dr. Anderson's accounts, I cannot believe that the committee is genuinely interested in learning about the safety, efficacy, or therapeutic application of the substances under review.

The history of modern synthetic pharmaceuticals includes many examples of products that have caused deleterious side effects and even toxicities. The existence of alternative options is a threat to pharmaceutical drugs and the industry's creation of proprietary versions of herbal products. We are witnessing the enactment of governmental policies that restrict our access to safe and gentle herbal medicines and shift control of their use to the auspices of giant companies. This trend also reflects a disturbing underlying philosophy: If a natural medicine is really *that* effective and valuable, then it shouldn't be left in the hands of alternative medicine clinicians, who are viewed as inferior and with suspicion. Instead, it should be moved into the capable hands of a drug company that, incidentally, has the capacity to develop the substance into a patentable drug that serves as a valuable adjuvant to the primary, essential pharmaceutical therapy. Herbalists are keenly interested in maintaining organically grown, high-quality, unadulterated herbal products, but pharmaceutical companies do not concern themselves with such purity issues. The goal of the drug companies is to create patentable compounds: manipulated molecules inspired by the natural plant constituents or entirely synthetic versions. Such molecules can be made into proprietary, branded, and patented products. We have seen some such molecules reach the market already, reporting enhanced bioavailability or targeted tissue delivery and sold at great cost compared to the whole crude herbs from which they are derived. Clinical studies are carried out using these molecules to support their medical applications, while research on whole herbs continues to be suppressed or fails to secure funding. Large pharmaceutical and regulatory entities can claim that the proprietary molecules have evidence of safety and efficacy and that the same level of research does not exist for the whole crude herbs. The whole herb products that herbalists and naturopathic physicians favor can be said to be inferior or not backed by research, even though centuries-old traditional usage inspired the research on the isolated compounds. Naturopathic physicians and herbalists are highly concerned that pharmaceutical versions of herbal products will not have the same safety profile or efficacy as traditional whole herbs. As such, they face the double dilemma of potentially being restricted from compounding formulas using high-quality, organic, whole herbs that they prefer. Many herbalists feel that, while the "active" compounds in herbs are certainly important, every other compound, enzyme, trace mineral, and nutrient in the plants are synergistic and that the removal of a single isolated compound robs the medicine of its nutrient base, delivery system, and energetic signature. Big Pharma is presently shaping the landscape of herbal medicine, and herbalists feel it is a detriment to the integrity and quality of the medicine.

It may be hyperbole to say, *"Learn about plants now, while it's still legal!"*, but there is cause for concern. The evidence keeps accumulating that natural medicine is being suppressed. How can we protect ourselves? By supporting our valued professional organizations: Together, we are better able to fight for our traditions and our rights.

About This Book

This text is the last in a set of five comprehensive volumes aimed at sharing my own clinical experience and formulas to assist herbalists, physicians, nurses, and allied health professionals in creating effective herbal formulas. The information in this book is based on the folkloric indications of individual herbs, fused with modern research and my own clinical experience.

I have organized this set of volumes by organ systems. Volume 1 features the organs of elimination—the gastrointestinal system, the liver and biliary system, the urinary system, and the skin. Herbalists know these organs are foundational to the health of the entire body. The treatment of many inflammatory, infectious, hormonal, and other complaints will be improved by optimizing digestion and elimination. Volume 2 covers respiratory, pulmonary, and vascular issues, including both cardiovascular and peripheral vascular complaints. Volume 3 focuses on metabolic and reproductive endocrinology: adrenal and thyroid disorders, diabetes and metabolic dysfunction, and male and female reproductive disorders. Volume 4 addresses headaches and pain management in a variety of organ systems, neurologic conditions ranging from neuropathy to Parkinson's disease and Alzheimer's disease, and psychologic issues from mood disorders to hyperactivity to addictions. In this volume, we explore herbal medicines and natural compounds to reduce allergic and autoimmune reactivity; share effective formulas and protocols to address infectious and inflammatory conditions of the eyes, ears, nose, mouth, and throat; and review traditional medicines for the most common musculoskeletal complaints.

Each volume in this set offers specific herbal formulas for treating common health issues and diagnoses within the selected organ system, creating a text that serves as a user-friendly reference manual as well as a guide for budding herbalists in the high art of fine-tuning an herbal formula for the person, not just for the diagnosis. Each chapter includes a range of formulas to treat common conditions as well as formulas to address specific energetic or symptomatic presentations. I introduce each formula with brief notes that help to explain how the selected herbs address the specific condition. At the end of each chapter, I have provided a compendium of the herbs most commonly indicated for a specific niche, a concept from folklore simply referred to as *specific indications*. These sections include most herbs mentioned in the corresponding chapter and highlight unique, precise, or exacting symptoms for which they are most indicated. Please note that these listings do not encompass *all* the symptoms or indications covered by the various herbs, but rather only those symptoms that relate to that chapter—the indications for autoimmune or allergic conditions, for conditions of the musculoskeletal system, or for issues of the ears, eyes, nose, mouth, and throat. You'll find certain herbs repeated in the specific indications section of more than one chapter of this book, but in each instance, the description will feature slightly different comments. Readers are encouraged to refer back and forth among the chapters to best compare and contrast the information offered.

The Goals of This Book

My first goal in offering such extensive and thorough listings of possible herbal therapies is to demonstrate and model how to craft herbal formulas that are precise for the patient, not for the diagnosis. It is my hope that after studying the formulas in this book and other volumes in the set and following my guidelines for crafting a formula, readers will assimilate this basic philosophic approach to devising a clinical formula. As readers gain experience and confidence, I believe they will find that they rely less and less on these volumes and more and more on their own knowledge and insight. That's what happened to me over the years as I read the research and folkloric herb books and familiarized myself with the specific niche-indication details of a wide range of healing plants. I now have this knowledge in my head, and devising an herbal formula for a patient's needs has become second nature and somewhat intuitive. But from talking with my herbal students over several decades of teaching, I have come to understand that creating herbal formulas is one of the most challenging leaps between simply absorbing information and using it to treat real, live patients. Students often feel inept as they try to sift through all their books, notes, and knowledge and struggle to use "information" to devise a single formula that best addresses a human being's complexities. Thus, I felt that it was high time that I created a user-friendly guide to help students refine their formulation skills and to help all readers develop their abilities to create sophisticated, well-thought-out formulas.

Another goal I aim to achieve through this set of herbal formularies is to create an easy-to-use reference that practitioners can rely on in the midst of a busy patient day. In this "information age," it is not hard to track down volumes of information about an herb, a medical condition, or even a single molecule isolated from a plant. The difficulty lies in remembering and synthesizing it all. While this text doesn't pretend to synthesize the "art" of medicine in one source, I believe it will help health professionals quickly recall and make use of herbal therapies they already know or have read about by organizing them in a fashion that is easy to access quickly.

Naturopathic physicians are a varied lot. Add in other physicians and allied health professionals, and the skill sets are varied indeed. I rely on my naturopathic colleagues to inform me about the latest lab tests, my allopathic colleagues to inform me about new pharmaceutical options, and my acupuncture colleagues to inform me as to which conditions they are seeing good results in treating. This text allows me to share my own area of expertise. I have included a large number of sidebars that feature some of the in-depth research on the herbs and individual molecular constituents, helping to provide an evidence-based foundation for the present era of medical herbalism.

I realize that not all clinicians, not even all naturopathic physicians, specialize in herbal medicine. I hope that this formulary will serve as a handy reference manual for those who can benefit from my personal experience, formulas, and supportive discussions.

Creating Energetically Fine-Tuned Formulas

Much like a homeopathic *materia medica*, this set of formularies aims to demonstrate to clinicians how to choose herbs based on *specific indications* and clinical *symptoms* and *presentations*, rather than on diagnoses alone. For example, this volume does not offer one-size-fits-all arthritis formulas, but rather details distinct approaches to creating herbal formulas for arthritis with digestive contribution, arthritis with poor connective tissue repair, arthritis with systemic inflammatory symptoms, and arthritis with immune activation. I include supportive research on herbs that helps to explain why a particular herb is chosen for a particular formula, as well as endnote citations that provide details of specific studies for those interested. I also provide findings from

research on individual herbs that are essential to the treatment of the various conditions featured in a chapter. To make the text as useful as possible for physicians and other clinicians, I also offer clinical pearls and special guidance from my own experience and that of my colleagues—the tips and techniques that grab attention at medical conferences year after year.

The Information Sourced in This Book

The source of the information in these volumes is based on classic herbal folklore, the writings of the Eclectic physicians, modern research, and my own clinical experience. Because this book is designed as a guide for students and a quick reference for the busy clinician, the sources and research are not cited rigorously, but enough so as to make the case for evidence-based approaches. When I offer a formula based on my own experience, I say so. I also make note of formulas I've created that are more experimental, either because of lack of research on herbs for that condition or my lack of clinical experience with it.

My emphasis is on Western herbs, but I also discuss and use some of the traditional Asian herbs that are readily available in the United States. In some cases, formulas based on Traditional Chinese Medicine (TCM) are featured because of a significant amount of research on the formula's usage in certain conditions. I readily admit that TCM creates formulas *not* for specific diagnoses, but rather for specific energetic and clinical situations. However, I have included such formulas, perhaps out of context but with the overall goal of including evidence-based formulas, with the expectation that readers and clinicians can seek out further guidance from TCM literature or experienced clinicians where possible. In reality, TCM is a sophisticated system that addresses specific presentations, and I have borrowed from this system where I thought such formulas might be of interest or an inspiration to readers. I admit that listing just one formula for a certain condition based on the fact that there have been numerous studies on it is somewhat of a corruption of the integrity of the TCM system, which is aimed at precise patterns and energetic specificity. Nonetheless, I chose to do so with the goal of creating a textbook to help busy clinicians find information quickly, while still encouraging individualized formulas for specific presentations.

How to Use This Book

Each chapter in this book details herbal remedies to consider for specific symptoms and common presentations of various diagnoses. Don't feel that you must be a slave to following the recipes exactly. When good cooks use a food recipe, they are always at liberty to alter the recipe to create the flavor that best suits the intended meal—the big picture. A formula listed should not be thought of as *the* formula to make, but rather as a guide and an example, inviting the clinician to tailor a formula for each individual patient.

To create an herbal formula unique to a specific person, the clinician should first generate a list of actions that the formula should perform (respiratory antimicrobial, expectorant, bronchodilator, mast cell stabilizer, and so on) and then generate a list of possible herbal *materia medica* choices that perform the desired actions. If these ideas are new to you, you may want to begin by reading chapter 1, "The Art of Herbal Formulation," before you start generating lists.

Unity of Disease (Totality of Symptoms)

The concept that any health issues a person may experience are actually one disease, as opposed to a number of disparate diagnoses to be treated individually, is a core tenet of naturopathic medicine and the philosophical underpinning of holistic medicine in general. Any one symptom does not provide the full story, and just because you can label the symptoms with a Western diagnosis and offer the established therapy for that diagnosis does not mean you are really helping a person to *heal*. A careful consideration of the totality of all symptoms is important to reveal underlying patterns of organ strength or weakness, excess or deficiency states, nervous origins versus nutritional origins, and, of course, a complex overlap of all such issues. The most effective therapies will address *all* issues in their entirety and involve an understanding of the entire energetic, mental, emotional, nutritional, hereditary, situational, and other processes, creating a complex web of cause and effect—the unity of any given individual's "dis-ease."

Look to the formulas in chapters 2 through 4 that address specific symptoms for guidance and inspiration. (These formulas are grouped within the chapter by a general diagnosis, such as "Formulas for Lupus" or "Formulas for Macular Degeneration.") Regard the lists and formulas I have provided as starting points and build from there. In my commentary on the individual formulas and in sidebars that focus on specific herbs, I offer further guidance as to whether the formula or individual herbs are safe in all people, possibly toxic in large doses, intended for topical use only, or indicated only in certain cases of that particular symptom. Once herb and formula possibilities have been identified, the reader should then review "Specific Indications" at the end of the chapter to focus on which herbs would be *most* appropriate to select and to learn more about how those herbs might be used. Herbalists can narrow down long lists of herbal possibilities to just a few *materia medica* choices that will best serve the individual. In many cases, the reader/clinician will be drawing upon herbal possibilities from a number of chapters and organ systems as the clinical presentation of the patient dictates. Thus, you are not making a formula by throwing together all the herbs listed as covering that symptom or symptoms, but you are studying further and narrowing down the list of possibilities to consider, based on the totality of the symptoms. In some cases, you will rule out herbs on the list for a particular symptom after reading the specific descriptions of those herbs at the end of the chapter. In some cases, you might decide to put one herb in a tea and another in a tincture because of flavor considerations. In other cases, you might decide that you will prepare only a topical remedy. And in other urgent situations, you might come up with a topical, a pill, an herbal tea, *and* a tincture to address the situation as aggressively as possible. Aim to select the best choices and avoid using too many herbs in one formula. Larger doses of just a few herbs tend to work better than smaller doses of many herbs, which can confuse the body with a myriad of compounds all at once. The use of three, four, or five herbs in a formula is a good place to start; this approach also makes it simpler to evaluate what works when the formula is effective as well as what is poorly tolerated, should a formula cause digestive upset or other side effects.

Learning from the Formulas in This Book

In reviewing the formulas in this book, notice how specific herbs are combined with foundational herbs to create different formulas that address a variety of energetic presentations. There are a handful of all-purpose immunomodulators, all-purpose alterative herbs, and all-purpose anti-inflammatories that can be foundational herbs in many kinds of formulas. Such foundational herbs can be made more specific for various situations by combining them with complementary herbs that are energetically precise. Notice how the herbs are formulated to be somewhat exacting to address specific symptoms and make a formula that is warming, drying, cooling, or moistening and so on. Also, note how acute formulas may have aggressive dosages and include some strong herbs intended for short-term use, while formulas attempting to shift chronic tendencies are dosed two or three times a day and typically include nourishing and restorative herbs intended for long-term use. Also notice how some potentially toxic herbs are used as just a few milliliters or even a few drops in the entire 2-ounce (60 ml) formula. These dosages should not be exceeded, and if this is a clinician's first introduction to potentially toxic herbs, further study and due diligence are required to fully understand the medicines and how they are safely used. Don't go down the poison path without a good deal of education and preparation. I am able to prepare all the formulas in these texts upon request, but I can offer those containing the "toxic" or low-dose herbs (*Atropa belladonna, Aconitum, Convallaria,* and so on) only to licensed physicians.

It is my sincere hope that this book helps you in your clinical work and efforts to heal people.

DR. JILL STANSBURY

CHAPTER ONE

The Art of Herbal Formulation

Creating an effective and sophisticated herbal formula is somewhat of an art, and like all art, it is difficult to put into step-by-step directions; however, this is my attempt to do just that. This book aims to explain how to create specific formulas for *presentations* rather than *diagnoses*—rather than offering a single formula or two for a general condition, such as hypertension or irritable bowel syndrome (IBS), this text offers exacting formulations to best address the precise presentation of the person. I have personally seen many *different kinds* of asthma, hay fever, arthritis, and other conditions that conventional medicine tends to treat with one across-the-board medication or therapy. My aim is to coach readers on how to create numerous finely tuned herbal options for treating the person and not the diagnosis—a core tenet of natural medicine.

Creating effective herbal formulas and treatment plans requires many skills: knowledge of the herbs; herbal combinations best for a particular situation; the proportions to use in a formula; the starting dose, frequency of dose, and how long to dose; what form of medicine is best, such as a tincture, a tea, a pill, or a topical application; broader protocol options that may include diet, exercise, nutritional supplements, or referral to allied health professionals; and the follow-up plan.

Hippocrates said, "It is more important to know what kind of person has a disease than to know what kind of disease a person has." To know what sort of person has a disease requires careful listening and skilled and nuanced questioning in a safe and comfortable setting. (See "Asking the Right Questions" on page 12.) Listen for underlying causes, for overarching emotional tone, for what a person is able to do for themselves (sleep more, exercise more, eat better, brew a daily tea), and for what they are reluctant to do (sleep more, exercise more, eat better, brew a daily tea). Address underlying causes and start where people show some interest and capacity. Cheerlead to instill enthusiasm if required, so that people become better educated and thereby better motivated to make important changes or adopt valuable healing practices. Giving the right medicine is only one aspect of doing healing work with a person; creating a sacred space that invites truth and sharing is key to getting to the point where you know what sort of person has a disease.

Healing also stems from understanding pain, suffering, challenges, and unique situations, from non-judgmental listening to the stories so often linked to our physical ailments. It comes from giving encouraging words and sympathy, congratulating and complementing people's efforts and accomplishments, and giving people the tools, resources, and support to succeed. This kind of true caring plus an earnest effort to provide real support is among our best medicines.

The Importance of Symptoms

Naturopathic medicine has a different philosophical stance on physical symptoms than allopathic medicine, especially infectious and eruptive symptoms, and that view is worth briefly describing here for those who may be unfamiliar.

The Biochemical Terrain

The terrain of the human body invites microbes specific to the chemical composition of the ecosystem, and when infectious microorganisms have consumed those specific chemical substrates, the disease-producing microbes are no longer supported. The microbes themselves change the chemical composition of the tissue by consuming the "food" that invited them, and when those nutritional resources are exhausted, the biochemical makeup of the tissues is changed for the better. Antibiotics are not the best way to treat chronic infections such as ear, throat, or sinus infections. Instead, optimize the ecosystem to support the desired beneficial flora. As when bacteria start to stink up the compost bucket on the kitchen counter, we don't spray the compost with a germicide—we clean the bucket! For example, we may be more susceptible to colds and infections leading to sinusitis when we eat too much dairy or are exposed to smoke or other triggers of congestion in upper respiratory mucous membranes. We might also suffer more upper respiratory congestion due to allergic reactivity increasing susceptibility to infection because the moist, boggy tissues support pathogenic microbes. Similarly, we might experience more joint pain and stiffness when the intestinal ecosystem is unbalanced, because the gut will trigger inflammatory reactivity that contributes to systemic inflammation, joint pain, and even autoimmune hypersensitivity reactions. One might take pain relievers or get a prescription to palliate stiffness in the joints, but such medicines would need to be consumed over and over again. The more deeply corrective approach may be to undertake the effort of improving the diet, avoiding inflammatory foods, nourishing the connective tissue, and utilizing alterative and broad anti-inflammatory agents that reduce the propensity to painful inflammatory processes over time.

The Role of a "Healing Crisis"

An acute infection, such as a simple cold, is a classic example of a "healing crisis"—meaning the acute symptoms are actually a part of the body's attempt to heal itself, allowing the infectious organisms to consume the "morbid matter" and thereby restore a healthier ecosystem. The symptoms of the cold—runny nose, fever, loss of appetite—are part of the body's process to heal itself.

Naturopathic philosophy embraces the symptoms of a healing crisis as a triumph of the vital force. Our symptoms serve us and call attention to the imbalances requiring changes. Be thankful when the body has the vitality to manifest a healing crisis. Be concerned when infections, eruptions, and discharges stop, and allergies, autoimmunity, joint pain, ulcers, blood pathology, and so on emerge. These are not healing crises, but rather signs that vitality is being damaged, and pathology is becoming deeper and more serious. Autoimmune disorders are among the most difficult of all conditions to "cure" because the body begins attacking itself, having become confused in its self-correcting healing efforts.

Because the symptoms of a healing crisis are the way in which the body can heal itself, such symptoms should not be suppressed, but rather supported and made as tolerable as possible. Only when such symptoms are so severe as to threaten damage to the body should they be suppressed. This is typical of autoimmune diseases in which immunosuppressive drugs are the mainstay of pharmaceutical therapy, sometimes necessary to protect the joints and connective tissue from severe damage. In less urgent circumstances, herbs can protect and even support the repair and regeneration of joints, for example in cases of osteoarthritis. They can be helpful complements to dietary and other treatments aimed at improving nutrient assimilation, building bone through hormonal support, or reducing inflammation through strengthening gut barriers. In numerous allergic disorders as well, optimizing gut barriers will reduce systemic inflammation that triggers immune and allergic reactivity.

The Harm in Suppressing Symptoms

Habitual suppression of symptoms over time can force the body to give up its struggle for health. For example, laxatives may relieve constipation temporarily, but ultimately damage normal peristalsis. Stimulants may initially provide energy, but ultimately exhaust the adrenals and nervous system. A pot of coffee may jolt an exhausted person awake, but it will not improve core energy status in the long run and will only further deplete it. Antibiotics can kill infectious pathogens in cases of acute upper respiratory infections, but unless the underlying ecosystem that supports such microbes is significantly improved, the same pathogens will be supported a second, third, and fourth time and become resistant to antibiotics if they are given repeatedly, such as for chronic otitis, pharyngitis, or sinusitis. Steroids can improve chronic joint pain, but can suppress the entire immune system and even contribute to joint damage over time. Many patients ultimately seek the help of herbalists and naturopathic physicians in such cases, as they realize their health is worsening rather than improving.

Asking the Right Questions

Learning how to ask questions that will elicit relevant information is as much an art form as creating an herbal formula. Aim your questions to gather information in several categories. What follows here is a broad, but not exhaustive, list of questions that can yield helpful information.

Etiology

How long has this been happening?
How did it begin?

How has it evolved over time?

What else was going on in your life at the time this began?

How was your very first infection or joint pain treated? Was it successful?

How long before the second infection or the next joint flare-up occurred?

Have any other symptoms, complaints, or problems accompanied this complaint?

What is your health history?

Any previous episodes? Related pathologies? Family members with this complaint?

What is the predominant emotion associated with this complaint? Did the complaint begin during a period of grief? Anxiety? Ambition?

Did the complaint begin following hard labor or an injury, stress, eating a new food, traveling out of the country, starting a new medication (and so on)?

Quality and Occurrence of the Complaint: "PQRST"

Provocation. Does anything seem to bring on the complaint? Does anything alleviate it? What makes the complaint better or worse?

Quality. What is the character of the complaint: burning, throbbing, dull, sharp, shooting, aching, and so on?

Radiation. Does the pain travel? Does this symptom affect any other organ? Is this complaint associated with any other symptom?

Severity. On a scale of 1 to 10, how severe is this? Does it interfere with sleep, activities, work, sex, relationships, child rearing, creativity, and hobbies?

Time. Timing throughout the day, throughout the month (hormonal fluctuations?), throughout the year (seasonal allergies? seasonal affective disorder? Asian or Native American concepts of seasons?). Timing may involve an association with eating food or going without eating, an association with anxiety/relaxation, an association with menstrual cycle, or an association to a certain environment or allergen exposure. Is the complaint any better or worse with sleep?

Concomitant Symptoms

Do you feel hot or cold when this occurs?

Is there lethargy or anxiety, heart palpitations, or weakness?

Does it occur during times of stress and activity, during sleep, or after prolonged sitting?

Is there fear or fatigue, mania or depression, weakness or restlessness (and so on)?

Is there a desire to be consoled or to be left alone?

Constitution and Energetic Considerations

Note the constitution by asking the right questions and observe the person to get additional clues as follows:

Complexion. Pale? Yellow? Cyanotic? Flushed? Haggard? Dry? Damp? Inflamed? Quick or slow to perspire? Oily or dry skin?

Pulse. Strong or weak? Fast or slow? The quality? The variability?

Tongue. Large or small?

Muscles. Well developed, overdeveloped, underdeveloped? Spastic, atrophic? Soothed by pressure and massage or aggravated by pressure or massage?

Senses. Hyperacute includes sensitivity to odors, noises, bright light; pain; racing thoughts. Hypoacute includes loss of taste, hearing, sensation; poor ability to concentrate, remember, respond.

Diet and appetite. Hungry, anorexic? Can eat large quantities or tolerate only small amounts? Hypoglycemic symptoms? Unusual cravings for a particular flavor or food? Unusual aversion to or aggravation by particular foods or flavors? Are the symptoms better following meals or between meals when the stomach is empty? What is the general diet, nutrient intake, bowel habit, and quality of the stool?

Thirst. Large thirst versus small or no thirst? Thirst due to dry mouth? Thirst due to compulsion? Thirst for large gulps or thirst for small sips? Thirst for warm fluids or thirst for cold fluids?

Sex drive. High, low? Markedly cyclical?

Sleep. Requires more than 8 hours of sleep to feel rested, able to function with little sleep? Sleeps soundly all night, or wakes many times? Takes a long time to fall asleep but then sleeps well, or falls asleep readily but wakes at a particular time unable to sleep further? Eating before bed disturbs or improves sleep? Restless during sleep or wakes with a jolt? Has nightmares or difficult dreams?

The "Triangle" Exercise

Once you have listened thoughtfully and well to a description of symptoms, I recommend starting with a traditional and simple method of thinking through the choice of the components of a formula: *the triangle*. Visualize a triangle with a horizontal base and two slanting sides that meet at a point at the top of the triangle. Like the triangle, your formula needs a *base* on which to rest—that's where you'll start—and it also needs two axes that "point" the formula in the right direction.

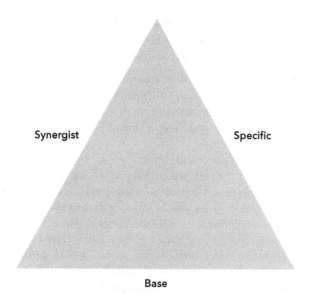

Figure 1.1. The Triangle—A Pragmatic Tool for Crafting Herbal Formulas

In general, the herb or herbs chosen as the base should be nourishing and nontoxic—something tonifying, restorative, alterative, adaptogenic, or nutritive to the main organ system, tissue, or issue of concern. For example, cardiovascular formulas might have *Crataegus* as a base herb on which to rest. A formula for insomnia with exhaustion might rest on *Withania*, and a skin condition formula might rest on *Calendula* or *Centella*. There is an old Wise Woman saying that all healing begins with nourishing. Everyone needs nourishment and tonification, so such herbs are always appropriate. Simple, right?

From there, herbs for the other two axes of the triangle are selected to point the formula in a more specific direction. The base herbs tend to be nonspecific and indicated in a very wide number of clinical situations, so to make your formula more specific for an individual patient, you will need to drive it in the right direction. Through the ages, herbalists have used varied vernacular to refer to such herbs as *synergists*, *specifics*, *energetics*, *kickers*, *directors*, and so on. The choice of names does not matter, but for the sake of discussion, I use the words *synergist* and *specific* to apply to the other two axes of the herbal formula triangle. In Traditional Chinese Medicine (TCM), royal terms such Emperor, Minister, and Assistants may be used.

The choice of a *specific* requires a detailed understanding of *materia medica*, such that one or two energetically and symptomatically precise herbs can be selected. The lists of herbs offered in "Specific Indications" at the end of each chapter help provide exacting details of symptoms for which individual herbs are most indicated. If there is more than one specifically indicated herb, use them both! If one tastes good and one tastes bad, put the tasty one in a tea and put the less palatable option in a tincture or pill. The point is to identify one or two specifics that best address all symptoms possible as well as the constitution, energy, and the entire body's strengths and weaknesses.

As an example, if your case mainly involves hypertension, you may choose to rest the formula on *Crataegus*. The unique identifying symptoms are that the woman's blood pressure began going up during menopause and she is also experiencing episodes of tachycardia and heart palpitations. After reviewing the specific indications for various herbs, you might choose *Leonurus* as your specific to point or drive the formula in the right direction and to the right place. If, on the other hand, the patient is a long-term smoker with high cholesterol and vascular inflammation, your specific may be *Allium*. And if the hypertension formula is for a highly stressed type A personality who also suffers from alarming hypertension accompanied by a throbbing headache, an appropriate choice of specific might be *Rauvolfia*.

A *synergist* is less specific but takes into account underlying contributors, such as other organ systems involved and other energetic considerations that contribute to the overall case. To follow up on the hypertension examples above, the menopausal hypertension formula might include *Actaea* as a synergist to also offer hormonal and nervine support. The long-term smoker may benefit from *Ginkgo* as a synergist to combat circulatory stress, and the type A severe hypertensive might benefit from *Piscidia* as a vasodilating synergist.

The "Specific Indications" in chapters 2, 3, and 4 offer details about individual *materia medica* options to help choose specifics and synergists to best address the sum total of the symptoms. A synergist may be very "specific"—it's all semantics. The point is to take into consideration various individual factors and unique presentations that contribute to the case.

Returning to the example of the menopausal woman with hypertension and episodic heart palpitations, if she also was suffering from stress and poor sleep as underlying contributors, you might select *Withania* as a synergist. If, on the other hand, she was sleeping fine and did not feel particularly stressed but was having some difficulty with constipation and hemorrhoids, you might select *Aesculus* as the synergist. Or if she was having constipation, hemorrhoids, and episodes of heavy menses as part of the menopausal transition, you might select *Hamamelis* as the synergist. Review "Specific Indications" at the end of each chapter to start developing a familiarity with each individual herb and its "personality" or the specific symptoms that it best addresses.

Whatever the details you are presented with, you will use the model of the triangle and select at least three herbs: a nourishing base (tonic), a synergist, and a specific. There is no reason not to select more than one nourishing base or more than one specific or synergist if something very appropriate pops out. If, for example, the menopausal woman with hypertension had a long-standing history of anxiety, stress, mood swings, and emotional lability well before the onset of the recent hypertension, then her nervous system may need nourishment and tonification as much as, or more so than, her cardiovascular system. In that case, you might choose two nourishing herbs upon which to rest the formula—one such as *Crataegus* for the vascular system and one such as *Avena* for the nervous system. Or, if she had a history of these stress-related symptoms and they also tended to cause irritable bowel and diarrhea on many occasions, then you might choose *Crataegus* and *Matricaria*, which provide a nourishing base for the vascular, nervous, and digestive systems.

Therefore, when you reflect on all the details of a case, other symptoms often emerge that are not the chief complaint but are highly important accompanying considerations that should not be overlooked. A simple presentation of perimenopausal hypertension with episodic heart palpitations and no other symptom details may lead to the formula as described above: *Crataegus*, *Leonurus*, and *Actaea*. However, if you learn that this woman has had gallstones, a history of fat intolerance, and many digestive symptoms and that the heart palpitations are worse after meals, you might choose *Curcuma* or *Silybum* as synergists because biliary congestion may contribute to hypertension and, for that matter, may deter the processing of hormones and postprandial lipids. This is another example of an important underlying factor that should not be overlooked. Or, if you find that she is a highly allergic person who sometimes has wheezing with chemical exposure, occasional hives, and chronic low-grade eczema on her hands that flares up after a day of heavy cleaning with exposure to a lot of water and cleaning products, you might choose *Angelica* as a second specific because it is specific for hypertension, asthma, eczema, and hives.

More about Formula Components

Base. Also called the lead herb or the Director, the base should be a tonic that has a nourishing and restorative effect on the main organ system affected. This herb does not require a great amount of skill to choose, as it is often among the herbs best known for having an affinity to a particular organ system.

Synergist. Also called the Adjuvant, the Balancer, or the Assistant, this component helps correct or complement the action of the lead herb and helps drive it to the desired tissues. The selection of these herbs requires an in-depth understanding of the case and person being treated in order to address underlying causes and give the formula the needed energetic specificity.

Specific. Also called the Kicker or the Energetic Specific, this component is selected not just for a specific condition or diagnosis but also for a specific quality, essence, or expression of any given disease or disorder. Such qualities include the pulse, the tongue, the affect, the pathology, the etiology, and the person themselves to guide the selection of the most specific medicine. The practice of learning and using specific medications is gleaned from careful study, observation, and clinical practice.

Mastering the Actions of Herbs

Herbal clinicians should have an excellent grasp of primary actions while gaining a solid knowledge of *materia medica* and specific indications. Clinicians should know which herbs are the best antispasmodics for a variety of situations, the best anti-inflammatories, the best vulneraries, the best nervines, the best antimicrobials, and so on. Such actions of herbs are also foundational considerations when creating an herbal formula or when considering which herbs to select as a tonic base, synergist, and specific of a formula triangle.

Supporting Vitality Instead of Opposing Disease

Western medicine has its "differential diagnosis," where the presenting symptoms can generate a list of possible (differential) causes, ultimately leading to the diagnosis. Although this approach has some value, herbal medicine is less concerned with the formal diagnosis. Instead, herbal medicine carefully considers organ system strengths and weaknesses, underlying causes (stress, toxicity, poor nutrition, poor sleep, circulatory weakness, inflammatory process, allergic hypersensitivity, and so on), and how to support and nourish basic organ function and systemic vitality. Medicines employed by herbalists are typically aimed at restoring function and supporting the innate recuperative powers of the body. The intelligent wisdom of the body to heal itself is sometimes referred to as the "vital force," and herbalists strive to support the vital force more so than to oppose the symptoms of disease. Almost all herbs offer at least some nutrition, being more like foods than drugs, and by nourishing the body and stimulating the vital force, they support the body in healing itself.

Another exercise that I often encourage my students to undertake is learning basic categories of actions of herbs. Actions include antispasmodics, antimicrobials, carminatives, alteratives, adaptogens, demulcents, vulneraries, and so on. I encourage my students to type up pages or create a "little black book" that helps to remind them of the hundreds of herbs they are learning, organized by their categories of action. And from there, they can go deeper. For example, individual antispasmodic herbs might be categorized as having an affinity for a certain organ system or being best suited for a particular quality of spasm. Consider the following antispasmodics: *Lobelia* is especially indicated for respiratory smooth

muscle and cardiac muscle spasms, *Dioscorea* is specific for twisting and boring muscle spasms about the umbilicus, *Piper methysticum* can allay acute musculoskeletal pain and urinary spasms, and *Viburnum opulus* or *V. prunifolium* is especially effective for spasms of the uterine muscle. Sidebars throughout this book offer quick reference lists that summarize various actions.

It is best to avoid a "What herbs are good for the eyes or the throat?" or "What herbs are good for pain?" style of creating herbal formulas. When treating chronic sore throats, for example, the condition may be associated with frequent colds, with smoking, with eating sugar, or with reflux of stomach acid. I encourage clinicians to ask themselves, "What *actions* do I need this formula to perform?" In this case, we need to tailor the formula to the individual and select herbs with specific desired actions. We may want to reduce the susceptibility of the oral mucosa to infections by drying the throat membranes with *Thymus* or *Allium* or N-acetylcysteine (NAC) supplements. Or we may want to support a person to quit smoking with nervine herbal formulas and motivational support. For those with chronic postnasal drip and chronically congested tonsils, we may offer education to improve dietary choices and reduce exposure to mucus-forming foods or allergens. Or we may need to treat underlying gastroesophageal reflux disorder (GERD) and offer demulcent combinations such as *Glycyrrhiza* and *Ulmus* lozenges to offer palliative comfort.

When thinking through an herbal formula, especially for difficult or complex cases, it is useful to write down what actions you wish the formula to perform and then list several herbs that perform this action. The next step is to consider any specific indications for the listed herbs to help home in on the best choices. And finally, the most nourishing herbs may be chosen as supporting the foundation of the triangle, a base on which the formula may rest. Other herbs from the action lists may then be chosen as specifics or synergists to best offer all the needed actions while creating energetically specific, finely tuned formulas. For the elderly with osteoarthritis and poor digestion, we may need to offer herbs in vinegar or take care that nutrients in capsule form will be assimilated. For children with ear or throat infections, we may need to prepare oral formulas in glycerin or provide other palatable options, especially for those too young to swallow capsules.

Energetic Fine-Tuning

Using the triangle method and an awareness of the actions of herbs, as detailed above, will assist you in selecting three or more herbs that address a case in its entirety, and as such, the formulas are likely to be effective and successful. To fine-tune your formulas even further, an added tier of specificity is the energetic state of your patient. TCM philosophy often depicts the energetic state as a mixture of polar opposites in keeping with the Taoist philosophy of yin and yang polarities. For example, is your patient hot or cold? Tight and constricted or loose and atrophic? Excessively damp in the tissues or excessively dry? Tired and lethargic or energized and manic—and so on. Ayurveda, the traditional medical system of ancient India, sets up a three-pronged system of doshas—vata, pitta, and kapha—rather than the two-pronged polar opposites of Taoism but is similarly aimed at addressing differing constitutions and energetic presentations. The four-elements theory of ancient Western herbalism looks for symptoms or presentations categorized into earth-, air-, fire-, and water-related symptoms, with herbal therapies chosen accordingly. Again, the precise system, vernacular, and approach do not matter as long as you are aware of some sort of energetic presentation. Whether or not you take it upon yourself to learn the doshas, TCM, or four-elements thinking and prescribing, you can still begin to notice whether a patient is hot or cold, damp or dry, for example, and choose herbs based on the specific clues or symptoms that you discern through thorough questioning and from simple and obvious physical exam findings.

For example, individuals may present with different qualities of hay fever. One woman might experience watery eyes and nose with frequent sneezing, quickly using up a whole box of tissues blowing her nose. *Euphrasia* could be offered to reduce her allergic symptoms, mixed with *Thymus*, which has a drying effect on the "damp" mucosa. A man who has hayfever symptoms might suffer from thick mucus in his nose and sinuses that won't move and tends to lead to sinus infections. This man might also be offered *Euphrasia* to reduce allergic reactivity, but in his case, mixed with *Allium* and *Armoracia*, which can thin and expectorate stuck mucus and reduce the tendency to development of a sinus infection.

As another example, a woman may present with general anxiety that is worse premenstrually and involves irritability and angry outbursts and a tendency to tightness in the neck and shoulders that progresses to muscle tension headaches two to three times a month. She relies on OTC Tylenol at these times and is experiencing worsening intestinal distress that includes episodes of bowel spasms and an urgent need to rush to the toilet to pass loose stools with increasing frequency and made worse by drinking coffee. This fairly common scenario could be described as heat and excess in the body: The nervous system is "wound up tightly" and would likely respond to calming, cooling remedies, such as *Avena*, *Matricaria*, *Scutellaria*, or *Lavandula*. The muscles become spastic as heat and tension build and may relax with the inclusion of *Actaea* or *Passiflora* in the formula. And the worsening of symptoms premenstrually suggests that hormonal excesses fan the flames of the underlying heat condition and that alterative herbs may help the liver to conjugate hormones and cool the mental-emotional fire. *Taraxacum*, *Arctium*, *Silybum*, or *Berberis* may be helpful when consumed every day for several months. And the intestinal system, too, displays heat symptoms, worsened by the irritating effects of coffee and regular Tylenol use, and may benefit from cooling demulcents, such as *Ulmus*, or the cooling nervines that also soothe the bowels, such as *Matricaria*. Given this scenario we might choose *Matricaria*, *Avena*, and *Ulmus* to be the trophorestorative base of an herbal tea, perhaps combined with additional nervines tasty enough to use in a tea, such as *Scutellaria* or *Passiflora* or both. *Actaea* may be one of our most specific *materia medica* choices in this scenario, being specifically indicated for anxiety, hormonal imbalance, and tightness in the neck and shoulder and muscle tension. Because *Actaea* is not tasty as a tea, it may be better employed as a tincture or encapsulation. The alterative herb choices might be combined with *Actaea* in a tincture or employed as a separate liver supportive encapsulation, many of which are commercially available. Thus, we can cover all the desired actions with several different medicines and follow up with our patient in several months to assess progress.

When I teach, I often present sample cases and lead discussions with my students to explore how to think through the choices of herbs for the base, synergists, and specifics, based on details of the particular patient. It can be helpful to work through two examples of people with the same "diagnosis," such as insomnia or acne, and discover how differently a formula evolves based on the patient's unique constitution, energetics, and symptoms. In the first sample case, I offer an extensive discussion that arrives at a specific formula. Following that, I present a more condensed sample case, with several options for base, synergist, and specific herbs. See how you do at finishing the selection process for the sample cases described below. There is no single right or wrong answer.

Pharmacologic versus Physiologic Therapy

Pharmacologic Therapy. Pharmacologic prescribing is the use of a potent medicine to force a rapid pharmacological response. The energy to catalyze changes, movement, and homeodynamic balance seems to come from the medicine itself. The chemical constituents have strong actions in the body and act as cardiosedatives, emetics, antibiotics, vasoconstrictors, antispasmodics, diuretics, and so on. Pharmacologic medicine can be used heroically and can save lives, but it doesn't build the vital force and restore organ function, plus its use may be needed repeatedly. Most pharmaceutical drugs are pharmacologic in nature and are often suppressive to the body's vital force—for example, acetaminophen to suppress a fever, antibiotics to kill pathogens, or bisphosphonate drugs to halt bone cell turnover rates.

Some of the more toxic herbs can have pharmacologic activity, but in general, herbs are more like foods, offering nutrition and physiologic support. Pharmacologic medications do very little to nourish, tone, or deeply "cure" anything; the symptoms return as soon as the medication is removed. For example, steroids may suppress wheezing in the lungs or suppress eczematous skin lesions, but because they do not alter the underlying condition, wheezing and itching will recur as soon as the medications are stopped. At times, for serious and acute situations, an herb with a pharmacologic action may be chosen as a synergist or a specific in an herbal formula, but never as a lead herb upon which to base a formula.

Physiologic Therapy. Physiologic prescribing involves the use of gentle medicines over an extended period of time to nourish organs and restore normal tone and function. Physiologic medications do not have rapid or strong pharmacological actions and even if prescribed inappropriately would be unlikely to push the limits of homeostasis or cause undesirable side effects. Physiologic medications balance, nourish, and tone, and they help restore optimal physiology through gentle support of assimilation, elimination, detoxification, metabolism, perfusion, and nerve function. Most herbs are physiologic in nature, capable of restoring normal functioning, organ tone, and homeostatic balance. In contrast to the above pharmacologic examples, herbs may be used to support a fever when needed to allow the body to fight infections, or to encourage the body to build new bone cells rather than halt all bone cell turnover. Because physiologic medications repair and restore tissue and organ function, they can usually eventually be stopped as the body becomes capable of maintaining balance without them. Herbs with nutritive physiologic actions should be those tonics upon which all formulas are based.

SAMPLE CASE: VARIATIONS OF INSOMNIA

Patient 1: This insomnia patient has been exhausted for years, with an accompanying history of chronic hay fever and occasional respiratory infections requiring medical attention. The insomnia is of recent onset and involves not being able to fall asleep for many hours, lying in bed very tired and exhausted, but awake. The person will finally fall asleep and wake in the morning groggy and still exhausted. She is cool with cold hands and feet in bed at night and even during the day. She experiences gas and bloating if she eats raw broccoli and onions. For such a patient, you might consider basing a formula on "energy" or "chi" tonics, such as *Panax, Eleutherococcus, Astragalus, Rhodiola, Ganoderma, Cordyceps,* or *Oplopanax.* You might settle on a base of *Panax* because it is more warming than some of the others. For the specific, you might select *Astragalus* because it is also a chi and immune tonic, and it is also specific for allergies and respiratory infections that linger and exhaust. The synergist might be *Zingiber* because it enhances circulation in the cold extremities and improves digestion by warming and stimulating core organs. It could help drive the other ingredients in the formula where they need to go by being a heating, stimulating, and moving herb.

Patient 2: This second patient has suffered from restless sleep for many years, but it is presently worse. The pattern has been to fall asleep readily but to wake after a few hours and then fall back to restless sleep for the remainder of the night, waking briefly every half hour or so, tossing and turning with stiff muscles, and going back to sleep for short stretches only. She has occasional episodes of feeling too hot in bed and rolling away from her partner, feeling uncomfortably warm, and taking a while to fall back asleep. The person has come for a consultation because recently, instead of frequently waking and rolling over and going back to sleep, she now lies awake for half an hour, sometimes several hours, before falling back to sleep. She also experiences back and joint stiffness at night in bed and for a short while in the morning upon waking. She reports having only a few minor colds per year, recovers quickly without needing treatment, and doesn't even feel terribly ill with them. She reports having oily skin, still prone to breakouts on the central face, especially premenstrually, and having some minor PMS symptoms, primarily emotional lability and crankiness.

There are many relevant details here. For this restless, stiff insomnia patient, who is excessive in movements, muscle tension, nervous tension, and heat, you want to rest a formula on something cooling and relaxing, particularly to the nervous and musculoskeletal systems, such as *Avena*, *Passiflora*, or *Scutellaria*. While *Valeriana* and *Piper methysticum* are two of the most powerful herbs specific for both sleep and muscle relaxation, you might choose not to "rest" the

formula on them, because they are rather hot and could be a problem long term or if put in the formula in too great a proportion. They are not only heating but also not particularly nourishing and never listed in the folkloric literature as nervous system restoratives or daily long-term-use herbs. *Valeriana*, *Actaea*, and *Piper methysticum* could certainly be used in the formula, as long as care is taken to use them as synergists or

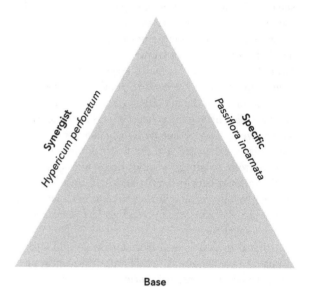

Base
Curcuma longa

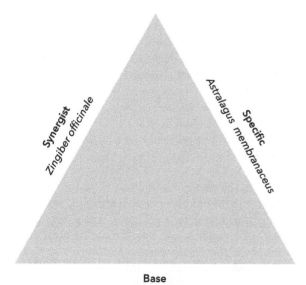

Base
Panax ginseng

Figure 1.2. Formula for Insomnia Patient with Exhaustion

Base
Avena sativa

Figure 1.3. Two Formulas for Restless Insomnia

specifics in smaller proportion than the other ingredients in the formula, while basing the formula on something more restorative to the nervous and musculoskeletal system and cooling to the body, such as *Avena* or *Scutellaria*. Another approach might be to consider adding *Valeriana* or *Piper methysticum* to a formula for short-term use or to be taken only before bed to reap the benefits of the more powerful muscle-relaxing effects of these herbs. In addition, you would create a separate formula to be used during the day and over the long term, attempting to restore the nervous and muscular tone so that the more powerful muscle relaxers and before-bed formula are no longer needed. *Scutellaria*, *Matricaria*, *Eschscholzia*, *Hypericum*, *Passiflora*, *Avena*, and *Melissa* are cooling, soothing, nourishing, and restorative to the nervous system, with *Scutellaria* and *Passiflora* having the greatest effects on muscle tension as well. Many people with insomnia, who are not particularly exhausted like the first insomnia patient, benefit from sedating, relaxing herbs like *Valeriana*, while patients with insomnia and long-term exhaustion might only become further tired or even lethargic and depressed with strong sedatives like *Valeriana* or *Piper methysticum*.

Other considerations or clues for the second insomnia patient are her oily skin, skin eruptions, aggravation by elevated premenstrual hormones, and hormone-related emotional tension. Although these are all very minor and very common in the general population, they are all heat symptoms, as well. These symptoms also suggest that the liver and hormonal metabolism may be contributing underlying factors in this person's overall constitution and balance. The liver not only processes hormones and removes waste products from the blood that can otherwise contribute to acne and skin eruptions, but in TCM, the liver is also said to "rule" the joints and tendons. In many traditions, including TCM, liver herbs are often said to be specific for vague muscle stiffness that does not represent tendonitis, arthritis, fibromyalgia, or other conditions. Therefore, a good synergist for this case might be a cooling liver herb noted to improve skin and hormonal balance. Some choices here might be a simple alterative such as *Taraxacum*, *Arctium*, *Silybum*, or *Curcuma*. Thus, for the second insomnia patient, you might end up with a formula such as *Hypericum*, *Passiflora*, and *Curcuma*, to be taken multiple times per day for many months, with a before-bed formula of *Avena*, *Piper*, and *Valeriana*. Yes, there is some arbitrariness in the selection

of herbs, but only to a degree. The arbitrariness stems from a somewhat capricious choice between herbs that have very similar energies and specific indications, or mixing and matching formulas in such a way that the overall base and specific energy is the same. It is possible to create several variations of a formula having nearly identical action. For example, both *Taraxacum* and *Arctium* roots may be somewhat interchangeable as an alterative base in a formula. Either fennel or caraway seeds may offer interchangeable carminative effects in an IBS formula. *Valeriana* or *Piper methysticum* might be effective in the previous insomnia case. Or a synergist of *Curcuma* or *Silybum* might be logical to help the liver process hormones.

SAMPLE CASE: VARIATIONS OF ACNE

Patient 1: The first case is a 25-year-old woman with acne. History reveals frequent heavy menses, PMS, including menstrual headaches, breast tenderness, and significant anxiety, along with chronic constipation as concomitant complaints. She rarely suffers from infections but when she does, they will come on quickly with painful sore throats, high fevers, and acute illnesses. She will recover quickly in a day or two. She also has frequent episodes of insomnia and muscle pain in the neck and upper back, most often premenstrually.

Figure 1.4 shows possible herbal combinations based on these considerations of the patient's constitution, action of herbs, and energetics.

4 Elements	Actions of Herbs	Energetics
Seek water to balance fiery symptoms	Seek alteratives, cholagogues	Seek cooling
		Seek sedating
	Seek hormonal-balancing agents	Seek yin tonics
Seek earthy therapies and herbs to ground the volatile tendencies	Seek nervines	
	Seek detox	

Patient 2: The second case is also a 25-year-old woman with acne, but her history reveals minor allergies, hay fever, frequent upper-respiratory infections, occasional yeast vaginitis, and tetracycline use for several years. She is chilly, has a damp constitution with much mucus production, tends toward loose stools, and has low-grade infections that linger a long time. Figure 1.5 shows possible formula combinations appropriate for this case.

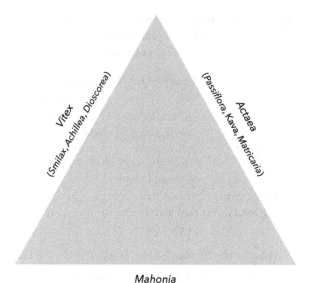

Vitex
(Smilax, Achillea, Dioscorea)

Actaea
(Passiflora, Kava, Matricaria)

Mahonia
(*Taraxacum, Arctium*)

Figure 1.4. Formula Possibilities for Acne with Fiery Symptoms

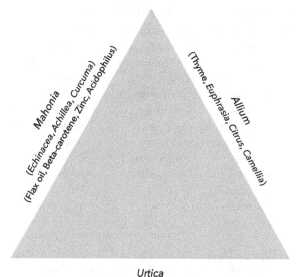

Mahonia
(Echinacea, Achillea, Curcuma)
(Flax oil, Beta-carotene, Zinc, Acidophilus)

Allium
(Thyme, Euphrasia, Citrus, Camellia)

Urtica
(*Taraxacum, Calendula*)

Figure 1.5. Formula Possibilities for Acne with Dampness and Infections

4 Elements	Actions of Herbs	Energetics
Seek warming herbs to balance cold	Seek immune stimulants	Seek chi tonics
		Seek yang tonic
	Seek antimicrobials	Seek warm and drying tonics
Seek drying herbs to balance damp	Seek intestinal, vaginal, and respiratory astringents	Seek to purge fluid and dampness
Seek fiery herbs	Seek antiallergy herbs	Seek to move medicine to head, skin, upper respiratory tract

SAMPLE CASE:
VARIATIONS OF RHEUMATOID ARTHRITIS

Patient 1: Our first example is a 50-year-old woman with a chief complaint of rheumatoid arthritis, primarily in the hands, wrists, neck, and shoulders. Onset was associated with conflicts and issues with children, loss of control in influencing children's lives, and disappointments. She reports minor anxiety and frequent bouts of insomnia. She also suffers from frequent constipation, a chronic cough due to a dry scratchy throat, and occasional brief episodes of cystitis. Her hands become red and swollen in acute episodes that are experienced as aching, burning, sore, and tender and then settle down over a month's time. Her neck becomes tight and spastic,

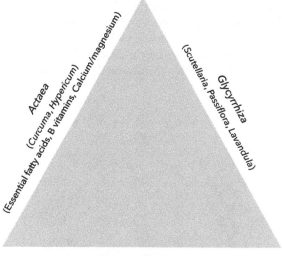

Actaea
(Curcuma, Hypericum)
(Essential fatty acids, B vitamins, Calcium/magnesium)

Glycyrrhiza
(Scutellaria, Passiflora, Lavandula)

Piper methysticum
(*Avena,* Gotu kola, *Althaea, Symphytum*)

Figure 1.6. Formula Possibilities for Rheumatoid Arthritis with Stress

with a throbbing headache and burning sensation on the skin. Symptoms wax and wane independently of diet or activity, but perhaps correlate to stress. She has a trim build, is often warm, is often thirsty, has a big appetite and a fast metabolism, and is very active. See figure 1.6 for formula possibilities for this case.

4 Elements	Actions of Herbs	Energetics
Seek to cool fire	Seek nervines	Seek to cool
Improve dryness, heat with watery, cooling herbs	Seek anti-inflammatories	and moisten
	Seek antispasmodics	Aim to soften, lubricate, and smooth
Improve insomnia, restlessness, and worry with grounding, earthy herbs	Seek tissue lubricants	Seek to quiet and calm energy
	Seek nerve and connective tissue tonics	Aim to move medicines to nerves
Aim to ground with moist earthy herbs		

Patient 2: The second example is a 50-year-old woman with a chief complaint of rheumatoid arthritis—the arthritic pain is in multiple joints including the low back, hands, wrists, shoulders, and hips. There is a family history of rheumatoid arthritis and osteoarthritis. Constant mild to moderate stiffness is worst after sleeping or prolonged inactivity and also is aggravated with exertion. She has some minor arthritic and degenerative changes in the low back and some bony deformity beginning in the finger joints. The sensation is heavy, stiff, and aching and is better when resting the affected limb. She tends to be chilly, has chronic postnasal drip, is on hormone

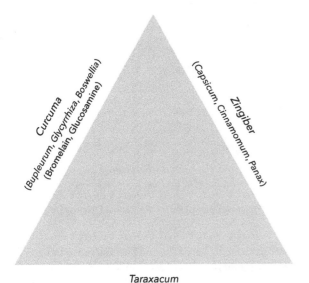

Taraxacum
(Gotu kola, Nettles, *Equisetum*)

Figure 1.7. Formula Possibilities for Rheumatoid Arthritis with Cold Constitution

replacement therapy, retains fluid, has mild constipation, is slightly overweight, and reports low energy. Figure 1.7 shows formula possibilities for this case.

4 Elements	Actions of Herbs	Energetics
Seek to dry out excess water	Seek anti-inflammatories, anodynes	Seek warming and drying
Seek to warm, fire	Seek to move fluid, diurese	Seek stimulating, moving therapies
Seek to lift, lighten with airy or volatile compounds	Seek connective tissue tonic	
	Seek alterative detox therapies	

The Use of "Toxic" Herbs in Formulas

In the sample case of insomnia with exhaustion where *Zingiber* is used as a synergist because of its heating and moving energetic qualities, *Zingiber* might be used in a lesser proportion than the other two herbs, in order to ensure the formula is overall nourishing, restorative, and tonifying. *Zingiber* is not particularly toxic, but because of its strong energy, there is a minor concern over its proportion in a formula. This is especially true for caustic, irritating, or outright toxic botanicals. Some such toxic or otherwise powerful herbs have strong moving or driving action: *Phytolacca* is a lymph mover, *Gentiana* is a bile mover, *Rhamnus purshiana* is an intestinal smooth muscle mover, *Sanguinaria* is a tissue mover, and *Pilocarpus* is a secretory stimulant. All are used in formulas in a small or lesser quantity than the lead herbs, and yet their presence is equally contributory and as powerful as the other primary ingredients.

The case is the same when energy or motion is excessive, and a goal of the formula is to calm and soothe or to quiet down an excessively hot or hyperfunctioning action in a tissue or organ. *Veratrum, Aconitum, Atropa belladonna, Digitalis, Lobelia,* and *Conium* all have powerful relaxing, sedating, diminishing effects on nerve, cardiac, respiratory, and musculoskeletal tissues, appropriate only in small amounts in the overall formula.

In many cases, herbs with such extremely strong energies are added to formulas only when extremely strong imbalances are occurring. In general, the stronger and more potentially toxic or harsh the herb, the smaller the dose in the formula. And yet even that milliliter, or as little as five drops, drives the formula and contributes equally to the other herbs occurring at a dose of ten or even one hundred times the amount. The smaller the degree of atrophy/

hypertrophy or the more minor the degree of imbalance of hot/cold or excess/deficient status of the body, the less the need for any strong-energy herbs, and the more the formula can be based on purely nourishing and restorative ingredients for the base, synergist, and specific alike.

In the previous example, the insomnia patient is so exhausted and so cold that a little bit of *Zingiber* to warm up our formula is appropriate. *Zingiber* is quite hot but overall nontoxic and safe, even long term. You would not include *Podophyllum peltatum* or *Iris* right off the bat to warm the bowels, and you would not choose *Thymus* or *Allium* (she is sensitive to onions and might react) to warm the lungs and reduce infections. None of those herbs are as indicated or specific as *Zingiber* to drive the formula into the proper locations. Thus, for this insomnia patient, you might end up mixing a formula of 30 milliliters *Panax*, 20 milliliters *Astragalus*, and 10 milliliters *Zingiber* to fill a 60-milliliter (2-ounce) dropper bottle of tincture.

Further Guidance on Creating Warming and Cooling Formulas

The actions of herbs—both traditional folkloric and mechanistic—can be organized into simple categories of warming or cooling. Vasodilators, for example, and the hot spicy "blood movers" are generally warming, and demulcent herbs are generally cooling.

Warming stimulants. Here are some general guidelines for the use of warming stimulants in treating acute conditions.

- Use for colds, fever, and chills with acute onset
- Use for abundant mucus, phlegm in the throat and lungs
- Use as diaphoretic for those who fail to mount a useful fever in acute illness
- Use for many common infections of childhood in small doses
- Discontinue therapy when improvement is achieved
- Contraindicated in yang constitutions, for those who feel uncomfortably hot
- Contraindicated in hemorrhage, free sweating, or night sweats

In cases of chronic disease, observe these guidelines for warming stimulants.

- Use for circulatory enhancement, excessive clotting, blood stasis
- Use for fluid accumulation, poor circulation to organs

Warming Stimulants

ACTION	HERBS
Peripheral vasodilators	*Achillea millefolium*
	Allium sativum
	Cinnamomum spp.
	Ginkgo biloba
	Zingiber officinale
Secretory stimulants	*Armoracia rusticana*
	Capsicum spp.
	Iris versicolor
	Pilocarpus jaborandi
Diaphoretics	*Achillea millefolium*
	Capsicum annuum
	Zingiber officinale
Warm antimicrobials	*Allium sativum*
	Curcuma longa
	Origanum spp.
	Thymus vulgaris
Chi tonics	*Eleutherococcus senticosus*
	Panax spp.
	Withania somnifera

- Use for chronic inflammation with stiff and swollen character that feels better in hot weather or after a hot bath
- Use for cold hands and feet
- Use for atonic, feeble, sluggish constitutions
- Don't use for high fevers due to chronic debility, rather for weakness and sense of chill

Cooling remedies. These are guidelines for the use of cooling remedies in treating acute conditions.

- Use for acute mucosal inflammations, cool tissues with demulcent anti-inflammatories
- Use for burning sensations in the throat, intestines, or skin
- Use for tight, dry sensations in the mucous membranes or joints
- Use for acute infections with fever or sensation of heat; cool with antimicrobials and alteratives
- Use for acute toxicity states, joint pain, or headache; cool with bitters, alteratives, diuretics
- Contraindicated for acute disease with chills, abundant watery mucus, consolidation in the lungs

Cooling Remedies

ACTION	HERBS
Demulcents	Aloe vera, A. barbadensis Althaea officinalis Ulmus fulva Verbascum thapsus
Bitters/alteratives	Arctium lappa Berberis aquifolium Rumex crispus Taraxacum officinale
Diuretics	Apium graveolens Equisetum arvense Galium aparine Petroselinum crispum
Cooling antimicrobials	Echinacea purpurea Hydrastis canadensis Lomatium dissectum Mentha piperita
Astringents	Geranium maculatum Hamamelis virginiana Rubus idaeus
Energy dispersants	Achillea millefolium Galium aparine Iris tenax, I. versicolor Mentha piperita

In cases of chronic disease, follow these guidelines when using cooling remedies.

- Use for dry, hot constitutions; use bitters, demulcents, alteratives, and diuretics
- Use for general yang states with warmth, redness, and heat in the body
- Use for tendency to acute fevers, infections, inflammation, and joint pain
- Contraindicated for chronic diseases associated with excessive dampness, coldness, deficiency, and fluid stasis

Types of Herbal Medicines

Herbs are available as teas, tinctures, powders, encapsulations, syrups, solid extracts, and other forms. Following are pros, cons, and indications for the most common types of these herbal preparations.

Herbal Teas

Teas are especially indicated for individuals with digestive and bowel weakness to avoid irritation by pills or sometimes tinctures and to improve absorption of desired substances. Teas are also indicated for urinary ailments where substances in water quickly reach the urinary passages. Demulcents to soothe inflamed mucosal surfaces, such as the esophagus, stomach, intestines, and bladder, are best delivered via teas. Teas are also relatively inexpensive and, when of good quality, can be a very effective method of getting nourishing herbs into the body, such as *Urtica, Centella, Medicago,* and *Equisetum,* which are all high-mineral herbs that support connective tissue and bone. When the desired herbs are particularly unpalatable, tinctures or pills would be friendlier. Or when a high dosage of an herb, action, or chemical is desired, it may be easiest to accomplish that with pills or tinctures.

Tinctures

Tinctures are especially handy when fresh plant juices are desired for preservation. For the home herbalist, tinctures are a practical way to stock the medicine chest with inexpensive, valuable medicines with a long shelf life. Tinctures will store for many years, whereas dried herbs age as the months go by and should be replaced about every two years to be of medicinal quality.

To any clinical herbalist, another important virtue of tinctures is that they can be combined into formulas as precisely indicated for a given individual, where commercially available pills are a one-size-fits-all formula. Tincture prescriptions can be formulated in an exacting manner to reduce the need to take many different medications to address many different organs, imbalances, and pathologies. Tinctures, when well thought out, can address many different pathologies, levels, and energies all at once.

Pills

Tablets and capsules can be convenient and practical ways to combine herbs with vitamins, minerals, amino acids, protomorphogens, bile, and numerous other diverse substances. Formulas designed for treating everything from prenatal nutrition to congestive heart failure or hypothyroidism are available, and when well indicated they are very helpful. However, unless you are producing your own encapsulations, you are usually limited to prescribing the pill combinations as the proprietor has formulated them. Because they might not be as specific for particular presentations as your own formulas, you might complement proprietary encapsulations with more specific herbs in tea or tincture form as needed to round out the herbal prescription.

Pills are also useful when high dosages are being pushed for urgent health problems or when numerous different medications are being employed at once. Herbal pills may be simply dried and powdered plant material, or they may contain purified and concentrated plant constituents, such as curcumin, manipulated to boost absorption and assimilation.

Knowing When to Use Which Form of Medicine

When there are urgent circumstances, standardized or concentrated medicines might be desirable for their concentration and known potency, such as using standardized *Ginkgo* for serious ischemic disease or *Convallaria* for heart failure or silymarin concentrate for acute *Amanita* toxicity. On the other hand, there is no reason to take expensive standardized *Matricaria* when a dollar's worth of good-quality chamomile tea would likely do the trick for a flare-up of indigestion or irritable bowel.

When treating complex and chronic disorders, it is often appropriate to use many different herbs or supplements at once. For example, a protocol might call for ginkgo, garlic, milk thistle, passionflower, and echinacea, along with a multivitamin, flaxseed oil, and some additional beta-carotene, zinc, magnesium, and calcium. A person on a budget (which is almost everyone) would not be able to keep up with such a program for very long if each item had to be purchased as a separate bottle of pills. Not to mention how difficult it is for many people to take handfuls of pills day in and day out. However, if the herbs could be combined in a tincture and a tasty tea and supported with the perfect vitamin and herbal/ nutritional combination, the cost and convenience could both be improved.

Preparing Formulas for Use

Teas and tinctures are used throughout this formulary. With regard to herbal teas, as a general rule, delicate plant parts, such as flowers, are steeped rather than boiled, because the aromas and flavors can be lost or destroyed with vigorous boiling. Green leafy herbs, such as nettles, alfalfa, and mint, are also gently steeped, but harder, denser plant parts, such as roots, barks, dried rose hips or hawthorn berries, and seeds, are best simmered to extract the medicinal components. Simply steeping herbs is referred to as preparing an infusion, while gently simmering herbs is referred to as a decoction. Unless otherwise specified, it is ideal to infuse teas for 10 minutes. For herbs that are decocted, simmer them gently for 10 minutes and then remove from the heat to stand 10 to 15 minutes more. In a few cases in this text, I recommend steeping or simmering for a longer period, when the intention is to liberate minerals or other compounds that are not readily released with simple infusions or decoctions. In some cases, it is recommended that vinegar or lemon juice be added to the water to best extract minerals. In other cases, it is recommended that mucilaginous herbs be macerated in cold water for many hours or overnight before bringing to a brief simmer to best extract mucilage.

A general dosage for a tea is a minimum of 3 cups (720 ml) a day. For an acute or urgent situation, however, such as a chest cold threatening to turn into pneumonia, a more aggressive approach would be recommended, such as preparing an 8- or 10-cup (1,920 to 2,400 ml) pot of tea with a goal of drinking 1 cup (240 ml) every hour for several days.

Herbal tinctures are also used throughout this formulary with recipes given for the proportions and ratios. Because commercial tinctures are readily available, I do not go into detail discussing the making of tinctures. However, in some cases, herbal oils or vinegars that are not readily available are used in formulas, and in these cases, I offer brief details on how to prepare them. I also provide a few unique methods of preparing formulas, such as using solid extracts to thicken formulas to help them cling to the esophagus or placing the formula in a spray bottle to use as a throat spray.

A general dosage strategy for tinctures is to use 1 dropperful (½ to 1 teaspoon) three times a day when treating chronic conditions. Most people have difficulty taking medicines more often than this, unless they are acutely ill or uncomfortable and motivation is high. In acute situations, tinctures may be taken as often as hourly for pain or infectious illnesses, or even as often as every 10 minutes. When tinctures or other herbal medicines are taken at this aggressive dosage, it is always for a very limited length of time, such as every 10 minutes for an hour, or every hour for a day, with instructions to reduce the frequency as symptoms improve.

Although the focus of the book is to create effective herbal formulas, in some cases, protocols are offered to create comprehensive therapies for chronic conditions such as osteoarthritis. In such cases, I may suggest the use of a tea, a tincture, medicinal foods, and herbal encapsulations to get as many bone-building nutrients into the body as possible, supported by digestive-support supplements to ensure proteins and minerals are well assimilated, especially in the elderly, whose stomach acid or pancreatic enzymes may be deficient. When herbal capsules are employed, they are often dosed as little as two per day or as much as two or three at a time, three times a day. While teas and tinctures can be taken with or without food, herbal capsules are most often taken with meals to enhance their digestion and to prevent them from being nauseating on an empty stomach.

Extending the Triangle Philosophy

As explained, herbal formulas exemplified throughout this book aim to use nourishing herbs as base ingredients in all formulas, complementing them with specifics and synergists. All the formulas include an introductory sentence or two explaining why the herbs are chosen, cite any research to support their use in specific conditions and situations, or comment on why they are appropriate for a specific presentation of the condition being discussed. I encourage readers to refer to "Specific Indications" at the end of each chapter to help master *materia medica* knowledge. Staying focused on the desired actions can help you avoid creating a formula that is too broad. With so many available herbs to choose from, knowledge of niche indications for various herbs helps prevent a kitchen-sink approach. Instead of choosing every herb you know of that may be effective for hypertension or respiratory viruses, you can select the best choices with a solid knowledge of *materia medica*. It is a common mistake to place too many herbs with the exact same function in a formula and miss including herbs that address the underlying cause or other special considerations. For example, beginning herbalists may attempt to put half a dozen herbs known

to help regulate blood glucose in a tincture, where some of those herbs would actually be more effectively given as medicinal foods (garlic, berries, fiber). The more medicine-like herbs could thereby be taken at a higher dose when just two or three are chosen for a tincture. In other cases, herbalists may place two or three adaptogens in a formula but overlook and omit hepatotonics, or use two blood movers that have similar action and forget the high-flavonoid vascular protectants. Another common mistake is to choose herbs based on a research study and be wooed by the newness of the information, thereby paying top dollar for a proprietary product when common alterative herbs and comparatively mundane herbs may be just as effective at a fraction of the cost. By choosing herbs with a desired action in mind or by listing all the actions that one wishes an herbal formula to perform, such redundancies and omissions are less likely. And by considering which herbs with the desired action best fit the person and energetic state—warming or cooling, stimulating or sedating, moistening or drying, and so on—one can craft a finely tuned formula that addresses the individual person rather than making a kitchen-sink attempt to treat the diagnosis. Once one has a solid knowledge of the *materia medica*, the specific indications of the herbs, prescribing and formulating become somewhat intuitive. This becomes easier, and even second nature, the longer one focuses on making formulas and the more experience one obtains.

Many herbalists believe that the plants are their teachers and that they learn from the plants themselves over time, as the plants teach us what they are capable of. Many herbalists are deeply aligned with nature and often have cultivated nature-based spirituality. Herbalists may embrace the notion that clinical intuition is evidence of a connection with the plants, and that this is the real root of herbal medicine as a healing discipline.

— CHAPTER TWO —

Creating Herbal Formulas for
Allergic and Autoimmune Conditions

Allergies are hypersensitivity disorders in which the immune system is hyperresponsive to fairly innocuous irritants such as dust or cat dander, triggering excessive activation of mast cells and release of histamine. Autoimmune diseases are analogous in that the body produces antibodies inappropriately, which initiates inflammatory cascades that can damage the tissues and cause numerous chronic symptoms. Exposure to toxins, mold, and chemicals may be underlying contributors to both allergies and autoimmune diseases.

Similar underlying molecular disorders contribute to both allergic and autoimmune disorders. Furthermore, many herbs that have broad immunomodulating effects are used to address both allergic and autoimmune conditions. Because of this, I've grouped these categories together in this chapter.

Allergic and autoimmune disorders can affect every organ system, and thus the conditions covered in this chapter overlap with those discussed in other volumes of *Herbal Formularies*. For example, irritable and inflammatory bowel diseases are discussed in Volume 1, with mention given to possible allergic reactivity and immunomodulatory disorders. In this chapter, I provide greater detail regarding the precise mechanisms of auto-inflammatory processes, as well as the research on herbal medicines that address those upregulated immune pathways. Similarly, asthma is discussed in chapter 3, Volume 2, with a discussion of herbs that have bronchodilating, expectorating, and antimicrobial effects on the airways. Here, I offer greater detail on allergic mechanisms that contribute to allergic airway disorders. Eczema is discussed in chapter 5, Volume 1; here, the focus is on allergic phenomena that contribute to atopic disorders of the skin, including eczema and hives. Also, some conditions discussed in other volumes, such as intestinal dysbiosis and leaky gut (discussed in detail in chapter 2, Volume 1) may contribute to both allergies and autoimmune disease, provoking hypersensitivity reactions and immune activation. I point out such overlap and include cross-references in this chapter where appropriate to alert readers to additional sources of information in other volumes.

Within this chapter, I begin with an overall discussion of allergic disorders, followed by groups of formulas for allergic conditions ranging from allergic airway disorders to dermatitis to vascular reactivity. The second half of the chapter focuses on the phenomenon of autoimmune disease, beginning with the section "Formulas for General Autoimmune Conditions," and provides formulas for conditions such as Sjögren's syndrome, lupus, IBS, and more.

Understanding Allergic Disorders

Atopy, the predisposition to be hypersensitive to various environmental antigens, involves the release of immunoglobulin E (IgE), which in turn triggers the release of vasoactive substances from leukocytes and mast cells. Antiallergy herbs may reduce allergic phenomena via direct effects on mast cells, platelets, and white blood cells such as eosinophilic granulocytes and reduce the release of histamine and inflammatory mediators. Common allergens include mold, dust, cat dander, pollen, and chemical antigens, and exposure to such substances can trigger the release of specific inflammatory mediators and antibodies in atopic individuals. In many cases the balance of T helper lymphocytes becomes altered and can drive long-term hypersensitivity and disease. Although vaccines to protect against the development of excessive type 2 T helper cells (Th2) are being developed, they are controversial at the time of this writing, and allergen-specific immunotherapy is another primary form of treatment of allergies. This approach is further discussed in the "Formulas for Inflammatory Bowel Disease" section on page 78, where the use of intestinal worms for similar purposes is discussed.[1]

On the other hand, some microbes and parasites, such as *Giardia* protozoa, may injure the tissues and promote allergic phenomena.[2] Damp indoor work environments foster mold growth and can activate the immune system via numerous mechanisms and lead to allergies such as allergic rhinitis.[3] The respiratory, digestive, and even urinary mucosa can become hypersensitive to various antigens and should be suspected when an individual with a history of any previous type of allergic symptoms develops new or chronic urinary, respiratory, or digestive complaints. In a chicken-and-egg cycle, impaired digestive health and particularly leaky gut syndrome can initiate allergic hyperreactivity in those without a significant personal or family history of atopic conditions previously. As the formulas in this chapter exemplify, herbs that support digestive and liver health and strengthen gut barriers are appropriate in formulas for allergies, especially when underlying leaky gut or intestinal dysbiosis are suspected to contribute to chronic allergic reactivity.[4]

Emotional stress may also increase allergic sensitivity in atopic individuals; hence, some physicians may offer B vitamins and include nervines in herbal formulas for asthma or for stress that induces hives or exacerbates inflammatory bowel conditions. Even though corticosteroids are used to suppress allergic symptoms, stress-induced cortisol surges can exacerbate immune hypersensitivity,[5] making adaptogenic and immunomodulating herbs helpful to include in some protocols for allergic issues. *Astragalus membranaceus, Ganoderma lucidum, Panax ginseng,* and other adaptogenic herbs are featured in many traditional Chinese formulas for allergic and autoimmune diseases, and these herbs appear in formulas throughout this chapter. While chronic stress is well known to impair the immune system and increase the risk for simple and complex infections, stress of all types can also impair cellular immune responses and promote excessive humoral immunity, thereby exacerbating some inflammatory and hypersensitivity states. Stress can, for example, alter the effects of glucocorticoids and catecholamines on cells, and especially alter the ratio of the T helper cells (Th1 to Th2 ratios, further discussed below) and the types and quantities of cytokines released.[6] Furthermore, the psychic effects of acute stress may provoke inflammation via effects on corticotropin-releasing hormone pathways that ultimately activate mast cells and promote the release of histamine. Most adaptogenic herbs also reduce inappropriate mast cell activation.

Various polyphenols in foods and medicinal herbs have potent antiallergic activities; catechins, anthocyanidins, flavanones, procyanidins, and phytoalexin improve a skewed ratio of Th1/Th2 cells and suppress formation of antigen-specific IgE antibodies. Clinicians can brainstorm with patients about how to make diet regimens as high in polyphenols as possible, such as consuming colorful fruits and berries, beets, carrots, dark green leafy vegetables, and mushrooms. It can also help to add supplements such as green tea catechins, resveratrol, bilberry extracts, and so on to the medical protocols to complement specific herbal formulas. Essential fatty acids (EFAs) are another foundational therapy used by alternative medical practitioners to treat atopy because EFAs play important roles in cellular reactivity and response. Supplementation with fish oil, flaxseed oil, evening primrose oil, and quality nut and other seed oils helps to provide important fatty acids. For more about the anti-inflammatory mechanisms of nut and seed oils, refer to chapter 2, Volume 4.

The Role of T Cells in Allergy

Much of the research on herbs for treating allergic disorders involves complex inflammatory cascades that originate in groups of white blood cells, in particular T helper cells, which are a type of lymphocyte. T cells play key roles in allergic inflammation, and atopic individuals show a bias toward T helper type 2 (Th2) cell regulation, overpowering T helper type 1 cells (Th1). Allergic diseases involve aberrant Th2-type immune responses against common innocuous environmental antigens such as pollen or dust; such responses are exacerbated by infection-triggered shifts in the balance of the Th1 to Th2 cells and their ease of activation. Th2 domination over Th1 contributes to rhinosinusitis, asthma, atopic dermatitis, chronic hives, severe insect venom reactivity, and other allergic phenomenon.[7] In these allergic disorders, excessive numbers of Th2 cells are excessively active and release large quantities of cytokines, particularly interleukins, which in turn activate mast cells and eosinophils to release histamine and other inflammatory mediators. T helper cells release various types of interleukins, and imbalances between Th1 and Th2 cells may also predispose to various diseases. A full discussion of the types and actions of inflammatory mediators such as the various cytokines and interleukins is outside of the scope of this text. Suffice it to say, some types of interleukins are triggered by bacterial and viral infection,[8] and in this way early childhood infections and significant intestinal dysbiosis can contribute to chronic allergic reactivity, especially in those genetically predisposed to atopy.

A strong association between childhood seborrhea and adult eczema, and childhood rhinosinusitis and adult asthma, has long been clinically observed in which early shifts in lymphocyte balance persist into adulthood. It is also noted that some lactose- or peanut-intolerant children "outgrow" their food allergies, and some beekeepers become somewhat immune to bee venom following repeated stings. The use of miniscule doses of allergens to slowly build tolerance is a time-honored

Types of Immune Responses

The following categories of immunoreactivity are common terms in medicine describing types of healthy, normal immune responses in the body as well as excessive or pathologic immune responses.

The innate immune response is a nonspecific response including epithelial barriers, phagocytes, and the complement system.

The adaptive immune response is a specific response dependent on the pathogen and can include the humoral immune response or the cell-mediated response.

The humoral immune response involves the production and release of antibodies by B-cells (IgM, IgD, IgG, IgE, and IgA).

The cell-mediated response involves the activation of T-cells, including CD4, CD8, and cytotoxic T-cells.

The hypersensitivity response involves unbalanced or excessive production of histamine, leukotrienes, prostaglandins, and platelet-activating factor and is the underlying phenomena of allergies.

Anaphylaxis is an acute, rapid, severe hypersensitivity response usually to food, drug, and insect antigens, such as severe peanut or shellfish allergy.

The cytotoxic response involves the binding of antigen-antibody complexes to a cell, damaging or destroying it. Some severe examples are hemolytic anemia and idiopathic thrombocytopenic purpurea. Cytotoxic immune responses can also include processes that the body uses to fight intracellular pathogens (such as chlamydia and TB) and cancer.

therapeutic approach to treating allergies—by injecting tiny amounts of the triggering substance into the skin, referred to as desensitization therapy, allergic reactivity is reduced slowly over time. Some people claim that the oral consumption of local pollen each winter reduces their spring hay fever, and some allergists have reported that consumption of food allergens in escalating doses is sometimes effective in reducing allergic reactivity to foods.[9] Due to these observations, desensitization therapy became one important prong of treating allergic disorders. In 1911, British scientist Leonard Noon published his success with injecting grass pollen into hay fever sufferers, and modern immunologists report that recovery from hypersensitivity states occurs when the balance between Th1 and Th2 ratios is reestablished.

The recommended therapy for hypersensitivity disorders therefore can range from strict avoidance of allergens to allow the immune system to calm down, to the use of tiny doses of the offending substance to build tolerance. Desensitization therapy with allergy shots is not uniformly effective, and some alternative medicine practitioners avoid this as a first line therapy. Increased understanding of the role of T regulatory (Treg) cells, another type of lymphocyte, further suggests that genetic and/or environmental factors can blunt the development of atopy. Immunologists propose that a defect in Treg function may allow inflammatory interleukins and T-cell activation to increase rather than decrease in some people, which may explain why some people fail to respond to allergy shots and desensitization therapy. No matter what treatment approach they choose, many herbalists aim to simultaneously support the intestinal microbiome and offer anti-inflammatory/ antiallergy herbs and nutrients.

Vitamin D is helpful to many allergic conditions and has long been used to treat dermatitis. Vitamin D has many immunoregulating properties, and activated lymphocytes possess receptors for vitamin D. When bound, these receptors downregulate interleukin 2 signaling and cytokines, enabling vitamin D to temper excessive allergic inflammation. Vitamin D is appropriate in protocols for multiple sclerosis, atopic dermatitis, and all allergies and autoimmune diseases.

Even though steroids are a mainstay drug therapy to suppress allergic reactivity, the side effects are numerous and usually not worth the consequences in any but the most severe cases, or in situations of life-threatening allergic reactivity. Most alternative health physicians avoid the use of steroids, both topical and oral, as a primary therapy for chronic allergic conditions, aiming higher by working to correct immune system imbalance, gut health, and nutrition and to reduce stress and exposure to environmental triggers.

The Hygiene Hypothesis and Allergy Etiology

The hygiene hypothesis acknowledges that epidemics of allergy-prone populations have been observed in the developed world, yet are relatively uncommon in some tropical areas with very poor sanitation. While a certain level of control over bacterial populations is essential for public health, the lack of exposure to basic microbes also appears to be detrimental to health as well. For example, exposure to livestock and basic farm microbes in the early years of life appears to guard against the development of allergies. It appears that excessive sanitization and living in relatively microbe-free environs impairs not only the development of a healthy, balanced immune system but also its function throughout life. The hygiene hypothesis posits that a lack of early microbial

---

Factors Contributing to Allergic Disorders

The atopic tendency is first and foremost a genetic phenomenon; genetic predisposition remains the single most important risk factor.[10] Exposure to antigens can also contribute to the onset, severity, and chronicity of allergic disorders. Here is a summary of agents and circumstances that may increase susceptibility to chronic allergic disease.

- Exposure to multiple allergens and respiratory viruses early in life [11]
- Allergic rhinitis early in life (associated with the development of asthma in adulthood)[12]
- Air pollution[13]
- Exposure to antibiotics; intestinal dysbiosis[14]
- Sensitization to dust mites, cats, dogs, and mixed grasses (all independently associated with asthma)[15]
- Lack of breastfeeding[16]
- Neonatal use of infant formulas[17]
- *Giardia, Blastocystis*[18]
- Maternal smoking during pregnancy

stimulation results in aberrant immune responses to innocuous antigens later in life. The hygiene hypothesis also extends to autoimmune disorders; lack of exposure to basic common microbes increases the incidence of rheumatoid arthritis, type 1 diabetes, and chronic inflammatory intestinal disorders.[19]

Further research has demonstrated the importance of Treg cells on dampening allergen-specific Th2 responses, but the evidence is complex and still poorly understood. Some microbes can suppress Th2 and are being explored as therapies for allergies, depending on their pathogenicity and risk-to-benefit ratio. *Pseudomonas aeruginosa*, for example, suppresses allergic airway eosinophilia, and *Bacteroides fragilis* is shown to correct Th1/Th2 imbalances in microbially deficient animals. Sensitivity to various allergens such as ragweed pollen and dust mites may be reduced in human subjects following treatment with synthetic DNA meant to mimic microbial nucleotides, and such products are being explored as a means of producing effective vaccines against development of allergies. There is even sound evidence that introducing hookworms or several other types of worms and parasites into the gastrointestinal system may be therapeutic to those with severe allergic hypersensitivity and autoimmune diseases. It has long been noted that people who live in places where tropical parasites abound suffer less asthma and autoimmune disease compared to other populations. For example, asthma is noted to be relatively rare in populations where intestinal helminths are endemic,[20] and treating intestinal helminths increases allergic skin reactivity.[21]

The Role of Mast Cells and Histamine

Mast cells are part of the immune system, harboring histamine and other compounds that are released in response to the presence of antigens binding to IgE, triggering the inflammatory cascades that are the hallmark of allergic disorders. Mast cells are typically found immediately underlying mucous membranes and reside in close proximity to small submucosal blood vessels. Mast cells respond to the entry of any potentially harmful and toxic substances. However, in the case of an allergic response, mast cells already primed with IgE degranulate when the antibody encounters the specific antigen. Thus, the histamine and other inflammatory mediators that mast cells release can readily enter the bloodstream. If the integrity of the intestinal mucosa is compromised, as in leaky gut, then incompletely digested proteins can trigger IgG-mediated hypersensitivity reactions, and

---

Herbs to Stabilize Mast Cells

The following herbs are included in folkloric remedies from around the world for treating hay fever, hives, and atopic dermatitis. Modern research has uncovered their mast cell–stabilizing effects.

Albizia lebbeck	*Grindelia squarrosa*
Allium cepa	*Inula racemosa*
Ammi visnaga	*Magnolia officinalis*
Azadirachta indica	*Mentha piperita*
Bacopa monnieri	*Ocimum sanctum*
Bidens parviflora	*Picrorhiza kurroa*
Cedrus deodara	*Solanum*
Clerodendrum serratum	*xanthocarpum*
Cnidium monnieri	*Tephrosia purpurea*
Coleus forskohlii	*Terminalia chebula*
Crinum glaucum	*Tinospora cordifolia*
Curcuma longa	*Tylophora asthmatica*
Drymaria cordata	*Vitex negundo*

---

microbial endotoxins can trigger the mast cells and contribute to allergic reactivity. (For more on this topic, see chapter 2, Volume 1.)

In other cases, mast cells are activated by immunoglobulin-bound allergens that attach to the cell surface, triggering the release of histamine, mast cell granule proteins, and various inflammatory mediators such as interleukins and tumor necrosis factor. Due to the important role that mast cells play in allergic phenomena, agents that reduce mast cell reactivity—known as mast cell stabilizers—can be valuable therapeutic tools.

Histamine is plentiful in central nervous system neurons, gastric mucosal parietal cells, mast cells, basophils, and other cells throughout the body. Once released, histamine binds to at least four different types of receptors and mediates allergic reactivity. It also plays roles in cell proliferation and differentiation, cell regeneration, and wound healing, in addition to having important neurologic and antiseizural effects.

Antihistamine drugs may block either the binding of histamine to its receptors, or the conversion of histidine into histamine. Pharmacology refers to "first-generation" and "second-generation" antihistamine drugs; first-generation antihistamines crossed the blood-brain barrier, leading to sedation and impaired cognitive and psychomotor performance. Second-generation antihistamines such as cetirizine, fexofenadine, and loratadine

Histamine-Blocking Botanicals

Histamine is a key contributor to rapid-onset allergy symptoms such as itching, swelling, watery secretions from mucous membranes, and formation of fluid-filled pruritic vesicles in the skin. All botanicals high in rutin and quercetin can reduce the rate of histamine release and decrease histamine levels. The following herbs also inhibit the release of histamine—from basophils and mast cells—and impair its tendency to increase membrane permeability.

Allium cepa
Anemarrhena asphodeloides[22]
Artemisia asiatica[23]
Capsella bursa-pastoris
Ephedra sinica[24]
Malus domestica[25]

Panax ginseng[26]
Picrorhiza kurroa
Scutellaria baicalensis
Tanacetum parthenium[27]
Tylophora asthmatica
Vaccinium species[28]

cross the blood-brain barrier to a significantly lesser extent than their predecessors and are increasingly prescribed to help treat hay fever and hives.

Some plant-derived natural compounds may block histamine receptors; such compounds include alkylamines, ethanolamines, ethylenediamines, phenothiazines, piperazines, and piperidines. The flavonoid khellin found in Ammi visnaga is one of the most researched such compounds and it has significant mast cell–stabilizing effects. Investigation into khellin led to the development of several important drugs, including nonsteroidal bronchodilating inhalers. Some antihistamine herbs may be consumed orally to achieve antiallergy effects over time, while other herbs may be prepared into eyewashes and nasal rinses to help relieve acute allergic rhinitis, sinusitis, and conjunctivitis symptoms acutely.

Treating Allergies with Botanicals

Most allergic individuals will remain so their entire lives, but therapies including herbs, diet management, nutritional supplements, and stress management can be helpful in reducing the severity, frequency of flares, and duration of allergic issues. The primary pharmaceutical approaches to treating allergic disorders are antihistamine drugs, corticosteroids, leukotriene inhibitors, and mast cell stabilizers. All of these medications are at least somewhat effective in alleviating allergic symptoms,

but none actually cure allergic disorders and may even worsen disorders over time by further injury to the gut and other undesirable long-term effects. Furthermore, many such pharmaceuticals have side effects. Antihistamines can cause dry mouth, drowsiness, gastrointestinal disturbances, headache, agitation, and confusion. Corticosteroids injure intestinal barriers, suppress the immune system and adrenal function, and when inhaled can lead to oral candidiasis.

Alternative medicine provides a much-needed complement to pharmaceutical approaches, and comprehensive protocols often involve dietary and nutritional efforts, supporting gut health, and using herbs and nutraceuticals to stabilize mast cells, improve T-cell balance, and increase levels of flavonoids and EFAs in the tissues. Herbal formulas to treat acute allergic conditions are usually based on herbs to reduce histamine, mucous secretions, activation of platelets and lymphocytes, and inappropriate and excessive immune activation. Ephedra sinica, Euphrasia officinalis, Tanacetum parthenium, Petasites hybridus, and Scutellaria baicalensis are examples of herbs mentioned folklorically to treat hay fever and allergies for which scientific research also shows a mechanism of action—these plants reduce platelet and mast cell activation. Many such antiallergy herbs are also shown to affect leukotrienes, thromboxanes, and other inflammatory mediators. Resveratrol, quercetin, and magnolol for example, offer numerous antiallergy effects. Rosmarinic acid found in mint family (Lamiaceae) plants has significant antiallergy and immunomodulating effects, limiting the release of inflammatory cells and Th2 cytokines in triggered airways and ameliorating airway hyperresponsiveness.[29] Metabolic wastes and toxins in the body can act as allergens and trigger allergic reactions, and alterative herbs can reduce these abnormal immune reactions by improving digestion and supporting the elimination of wastes via the liver and bowels. Immunomodulating herbs such as Astragalus,[30] Ganoderma, and Panax might be included to help reduce hypersensitivity.

In addition to basic antiallergy herbs and alteratives, a third or fourth herb might be included in allergy formulas as specifically indicated. For example, Euphrasia is specific for a hay fever sufferer who has itchy, watery eyes. If a person with airway reactivity suffers from thick mucus in the bronchi, causing a spastic cough, Thymus vulgaris might be added to the formula for antispasmodic and drying actions, and Lobelia inflata could be added temporarily for acute wheezing. When leaky gut

Antiallergy Herbs from Many Traditions

Although herbs always have numerous mechanisms of action, this summary highlights at least one important antiallergy constituent and its mechanism of action.

HERB	MECHANISM OF ACTION
Albizia lebbeck	Saponins have immunomodulating activity.
Alisma plantago-aquatica subsp. *orientale*	Triterpenoid compounds are credited with broad lipid-regulating, anti-inflammatory, and antimicrobial activity.
Angelica sinensis	Ligustilides have shown many antiallergy effects including inhibition of nitric oxide and cyclooxygenase via NF-κB pathways.
Camellia sinensis	Epigallocatechins are strongly antioxidant and inhibit a variety of cytokine signals.
Chelidonium majus	Chelerythrine inhibits prostaglandin 2 via regulating cyclooxygenase activity.
Citrus aurantium	Nobiletin inhibits pro-inflammatory mediators by blocking NF-κB signaling pathway.
Cnidium monnieri	Osthol has numerous stabilizing effects on blood cells and vessels.
Coptis chinensis	Berberine downregulates T helper cells and reduces the release of cytokines.
Corydalis species	Pseudocoptisine inhibits NF-κB activation and reduces inflammatory mediators.
Crinum glaucum	In addition to the anticancer and cognition-enhancing effects, alkaloids lycorine and crinamine have numerous immune-modulating and neuron-stabilizing actions.
Curcuma longa	Curcuminoids, such as curcumin, have broad anti-inflammatory and immunomodulating activities.
Desmodium adscendens	The triterpenoid saponins have immunomodulating effects.
Ephedra sinica	Ephedrine and related alkaloids have mast cell–stabilizing effects.
Euphrasia officinalis	Constituents may reduce allergic reactivity in the cornea and upper airways.
Ginkgo biloba	Ginkgolide flavonoids stabilize mast cells.
Glycyrrhiza inflata	Licochalcone inhibits pro-inflammatory cytokines including interleukins, tumor necrosis factor (TNF)-α, and NF-κB.
Humulus lupulus	Xanthohumol and related humols inhibit nitric oxide production.
Hydrangea macrophylla	Glycosides may have an affinity for the urinary tissues and the coumarin skimmin is shown to protect tissues from heavy metal–induced renal reactivity.
Inula racemosa	The sesquiterpene lactone inulolide has antiallergy effects on the airways.
Justicia adhatoda (also known as *Adhatoda vasica*)	Vasicinol and vasicine have broad anti-inflammatory effects.
Leonurus japonicus	Leonurine downregulates pro-inflammatory cytokines and upregulates interleukin-10.
Ligusticum striatum	Tetramethylpyrazine inhibits pro-inflammatory cytokines and macrophage chemotaxis.
Lonicera japonica	The flavonoid luteolin decreases the release of numerous inflammatory mediators.
Magnolia officinalis	Honokiol and magnolol have numerous immunomodulating activities.

Antiallergy Herbs from Many Traditions (*continued*)

HERB	MECHANISM OF ACTION
Perilla frutescens	Perilla is a traditional antiallergy herb in TCM and used in cooking to reduce the allergenicity of shellfish.
Petasites hybridus	Used for inflammatory pain and migraines and long emphasized for asthma, cough, and allergic reactivity in the lungs.
Piper longum	Piperine reduces the levels of pro-inflammatory cytokines.
Pueraria species	Daidzein and puerarin decrease TNF-α, interleukin-1β, and nitric oxide.
Ruta graveolens	Rutin inhibits leukocyte migration and suppresses TNF-α and interleukin-6 production.
Scutellaria baicalensis	Baicalein and oroxylin inhibit nitric oxide, cyclooxygenases, and TNF-α.
Tanacetum parthenium	Parthenium and sesquiterpene lactones have mast cell–stabilizing effects.
Terminalia chebula	Ellagic acid, tannins, and chebulagic acid are all credited with a broad range of anti-inflammatory and immunomodulating effects.
Uncaria rhynchophylla	Rhynchophylline inhibits the phosphorylation of protein kinases that may be upregulated in immune and allergic disorders.
Vitex negundo	Casticin, chrysophanol, and luteolin have many anti-inflammatory effects.

or intestinal dysbiosis contribute to allergic reactivity, consider alterative herbs such as *Arctium lappa*, *Taraxacum officinale*, or *Berberis aquifolium*. And for an allergic person experiencing acute irritable bowel symptoms of pain and diarrhea, herbs might be delivered as a tea and include a soothing intestinal demulcent such as *Ulmus* species or *Althaea officinalis* to achieve direct surface contact with the intestinal mucosa.

Allergic Conditions of the Skin

As noted in chapter 1, allopathic medicine frequently blames certain pathologies as being caused by associated microbes, such as *Helicobacter pylori* as causing gastric reflux, or *Streptococcus* as causing strep throat. Many alternative practitioners view such pathogens as only one prong of such illnesses and regard the ecosystem that supports the microbes as being just as important or more so. Along this line, atopic dermatitis has long been associated with *Staphylococcus aureus* skin colonization, but because this is a ubiquitous microbe that is ever present and typically flares due to opportunistic shifts in the underlying ecosystem, many naturopathic physicians do not believe that *Staph* or any other microbes alone "cause" dermatitis. Rather, atopic dermatitis may involve *shifts* in microbes, as opposed to *causative*

microbes, and healthy skin typically supports a broad collection of fungal and bacterial members, which in a state of balance do not act as pathogens. Common commensal bacteria of the dermis include *Staphylococcus* species, *Streptococcus* species, *Propionibacterium* and *Corynebacterium* species, and common fungal spores. These microbial communities are linked and typically one single microbe doesn't proliferate and "cause" allergic inflammation. Shifts in the communal microbiota due to diet, leaky gut, stress, the use of antimicrobials, high sugar load or diabetes, and many other factors may trigger inflammatory disease and immune dysregulation. Viral infections, especially with herpes simplex virus, may also aggravate atopic dermatitis. While some intestinal microbes are possibly protective, the presence of *Giardia lamblia* and *Blastocystis hominis* may increase the prevalence of chronic urticaria (hives), atopic dermatitis, and idiopathic pruritus.[31]

"Dermography," in which white lines persist when the skin is lightly scratched, is a hallmark of atopic dermatitis and occurs due to rapid constriction of the blood vessels. Most patients with atopic dermatitis tend toward dry skin, and dietary fats are often a valuable aspect of the therapeutic protocol. Dry skin leaves the dermis more susceptible to external allergens and triggers, from *S. aureus* exotoxins to chemical irritants. EFAs are considered to be "essential" because the body cannot synthesize them, and humans must obtain linoleic, linolenic, and other fatty acids from the diet. ("Plant EFAs for Allergic Disorders" on page 48 details a few of the readily available food sources of EFAs.) Ceramides, a waxy fat found in small amounts in healthy epidermal cell membranes, are also crucial for skin health. In the outermost layer of skin cells, the stratum corneum, ceramides may comprise 50 percent of the total lipids in the cell and support the barrier function of epithelial cells. Low ceramide levels are associated with atopic dermatitis; both oral consumption and topical application of ceramides may improve eczema and other types of dermatitis. Niacinamide increases epidermal ceramide and other intercellular lipid levels, and the application of niacinamide-containing creams and lotions may reduce inflammation, decrease transepidermal water loss, and increase stratum corneum thickness.[32] Palmitoylethanolamide (PEA) is another endogenous lipid with anti-inflammatory properties and has been shown to improve atopic dermatitis when used topically.

In this chapter, I provide formulas for specific allergic skin conditions using some of the most common antihistamine and anti-inflammatory herbs. I also delve into detail on precise antiallergy research and mechanisms of action, beginning with the "Formulas for Contact Dermatitis" section on page 52. Take note that chapter 5 of Volume 1 covers dermatologic conditions and also provides valuable formulas for eczema, hives, and allergic dermatitis; I do not repeat those formulas here.

Leukotrienes

Leukotrienes, a group of inflammatory mediators, are synthesized in white blood cells from the fatty acid eicosapentaenoic acid (EPA), and they play many immunomodulating roles. Levels of various types of leukotrienes are increased in allergic, inflammatory, and autoimmune disorders. Lipoxygenase is an enzyme in mast cells, eosinophils, macrophages, and other cells that converts fatty acids into leukotrienes. Therefore, lipoxygenase inhibitors will reduce endogenous synthesis of leukotrienes. Leukotrienes have potent pro-inflammatory activity and increase vascular permeability, mucus secretion, bronchial hyperresponsiveness, and smooth muscle contraction that contribute to asthma, allergies, and airway inflammation. Antileukotriene drugs target elevations in pro-inflammatory leukotrienes. Antileukotriene drugs include the lipoxygenase inhibitors and directed leukotriene inhibitors or receptor blockers. Leukotrienes bind to specific receptors on target cells, and drugs such as montelukast may interfere with receptor binding and activation and may block allergen reactivity following a single dose.

Formulas for Allergic Hypersensitivity

The following basic formulas may complement more specific formulas to address specific diagnoses. As such, these formulas do not fall under tidy diagnostic headings, but rather are general or nontypical categories.

Tincture for Sensitivity to Medication

Numerous patients experience sensitivity to required medications. In allopathic medicine, reactivity to drugs is often treated with additional drugs—such as antihistamines, corticosteroids, leukotriene inhibitors, and mast cell stabilizers—sometimes with a lengthy struggle to find the right "cocktail" that an individual will tolerate. Many such patients find their way to herbalists and

alternative practitioners, wishing to reduce the number of drugs they use or possibly treat the underlying issue in a nonpharmaceutical manner. These antiallergy herbs can reduce drug reactivity and wean from secondary medications that are used to manage drug side effects.

Perilla frutescens
Picrorhiza kurroa
Scutellaria baicalensis
Astragalus membranaceus

Combine in equal parts. Take 1 to 2 dropperfuls at a time, 3 or 4 times a day, continuing for several months. Reduce the dosage and frequency as symptoms improve.

Tincture to Support Weaning Off of Steroids

Steroids are often prescribed for years on end to manage eczema, asthma, and chronic allergies. While they may be effective in reducing symptoms, steroids do not correct underlying allergic hypersensitivity, and worse, may promote allergies by contributing to leaky gut syndrome. Long-term steroid use will also downregulate the adrenal glands. This formula contains four herbs known to upregulate the adrenals to assist in weaning from steroids.

Glycyrrhiza glabra
Ganoderma lucidum
Ocimum tenuiflorum
Panax ginseng

Combine in equal parts. Take 1 teaspoon, 3 to 4 times daily for 3 to 6 months.

Antifungal House-Cleaning Spray

Patients with allergies should avoid solvents and yard and household chemicals wherever possible—and avoiding such products is recommended for everyone as being best for the environment. This formula is a cleaning product that will kill mold and bacteria, yet is unlikely to trigger hypersensitivity reactions. Apple cider vinegar or white vinegar, both relatively inexpensive, are suitable options for this formula.

Water	2 cups (480 ml)
Vinegar	1 cup (240 ml)
Melaleuca alternifolia essential oil	100 drops
Lavandula angustifolia essential oil	100 drops
Origanum vulgare essential oil	100 drops

Combine all ingredients in a 1-liter spray bottle and use as a cleaning spray.

Formulas for Gastrointestinal Hypersensitivity

Gastrointestinal health is foundational to the health of the entire body and can play a role in initiating or exacerbating allergic and autoimmune disorders. Immune cells are plentiful within and just beneath the gastrointestinal mucosa, a handy location for dealing with any pathogens or toxins orally ingested. Food additives such as synthetic ingredients and preservatives, drugs such as steroids and nonsteroidal anti-inflammatories (NSAIDs), and poor nutrition may weaken the integrity of the mucous membrane barrier and allow undigested proteins and carbohydrates to be absorbed and trigger immune responses. Beneficial intestinal microflora help protect against food allergies, and intestinal dysbiosis may make the intestinal mucosa more susceptible to sensitization, helping to explain the use of antibiotics in early life as a risk factor for developing atopic conditions.[33] (For more about this, see chapter 2, Volume 1.)

True food allergies are relatively rare, but the term *allergy* is sometimes used indiscriminately to refer to any reactivity to a particular food. The term *intolerance* more accurately applies in this case. True food allergies involve an excessive release of histamine and immunoglobulins in response to specific ingested proteins, lectins, and other components of foods. Such allergies may trigger mild, acute, or anaphylactic reactivity involving IgE antibody release and trigger cell-mediated eosinophilic responses. Food intolerances may involve local digestive symptoms or may promote systemic reactivity manifesting as joint pain, headaches, or general malaise.[34] Many patients with food reactivity are treated as having irritable bowel syndrome without underlying mechanisms being addressed or corrected. And many patients, especially women, may be diagnosed as having anxiety-induced symptoms and offered anxiolytic medications and/or referred to psychiatrists. True allergies are mediated by IgE immunoreactivity and can be rapid in onset, while many nonimmune globulin-I-mediated types of food "intolerance" reactivity in the gut emerge more slowly. More precisely "food allergies" may be classified as food-protein-induced enterocolitis, enteropathy (such as gluten enteropathy or celiac disease), proctocolitis, and allergic eosinophilic gastroenteritis, diagnoses all reported to be on the rise in many areas of the world.[35]

Allergic sensitization to food in the pediatric population is increasingly common and increasingly severe as with anaphylactic disorders,[36] and a common emergency room presentation around the world. Eggs, shellfish, milk, and nuts—especially peanuts and cashews—are reportedly the most common offenders, followed by wheat and soy and, to a lesser extent, nightshades (tomatoes, potatoes) and citrus. While not an actual allergen, sugar is a common contributor to bowel reactivity due to the manner in which it supports undesirable intestinal flora, increases inflammation, and harms innate immune responses. Around 10 percent of children outgrow tree nut allergies,[37] but lifelong avoidance of the offending food is the only cure for some people.[38] Herbal formulas may reduce anaphylactic reactivity and even reestablish tolerance.[39]

Food "allergy" symptoms are not confined to the gut and can manifest as urticaria, angioedema, anaphylaxis, atopic dermatitis, or respiratory symptoms. Many individuals are not aware of respiratory or skin symptoms being associated with the ingestion of particular foods, often because the offending food is eaten so regularly that it obscures an observable cause and effect. A common approach embraced by many clinicians is the immediate implementation of an elimination diet in which all the common allergens are avoided for a minimum of 6 weeks. EFAs, probiotics, antioxidant nutrients, herbal formulas, and anything specific for an individual patient are prescribed. This protocol can be a lot to ask of patients, but I prefer this approach over expensive laboratory testing. The elimination diet has the further benefit of supporting patient compliance; after a period of 6 to 8 weeks of avoiding all allergens, foods are added back one at a time. If migraines, wheezing, pruritus, or diarrhea occur with a certain food, the cause and effect are then more obvious to the person and this confers motivation for long-term dietary improvements.

Tea for Allergic Reactivity with Diarrhea

This formula will treat diarrhea due to allergic reactivity. It includes the bowel anti-inflammatories *Glycyrrhiza*, *Matricaria*, and *Filipendula* to reduce allergic reactivity, combined with the mucous membrane astringents *Quercus* and *Salix* to reduce excessive secretions and help control the diarrhea. *Atractylodes macrocephala* is used in Asia to treat allergic diarrhea and has been shown to support T-cell balance.[40] This formula is best prepared as a tea to ensure surface contact with the intestinal mucosa. While treating bowel hypersensitivity, it is also helpful to fast for 1 to 2 days or consume an allergen-free, anti-inflammatory diet.

Glycyrrhiza glabra	2 ounces (60 g)
Matricaria chamomilla	2 ounces (60 g)
Filipendula ulmaria	1 ounce (30 g)
Quercus rubra	1 ounce (30 g)
Salix spp.	1 ounce (30 g)
Atractylodes macrocephala	1 ounce (30 g)

Steep 1 tablespoon of the combined herbs per 1 cup of hot water and then strain. Prepare 5 or 6 cups (1,200 to 1,440 ml) at once and sip frequently throughout the day to keep a constant trickle moving through the bowels. Use the relatively large 6 cups/day dosing until diarrhea is controlled, typically over 2 or 3 days. The tea may then be continued at half the dose of around 3 cups (720 ml) per day for several weeks and then discontinued altogether, or consumed several times a week as part of a long-term maintenance protocol as needed.

Tincture for Bowel Reactivity Associated with Toxins

The presence of toxins and metabolic wastes in the blood and tissues can act like antigenic material and trigger immune reactivity. The bitter alterative and liver-supportive herbs in this formula may promote removal of antigenic substances from the body and help bolster the liver's detoxification capacity. This tincture includes *Echinacea*, a popular herb that excels in reducing septic conditions and supporting immune homeostasis. Chelerythrine in *Chelidonium* is credited with numerous anti-inflammatory effects, including inhibition of nitric oxide and cyclooxygenase cascades in animal models of bowel toxicity.[41] A formula such as this may be used in between episodes of acute allergic flares to help reduce chronicity.

Echinacea angustifolia
Achillea millefolium
Silybum marianum
Andrographis paniculata
Schisandra chinensis
Chelidonium majus

Combine in equal parts. Take 1 to 2 teaspoons, 3 to 6 times daily for several weeks, reducing as symptoms improve. Patients with chronic issues may benefit from continuing at a lower dose of 1 teaspoon, 2 or 3 times a day for 3 to 6 months.

Tea for Chronic Food Reactivity

This formula is similar to the Tea for Allergic Reactivity with Diarrhea, but substitutes some of the herbs for acute pain and bloating with *Camellia* and *Zingiber*, both of which have broad anti-inflammatory and antioxidant effects. Berberine, found in *Berberis* (Oregon grape), is noted to reduce IgE-mediated bowel reactivity[42] while enhancing liver function and the digestion of fats and carbohydrates. Reishi mushrooms also offer antiallergy effects via immunomodulation. Reishi triterpenoid saponins inhibit the production of inflammatory mediators and cytokines.[43] The wood-like slices of this medicinal mushroom can be decocted, or the powdered mushroom may be included in teas or blended into smoothies. The following tea should be used in conjunction with an anti-inflammatory diet and the avoidance of food allergens. Supplements such as antioxidant nutrients, quercetin and other flavonoids, and EFAs such as fish oil, flaxseed oil, or evening primrose oil may be complementary.

Matricaria chamomilla
Glycyrrhiza glabra
Camellia sinensis
Zingiber officinale
Ulmus rubra
Berberis aquifolium
Ganoderma lucidum mushroom powder

Combine in equal parts, adding slightly more *Glycyrrhiza* to sweeten if desired. Gently simmer 1 tablespoon per cup for 1 minute, then cover and let stand for 15 minutes before straining. Drink 3 cups (720 ml) or more daily for 3 to 6 months.

Tincture for Chronic Food Reactivity

This formula has the same goal in mind as the Tea for Chronic Food Reactivity, but the incorporation of less-tasty herbs makes it better suited to tincture form. Tinctures may be preferred by people who do not take well to teas, or both formulas may be used in tandem for urgent or severe cases. The herbs in this tincture are expected to work systemically. *Silybum* may help the liver and kidneys eliminate toxins and improve digestion, reducing bowel inflammation. The other herbs have antiallergy effects that can help make the intestines less reactive over time. *Tanacetum*'s sesquiterpene lactones have been repeatedly shown to reduce vascular inflammatory cascades. *Scutellaria baicalensis* is a traditional anti-inflammatory herb and may reduce food allergies through down regulation of T lymphocytes.[45] The constituent wogonin is credited with stabilizing interleukin and immunoglobin E release,[46] which is frequently excessive in those with food allergies. *Bupleurum falcatum* is a traditional medicine for inflammation in Asia, and saikosaponins in the plant are credited with reducing

Immune Dysregulation in the Brain

The notion of a "brain allergy" refers to neurological effects that result from dysregulated immune response and related inflammation within the central nervous system (CNS). For example, some conditions in the central nervous system may be due in part to hypersensitivity-driven inflammation that leads to neurologic symptoms and neurodegeneration. Until very recently, the brain was thought to be one of the few immune-free regions of the body, which is part of the reason why meningitis is so deadly. However, it is now recognized that allergy-like inflammatory processes can occur within the brain and central nervous system when cytokines and other inflammatory mediators in cerebral circulation and tissue have neurologic impact. Dysregulated immune responses in the brain may cause mental symptoms and concentration difficulties. Histaminergic neurons can be excessively stimulated and may be involved with parkinsonism, schizophrenia, Tourette syndrome, and other neurologic pathologies, due to inadequate histamine clearance and histamine transport.[44] Consider some of the antiallergy approaches discussed throughout this chapter in such situations, especially in individuals with other obvious atopic tendencies.

Chemicals, endotoxins, elevated cortisol levels, and insufficient antioxidant levels may all promote excessive inflammation and contribute to neuronal degeneration. Protective nutrients include vitamins C and E and beta-carotene, as well as zinc and selenium. The presence of heavy metals, including iron, contributes to brain inflammation and damage. Exposure to aluminum and mercury should be avoided wherever possible, and children should not take multivitamins that contain iron unless they are truly iron deficient. Herbal medicines that reduce allergic sensitivity may improve cognitive function by reducing inflammation in the brain, especially in children with known allergies, hay fever, eczema, hives, or asthma.

food reactivity via effects on white blood cells in a manner that reduces inflammatory signaling cascades.[47]

Bupleurum falcatum
Silybum marianum
Tanacetum parthenium
Glycyrrhiza glabra
Angelica sinensis
Scutellaria baicalensis

Combine in equal parts. Take 1 teaspoon or more, 3 to 6 times per day, 5 days per week, for several months. Continue for up to 6 months if symptoms persist.

Mushroom Powder for Allergies

Many medicinal mushrooms have immunomodulating effects that may reduce hypersensitivity reactions and improve basic immunity when it has been suppressed by long-term corticosteroid use. *Lentinula edodes* (shiitake), *Ganoderma lucidum* (reishi), and *Grifola frondosa* (hen of the woods or maitake) all contain immune polysaccharides, a well-studied group of plant constituents shown to optimize immune reactivity.[48] *Lentinula* and *Grifola* are edible and can be consumed daily or taken in capsule or tincture form. Reishi mushrooms are woody and inedible, but can be used in decoctions and medicinal soups, prepared into tinctures, or powdered. This formula is just one possible combination, utilizing the most readily available, affordable medicinal mushroom powders.

Ganoderma lucidum
Lentinula edodes
Grifola frondosa

Use 1 teaspoon to 1 tablespoon of single or combined mushroom powders. Combine with ground coffee, stir into nut butter to make medicinal truffles (as described in chapter 3, Volume 3), or blend into smoothies, oatmeal, yogurt, applesauce, and so forth. Use once a day for general health support and prevention, or as often as 3 to 4 times a day for acute allergic reactivity, autoimmune disease, or cancer.

Wu Mei Wan for Food Allergies

Wu Mei Wan is a classic TCM 10-herb formula used to treat intestinal parasite infections and gastrointestinal disorders with symptoms similar to those of food allergies and gastroenteritis. The formula was reportedly effective in improving lung and digestive symptoms in allergic subjects without toxicity or side effects.[49] One group of researchers adapted the traditional formula,

Herbs for Acute Allergic Pain in the Gut

Teas and encapsulated powders are usually more effective than tinctures when treating gastrointestinal complaints due to their ability to achieve more direct surface contact with the digestive mucosa. Many of the herbs in this list can be prepared as foods; for example, slippery elm (*Ulmus*) can be prepared into porridge and freshly grated ginger root (*Zingiber*) can be included in cooking wherever possible. Alteratives and liver herbs are appropriate for those with chronic digestive symptoms and reactivity, but are usually best avoided in acute situations. Strong, bitter alterative roots may promote peristalsis and digestive secretions in those with diarrhea, pain, or irritable bowel, so they are best left out of acute formulas or used in small amounts combined with more soothing demulcents in formulas for sensitive individuals.

Althaea officinalis	*Matricaria chamomilla*
Angelica sinensis	*Mentha piperita*
Foeniculum vulgare	*Perilla frutescens*
Ganoderma lucidum	*Zingiber officinale*
Glycyrrhiza glabra	

Nutritional Supplements for Bowel Reactivity

In addition to the herbal formulas in this section, the following nutrients and supplements are helpful for patients with general bowel reactivity. Those with Crohn's disease or ulcerative colitis may require an aggressive protocol that combines various options. Those with milder issues may benefit from one or more choices as individual cases warrant. A several-month-long elimination diet is also a highly valuable initial therapy for bowel reactivity.

Aloe vera gel in juice, water, or smoothies	vitamins A, E, D, and C; beta carotene)
Glutamine	Probiotics
Essential fatty acids	Fiber and prebiotics
Antioxidant nutrients (zinc; selenium;	Quercetin and other bioflavonoids
	Resveratrol

adding *Ganoderma* and removing aconite, and evaluated the effects for patients with food allergies and asthma in the United States. Researchers reported the formula to "completely block" peanut-induced anaphylaxis in a murine model of peanut allergy—an effect that endured for at least 6 months after discontinuation of the formula. Human clinical trials show beneficial immunomodulatory effects on peripheral blood mononuclear cells in children with peanut and multiple other food allergies.[50]

Prunus mume
Zanthoxylum bungeanum
Angelica sinensis

Zingiber officinale
Cinnamomum cassia
Phellodendron amurense
Coptis chinensis
Panax ginseng
Ganoderma lucidum

Combine in equal parts. Gently simmer 2 heaping tablespoons in 8 cups (1,920 ml) of water for 30 minutes. Let stand until cool enough to strain. Drink 3 cups (720 ml) or more each day for 3 months or more, reducing as symptoms improve. Take weekends off or skip days here and there after a month of consistent use.

Formulas for Respiratory Allergies

Asthmatics will often improve when they move from a damp or humid climate to an arid location. Providing adequate ventilation to clear buildings of indoor air pollution, mold, and harmful chemicals and volatiles is also essential. Living in a damp, moldy, mildewy environment may sensitize infants and lead to allergic airway disorders.

Chronic mucous congestion in the respiratory passages can support opportunistic microbes and lead to secondary bacterial rhinitis, sinusitis, otitis, and bronchitis. Reducing underlying hyperreactivity is often an important aspect for improvement. It is often helpful to have a broad protocol aimed at reducing allergic hypersensitivity, while keeping a specific herbal antimicrobial formula on hand for those prone to secondary bacterial infections. Smokers or those with superimposed pulmonary disease can suffer severe lung infections when allergic reactivity promotes bronchial congestion, sometimes necessitating hospitalization when occurring in the elderly or debilitated persons. Aggressive therapies are required in such patient populations.

Basic Tea for Allergic Rhinitis

Euphrasia, with its bland, grassy flavor, is a base herb for this tea. The tea is not powerful enough to alleviate acute symptoms on its own. However, it can be a valuable part of a long-term, broad protocol that includes a tincture, antioxidants, and EFA supplements. It is most effective to begin drinking this tea in mid to late winter so it is consumed for at least 1 month prior to the onset of the typically high-pollen months of spring.

Euphrasia officinalis leaf	3 ounces (90 g)
Urtica spp. leaf	2 ounces (60 g)
Ephedra sinica stems	2 ounces (60 g)
Glycyrrhiza glabra root, shredded	2 ounces (60 g)
Rosmarinus officinalis leaf	1 ounce (30 g)

Combine the ingredients and store in an airtight bag or container. Steep 5 to 6 tablespoons in 5 or 6 cups (1,200 or 1,440 ml) of water for 10 minutes, then strain. Drink 3 cups (720 ml) or more each day for a month prior to the typical onset of hay fever when possible, and as much as possible for acute symptoms.

Dietary Protocol for Respiratory Allergies

The supplements listed in "Nutritional Supplements for Bowel Reactivity" on page 39 are also appropriate for most respiratory allergies. In addition, the following foods are recommended for the daily diet while attempting to control severe reactivity of the airways. Dairy products should be avoided because they are mucus forming, and common food allergens such as wheat, sugar, soy, and chemical additives, especially sulfites, may also trigger or exacerbate symptoms.

- Omega-3 fatty acids (fish, almonds, walnuts, pumpkins, evening primrose, flaxseed oil, etc.).
- Onions and garlic provide quercetin; in addition, they are expectorating to excessive mucus and deter infections.
- Fresh fruits and vegetables (strive to consume 8 different kinds per day) to support gut health and provide fiber and anti-inflammatory nutrients.

Petasites for Inflammation

Petasites hybridus is a traditional medicine that has been used for centuries for all manner of painful and inflammatory conditions and has a long history of use for respiratory inflammation and migraine headaches. One clinical trial reported that *Petasites* was as effective as antihistamines in reducing the symptoms of allergic rhinitis compared to placebo.[51] A review of clinical trials on *Petasites* suggest efficacy for allergy and asthma.[52] One clinical trial reported that *Petasites* alleviated symptoms of allergic rhinitis in a manner equal to antihistamine pharmaceuticals,[53] and another reported equal efficacy to fexofenadine.[54] An asthma trial reported a reduction in both frequency and severity

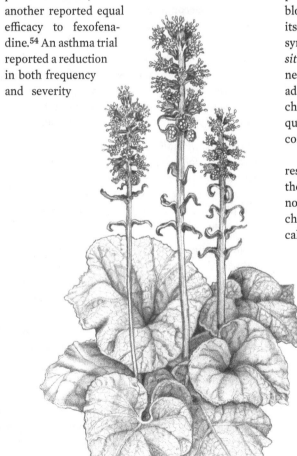

Petasites hybridus,
butterbur

of asthma attacks, and improved spirometry readings following 2 months of *Petasites* use.[55] Researchers have also reported *Petasites* preparations to reduce the occurrence of migraines.[56]

Petasites' mechanisms of action include inhibition of cyclooxygenase and inflammatory prostaglandins[57] and reduction of inappropriate vasoconstriction via a mechanism that involves blocking calcium-gated channels in the vasculature.[58] *Petasites* may also relax bronchial smooth muscle via activity at muscarinic sites.[59] In asthma patients, *Petasites* reduces elevated eosinophil and blood nitric oxide levels.[60] *Petasites* also inhibits cyclooxygenase, an enzyme involved with the synthesis of inflammatory prostaglandins.[61] *Petasites* species may limit synthesis in eosinophils and neutrophils[62] and corticosterone release from the adrenal gland, both deterring vascular and bronchial smooth muscle contractions.[63] Petasin, a sesquiterpene in *Petasites*, is believed to be one active constituent responsible for some of these effects.

Allergic rhinitis, sinusitis, hay fever, and other respiratory allergies may be improved by herbal therapies,[64] and the digestive support and immunomodulating therapies discussed throughout this chapter can help a person wean from pharmaceuticals and even reduce the need to rely on respiratory inhalers. *Euphrasia*, *Petasites*, and *Tanacetum* are among the key herbal options for hay fever. *Petasites* reduces allergic phenomena by a number of different mechanisms and clinical trials show efficacy in alleviating the symptoms of allergic rhinitis. When allergic symptoms tend to progress into sinus or ear infections, *Achillea*, *Berberis*, and other antimicrobials would be appropriate additions to tea and tincture formulas. When there is copious watery mucus, drying agents such as *Thymus* or *Salvia officinalis* might be added to herbal formulas. See also chapter 3 of Volume 2 for additional guidance in creating formulas for allergic conditions of the airways.

- Shiitake, maitake, and other mushrooms in soups, stir-fries, and casseroles.
- Berries to provide antioxidant flavonoids such as anthocyanins and the phenolic compound resveratrol.

Tincture for Chronic Allergic Rhinitis

When upper respiratory allergies are chronic, the following herbs may help reduce hypersensitivity reactions. This tincture is most effective when used in conjunction with antioxidant nutrients, resveratrol, EFAs such as black cumin (*Nigella sativa*) or evening primrose, and diet optimization as detailed in the Dietary Protocol for Respiratory Allergies on page 40. Also, reduce exposure to mold, chemicals, and other antigens.

Petasites hybridus
Tinospora cordifolia
Perilla frutescens

Combine in equal parts. Take 1 dropperful, 3 times a day for 3 to 6 months.

Eye Drops for Hay Fever Symptoms

This formula employs a dry herb blend mixed with rosewater and saline to form eye drops at the time of each use. *Euphrasia* is effective in alleviating the itchy, watery eyes that accompany hay fever[65] and in improving catarrhal conjunctivitis. When eye symptoms are severe and uncomfortable, this formula can be used directly in the eyes for fast-acting relief. Use these eye drops as a soothing palliative agent, while treating underlying hay fever or chemical sensitivity with a systemic formula and nutritional support. See chapter 4 of this volume for more information on treating eye symptoms.

Euphrasia officinalis	1 ounce (30 g)
Calendula officinalis	1 ounce (30 g)
Foeniculum vulgare seeds, powdered or crushed	1 ounce (30 g)
Sterile saline	1 cup (240 ml)
Rosewater (optional)	

Combine the dry herbs in a small jar or ziplock bag. Prepare the eye drops by gently simmering 1 tablespoon of the herb mixture in the sterile saline for 10 minutes. Remove from the heat and let stand for 10 minutes. Take care to remove all particulate by using a fine strainer, such as a muslin-lined strainer or a coffee filter. Once cool, the eye drops can be used as is, or the solution can be combined with rosewater in equal parts or 2 or 3 parts tea to 1 part rosewater. Use an eye dropper or eyecup to instill in the eyes at least 4 times per day, and as often as hourly if possible. Store the unused formula in the refrigerator and use within 48 hours, discarding what remains after that time and making a new batch.

Nasal Spray for Allergic Rhinitis

Euphrasia reduces allergic reactivity in the upper respiratory mucous membranes and the eyes. *Hydrastis* is specifically indicated for abundant mucus and chronic

Herbal Specifics for Hay Fever

The following herbs are among the folkloric classics recommended to help alleviate the discomforts of hay fever, as well as to reduce allergic and reactive tendency.

Achillea millefolium. With its alterative and broad-acting anti-inflammatory effects, *Achillea* can act as a supportive herb in formulas.

Euphrasia officinalis. *Euphrasia* can often serve as a base foundation herb in allergy formulas due to its affinity for ear, nose, and throat mucous membranes.

Ephedra sinica. Although *Ephedra* has general antiallergy effects, it also has some limitations due to caffeine-like stimulating effects.

Petasites hybridus. *Petasites* has broad anti-inflammatory and antiallergy effects on blood cells and mucous membranes.

Curcuma longa. Its effects are more anti-inflammatory than antiallergy effects, but *Curcuma* is often useful as a supportive alterative.

Thymus vulgaris. When abundant mucus leads to secondary infections, *Thymus*'s drying and antimicrobial effects are useful.

Armoracia rusticana. Due to its affinity for the sinuses and ability to thin stuck mucus, *Armoracia* makes an excellent synergist.

catarrh; it is effective in drying the nasal passages and alleviating profuse congestion.

Euphrasia officinalis	½ tablespoon
Hydrastis canadensis powder	½ tablespoon
Sterile saline	1 cup (240 ml)

Heat the sterile saline (or, when unavailable, use tap water) in a small saucepan, adding the herbs when simmering. Remove from the heat immediately and let stand for 15 minutes before straining well. Administer with a neti pot or bulb syringe to lavage congested nasal passages.

Licorice Rinse for Nasal Polyps

Polyps can result from chronic congestion, edema, swelling, and allergic reactivity in the nasal mucosa. Restoring mucociliary clearance of the nasal epithelium when it has been altered thus represents a key therapeutic tool against rhinosinus chronic diseases. The glycyrrhetinic acid in *Glycyrrhiza glabra* has been shown to improve vasomotor rhinitis in human subjects when used as a nasal spray in combination with mannitol and glycerol.[66] A simple licorice tea may also be used as a nasal spray in place of this formula as part of a broad protocol for treating respiratory allergies.

Glycerine	1 tablespoon
Glycyrrhiza glabra	1 teaspoon
Mannose powder	¼ teaspoon

Prepare ½ cup (120 ml) of *Glycyrrhiza* tea by gently simmering the dried herb in ½ cup (120 ml) of water for 10 minutes. Strain and combine with the mannose powder and glycerine in a small squeeze bottle. Use the nasal spray twice a day for 30 to 60 days, and then evaluate the results. For optimum effect, combine with a nutritional protocol, allergen avoidance, and the systemic antiallergy approaches described throughout this chapter.

Tincture for Hay Fever with Spastic Cough

For some patients, the amount of mucus produced due to respiratory allergies leads to postnasal drip and coughing. Some patients with allergic airways may also develop an allergic cough that is a variant of asthma. This formula emphasizes respiratory anti-inflammatories and antispasmodics, addressing both the lower airway irritation and the underlying allergy. Because *Petasites* contains a small amount of pyrrolizidine alkaloids, limit its use to 10 to 15 days and then omit from the formula. This formula can be altered to address different presentations. For tight spastic wheezing, *Lobelia inflata*

Alliaceous Allergy and Migraines

Although it is an uncommon reaction, some atopic individuals are sensitive to garlic and onions and experience dermatitis, gingival reactions, eye sensitivity, asthma, and migraines. Some tincture formulas in this chapter include *Allium* due to garlic's stellar ability to thin and expectorate mucus, but don't overlook that some atopic individuals do not tolerate alliaceous herbs.[67]

Allium sativum, garlic

may be added or substituted for part of the *Foeniculum*. When there is thick stuck mucus deep in the lungs, the amount of *Thymus* in the formula may be increased and complemented with *Allium sativum* or *Eucalyptus globulus*. Where abundant mucus leads to bronchial infection, *Lomatium dissectum* or *Ligusticum striatum* may be substituted for part of the *Euphrasia*.

Euphrasia officinalis	20 ml
Thymus vulgaris	10 ml
Ammi visnaga	10 ml
Foeniculum vulgare	10 ml
Petasites hybridus	10 ml

Take 1 dropperful of the combined tincture as often as hourly, reducing as symptoms improve.

Tincture for Hay Fever with Sinus Pain

This formula employs herbal drying agents and horse-radish (*Armoracia*), which can effectively thin and reduce mucus and move it out of the sinuses. This formula is best used acutely and swapped with antiallergy formulas once the acute pain and pressure symptoms are improved. Patients with excessive respiratory mucus should also avoid dairy products.

Euphrasia officinalis	15 ml
Thymus vulgaris	15 ml
Achillea millefolium	15 ml
Armoracia rusticana	15 ml

Combine in equal parts to fill a 2-ounce (60 ml) tincture bottle, or double the recipe to yield 4 ounces (120 ml). Take 1 dropperful or ½ teaspoon as often as hourly, reducing as symptoms improve. Once acute symptoms have been reduced, the formula may be continued at a lower dose, such as 2 to 4 times daily to maintain improvement. Steam inhalation 3 or more times a day can also help thin mucus and prevent sinus congestion from turning into a sinus infection and be complementary to the use of this tincture.

Tincture for Febrile Allergic Reactions

The term hay "fever" reflects the febrile state that some severe allergic reactions may manifest. According to folklore, the herbs in this tincture are especially appropriate for an allergic fever.

Curcuma longa	15 ml
Filipendula ulmaria	15 ml
Petasites hybridus	15 ml
Tanacetum parthenium	15 ml

Take ½ teaspoon every 30 minutes for several hours, reducing gradually over the first 24 hours. Thereafter, continue taking 6 times a day for several days, until the fever subsides.

Formulas for Allergic Airway Disorders

Numerous medicinal herbs have been documented as useful bronchodilators, mast cell stabilizers, and immunomodulating agents via inhibiting the release of inflammatory mediators including leukotrienes, lipoxygenase, cyclooxygenase, platelet-activating factor, phosphodiesterase, and cytokines.[68] Methylxanthines such as theophylline and caffeine have been used to promote bronchodilation since at least the 1930s, with *Ephedra* and the ephedrine it contains used for wheezing long before that. Such methylxanthines improve airway reactivity by inhibiting the release of mediators from mast cells and leukocytes, along with other mechanisms. Numerous plants have mast cell–stabilizing effects, thereby reducing the release of histamine (See "Herbs to Stabilize Mast Cells" on page 31). *Boswellia serrata*, *Glycyrrhiza glabra*, and *Curcuma longa* are shown to inhibit leukotriene, and clinical trials showed that an encapsulation containing these herbs significantly reduced plasma leukotriene levels and improved pulmonary function tests.[69]

Note that asthma and bronchodilators are also covered in chapter 3, Volume 2. In this section, I focus on those herbs specifically noted to improved allergic reactivity in the airways.

Ban Xia Hou Po Decoction for Asthma

Perilla leaf is a traditional antiallergy ingredient in many Chinese formulas, such as Ban Xia Hou Po. This classic decoction is used for resolving phlegm, relieving cough and asthma, and loosening the bowel to relieve constipation. The ingredient amounts in this formula are listed in grams rather than milliliters or ounces, as is typical for Chinese granules. If individual granules for this precise formula are unavailable, this blend may also be prepared as a decoction of crude herbs or a tincture. These herbs are all commonly available from purveyors of Chinese herbs.

Pinellia ternata	½ ounce (15 g)
Magnolia officinalis	½ ounce (15 g)
Poria cocos	½ ounce (15 g)
Zingiber officinale	⅓ ounce (10 g)
Perilla frutescens	⅓ ounce (10 g)

Because granules are prepared by concentrating a tea down to a dry residue, just 1 teaspoon of the combined granules are reconstituted per 1 to 2 cups (240 to 480 ml) of hot water to prepare a tea. If using dry herbs instead of granules, then simmer about 1 tablespoon of the herb mixture gently in 3 cups (720 ml) of water for 15 minutes,

Perilla for Almost Any Allergy

Perilla has been used as an ornamental, food, and medicine in East Asia since ancient times. The anthocyanin-rich purple forms of *Perilla* are frequently used as food colorants in Japan and China, and all varieties have been used medicinally. The leaves are used in cooking, especially to reduce the allergenicity of fish and shellfish. The seeds have been processed into oils used for culinary and medicinal purposes, including in massage blends and ear oils. As with most common mint-family plants, *Perilla* has numerous uses in folklore, including as treatment for many allergic disorders such as asthma, hay fever, and food reactions. It is also used for abdominal pain and indigestion and respiratory congestion. Numerous anti-inflammatory constituents have been identified, including phenolic acids, flavonoids, essential oils, triterpenes, carotenoids, phytosterols, fatty acids, tocopherols, and policosanols. Individual phenolic acids and flavonoids include rosmarinic acid, perillaldehyde, luteolin, apigenin, tormentic acid, and isoegomaketone, all credited with antiallergy and immunomodulating properties. Carotenoid levels in *Perilla* are five times higher than those of carrots.[70] *Perilla* contains a wide range of essential oils credited with antibacterial, antiviral, antifungal, anti-inflammatory, antimutagenic, anticarcinogenic, antidiabetic, antiprotozoal, and antioxidant activities.

Animal research shows *Perilla* to have positive effects against atopic dermatitis, anaphylaxis, and allergic airway disorders, with some activity credited to rosmarinic acid. *Perilla* glycoproteins are shown to inhibit mast cell degranulation and hyaluronidase release. Additional animal studies show the daily consumption of *Perilla* to prevent the increase in the

Perilla frutescens, shiso

numbers of eosinophils in bronchoalveolar lavage fluids and inhibit the expression of interleukins and eotaxin in allergen-sensitized lungs.[71] *Perilla* is also indicated for inflammatory bowel diseases[72] and may be included in tinctures and teas both for prevention and for acute flares. *Perilla* is also prokinetic to the gastrointestinal tract and may be suitable for patients with gastroparesis.[73] One human clinical trial investigated a nutraceutical containing *Perilla*, quercetin, and vitamin D_3 in children with allergic rhinitis and reported the formulation to reduce acute episodes and exacerbations by 50 percent without adverse events or side effects.[74]

and then strain. Consume 3 cups (720 ml) per day for a month and then reduce to several cups at a time a few times a week as a maintenance therapy.

Chinese Trio Tincture for Asthma

Clinical trials have suggested this simplified version of a traditional Chinese asthma formula may significantly improve lung function and reduce symptom scores in asthma patients.[75] *Sophora flavescens* is an herb with one of the highest concentrations of quercetin, which has antiallergy and anti-inflammatory effects. The prepared root is called Radix *Sophorae flavascentis* and is available from Chinese herb suppliers. Researchers report the therapy to be well tolerated, suggesting it may offer an alternative to standard corticosteroid therapy, without the undesirable side effects. Animal studies of the formula report a broad spectrum of therapeutic mechanisms, including inhibition of airway hyperreactivity, pulmonary inflammation, and airway remodeling, as well as downregulation of Th2 responses and direct modulation of airway smooth muscle contraction.

Ganoderma lucidum
Radix *Sophorae flavescentis*
Glycyrrhiza uralensis or *G. glabra*

Combine in equal parts in a 2-ounce (60 ml) tincture bottle. Take 2 dropperfuls, 2 or 3 times a day, reducing dose and frequency as airway reactivity subsides.

Saiboku-To Tincture for Asthma

Saiboku-To is a traditional Japanese Kampo herbal medicine for bronchial asthma. Several individual molecules found in the herbs in this formula, such as liquiritigenin, scutellarin, glycyrrhizin, wogonin, baicalein, and magnolol, have been studied for their significant anti-inflammatory and antihistamine effects.[76] Saiboku-To may enhance serum prednisolone concentrations via inhibition of 11β-hydroxysteroid dehydrogenase.

Glycyrrhiza glabra
Magnolia officinalis
Scutellaria baicalensis

Combine in equal parts in a 2- to 4-ounce (60 ml to 120 ml) bottle. Take 2 dropperfuls, 2 or 3 times a day, reducing the dose and frequency as asthmatic reactivity subsides.

TCM Nebulizer Formula

Xiao-Qing-Long-Tang is a traditional Chinese formula used in the treatment of asthma. Modern research has shown nebulized forms of this medicine to reduce inflammatory markers of allergic activation.[77] The formula listed here is the complete traditional recipe, and most of the herbs are commercially available.

Ephedra sinica	⅔ ounce (20 g)
Paeonia lactiflora	⅔ ounce (20 g)
Glycyrrhiza uralensis	⅔ ounce (20 g)
Cinnamomum cassia	⅔ ounce (20 g)
Zingiber officinale	⅔ ounce (20 g)
Prunus armeniaca	⅔ ounce (20 g)
Perilla frutescens	⅔ ounce (20 g)

Combine the herbs. Bring a pot of water or saline solution to a boil, then add the herbs and simmer for 15 minutes before straining through a fine mesh filter. While it may not be a simple matter for all clinicians to prepare nebulized therapies in their offices, this formula can also be simplified and prepared as a steam, a nasal lavage, or a tea.

Cromolyn Sodium Inhaler for Allergic Airways

Cromolyn sodium (sodium cromoglycate) is a derivative of khellin, a flavonoid occurring in *Ammi visnaga* (khella). The herb *Ammi visnaga* may be included in teas and tinctures for allergic airway disorders, while cromolyn sodium is available as a prescription inhaler from pharmacies and is used for maintenance. Cromolyn is a mast cell stabilizer with various inhibitory effects on macrophages and eosinophils, but it must be taken regularly to block both the early response to allergens that may be mediated by mast cells and the later response involving inflammatory cascades that contribute to bronchial hyperresponsiveness. Cromolyn sodium inhalers are often a first-choice medication for children with asthma, because they have few adverse effects and are considered safer than corticosteroid inhalers.

Cromolyn sodium inhaler (trade name Intal)

Take 2 puffs per day as a maintenance therapy. Improvements are generally seen after 3 to 4 weeks of regular use. Cromolyn sodium may be used long term as part of maintenance therapy.

Theophylline for Allergic Airway Disease

Theophyllline is a naturally occurring caffeine-like molecule found in green tea and chocolate, and has been available in a purified form for over 80 years. Even though β-agonist inhalers (such as albuterol) and steroids are the preferred treatment for rapid bronchodilation from acute asthma, theophylline remains popular because it is inexpensive and complementary to other therapies. Theophylline is a not a fast-acting bronchodilator, but it inhibits the release of mediators from mast cells, increases mucociliary clearance, and prevents microvascular leakiness and excessive T-lymphocyte activation. It also reduces airway excitability and chronic cough by decreasing the excessive stimulation of sensory nerves through activation of calcium-activated potassium channels.[78] Theophylline has a narrow therapeutic window because large or repetitive doses can have side effects similar to caffeine overdose, such as heart palpitations, upset stomach, insomnia, nervousness, and, rarely, seizures. Theophylline is available by prescription (aminophylline

Household Allergens as a Trigger

Household antigens, including mold,[79] mouse, or cockroach exposure,[80] can trigger chronic allergic symptoms. The ability of molds to promote allergic sensitivity has been described since at the least the 1860s, and mold-induced illness is referred to as mycotoxicosis and contributes to the development of allergic rhinitis, sinusitis, nasal polyps, and the enlargement of the adenoids or tonsils.[81] Mold exposure promotes nasal obstruction more commonly than the watery mucus of hay fever or other allergies, and many patients are also predisposed to bronchial asthma.[82] Many molds are toxic or colonizing, but many of their metabolic by-products are also harmful,[83] and some mold species emit volatile compounds that contribute to pneumonitis and other allergic reactivity symptoms.

While exposure to some amount of mold is unavoidable, heavy and persistent exposure to molds and their products can contribute to the development of asthma, allergic rhinitis, allergic bronchopulmonary aspergillosis, sinusitis, and hypersensitive airways.[84] *Stachybotrys chartarum* is one culprit and can promote severe illness. *Aspergillus*, long noted to live in some air-conditioning units, is another common mycotoxin.

Mold can be a trickster as a trigger of allergic phenomena, causing clinicians to run numerous tests and perform extensive diagnostic work-ups. A person may present with symptoms ranging from fatigue, nausea, headaches, or dizziness to cognitive issues, numbness and tingling, skin rashes, and many other common symptoms. And ultimately, testing may fail to reveal that mold exposure is the cause.[85]

This text cannot do justice to the vast pathophysiology, diagnostic techniques, or therapeutic protocols helpful to address mold-induced illnesses. At the very least, ask patients if they have any obvious exposure to molds at home, in the workplace, or elsewhere, such as damp basements, musty smelling rooms, barns, and other locations commonly susceptible to mold growth.

Among the most common mold-induced symptoms are rhinorrhea, sinus tenderness, and wheezing. Also worth noting are neurologic symptoms including the inability to stand on the toes or to walk a straight line with eyes closed and short-term memory loss. Avoiding further mold exposure, even if it means remodeling the home, moving, or changing jobs, is essential to recovery. Sometimes referred to as "sick building syndrome," an unrealized mold reservoir should also be searched for in any edifice where a number of occupants suffer from chronic illness.

---

or other trade names) to treat asthma, allergic cough, angiotensin-converting-enzyme-inhibitor-related cough, and chronic obstructive pulmonary disease (COPD).

Theophylline   100 to 600 mg capsule

Capsules are available in various strengths and may be taken once or twice a day for a month or more, as directed. Use with caution in patients with hyperthyroidism, seizure disorder, peptic ulcer, or cardiovascular disease. Theophylline clearance may decrease in patients with congestive heart failure, acute pulmonary edema, hepatic disease, cor pulmonale, acute hepatitis, hypothyroidism, cirrhosis, fever, or sepsis with multiorgan failure and shock.

Topical Compress for Asthma

Topical herbs, and even simple heat, can provide rapid relief for chest pain, tightness, and asthma concomitant with lower respiratory infections. *Lobelia* is one of the fastest-acting herbs that I have tried.

Hot water	½ cup (120 ml)
Lobelia inflata dry herb	¼ cup (12.5 g)
Apple cider vinegar	¼ cup (60 ml)
Apium graveolens essential oil	30 drops
Castor oil, for topical application at each treatment	

Soak the dry *Lobelia* in the vinegar for 10 minutes, then cover with the hot water and mix in the *Apium* essential oil. Place the entire mass inside a muslin cloth,

Plant EFAs for Allergic Disorders

Many seeds are excellent sources of EFAs, which are important components of cell membranes and play many important immunomodulating roles. The following oils can be applied to the skin and included in medicinal foods or consumed as a supplement to treat immune and allergic disorders.

Argania spinosa. Argan oil contains monounsaturated and saturated fatty acids and has been shown to improve skin elasticity and skin hydration and promote wound healing. Although argan oil has been used traditionally as a culinary oil, it is also a useful skin oil and may increase transdermal absorption of other ingredients in skin care creams.

Borago officinalis. Borage oil contains high levels of EFAs, including linoleic acid. It has therapeutic effects for atopic dermatitis and may improve seborrheic dermatitis in infants and children when topically applied. Borage oil capsules are available for oral use.

Cocos nucifera. Coconut oil contains saturated fat, medium-chain triglycerides, and numerous fatty acids. Applying coconut oil to the skin improves barrier function and speeds epithelialization. It also increases neovascularization, fibroblast proliferation, and collagen synthesis, and it may deter opportunistic skin bacteria by causing bacterial membranes to disintegrate.

Hippophae rhamnoides. Oral ingestion of sea buckthorn oil improves fatty acid composition of the skin in patients with atopic dermatitis,[86] speeds wound healing,[87] has significant antioxidant activity,[88] increases hydroxyproline in skin and connective tissue, and offers UV light protection.[89] Sea buckthorn is rich in vitamins A, C, and E and the minerals sulfur, selenium, zinc, and copper.[90]

Linum usitatissimum. Flaxseeds are a rich source of α-linolenic acid (ALA). ALA can be transformed into epoxyeicostetraenoic and eicosapentaenoic acids in the human gut and exert antiallergy affect in the intestines[91] and throughout the tissues. The consumption of flaxseed oil can improve the lipid profiles of cell membranes, such as red blood cell membranes, and help protect against various oxidative stressors.[92]

Nigella sativa. Black cumin seeds and pressed oil contain thymoquinone credited with anti-inflammatory, antioxidant, and chemopreventive effects, and now shown to increase motion of respiratory cilia, explaining *Nigella*'s traditional use as a respiratory expectorant.[93]

Oenothera biennis. Evening primrose seed oil is extremely high in linoleic acid and γ-linolenic acid (GLA), important precursors of anti-inflammatory eicosanoids. It has been shown to benefit atopic dermatitis, psoriasis, Sjögren's syndrome, and asthma.[94]

Rosa canina. Rose hip seed oil can be consumed orally, but due to its expense, is more commonly used topically. Rose hip seed oil contains linoleic acid, α-linolenic acid, as well as tocopherols and carotenoids. Rose hip oil is used in medicinal skin oils and high-end cosmetics, and may improve eczema, neurodermatitis, and cheilitis when topically applied.

Simmondsia chinensis. Jojoba oil is fairly shelf stable compared to the highly labile flaxseed and evening primrose oils. It is widely used in lotions and creams and shown to improve skin barriers in cases of seborrheic and eczematous dermatitis. Jojoba may also enhance the absorption of other ingredients in formulas.

Vitellaria paradoxa. Shea butter is composed of triglycerides with oleic, stearic, linoleic, and palmitic fatty acids. Because the "butter" is solid at room temperature it is more commonly used topically than orally. Shea butter is shown to inhibit nitric oxide, cyclooxygenase, and cytokines via the NF-κB pathway and shown to help treat allergic dermatitis when used topically.

bandana, or thin towel. Apply castor oil to the patient's chest or between the shoulder blades, then apply the *Lobelia* compress and cover with a heating pad or a hot, moist towel.

Classic Chinese Formula for Asthmatic Wheezing

The herbs in this traditional formula can be found in various forms, including fluid extracts, powders, and granules. Other species of *Asparagus* such as *A. racemosus* may be substituted if more readily available.

Rehmannia glutinosa
Paeonia suffruticosa
Schisandra chinensis
Asparagus cochinchinensis
Prunus armeniaca
Scutellaria baicalensis
Stemona sessilifolia

If using granules, combine in equal parts. Stir 1 teaspoon of the combined granules into 1 to 2 cups of hot water or other herbal tea and drink 2 or 3 times a day. The formula can also be prepared from dry herbs. Simmer gently 1 heaping tablespoon of the herbs in 3 or 4 cups of water for 15 minutes, then strain, and consume the entire amount over the course of the day. If using tinctures, combine in equal parts and take 1 to 2 dropperfuls, 2 to 5 times a day, reducing as symptoms improve.

Tincture for Acute Asthma with Wheezing

This formula can offer a strong antispasmodic effect on the bronchial smooth muscle in instances of acute wheezing. (Asthma and specific herbal formulas are further discussed in chapter 3, Volume 2.)

Lobelia inflata	15 ml
Thymus vulgaris	15 ml
Angelica sinensis	10 ml
Ammi visnaga	10 ml
Ginkgo biloba	10 ml

Take 1 dropperful of the combined tincture every 5 to 10 minutes, reducing as symptoms improve.

Formulas for Oral Allergic Reactivity

The tongue and buccal tissue may occasionally be affected by allergic phenomena, and autoimmune disorders can often affect the oral cavity. Surprisingly little is known about the complex workings of the immune system in the oral cavity, however, making a trial and error elimination of possible allergens a key clinical approach to suspected allergic phenomena in the mouth. It is known that epidermal dendritic Langerhans cells are important regulators of immune reactivity in the oral cavity, reacting to microbes, antigens, and tumors. These cells may initiate oral hypersensitivity reaction akin to contact dermatitis and trigger a variety of inflammatory cascades.[95] Allergic reactivity in the mouth may cause stomatitis, cheilitis, leukoplakia-like and lichenoid lesions, geographic tongue, and burning mouth syndrome. For additional information about and formulas for these conditions, see also sections later in this chapter as well as chapter 4 of this volume.

Despite being exposed to all manner of noxious stimuli, irritants, and allergens, the oral cavity is relatively resistant to allergic reactivity. The oral mucosa of atopic individuals may be sensitive to certain foods such as nuts, citrus, wheat, and other common allergens. Celiac disease and gluten intolerance are most commonly associated with bowel symptoms and enteropathy, but gluten sensitivity can also induce skin lesions, including oral manifestations.[96] Shifts in microbial populations of the oral cavity and increases in fungal strains, particularly *Malassezia* yeasts, may play a role in atopic disorders of the mouth.[97] In a chicken-and-egg cycle, patients with allergic reactivity may support yeast more readily, and patients with high fungal burden may develop atopic disorders. Metals are well-known triggers of contact dermatitis in the skin, and the metals used in dental bridges and crowns can act as contact irritants and trigger oral reactivity.[98] The metals used in dental amalgams have been associated with oral lichenoid reactions,[99] and exposure to mercury for just 20 days can result in oral lesions. Nickel is one of the most common triggers of cutaneous contact dermatitis and may cause oral reactivity to braces or other dental devices that contain nickel. Potassium dicyanoaurate (a gold salt that may be used in dental devices) may trigger burning mouth syndrome and other conditions. Cobalt (a chloride used in electroplating of various metals, including dental devices) can trigger perioral dermatitis. Lamentably, no dental material from gold, to palladium, to titanium, to ceramic resin is without evidence of triggering sensitivity. Oral hygiene products and flavoring agents, chemicals, and preservatives may be oral allergens,[100] and all such synthetic products may be harmful over time. Allergic individuals can also react to natural mouthwashes and chewing gums containing cinnamon, eugenol in clove oil, and menthol in peppermint essential oils, although such substances are very effective oral disinfectants and usually well tolerated.

Some of the herbs most useful for treating oral mucosal allergies are *Allium cepa*, *Allium sativum*, *Aloe vera*, *Althaea officinalis*, *Calendula officinalis*, *Euphrasia officinalis*, and *Glycyrrhiza glabra*. More information about these herbs can be found in the "Specific Indications" section of this chapter beginning on page 85.

Syrup for Allergic Glossitis

Allergic glossitis can be associated with food allergy or a general atopic constitution. Other causes include microcirculatory disease, hyperviscosity of the blood, collagen vascular inflammation, pemphigus, Behçet's disease, and lupus, which may affect both the oral cavity and the tongue. *Angelica sinensis* may improve allergic glossitis and glossodynia.[103] *Glycyrrhiza* solid extract has anti-inflammatory effects, and also provides a thick base to hold the *Angelica* against the tongue.

Glycyrrhiza solid extract	90 ml
Angelica sinensis tincture	30 ml

Combine the two ingredients in a small container and mix thoroughly into a thin syrup. Take ¼ teaspoon every several hours, reducing as symptoms improve.

Syrup for Oral Mucosal Reactivity

Allergic reactions to nuts, shellfish, citrus fruits, or food additives such as benzoates, sulfites, and artificial colors can extend to the oral mucosa. Alcohol and coffee can also trigger reactivity, as can pharmaceutical drugs. There is increasing evidence that specific HLA alleles influence the risk of drug reactions.[104] An underlying allergic constitution can be addressed systemically; this

Supportive Therapies for Stomatitis

Autoimmune diseases, particularly erythema multiforme, lichen planus, Behçet's disease, Stevens-Johnson syndrome, and lupus, may involve stomatitis and oral lesions. Any food may trigger a peculiar idiosyncratic reaction in an individual, but nuts and citrus are especially common triggers. Heavy metal toxicity is also associated with oral lesions, and mercury toxicity is especially noted to promote excessive salivation. Low-level laser therapies are being explored as a possible option for recurrent or severe cases.[101] Stevens-Johnson syndrome can be severe and has been fatal in 10 to 50 percent of cases depending on the severity.[102] The condition may cause large layers of skin and tissue to become necrotic, and therefore treatment requires expert guidance. Such severe cases are often treated with high doses of drugs such as prednisone and colchicine. Herbal formulas may provide pain relief, shorten the duration of an episode, and reduce the need for potentially harmful drugs. The following herbs and supplements may be used in aggressive protocols to address autoimmune diseases with oral lesions.

Folic acid
Vitamin A
Zinc
Glutamine
Glycyrrhiza solid extract
Aloe vera gel
Curcuma longa
Mucoadhesive agents such as *Althaea officinalis* or *Tamarindus indica*
Tannins
Flavonoids

As part of a protocol, consider taking the vitamins and minerals as supplements, using the herbs in mouth rinses, and preparing mouth pastes from glutamine, *Aloe vera* gel, and licorice solid extract as exemplified in formulas in this chapter.

formula addresses acute oral symptoms. *Glycyrrhiza* is the anti-inflammatory base; in a solid extract form, it helps the other herbs cling to the oral mucosa. *Hypericum*, *Commiphora*, and *Sanguinaria* are specific for oral pain and gingival and buccal lesions. Do not exceed the suggested amount of *Sanguinaria*, as it can prove irritating rather than helpful in higher quantities.

Glycyrrhiza glabra solid extract	30 ml
Hypericum perforatum	15 ml
Commiphora myrrha	15 ml
Sanguinaria canadensis	15 ml

Thoroughly mix the 3 tinctures into the *Glycyrrhiza* solid extract. Place ¼ teaspoon directly in the mouth 2 or 3 times a day. The formula can also be diluted in a bit of water and swished in the mouth for several minutes before swallowing. Use hourly in ⅛ teaspoon doses for acute symptoms.

Tea for Oral Mucosal Reactivity

As in the Syrup for Oral Mucosal Reactivity, *Glycyrrhiza* and *Hypericum* are included here for their effects on oral lesions and pain. *Salix* provides additional anti-inflammatory effects, and *Tabebuia* is gently astringent in the case of swelling, ulceration, or increased salivary and mucous secretions.

Glycyrrhiza glabra
Hypericum perforatum
Salix alba
Tabebuia impetiginosa

Combine in equal parts. More *Glycyrrhiza* can be added for flavor, if desired. Steep 1 tablespoon per cup of hot water. Drink 3 cups (720 ml) or more throughout the day for a week or until condition abates. For acute symptoms, attempt to sip as constantly as possible, reducing as symptoms subside.

Demulcent Tea for Oral Lesions

Teas are ideal for oral lesions because they can be gargled and swished about the mouth to prolong surface contact. This formula is slightly demulcent and appropriate for pain, burning sensations, and dry, irritated mucosal surfaces.

Ulmus fulva
Glycyrrhiza glabra
Calendula officinalis

Combine in equal parts. Steep 1 tablespoon of the combined herbs per cup of hot water for 10 to 15 minutes. Drink at least 3 cups (720 ml) per day or more as desired.

Collagen-Vascular Oral Lesion Tincture

Small pinpoint lesions and ulcerations of the gums and oral mucosa may occur in lupus, Behçet's disease, and other autoimmune conditions. Licorice tastes good, helps heal ulcers, and may act as an immunomodulator. *Hydrastis* and *Commiphora* are also specific for lesions and inflammation of the oral mucosa, helping to astringe tissues that are swollen or bleeding. *Althaea* provides a soothing base to help heal lesions and alleviate pain and burning sensations.

Glycyrrhiza glabra	20 ml
Calendula officinalis	10 ml
Althaea officinalis	10 ml
Hydrastis canadensis	10 ml
Commiphora myrrha	10 ml

Mix 1 to 2 teaspoons of the combined tincture in a small amount of water and use as a gargle, swallowing when finished. Take 3 or more times daily.

Tincture for Aphthous Stomatitis

Recurrent aphthous stomatitis involving painful ulcerative or erosive lesions of the oral mucosa may be idiopathic or secondary to an underlying immune disorder such as Behçet's disease. Colchicine at a dose of 1 to 1.5 milligrams daily for a minimum of 3 months may help to resolve or improve chronic lesions as effectively as 5 milligrams daily of prednisolone does. Colchicine is available by prescription only, but herbalists might also use the whole *Colchicum* plant, from which the colchicine alkaloid is derived, in tincture form. (See "*Colchicum*—An Herbal Immune Suppressant?" on page 73 for more information.) **Caution:** Do not exceed the dosage of *Colchicum* noted below due to its toxic potential.

Colchicum autumnale	40 ml
Glycyrrhiza glabra solid extract	20 ml

Combine the 2 ingredients in a 2-ounce (60 ml) bottle and shake vigorously. The tincture should dilute the solid extract to the extent that it can pass through the dropper of the tincture bottle. Take 1 dropperful of the tincture and hold in the mouth for at least 2 minutes before swallowing. Take 2 to 3 times a day for 3 months, then evaluate for efficacy, continuing as needed.

Mouthwash for Lupus

Oral lesions and ulcers occur commonly in patients with lupus, and may range from a minor nuisance to a significant interference to talking and eating. This formula uses *Glycyrrhiza* for its anti-inflammatory, immunomodulating, and ulcer-healing effects (it may also be used in the form of a solid extract). *Calendula* improves microcirculation to the skin and mucosa, and *Symphytum* may help allay burning pains and heal tissue.

Glycyrrhiza glabra
Calendula officinalis
Symphytum officinale

Combine in equal parts and take with a sip of water. Swish around the mouth for several minutes to prolong surface contact to increase efficacy.

Mouthwash for Behçet's Syndrome

Similar to mouthwashes for aphthous ulcers and lupus, this formula may help resolve oral lesions while allaying pain. This formula uses a *Calendula* tea rather than a tincture.

Calendula officinalis tea	¼ cup (60 ml)
Glycyrrhiza glabra solid extract	1 teaspoon
Aloe vera gel	1 teaspoon

Combine the ingredients, mix well, and use as a mouth gargle, swallowing when finished. Use 3 or 4 times daily until lesions are fully resolved, and follow with once a day as a maintenance or preventive therapy as needed.

Mouthwash for Sjögren's Patients

Licorice is demulcent and anti-inflammatory to the oral mucosa. The solid extract form makes the medicine thick and sticky to prolong surface contact with the buccal membranes. *Sanguinaria* acts as a counterirritant in this formula, stimulating circulation and saliva flow. *Pilocarpus* is a specific and a profound sialagogue.

Water	¼ cup (60 ml)
Glycyrrhiza glabra solid extract	25 ml
Aloe vera gel	25 ml
Pilocarpus jaborandi	5 ml
Sanguinaria canadensis	5 ml
Folic acid liquid, to deliver 500 mcg per rinse as per product label	

Combine the *Glycyrrhiza*, *Aloe* gel, *Pilocarpus*, and *Sanguinaria* in a 1-ounce (30 ml) bottle and shake vigorously. Add 1 to 2 dropperfuls to the water, then add the commercial liquid folic acid. Swish around the mouth for a full 2 minutes, swallowing afterward, repeating until the ¼ cup is gone. Repeat the procedure at least 3 times daily. The use of the mouth rinse immediately before meals may increase saliva to support comfort of eating and swallowing. The quantity of *Pilocarpus* in the formula may be increased if results aren't seen after several days. Because *Pilocarpus* is a systemic secretory stimulant, the mouth rinse can be expectorated rather than swallowed if increased sweating or intestinal stimulation occurs.

Formulas for Contact Dermatitis

Contact dermatitis is an atopic phenomenon in which the skin reacts to chemicals or natural substances when they come into contact with the skin. It is not a distinct diagnostic category from atopic dermatitis, but is rather a specific presentation. The patient notes dermatitis when sitting in the grass, when nickel touches the skin, or when certain cosmetics or perfumes are applied, as opposed to the spontaneous appearance of pruritic vesicles or urticarial wheals as with eczema. Also known as contact allergy or contact sensitization, this form of dermatitis is a common form of delayed hypersensitivity to molecules referred to as *haptens*—small molecules capable of triggering a sensitization reaction by binding body proteins of the skin and initiating an immunoglobulin response.[105] A hapten elicits an immune response only when attached to proteins and it is a more specific

term than *allergen*, which refers to anything that elicits an allergy-like symptom. Haptens can also form in the GI tract and the blood. For example, penicillin can cause autoimmune hemolytic anemia, and hydralazine can cause drug-induced lupus.

Patch testing is a diagnostic procedure performed by allergists to diagnose what substances are causing contact dermatitis or worsening general atopic dermatitis. Adhesive patches impregnated with various contact allergens are applied directly to the skin, usually by injuring the skin slightly first with a small scratch, hence the term "scratch testing." Typically, some 30 patches carrying the most common contact allergens are applied. Although this is a long-standing medical approach, I don't subscribe to this method. It is usually obvious that a person's skin has become sensitive, and finding out what specific

Common Contact Allergens

The following list of metals, drugs, and chemicals are among the common substances known to trigger allergic reactivity. The metals may trigger allergies when found in zippers and snaps on clothing, when used in dental materials, and even when present in eating utensils. Aluminium contact sensitization has gained some attention due to the observation of cutaneous or subcutaneous inflammation after injection of aluminium-adsorbed vaccines. Many hygiene products can promote hypersensitivity disorders. And many household cleaning products, aerosols, and propellants can be problematic, as can glue, dye, and bandaging materials.

2-hydroxyethyl methacrylate
Acrylates
Bacitracin
Balsam of Peru
Cobalt
Cocamidopropyl betaine
Essential oils, especially oils that contain psoralen and coumarin
Formaldehyde
Fragrance
Hair dye
Local anaesthetics, such as dibucaine and benzocaine
Makeup and personal hygiene products
Neomycin sulfate
Nickel and occasionally other metals, including gold salts
Potassium dichromate
Propylene glycol
Rubber contact allergens
Sesquiterpene lactones (components of aster-family plants such as goldenrod and chamomile)
Synthetic aromatic terpenes, particularly hydroperoxides of linalool and limonene
Thiomersal

---

substance they are sensitized to is less important than reducing the hypersensitivity reaction itself. If the only outcome of such testing is a recommendation to avoid pollen, or dust mites, or metals, or formaldehyde, that is no help at all. And if a certain type of shampoo or laundry soap is suspected to be the culprit, it doesn't really mean that the particular substance *causes* the problem; rather, the person's allergic nature is the cause. If we do not help to reduce the underlying atopic tendency, the person will likely become sensitive to an increasing number of compounds over time. Furthermore, the scratch test is traumatizing to some, and I have seen adult men quiver at the memory of their childhood allergy tests.

Note that any formula in this chapter for hives or other forms of atopic dermatitis would also be appropriate for treating contact dermatitis.

Licorice Compress for Contact Dermatitis

Glycyrrhiza glabra has numerous immunomodulating functions. Its glycyrrhetinic acid and licochalcones inhibit T-cell proliferation and inflammatory cytokine production. The topical application of licorice may improve atopic dermatitis. Licorice-containing ointments or a simple compress prepared by soaking a cloth in licorice tea may be applied. These applications may ameliorate dryness, pruritus, and inflammation accompanying dermatitis.

Glycyrrhiza glabra dried root   1 tablespoon

Gently simmer the dried licorice root in 3 cups (720 ml) of water for 10 minutes. Let stand until cool enough to comfortably handle, then strain. Soak a soft cloth in the liquid and fold into a shape and size convenient for topical application. Apply for 15 minutes every several hours, reducing as symptoms subside.

Colchicum Tincture for Chronic Urticaria

Colchicine, a well-known gout medicine derived from *Colchicum autumnale*, may reduce urticarial reactivity. See "*Colchicum*—An Herbal Immune Suppressant?" on page 73 for a discussion of the molecular research. **Caution:** Due to potential toxicity, do not exceed the amount of *Colchicum* used in this formula. The formula also uses *Urtica dioica* (nettle) root rather than leaves to

Topical Antipruritics

Mint and clove essential oils and oatmeal baths are the best palliatives I have tried for providing antipruritic effects. Dot *Mentha* or *Syzygium* essential oil on the skin or apply with a small gauze pad. Be sure to keep out of the eyes. Oatmeal (*Avena*) baths are very soothing. See chapter 5 of Volume 1 for detailed directions for oatmeal baths.

Herbal Specifics for Allergic Skin Eruptions

PRESENTATION	RECOMMENDED HERBS
Hot, red, stinging skin	*Aloe vera* *Apis mellifica* (homeopathic remedy) *Sanguinaria canadensis* (drop or homeopathic dosages)
Ulcerated skin with discharges	*Apium graveolens* *Betula* spp. *Calendula officinalis* *Geranium maculatum* *Phytolacca americana* *Quercus* spp.
Sense of crawling in the skin	*Apium graevolens* *Hypericum perfoliatum* *Scrophularia nodosa*
Dry, scaling dermatitis	*Avena sativa* (baths) *Centella asiatica* *Fucus vesiculosus* (and thyroid support) *Phytolacca americana*
Acute itching	*Ammi visnaga* *Angelica sinensis* *Apium graveolens* *Scutellaria baicalensis* *Salix* spp. *Urtica dioica*

best support renal clearance of possible antigens. *Urtica* tinctures are available as both root and leaf preparations.

Angelica sinensis	20 ml
Urtica dioica root	15 ml
Petroselinum crispum root	15 ml
Colchicum autumnale	10 ml

Take 1 dropperful of the combined tincture per hour, reducing to every 2 to 4 hours as symptoms improve. To maintain the effects once relief of acute hives has been achieved, the tincture may be taken by the dropperful twice per day for several months.

Tincture to Relieve Itching

This formula uses herbs recommended in folklore for alleviating itching. *Hypericum* may help pruritus by calming irritated nerve endings. Reishi mushroom, *Ganoderma lucidum*, inhibits pruritus via effects on histamine receptors and mast cells.[106] *Picrorhiza kurroa* is used in Ayurvedic medicine for allergy; it inhibits NF-κB signaling, thereby reducing the release of inflammatory mediators.[107] These antihistamine and anti-inflammatory herbs can be taken frequently in cases of acute eczema, hives, contact dermatitis, and reactivity to bee stings and bug bites.

Hypericum perforatum
Picrorhiza kurroa
Scrophularia nodosa
Tylophora indica
Ganoderma lucidum

Combine in equal parts. Take ½ to 1 teaspoon as often as every 15 to 30 minutes for acute itching, reducing dose and frequency as symptoms improve over the course of the day, or several days if necessary. Once acute pruritus has subsided, use the formula 3 times a day to reduce a chronic tendency.

Tincture for Seborrhea

Seborrhea most commonly affects infants and is referred to as cradle cap. However, it can also occur into adulthood, presenting on the eyebrows, behind the ears, and at the nasolabial creases. Cradle cap in a newborn may be a harbinger of an atopic constitution, and extra care should be taken introducing infants to foods and avoiding exposure to mold, chemicals, and environmental toxins that could promote further allergic sensitization. This formula features herbs emphasized in the folkloric

Topical Options for Skin Eruptions

The following are herbs specifically indicated in folklore for various types of skin lesions. See also chapter 5 of Volume 1 for more information on topical treatments.

PRESENTATION	RECOMMENDED HERBS
Moist, weeping eruptions	Betula pendula Geranium maculatum Hamamelis virginiana Quercus spp.
Ulcerated moist lesions	Commiphora myrrha Grindelia squarrosa Juglans nigra
Dry, fissured, bleeding eruptions	Aloe vera Calendula officinalis Symphytum officinale
Pustules, pimples, and boils	Achillea millefolium Echinacea purpurea Hydrastis canadensis Melalueca alternifolia

literature as specific for seborrhea, and especially the tendency to hard yellow crusts behind the ears. EFAs, zinc, and vitamin A are important to supplement in severe cases, including to supplement nursing mothers to ensure the breast milk is plentiful in these nutrients.

Juglans nigra	15 ml
Scrophularia nodosa	15 ml
Stillingia sylvatica	15 ml
Trifolium pratense	15 ml

Take ½ to 1 teaspoon of the combined tincture 3 to 6 times per day for several months. Continue at a lower dose of ½ teaspoon once or twice a day thereafter as a maintenance therapy and preventive if recurrent episodes occur.

Tincture for Pompholyx

Pompholyx is a less common type of eczema in which pruritic vesicles, and sometimes large bullae, emerge on the palms and soles. *Hypericum* and *Stillingia* are both emphasized in the folkloric literature for skin eruptions on the palms and soles. *Tanacetum, Curcuma, Calendula,* or *Scutellaria baicalensis* may make useful adjuvants to this simple formula depending on the presentation.

Hypericum perforatum
Stillingia sylvatica

Combine in equal parts and take several dropperfuls every ½ hour for several days to a week or more as needed for acute eruptions. Use the formula 3 times a day to reduce a chronic tendency.

Tea for Pruritus

Pruritus is an unpleasant cutaneous sensation associated with the immediate desire to scratch, sometimes so severe that it leads to intense scratching to the point of bleeding. Pruritus is the diagnostic hallmark of atopic dermatitis, and a chief symptom of urticaria. Itching sensations are caused by the release of inflammatory mediators from mast cells, such as histamine, serotonin, and substance P.

Perilla frutescens
Matricaria chamomilla
Apium graveolens
Lonicera japonica

Combine in equal parts—for instance, 4 ounces (120 g) of each herb to yield a pound of the tea. Steep 1 heaping tablespoon per cup of hot water, for 10 to 15 minutes then strain. Drink 3 cups (720 ml) or more daily for several weeks during episodes of hives or acute dermatitis. Once acute itching is controlled, continue to drink several cups throughout the day, several days per week, as a maintenance therapy.

Field Poultice for Skin Reactivity

Prunella vulgaris, or self-heal, contains triterpenes credited with antiallergy effects. This common meadow weed is a traditional simple for numerous allergic and inflammatory disorders, and can be included in teas and tinctures or prepared into a simple field poultice for topical application. Modern research has shown *Prunella* to exert an anti-inflammatory effect via inhibition of mast cell degranulation, reducing histamine release.[108]

Prunella vulgaris fresh leaves	1 tablespoon

Chew the fresh leaves into a moist bolus and apply directly to the affected skin. Replace every half hour for several hours to alleviate discomfort, or until other treatment options are available.

Formulas for Angioedema

Allergic angioedema involves acute and often gross swelling in the skin and upper airways as mucosal connective tissue swells, sometimes extending to the digestive system and causing abdominal pain. Angioedema accounts for around 100,000 emergency room visits in the United States each year. The edema is nonpitting, and research has revealed that both histamine and bradykinin pathways may underlie angioedema.[109] While the histamine-mediated forms present similarly to anaphylaxis and may be triggered by foods, medications, insect venom, or sunburn, bradykinin-mediated angioedema presents with greater face and oropharyngeal involvement and higher risk of progression to respiratory distress. Severe histamine-mediated angioedema can be treated with intramuscular injections of epinephrine and oral antihistamine medications or steroids. These medications are not effective for bradykinin-mediated forms. Clinicians may misdiagnose the digestive aspects of angioedema and offer antacids or irritable bowel therapy, or may suspect ulcers, appendicitis, and other gastrointestinal lesions when particularly acute.

The most common presentations of angioedema may be acute swelling of the lips or ears following a sunburn, or following exposure to topical chemicals or ingestion of an allergen. The swelling is not particularly painful but may be associated with tingling or itching sensations, and urticaria is concomitant in roughly 50 percent of cases. Although the condition usually subsides on its own within a weeks' time, botanical agents may hasten the resolution or may help reduce severity in those with swelling significant enough to compromise the airways. Furthermore, for those with repetitive or particularly severe episodes, botanical and nutritional therapies may reduce the tendency to allergic inflammation. *Crataegus* species, *Tanacetum parthenium*, *Curcuma longa*, *Centella asiatica*, and *Glycyrrhiza glabra* can be used long term as a preventive, or a mouth rinse or

Tanacetum parthenium for Migraines and More

Tanacetum parthenium is a classic herb for treating and preventing migraine headaches, asthma, rheumatism, skin inflammation, and allergy. Numerous mechanisms of action include an antihistamine effect due to mast cell stabilization[110] and inhibition of pro-inflammatory leukotriene release. One of the active constituents identified in feverfew is parthenolide, shown to limit macrophage-driven inflammation[111] and to exert antiproliferative effects on fibroblasts, making the plant appropriate for connective tissue pathologies such as rheumatoid arthritis.[112] *Tanacetum* also contains arabinogalactans, immune polysaccharides with numerous immunomodulating actions, including inhibition of excessive tumor necrosis factor (TNF),[113] an immune trigger that often initiates and drives inflammatory cascades. *Tanacetum*'s inhibition of TNF reduces blood cell activation and related inflammatory cascades,[114] and blocks vascular serotonin 5-HT receptors, contributing to the plant's well-known ability to reduce migraine headaches.

Tanacetum parthenium, feverfew

lip balm of *Hypericum perforatum*, *Calendula officinalis*, and *Aloe vera* may be used acutely.

Tincture for Acute Angioedema

The herbs in this formula may reduce acute allergic reactivity and speed the resolution of angioedema.

Perilla frutescens
Curcuma longa
Ginkgo biloba

Combine in equal parts and take 1 dropperful as often as every 15 minutes for several hours, then reducing to hourly for several days.

Mouth Rinse for Oral Angioedema

Angioedema of lips and tongue may be treated with an oral rinse that provides soothing and anti-inflammatory effects and reduces the risk of airway involvement.

Aloe vera gel	15 ml
Glycyrrhiza glabra solid extract	15 ml
Centella asiatica tincture	15 ml
Hypericum perforatum tincture	15 ml

Place 1 teaspoon of the combined ingredients in ¼ cup of water. Swish around the front of the mouth, then swallow. Repeat every ½ to 1 hour at the onset of acute swelling, reducing as symptoms abate.

Formulas for Allergic Reactivity of the Urinary Tract

The urinary epithelium can become sensitized and display allergic reactivity, a fact that can be overlooked when patients present with symptoms suggestive of a bladder or prostatic infection, but no infection present. Ultimately, no adequate therapies may be offered. Most such patients have a history of atopy or other concomitant allergic disorder. Initial allergic irritations of the bladder are often undertreated, while repeat episodes may be diagnosed as interstitial cystitis and chronic prostatitis. (See chapter 4 of Volume 1 for discussion of these complex conditions.) Such conditions may be in part autotoxic, in part inflammatory, and in part autoimmune and thus overlap with conditions discussed in this section. Such inflammatory disorders of the urinary tract are sometimes referred to as lower urinary epithelial dysfunction (LUED). Chemical reactivity can trigger acute inflammation in the bladder characterized by vasodilation and increased vascular permeability, leukocyte migration to the site of injury, and activation of biochemical cascades of inflammation causing release of mediators such as cytokines, histamines, kinins, complement factors, clotting factors, nitric oxide, and proteases. Autoimmune sensitization can occur where T-lymphocyte cells proliferate and activate B cells to produce immunoglobulins. There is frequently a great deal of overlap between those who experience allergic or autoimmune reactivity of the bladder, and other allergies and autoimmune disorders including fibromyalgia, endometriosis, and inflammatory bowel disorders.[115]

Tincture for Allergic Bladder Symptoms

Symptoms of cystitis but with a sterile urine can be due to interstitial cystitis or can be due to allergic reactivity in the urothelium. Herbs in the Apiaceae family (including *Ammi*, *Angelica*, and *Apium*) have systemic antiallergy effects, as well as antispasmodic effects on urinary muscle. The *Apis mellifica* homeopathy mother tincture is specific for stinging and burning urination.

Ammi visnaga	20 ml
Angelica sinensis	20 ml
Apium graveolens	18 ml
Apis mellifica	2 ml

Take ½ to 1 teaspoon of the combined tincture as often as every 15 to 30 minutes for new acute symptoms, reducing as symptoms improve. The tincture should be helpful over the span of 2 or 3 days; frequency can be reduced to hourly, then every 2 or 3 hours, and then to 3 or 4 times a day as symptoms abate. For those with recurrent episodes, this tincture can be continued long term at a dose of 2 or 3 times per day, or replace with an Apiaceae-containing tea formula.

Tea for Allergic Bladder Symptoms

Teas may be more effective for urinary allergies than tinctures due to better direct surface contact with the bladder mucosa. *Apium*, *Glycyrrhiza*, and *Filipendula* all have antiallergy and anti-inflammatory effects. *Hypericum* is traditionally used to calm urinary irritability, and is often suggested as a folkloric treatment for

Herbs for Urinary Hyperreactivity

Allergic phenomena of the urinary mucosa should be especially suspected in individuals who have other atopic and allergic conditions. Urinary allergy may mimic prostatitis in men. The following herbs can be used to treat urinary hyperreactivity.

Apiaceae-family herbs. *Ammi visnaga, Angelica sinensis, Apium graveolens,* and *Petroselinum crispum* may be used for inflammation and especially spasm of the ureters and bladder. They also promote sodium excretion and diuresis, and all contain coumarins.

Bupleurum falcatum. Consider *Bupleurum* in formulas for acute and chronic nephritis and for autoimmune reactivity in the kidneys. It may be used for inflammation of the kidneys following exposure to nephrotoxins.

Cordyceps militaris. This medicinal mushroom is best for long-term use in cases of autoimmune, autoinflammatory, and chronic allergic reactivity in the kidneys. *Cordyceps* may also be used for renal inflammation, renal failure, lupus, Sjögren's syndrome, and other autoimmune damage to kidneys.

Hypericum perforatum. *Hypericum* can help reduce nervous excitability in the bladder. It may improve bedwetting due to allergic sensitivity of the urinary mucosa and is noted to reduce urinary pain and spasm in inflammatory and infectious conditions.

Solidago canadensis. *Solidago* has trophorestorative effects on the urinary system. *Solidago* may help reduce allergic and inflammatory conditions in the urinary system, such as prostatitis and interstitial cystitis. *Solidago* flowers have an extremely high pollen content, and many types of pollen have been shown to be beneficial to the prostate.

As discussed throughout this chapter, hypersensitivity reactions often involve an increase in the permeability of an epithelial barrier. In this case, the barrier is the urothelium, which lines the interior of the bladder, and cellular injury contributes to sensations of burning or pain. The urothelium is also a sensory web that receives, amplifies, and transmits information as it senses the extracellular environment. It responds to chemical, mechanical, and thermal stimuli via the release of ATP, nitric oxide, acetylcholine, and other molecules. The urothelium possesses a variety of receptors and ion channels, including purinergic receptors, nicotinic and muscarinic receptors, and TRP channels.[116] Altered signals in any of these pathways can affect sensation, barrier efficacy, and immune functions of the urothelial tissues and contribute to allergic and autoimmune disorders. Damage to the claudin proteins of the tight junctions that support cellular adherence can occur due to inflammatory and antigenic triggers, and loss of the urothelial barrier allows urinary solutes to diffuse through the urothelium and sensitizes the bladder's afferent nerve fibers. Increased firing of sensory nerves alters the sensations normally involved with filling and emptying the bladder.[117] Increased cytokine release is associated with various inflammatory and autoimmune diseases and also contributes to overactive bladder, chronic prostatitis, and chronic pelvic pain syndromes such as lower urinary epithelial dysfunction and interstitial cystitis. Sensitization promotes the release of histamine from mast cells and contributes to pelvic pain and uncomfortable urinary symptoms and voiding dysfunction.[118] Purinergic signaling pathways may also be disrupted, and molecular and dietary purines may begin to trigger urothelial symptoms.[119] Estrogen supports the health and integrity of the urothelium and hormonal fluctuations or decline may alter the stability of the urothelium.[120] I have sometimes found a weekly or bi-weekly application of an estrogen cream to the external genitals to improve urinary irritability in elderly women.

bedwetting, especially when due to urinary irritation and allergic inflammation.

Glycyrrhiza glabra	2 ounces (60 g)
Apium graveolens	1 ounce (30 g)
Filipendula ulmaria	1 ounce (30 g)
Hypericum perforatum	1 ounce (30 g)
Urtica urens	1 ounce (30 g)

Combine the dry herbs and steep 1 tablespoon per cup of water for 10 to 15 minutes. It may be easiest to brew 4 or 5 cups (960 to 1,200 ml) at once to sip throughout the day. Consume at least 3 cups (720 ml) per day for several weeks, then reduce to 1 or 2 cups (240 to 480 ml) per day for another month. The tea may be consumed long term as a maintenance therapy to reduce allergic reactivity of the bladder in susceptible individuals.

Tincture for Allergic Bladder Tenesmus

Tenesmus is a medical term referring to cramping of the bladder. When urinary irritation is severe enough to trigger spasms in the bladder smooth muscle, cramping can result. *Lobelia* and *Piper methysticum* are very effective in palliating such symptoms, but when symptoms are due to chronic atopic and allergic phenomena, a separate systemic approach will usually be necessary.

Apium graveolens	20 ml
Lobelia inflata	20 ml
Piper methysticum	20 ml

Take 1 teaspoon of the combined tincture in ¼ cup of water every 5 to 15 minutes, reducing as cramping subsides.

Urinary Irritants

The foods most aggravating to the urinary system include hot spices, coffee, hot peppers, alcoholic beverages, tea, chili peppers,[121] carbonated beverages, artificial sweeteners, citrus fruits and juices,[122] chocolate, and tomatoes. Formaldehyde is ubiquitous in the environment and when absorbed is metabolized into formic acid in the liver and excreted via the urine and feces and respiratory system. Formaldehyde is toxic to the urinary system.[123] Tobacco products, including e-cigarette smoke and tobacco metabolites, are known to be damaging to the bladder and are urinary carcinogens. The damaging effects of tobacco on the urothelium may be due to oxidative stress on urinary cell DNA, leaving cells less able to repair cellular proteins and recover from other allergens and stressors.[124] Soap can also act as a local irritant that may trigger urogenital symptoms in some allergy-prone women.[125] Atopic individuals may be especially reactive to common foods, food additives, and environmental chemicals, with metabolites acting as allergens as they pass through the urinary system.

Formulas for Vascular Reactivity

Vascular reactivity may promote less obvious symptoms than hay fever or contact dermatitis, but hypersensitivity and autoimmune reactions can afflict the veins, arteries, and microcirculatory blood vessels. Chemical irritation of the blood vessels can occur with intravenous (IV) fluids or exposure to toxins that enter the blood via oral or respiratory routes. Mast cell activation and hypersensitization reactions elsewhere in the body may allow large amounts of cytokines, thromboxanes, and other inflammatory mediators to enter the bloodstream and lead to platelet activation, nitric oxide release, and endothelial damage. This short section features some of the most commonly encountered types of vascular hyperreactivity.

Tea for Cold Sensitivity

Cold sensitivity involves excessive vasoconstriction via sympathetic nerves and central thermoregulatory dysfunction, although the precise mechanism is not completely clear. Chinese herbalists report that cold hypersensitivity is more common in women and is sometimes associated with gynecological problems such as irregular menstruation, miscarriage, and infertility. (A more severe form of cold sensitivity—Raynaud's syndrome—is discussed in chapter 2, Volume 2). In TCM philosophy, treating cold sensitivity requires balancing yin and yang energy in the body. Accordingly, blood movers and yin tonics such as *Angelica* and *Cnidium*

Herbal Specifics for Vascular Allergy

All allergic reactions involve the release of immune mediators from blood cells. Here, the term *vascular allergy* is used to refer to allergic hyperreactivity of blood cells and microcirculatory vessels specifically. Raynaud's syndrome and hypercoagulability would fit this category. Allergic inflammation may also contribute to atherosclerotic plaques and arteriosclerotic loss of vessel elasticity.

Allium sativum. May be added to formulas to reduce blood reactivity in those with ischemic disorders, poor circulation, and tendency to clots. *Allium* is specific for those with hyperlipidemia, hyperglycemia, and hypertension.

Angelica sinensis. Consider for Raynaud's syndrome and as an all-purpose vascular anti-inflammatory for vasculitis, hives, and allergic phenomena in the blood. *Angelica* also resists vasospasms and reduces allergic reactivity in blood cells.

Capsicum annuum. Consider including *Capsicum* in formulas for vascular inflammation associated with stasis, cold extremities, poor circulation, clots, and phlebitis.

Centella asiatica. *Centella* may reduce vascular inflammation that can lead to permanent fibrosis, scarring, and loss of elasticity and also reduce autoimmune inflammation in the blood vessels, connective tissue, and skin.

Cinnamomum verum. Consider *Cinnamomum* in formulas for vascular inflammation associated with high levels of lipids, blood sugar, and blood pressure; cold hands and feet; poor microcirculation; and stasis. *Cinnamomum* may also be included in skin allergy formulas to help move energy to the outer body.

Crataegus spp. *Crataegus* is appropriate for all types of vascular reactivity and allergy, as well as phlebitis, angioedema, microcirculatory inflammation, and hives.

Curcuma longa. *Curcuma* has broad anti-inflammatory effects and may be used for acute and chronic vascular inflammation, phlebitis, clots, and angioedema. Because of liver supportive effects, *Curcuma* is also appropriate for liver and bowel congestion that leads to blood reactivity and allergy.

Ginkgo biloba. *Ginkgo* is well studied for numerous effects on platelets and blood cells. Consider *Ginkgo* for allergic reactivity of the blood and blood vessel, hives, Raynaud's syndrome, clots, phlebitis, and atherosclerosis.

Hypericum perforatum. Consider including *Hypericum* in formulas for hives, microcirculatory phenomena, easy bruising, traumatic inflammation, neuropathies, itching, and acute pain due to allergic phenomena.

Phytolacca americana. Consider *Phytolacca* as a possible synergist in formulas for vascular and tissue inflammation associated with lymphatic and fluid stasis. *Phytolacca* may also help with pain and congestion related to diabetic nephropathy.[126]

Tanacetum parthenium. *Tanacetum* is very useful for hyperreactivity of blood cells, hives, angioedema, sun sensitivity, clots, migraines, chemical sensitivity, contact dermatitis, and excessive vascular reactivity and histamine response.

Zingiber officinale. *Zingiber* is useful for vascular inflammation with coldness, poor circulation, tendency to clots, atherosclerosis, cold hands and feet, stasis, and poor digestion.

are appropriate in formulas. One human clinical trial showed a TCM formula to be effective in treating cold hypersensitivity in the hands and feet[127] and reducing the symptoms that occur when a patient is exposed to cold temperatures and speeding the rate of recovery after exposure. The complex formula below has been altered very slightly from the study cited to use herbs most readily available in the Western marketplace.

Atractylodes lancea	4 ounces (120 g)
Angelica sinensis	4 ounces (120 g)
Ephedra sinica	2 ounces (60 g)

The Blood Mover *Angelica sinensis*

Angelica sinensis, dong quai, is one of the most well-known blood movers in TCM, long used to treat congested blood and lymph, which may often result from blood reactivity and allergic phenomena. Studies show that *Angelica* has numerous molecular actions that benefit allergies,[128] including antihistamine and antiserotonin effects.[129] Ligustilide in *Angelica* has been studied for anti-asthmatic activity.[130] The coumarins in *Angelica* species

Angelica sinensis, dong quai

have platelet anti-aggregating effects equal to, or greatly surpassing, that of aspirin. *Angelica* coumarins reduce mast cell secretions via mechanisms involving blockade of the influx of calcium.[131] Other antiallergy mechanisms include cyclooxygenase and lipoxygenase inhibition.[132] *Angelica sinensis* may improve the cytokine profile, reduce excessive cytokine levels and provide overall antiallergy support.[133] Clinical studies have demonstrated *Angelica*'s effectiveness in treating allergies and respiratory complaints.[134] These numerous anti-inflammatory effects contribute to *Angelica*'s reputation as a blood mover, reducing blood cell reactivity and resulting inflammatory cascades, reducing inappropriate clotting and vascular stasis, and exerting antiallergy effects by direct effect on blood cells and the vasculature.

Citrus aurantium	2 ounces (60 g)
Pinellia ternata	2 ounces (60 g)
Zingiber officinale	2 ounces (60 g)
Cinnamomum verum	2 ounces (60 g)
Ziziphus jujuba	2 ounces (60 g)
Glycyrrhiza glabra	2 ounces (60 g)
Paeonia lactiflora	2 ounces (60 g)
Poria cocos	2 ounces (60 g)
Cnidium monnieri	2 ounces (60 g)
Magnolia officinalis	2 ounces (60 g)
Platycodon grandiflorus	2 ounces (60 g)
Cyperus rotundus	2 ounces (60 g)

This formula yields over 2 pounds (960 g), which should be stored in a cool, dark cupboard. Gently simmer 1 tablespoon of the combined herbs in 4 cups (960 ml) of water for 20 minutes. Let stand until cool, then strain.

Drink the entire amount over the course of the day, making a fresh batch each day.

Tincture for Allergic Migraines

Some types of migraines and headaches are considered allergic phenomena, as the release of histamine from mast cells and platelets may initiate vascular migraines. This formula is more effective for preventing chronic migraines than for abating a full-blown migraine. See chapters 2 and 4 of Volume 4 for more headache and migraine formulas.

Ephedra sinica
Ginkgo biloba
Tanacetum parthenium

Combine in equal parts and take ½ teaspoon, 3 to 4 times daily for at least 3 months, evaluating for efficacy and continuing long term if helpful.

Encapsulated Therapies for Vasculitis

The following are some of the most robustly researched anti-inflammatories having specificity for the heart, blood cells, and blood vessels. Each of them is readily available as an encapsulated supplement. Regular use may help prevent the loss of elasticity and functional endothelium of the blood vessels during bouts of acute vasculitis. Any of the following may be used in tandem during active episodes of vasculitis.

Crataegus spp. capsules
Resveratrol capsules
Vaccinium myrtillus or *V. corymbosum* capsules
Ginkgo biloba capsules

Take 1 to 2 capsules of each, 1 to 3 times daily as a convalescent recovery and rebuilding formula. In cases of acute vasculitis, 2 or 3 capsules of each can be taken 4 or 5 times a day, reducing back down to the maintenance dosage as active flare-ups subside. *Vaccinium* and *Crataegus* are also available as solid extracts and can be taken at ¼ teaspoon doses 3 times per day. Alternately, teas and tinctures of *Vaccinium*, *Ginkgo*, and *Crataegus* are readily available and can complement or replace capsules. Many options for this are discussed in chapter 2, Volume 2.

General Vasculitis Support Tincture

Vasculitis may accompany autoimmune diseases, contributing to general inflammation and tissue destruction. The herbs in this formula are chosen both for their general systemic anti-inflammatory effects and for their particular affinity for the blood vessels. All of these herbs are gentle, broad in action, and nourishing; thus, this formula may be thought of as a trophorestorative formula for the vasculature.

Ginkgo biloba
Angelica sinensis
Curcuma longa
Crataegus oxyacantha or *C. monogyna*

Combine in equal parts and take 1 to 2 teaspoons, 3 to 6 times daily, long term.

Apiaceae Family Plants for Allergy

Many Apiaceae family (carrot family) plants are prominent in world ethnopharmacology to treat allergies. *Angelica* species (dong quai) have long been used in China for allergies and asthma, and *Ammi visnaga*, or khella, is a traditional Ayurvedic herb for the same. Khella contains the flavonoid khellin, the inspiration for a nonsteroidal bronchodilating inhaler. *Cnidium monnieri* is another Chinese herbal medicine for allergy, asthma, and dermatitis,[135] and *Foeniculum vulgare* (fennel), *Apium graveolens* (celery),[136] and *Pimpinella anisum* (anise) are also traditional bronchodilators included in cough, asthma, and lung formulas for centuries. The aromatic and tasty seeds of fennel and anise are safe, nutritious, and mineral-rich, and featured in many of the teas in this chapter, and the tinctures of all the Apiaceae family herbs are included in the allergy tincture formulas. Osthol, a constituent in many Apiaceae family members, reduces platelet aggregation induced by adenosine diphosphate, arachidonic acid, platelet-activating factor, collagen, and thrombin via direct thromboxane inhibition.[137] Osthol can relax vascular and bronchial smooth muscle via inhibition of calcium channels,[138] one mechanism whereby such plants relax bronchoconstriction in cases of allergic airway disease and improve blood flow in various states of blood and vascular reactivity.

Ammi visnaga, khella

Tea for Vasculitis

As is often the case with herbal medicines for chronic vasculitis, the more the better. Using both a tincture and a tea, with the possible addition of the encapsulated herbs mentioned in Encapsulated Therapies for Vasculitis, helps deliver a larger dose of trophorestorative herbs than any single treatment option. The more palatable herbs have been chosen for this tea. Because vasculitis involves the conversion of functional vascular endothelium and muscle fiber to fibrous tissue, *Centella* is included to help maintain tissue integrity and reduce fibrosis. (Chapter 2 of Volume 2 details the mechanisms whereby herbs can maintain elasticity in arteriolar sclerosis and collagen vascular disorder.)

Rubus idaeus
Centella asiatica
Ginkgo biloba
Crataegus oxyacantha leaf and flower
Crataegus oxyacantha berry powder

Combine in equal parts, or to taste. Steep 1 tablespoon per cup of hot water for 10 to 15 minutes, then strain. Drink 3 cups (720 ml) or more per day, long term.

Decoction for Vasculitis

Berries and bark, as opposed to the more delicate leaves and flowers in the Tea for Vasculitis, are best decocted. Flavonoids, which help protect the delicate endothelium, are plentiful in colorful berries, including *Sambucus*, *Crataegus*, and *Rosa*. *Glycyrrhiza* sweetens the brew while offering general anti-inflammatory action. *Hamamelis* is specific in the folkloric literature for vascular inflammation.

Sambucus canadensis berries
Crataegus oxyacantha berries
Rosa canina, whole or coarsely chopped
Glycyrrhiza glabra, shredded or coarsely chopped
Hamamelis virginiana bark

Combine in equal parts, or to taste. Gently simmer 1 teaspoon per cup of hot water 10 to 15 minutes, then strain. Drink 3 cups (720 ml) or more per day, long term.

Compress for IV-Induced Phlebitis

Phlebitis is one of most common complications in patients receiving IV therapy. Some cases are trauma-induced, and others are reactions to the IV medication. Depending on the substance being administered, 20 to 70 percent of patients receiving IV therapy develop phlebitis. The application of *Aloe* gel or compresses may prevent infusion-related phlebitis.[139] *Aloe* is combined here with *Calendula* and witch hazel, which is an herb commonly available in hydrosol form.

Aloe vera gel	1 tablespoon
Calendula officinalis succus	1 tablespoon
Hamamelis virginiana hydrosol	1 tablespoon

Combine the ingredients in a small dish. Saturate a gauze pad or compress with the herb blend and apply topically for 10 or 15 minutes every 1 to 2 hours, reducing as the inflammation resolves.

Formulas for General Autoimmune Conditions

Autoimmune conditions, in which the body mounts an immune reaction to its own tissue, are among the most difficult of all medical conditions to treat. "Cures" are rare to nonexistent, and most of those diagnosed require lifelong use of immune suppressants such as steroids to control the condition and slow the destruction of joints, mucous membranes, and other tissues. It is not entirely clear what causes the body to attack itself or not recognize self-destructive inflammatory processes. One theory among many others is that the presence of toxins, heavy metals, and other foreign substances in the tissues act as immune triggers that initiate autoinflammatory cascades. Leaky gut syndrome, as mentioned earlier in this chapter, may also initiate autoimmune disease. Therefore, herbs that optimize intestinal health and barriers and promote detoxification are included in autoimmune formulas.

Another possible trigger of autoimmunity is chronic viral infections. The similarities between the genetic proteins (HLA compounds) in both human tissues and some infectious viruses is hypothesized to be a mechanism by which some viral infections such as Epstein-Barr (EBV) trigger autoimmune disorders. The body produces autoreactive T cells (this is a normal phenomenon), and the T cells are trained to react to a large number of antigens, including our own tissues. These do not initiate autoimmune disease unless an infectious illness induces higher numbers and activities of T cells. Viruses such as EBV and

Immunomodulating Herbs for Autoimmune Conditions

A common question that herbalists face is whether herbs commonly known to increase immune function in deficiency states are safe for those with autoimmune diseases. Is "stimulating" the immune response inappropriate, or even harmful, for someone with lupus, multiple sclerosis, or rheumatoid arthritis, for example? There has not been enough clinical research or scientific investigation to fully answer these questions, but the available research suggests that immune supportive herbs are *not* contraindicated and might in fact be helpful. Even the phrase "immune-stimulating herbs" may be a misnomer. Many such herbs may actually *modulate* rather than stimulate the immune response. The herbs most deeply researched in this regard are probably *Panax* and *Astragalus*. Studies have shown *Panax* and *Astragalus* to both enhance immune function in cases of immune deficiency and normalize immune reactivity in cases of allergy and immunoreactivity. In fact, there is very little concrete evidence that such herbs are capable of aggravating autoimmune or allergic disorders. *Astragalus* has been traditionally included in formulas for allergies and inflammation, suggesting an immunomodulating activity. Modern research suggests *Echinacea* to be useful in the management of autoimmune uveitis[140] and diabetic inflammation of the pancreas.[141] Furthermore, heavy metals in the tissues may trigger autoimmune activity, and *Echinacea purpurea* has been shown to assist the body in eliminating cadmium.[142] Also consider immunomodulators in formulas for collagen vascular arthritic disorders.

other infections such as pronounced intestinal dysbiosis can stimulate large numbers of T cells to overreact and increase the likelihood that autoreactive T cells will reach and interact with tissues. Misguided and hyperreactive T cells can propagate and remain active in the tissues even long past when infections have cleared. For example, EBV has been associated with juvenile-onset autoimmune arthritis and other autoimmune disorders in which the virus triggers cytotoxic T cells and the production of pro-inflammatory cytokines, which are then sustained long term due to the similarity of viral proteins to tissue proteins.[143] In this same way, immunizations that introduce microbial antigens into the body can also sometimes initiate autoimmune disease such as systemic lupus erythematosus (SLE) and rheumatoid arthritis (RA).[144]

This text and the research on autoimmune diseases focuses on the discrete diagnostic entities of RA, Sjögren's syndrome, Behçet's disease, and so on, but in reality, the diagnoses are not always so clear. There is often a great deal of overlap in the conditions, and patients are often worked up for one diagnosis that ultimately proves inconclusive, and then referred to a rheumatologist or immunologist to be further worked up for related diagnoses. Furthermore, many of the diagnostic tests are blood titers, antinuclear antibody (ANA) for example, and when the results are only mildly elevated, and below the diagnostic threshold, the test is considered negative. Such tests are performed when symptoms flare, for example when joint pains are rapid in onset. When the tests are negative, many patients receive no treatment and in most cases the symptoms abate over time. I have seen a hundred or so individuals who present with this pattern of being suspected to have RA or lupus, and while the ANA may be mildly elevated, it is not elevated to the degree of being diagnostic of an autoimmune disease. The symptoms abate over several months and recur a year or two later, and the tests are repeated with similar inconclusive results. Many patients have been through this cycle three or four times when they decide to seek help from the alternative medicine community. I am of the opinion that such cases are indeed low-grade autoimmune-driven inflammation and advise beginning treatment at once with the concern that full-blown RA or another entity will emerge in time. Initial therapies might include herbal formulas that address gut health, stress, and sleep, while searching for toxins, allergens, and immune triggers and protecting the vasculature and connective tissue.

In addition to the well-known glucocorticoid immunosuppressive steroids, several additional types of medications exist to suppress overreactive T lymphocytes, but such pharmaceuticals can have severe and life-threatening side effects, an example being the monoclonal antibody muromonab-CD3, approved by the FDA in 1986 but withdrawn from the market a few years later when significant toxicity became apparent. Monoclonal antibodies remain valuable pharmaceuticals, however, and such drugs may target tumor necrosis factor (TNF) and TNF receptors and are widely prescribed for RA, psoriasis, and inflammatory bowel diseases.

Ribosomally synthesized peptides are another category of immunosuppressant drugs for autoimmune diseases, favored for their enhanced selectivity and low toxicity compared to the steroidal and the monoclonal antibody drugs. Some plants produce unique peptides, such as cyclotides, presently being investigated for possible novel T-cell inhibitors. These are generally less toxic than the monoclonal antibodies, but such immunosuppressant herbs are not typically nutritive tonics and have their own cautions and toxicities. Cyclosporine derived from the *Tolypocladium inflatum* fungus is used to inhibit calcineurin, an enzyme involved with the activation of T cells; however, cyclosporine shows hepatotoxicity, nephrotoxicity, neurotoxicity, and cytotoxicity. Substances found in other fungi as well as cycloleonurinin from *Leonurus japonicus* and cyclolinopeptides from *Linum usitatissimum* are being researched for possible immunosuppressive properties, as are other peptides from bacteria, marine sponges,[145] and scorpion venom.

The autoimmune disorders known as collagen vascular diseases involve autoinflammation and destruction of the connective tissue of joints, tendons, and vasculature. *Curcuma*, *Centella*, and *Equisetum* are among herbs that may both help protect and help regenerate connective tissue. *Angelica*, *Ginkgo*, and *Crataegus* may be included in formulas where vasculitis and vascular inflammation are present because they can reduce histamine, platelet activation, and endothelial damage. Celastrol is a triterpene found in medicinal herbs of the Celastraceae family. It has numerous anti-inflammatory activities and may reduce inflammatory cytokines, correct imbalances between T helper and T regulatory cells and help to control the migration of inflammatory mediators into the joints and protect against bone damage.[146] Animal models of autoimmune disorders have shown celastrol to be a potential therapy for RA, multiple sclerosis, inflammatory bowel disease, and SLE. *Tripterygium wilfordii* is one such plant that contains celastrol, and extracts have been widely used in the treatment of autoimmune and inflammatory diseases, including RA, SLE,

Rosmarinic Acid for Allergy and Autoimmune

Rosmarinic acid is a caffeic acid derivative that occurs in *Rosmarinus*, as well as *Perilla*, *Prunella*, *Melissa*, and *Salvia*. Rosmarinic acid may help treat autoimmune disease and allergy due to beneficial effects on T cells,[147] and studies suggest efficacy for hay fever,[148] RA, organ rejection in transplant patients,[149] acute snake bite,[150] acute chemical reactivity,[151] and allergic disorders.[152]

Allergic reactivity in asthma, dermatitis, and chemical reactivity involve activation of T and mast cells and their release of pro-inflammatory interferon and interleukins. NF-κB and tumor necrosis factor-α (TNF-α) are several agents at the top of the inflammatory cascade that activate T cells. Rosmarinic acid limits the expression of TNF-α and NF-κB, offering broad effects in allergic and autoimmune disorders, limiting the release of cytokines and other inflammatory mediators.[153] Rosmarinic acid induces T-cell apoptosis[154] and reduces autoinflammatory activity. Rosmarinic acid may prevent lung inflammation following exposure to airborne irritants[155] via stabilizing effects on interleukins and inflammatory proteins and by limiting the infiltration of neutrophils in the lung tissue and local interstitial edema.

Rosmarinic acid may also offer protection in collagen vascular and autoimmune conditions, via stabilizing effects on T cells and limiting cytokine-driven breakdown of connective tissues,[156] including collagen,[157] and reducing enzymatic digestion by tyrosinase, hyaluronidase, elastase, and collagenase.[158] An animal model of collagen-induced arthritis has shown rosmarinic acid to reduce inflammation. Animals treated with rosmarinic acid displayed "remarkably reduced" synovial inflammatory markers, such as cyclooxygenase, compared to the control animals.[159] In RA, rosmarinic acid can induce apoptosis of the aberrant T cells.[160]

psoriasis, and dermatomyositis. However, this extract is not widely available in the United States and so not featured further in this text.

Because the amount of detailed information is voluminous and highly molecular, this section focuses on the autoimmune disorders most commonly encountered in general family practice, including collagen vascular diseases such as lupus, RA, Sjögren's syndrome, and Behçet's disease. RA, however, is covered in chapter 3 of this volume. Thyroid disorders that are considered to have autoimmune components are covered in chapter 2, Volume 3. Some autoimmune skin diseases such as vitiligo are discussed in chapter 5, Volume 1.

Gelsemium Water for Acute Immune Reactivity

Gelsemium has historically been used to treat neuralgia and pain, and folkloric homeopathic texts mention using *Gelsemium* for double vision and visual disturbances accompanied by mental fatigue and physical weakness, symptoms that commonly occur with multiple sclerosis. Due to its potent toxicity, *Gelsemium* is only used as a nanomedicine, or a highly diluted medicinal preparation. *Gelsemium* species contain the alkaloids gelselegine and koumine, which have been shown to inhibit T-lymphocyte proliferation and have other immunomodulatory benefits. Koumine can reduce the release of pro-inflammatory cytokines[161] and significantly reduce neuropathic pain via inhibitory effects on activated microglia and astrocytes. *Gelsemium* is shown to reduce symptoms and deter disease progression in animal models of RA.[162]

Gelsemium sempervirens tincture	100 drops
Water	4 ounces (120 ml)

Combine the *Gelsemium* and water, and take ½ teaspoon 3 times per day for several months, continuing at once or twice a day if proven effective.

Tea for General Autoimmune Conditions

The two herbs in this formula are basic immunomodulators that can be considered all-purpose for many autoimmune disorders.

Astragalus membranaceus
Rehmannia glutinosa

Combine in equal parts and gently simmer 1 heaping teaspoon per 1 cup (240 ml) of water for 15 to 20 minutes. Let stand in a covered pan for 20 minutes or more, then strain. Drink 3 cups (720 ml) or more per day, long term, evaluating efficacy after 6 months of regular use and continuing as needed.

Ganoderma Simple for Autoimmune Disease

Reishi, or *Ganoderma lucidum*, is a traditional immunomodulating agent included in numerous classic TCM formulas for inflammatory arthritis, fatigue, edema, and heart palpitations. Research shows that *Ganoderma* can significantly improve the ratio of Th1 to Th2 white blood cells and reduce inflammatory cytokines. This formula is based on the powdered form, but reishi can also be consumed as a tea, tincture, or encapsulation.

Ganoderma lucidum powder	1 teaspoon

Mix the *Ganoderma* powder into a liquid, whisking briskly, and consume 3 times a day. The powder can also be added to coffee before it brews, or stirred into smoothies. Note that the powder has an earthy taste and will not dissolve but this approach is inexpensive and practical for those who don't mind the flavor and texture.

"Cold" Tincture for Autoimmune Disorders

This formula and the "Warm" Tincture for Autoimmune Disorders are examples of formulas energetically specific for "hot" versus "cold" presentations. In this formula, all the herbs except the *Curcuma* are cooling and balancing, helping with fevers and acute hot episodes. Joint pains that are sudden in their onset and accompanied by red, hot skin that feels good under an ice pack are typical of hot and "excess" constitutions.

Tanacetum parthenium
Ginkgo biloba
Curcuma longa
Scutellaria baicalensis
Angelica sinensis

Combine in equal parts and take 1 to 2 teaspoons, 3 to 6 times daily for several months and evaluate for efficacy. The formula may be continued at 1 teaspoon 2 or 3 times a day, long term, once acute symptoms have subsided. Many herbalists may also alter the formula slightly as hot or cold symptoms resolve over time.

"Warm" Tincture for Autoimmune Disorders

This is a much warmer formula than the "Cold" Tincture for Autoimmune Disorders due to the inclusion of *Panax*, *Zanthoxylum*, and *Zingiber*. Patients who feel cold and weak and have low-grade symptoms that come

Rehmannia for Yin Deficiency

Rehmannia is a gently warming tonic herb used in TCM for blood and yin deficiency indicated by coldness, weakness, night sweats, heart palpitations, anemia, and fatigue. Modern molecular research shows that *Rehmannia* increases leukocyte production and balances T- and B-lymphocyte ratios and activities, which is beneficial for autoimmune and allergic reactivity. While immunosuppressive fractions[163] have been identified in *Rehmannia*, the herb is immunomodulating overall due to catalpol, an iridoid glucoside, as well as raffinose oligosaccharides. Because raffinose oligosaccharides are difficult to digest, *Rehmannia* roots are traditionally steam-processed,[164] allowing the starches to act as prebiotics[165] in the gut. Catalpol affects T-cell balances, modulates T helper to T regulatory cell ratios, and has numerous immunomodulating actions on pro-inflammatory pathways including NADPH

Rehmannia glutinosa,
di huang

oxidase enzymes,[166] NO production and the expression of inducible NO synthase, cyclooxygenase, prostaglandin E2, pro-inflammatory interleukins,[167] TNF-α, and activation of NF-κB.[168] It also promotes anti-inflammatory pathways for superoxide dismutase (SOD) and glutathione.[169]

Atopic conditions such as allergic airway diseases and eczema involve increased Th2 cell responses leading to increased serum immunoglobulin E and increased leukocyte infiltration. *Rehmannia* can help reduce atopic reactivity by reducing the secretion of cytokines, chemokines, and cellular adhesion molecules. Animal models of allergic dermatitis show *Rehmannia* to limit dermal infiltration of inflammatory cells when provoked by TNF-α and interferon.[170] *Rehmannia* decreases blood levels of TNF-α, interleukin-6, and other pro-inflammatory mediators and corrects elevated blood levels of SOD in diabetic mice.[171]

on slowly and persist for a long time may benefit from the more warming immunomodulators. In autoimmune diseases, joint pains that are cold and stiff and feel better with hot applications are typical of cold constitutions.

Panax ginseng
Zanthoxylum clava-herculis
Smilax ornata
Zingiber officinale
Astragalus membranaceus

Combine in equal parts and take 1 to 2 teaspoons, 3 to 6 times daily for several months and evaluate for efficacy. The formula may be continued at 1 teaspoon 2 or 3 times a day, long term, once acute symptoms have subsided. Many herbalists may also alter the formula slightly as hot or cold symptoms resolve over time.

Supportive Tincture for Connective Tissue

Many autoimmune conditions can deform, if not devastate, the joints due to steady degeneration of connective tissue. As blood and lymph supply to the joints becomes impaired, inflammation, loss of elasticity, and diffuse fibrosis can occur. *Phytolacca* and *Iris* may be useful lymphagogues, and are combined here with broad-acting anti-inflammatories such as *Curcuma*. As connective tissue in the skin may also be affected, *Calendula* is included to promote microcirculation in the skin. *Echinacea* is mentioned in many old formularies for prevention of tissue breakdown.

Calendula officinalis	13 ml
Curcuma longa	13 ml
Echinacea angustifolia	10 ml

Phytolacca americana	10 ml
Iris versicolor	10 ml
Pilocarpus jaborandi	4 ml

Take 1 teaspoon of the combined tincture 4 or more times daily for at least 3 months and continuing at a lower dose as a maintenance therapy long term.

Tincture for Toxin-Triggered Disorders

As discussed at the beginning of this chapter, toxins and metabolic wastes in the blood and tissues can act like antigenic material and trigger immune reactivity. The inclusion of bitter alterative and liver-supportive herbs in this formula may promote removal of antigenic substances from the body and help prepare the liver to detoxify foreign and endogenous toxins. Coated tongue, skin blemishes, and digestive difficulties indicate liver and/or digestive burden and suggest which specific herbs would be indicated in this formula.

Echinacea purpurea
Achillea millefolium
Silybum marianum
Andrographis paniculata

Schisandra chinensis

Combine in equal parts and take 1 to 2 teaspoons, 3 to 6 times daily for at least 3 months and continuing at a lower dose as a maintenance therapy long term.

Antiviral Tincture for Autoimmune Disorders

Viruses are among the many conditions that may initiate autoimmune reactivity. Viral symptoms such as low-grade fevers, sore throats, and enlarged and tender lymph nodes are clues that antiviral herbs should be included in an autoimmune infection formula. Three to 6 months of an antiviral and immune-support protocol may also prove helpful in some cases.

Curcuma longa
Glycyrrhiza glabra
Phytolacca decandra
Eleutherococcus senticosus
Andrographis paniculata

Combine in equal parts and take 1 to 2 teaspoons, 3 to 6 times daily for at least 3 months and continuing at a lower dose as a maintenance therapy long term.

Formulas for Sjögren's Syndrome

Sjögren's syndrome involves dysfunction of exocrine glands such as lacrimal and salivary glands, manifesting in dry mouth and dry eyes. Sjögren's syndrome affects roughly 7 per 100,000 people worldwide, but the rate is much higher in European and Asian populations, at 43 per 100,000 people. Sjögren's syndrome may present as an isolated condition but may occur in tandem with other autoimmune diseases, including RA and SLE; there is an increased risk of developing non-Hodgkin's lymphoma. Sjögren's syndrome typically progresses slowly and gradually. Herbal support of connective tissue is appropriate because joint pain and damage occurs in roughly one-third of cases. In addition to a 50-fold increased risk of lymphoma, a third of Sjögren's patients are anemic, and a quarter are leukopenic. Thus, immune support, blood builders, and possibly lymphatic herbs may be valuable. Furthermore, 20 percent of Sjögren's patients develop renal tubular acidosis, interstitial nephritis, and glomerulonephritis, and atrophy of the intestinal mucosa and hepatobiliary and pancreatic duct glands may also occur.

Long-term herbal formulas may feature lymphagogues, demulcents, and ocular-circulation enhancers, as well as renal, biliary, and pancreatic support and protection to restore connective tissue and joints, in addition to general anti-inflammatories and immunomodulators.

Because the secretory glands are targeted, Sjögren's syndrome patients often have dry eyes, dry mouth, vaginal dryness, and possible trouble swallowing. (In TCM such patients are considered to have "yin deficiency.") Dry eyes of Sjögren's syndrome may be treated with the topical application of immunosuppressant drugs, such as fluorometholone, tacrolimus, or cyclosporine, but oral pilocarpine, an alkaloid from *Pilocarpus*, may be more effective.[172] In the most severe cases of dry eyes, keratitis filiformis may emerge and moistening agents that improve circulation to the eye may help. Lactoferrin is a major tear protein known to be diminished in dry eyes and known to play a role in modulating oxidative stress and deterring microbial infections. Oral lactoferrin may improve corneal health but may need to be continued long term to maintain the benefits. As with all autoimmune diseases, Sjögren's is difficult to treat, and lifelong support and prevention of the above sequalae are reasonable goals of alternative treatment. When using herbal formulas for years on end, as may be necessary for Sjögren's patients, it is common to take occasional days off, or even a week's break, from taking a particular remedy.

Basic Moistening Tea for Sjögren's

Lycium barbarum may protect the collagen matrix in the skin and, due to its delicious taste, it can be added to oatmeal, teas, and medicinal soups.[173] *Ulmus, Glycyrrhiza,* and *Rehmannia* are all known for their moistening and lubricating effects and may support mucous membranes and prevent loss of secretions when consumed on a regular basis. *Althaea, Glycyrrhiza,* and *Calendula* are nourishing to the oral mucosa, while *Equisetum* and *Centella* may help maintain collagen and elastin, which can become fibrotic in Sjögren's and other autoimmune conditions. *Camellia sinensis* may protect salivary glands from inflammatory damage,[174] and the catechins may help reduce autoantigen production.

Lycium barbarum berries
Rehmannia glutinosa
Glycyrrhiza glabra
Ulmus fulva
Althaea officinalis
Calendula officinalis
Equisetum spp.
Centella asiatica
Camellia sinensis

Combine in equal parts—for example, 4 ounces (120 g) of each herb yields over 2 pounds (960 g) of dry mixture to store in a glass jar. *Althaea* offers the most mucociliary and moistening effects following long maceration. Thus, for best results place 2 heaping tablespoons of the herb blend in a saucepan and cover with 6 cups (1,440 ml) cold water and allow to stand overnight, or at least for 3 hours. In the morning, bring the mixture to a gentle simmer for 15 minutes, then remove from the heat and let stand 20 minutes, and then strain. Drink the entire amount over the course of the day. Continue for at least 2 months and evaluate efficacy; then continue long term if effective.

Prescription Pilocarpine for Sjögren's

Oral pilocarpine was shown to be superior to lubricating eye drops for the treatment of Sjögren's syndrome symptoms in a review of clinic trials. It induced gastrointestinal disturbance in only 5 percent of the study's participants,[175] and was reportedly more effective than the various eye drops and lacrimal duct plugs.[176] Pharmaceutical pilocarpine is available by prescription.

Pilocarpine	5 mg

Take 5 milligrams twice a day for at least 3 months, evaluating for efficacy.

Supportive Tincture for Sjögren's

Curcuma is the lead herb in this formula given its ability to protect connective tissue. It is combined with immunomodulating and circulatory-enhancing herbs.

Curcuma longa	30 ml
Panax ginseng	10 ml
Viscum album	10 ml
Salvia miltiorrhiza	10 ml

Take 1 teaspoon of the combined tincture 3 to 4 times daily, long term, following a schedule of 3 months on, 2 weeks off.

Eye Drops for Sjögren's Patients

Sjögren's syndrome patients often need lifelong eye drops. This formula is an example of an eye drop that is nourishing and supportive, rather than merely moistening. The *Aloe* gel is a soothing base and reduces the chance that the tinctures will sting the eyes. *Pilocarpus* tincture may be useful in promoting salivation and other secretions, as may the isolated alkaloid pilocarpine, which is available by prescription in liquid form from pharmacies. Research has shown that in both systemic and mouthwash formulas,[177] pilocarpine acts as a muscarinic agonist, stimulating the secretions from lacrimal glands and goblet cells. Sterile saline is available from any pharmacy. *Euphrasia* is specific for eye irritation. See also the Formulas for Eye Conditions in chapter 4.

Aloe vera gel	½ tablespoon
Euphrasia officinalis	10 drops
Calendula officinalis	10 drops
Pilocarpus jaborandi	10 drops
Sterile saline	¾ ounce (23 ml)

Sterilize an eye dropper, a small bowl, measuring spoons, a fork, and a storage bottle or container by pouring boiling water over them. Use the spoon to add the *Aloe* gel to the bowl, then add the tinctures and sterile saline and whisk with the fork until homogenous. Use out of the bowl immediately, or pour into the sterilized container. Place a dropperful into each eye, and repeat 2 or 3 times as desired, using an eye dropper, or use an eyecup to administer as an eyewash throughout the day, from 3 times per day or more frequently

to allay discomfort. Refrigerate the remaining solution, discarding after 2 days. If preferred, the eyecup wash can be warmed by placing the entire bottle in a pan of hot water before administering.

Formulas for Dermatomyositis and Polymyositis

Dermatomyositis and polymyositis are related and often overlapping conditions involving inflammatory degeneration of muscles, skin, and connective tissue

Psoriasis

Psoriasis is a T-cell-mediated autoimmune chronic inflammatory skin disease that affects from 1 to 3 percent of the global population. It is characterized by hyperproliferation of keratinocytes with a greatly accelerated cell turnover rate. The skin lesions are usually well demarcated red inflammatory patches covered superficially with fine silvery scales, commonly on the trunk, elbows, knees, and scalp; lesions may also affect fingernails and toenails. Psoriasis is often recalcitrant and treated with immunosuppressants, such as steroids, cyclophosphamide, or cyclosporine, all having toxicity and numerous side effects. The topical use of antiproliferative agents may be helpful, and the application of coal tars and ultraviolet light are long-standing therapies. Animal studies suggest that *Scutellaria baicalensis* and the anti-inflammatory flavonoid baicalin may improve the differentiation of keratinocytes.[178] Topical application of *Curcuma longa* can inhibit several inflammatory enzymes mainly involved in the inflammatory process of psoriasis,[179] and although it will stain fabric, *Curcuma* ointment can be part of a broader protocol for psoriasis. For more information on psoriasis, including treatment of the liver intestinal function and possible leaky gut, see chapter 5, Volume 1.

through the entire body. Symmetrical muscle weakness is common, and dysphagia may occur when pharyngeal and esophageal muscles are affected, and even breathing may be impaired when the diaphragm is involved. Malnutrition can occur due to impaired intestinal motility and serious pneumonias may occur secondarily to impaired respiratory excursion. Research suggests that picornaviruses or other viruses may initiate these conditions, or occult low-grade malignancies may be triggers. Formulas might include immunomodulating antivirals such as *Astragalus, Ganoderma, Cordyceps sinensis, Viscum album*, and other anticancer herbs. Skin and muscle inflammation in these conditions does not lead to fibrosis or deformity but more to muscle pain and weakness. Therefore, connective tissue herbs are less indicated, and immunomodulators and muscle builders may be more specific. *Panax, Smilax ornata, Rhodiola rosea*, and *Lepidium meyenii* (maca) are herbs with a folkloric reputation for muscle strength and stamina and can also be immunomodulating. *Eupatorium perfoliatum* is a folkloric specific for muscle pain due to viruses, although no modern research has suggested it for this condition. *Actaea racemosa* is another herb specific for muscle pain that may help keep the use of steroids down. *Hypericum* is used for neuralgic pain and is anti-inflammatory and antiviral, and *Boswellia* and *Yucca* species are folkloric herbs for arthritic and musculoskeletal pain.

Tincture for Myositis

Because dermatomyositis and polymyositis typically involve inflammatory pain in the muscles, the gentle immune-modulating herbs in this formula are chosen to offer multiple anti-inflammatory and tissue-supportive mechanisms of action.

Panax ginseng	20 ml
Rhodiola rosea	20 ml
Ganoderma lucidum	20 ml

Take 1 teaspoon of the combined tincture 4 or more times daily long term.

Tea for Dermatomyositis and Polymyositis

This formula has the same general effects as the Tincture for Myositis, but features the tastier herbal options for antiviral and immune support, along with a general anti-inflammatory action. The *Ulmus* may facilitate swallowing and lubricate the intestines in cases of weak peristalsis, and the amount used in the formula may be increased when this is a significant symptom.

Centella asiatica
Hypericum perforatum
Glycyrrhiza glabra
Astragalus membranaceus
Ulmus fulva
Zingiber officinale
Yucca spp.

Combine in equal parts, then steep 1 tablespoon per cup of hot water and strain. Drink 3 cups (720 ml) or more per day long term. If effective, continue long term; occasional breaks of a week or two are suggested.

Topical for Polymyositis Muscle Pain

Gaultheria, or wintergreen, has been used traditionally as a topical or oral treatment for joint pain and swelling. It is a source of methyl salicylates, which help suppress the inflammatory responses in macrophages, microglia, and astrocytes, and have a significant antiarthritic effect.[180]

Plant-based methyl salicylates are transformed into salicylic acid in the gut when ingested orally. Methyl salicylate is lipophilic, and readily penetrates the skin when topically applied. It is then hydrolyzed to salicylic acid and distributed throughout tissues and transcellular fluids. This formula is intended to complement oral therapies or more comprehensive protocols for polymyositis, and to reduce the need for harmful pharmaceutical anodynes.

Gaultheria fragrantissima essential oil   ½ ounce (15 ml)
Castor oil   1½ ounces (45 ml)

Combine in a 2-ounce (60 ml) bottle. Rub ½ teaspoon of the oil blend into painful muscles every 4 to 6 hours.

Tincture for Polymyositis Muscle Pain

This formula can be dosed frequently to help allay muscle pain and reduce the need for steroids or other pharmaceuticals. It features herbs noted for their anodyne effects on muscle pain and inflammation.

Actaea racemosa	15 ml
Eupatorium perfoliatum	15 ml
Panax ginseng	10 ml
Piper methysticum	10 ml
Boswellia serrata	10 ml

Take 1 to 2 teaspoons of the combined tincture 3 to 6 times daily, reducing as muscle pain improves.

Formulas for Reactive Arthritis (Reiter's Syndrome)

Formerly known as Reiter's syndrome, reactive arthritis results from infectious disorders that trigger cross-reactivity, manifesting as a triad of inflammatory reactivity in the large joints, the eyes, and the urethra in both genders or the cervix in women. People ages 20 to 40 are most commonly affected. Complicating the diagnosis, the inciting infection is often resolved at the time the joint pain emerges. Sometimes mucocutaneous lesions accompany, as well as psoriatic lesions. Eye inflammation typically involves uveitis and/or conjunctivitis, and the affected joints are most often the knees and the sacroiliac joint, usually asymmetrically, with additional joints becoming inflamed after the initial arthritic areas have resolved. Some texts cite the occurrence of "sausage digits," referring to gross swelling of single fingers or toes. Plantar fasciitis, Achilles tendinitis, and sudden onset of heel pain can also occur. Urinary symptoms are commonly part of the initial onset of reactive arthritis where burning pain and frequency may progress to prostatitis and cervicitis or vulvovaginitis. Penile lesions occur in 20 to 40 percent of affected men. Mucocutaneous ulcers on the mouth and penis can be painless, but in other cases are painful. About 10 percent of those affected by reactive arthritis, especially those with prolonged episodes, will develop cardiac manifestations that can include aortic regurgitation and pericarditis.

Reactive arthritis occurs in those genetically predisposed due to the presence of the HLA-B27 gene, and the most common inciting pathogens are the intestinal microbes *Salmonella*, *Shigella*, and *Campylobacter*; *Yersinia* species (from improperly cooked pork); the genitourinary pathogen *Ureaplasma urealyticum*; and the sexually transmitted infection *Chlamydia trachomatis*. The symptoms of reactive arthritis often begin between 1 and 3 weeks after the inciting infection.

Tincture for Reiter's Syndrome

Reactive arthritis (Reiter's syndrome) is triggered by bacterial and chlamydial infections, so the use of immunomodulators and antimicrobials—such as *Echinacea* and *Astragalus*—is appropriate in herbal formulas. Because the skin and connective tissue are also affected, connective tissue tonics such as *Curcuma* or *Centella* may support the healing of the joints and mucocutaneous lesions. See chapter 3 of this volume for further discussion of the various types of arthritis, as well as pain management and tissue repair formulas. The condition may resolve on its own, but this formula can help speed resolution and prevent the joint destruction that may occur with chronic exacerbations and remissions.

Echinacea purpurea	15 ml
Astragalus membranaceus	15 ml
Curcuma longa	15 ml
Phytolacca americana	15 ml

Take 1 to 2 teaspoons of the combined tincture 3 or more times daily, reducing after several weeks as symptoms improve.

Sitz Bath for Reactive Arthritis

Reactive arthritis patients may present with noninfectious urethritis, penile or vaginal discharge, cervicitis, cystitis, hematuria, and penile lesions in men. This sitz bath formula astringes suppurating tissue and provides healing and anti-inflammatory effects. This infusion could also be used as a vaginal douche in women, or prepared into an eyewash for conjunctivitis symptoms. See chapter 4 of this volume for additional eyewash formulas.

Hamamelis virginiana
Berberis aquifolium
Filipendula ulmaria
Achillea millefolium
Calendula officinalis

Combine in equal parts. Steep 2 heaping cups (100 g) of the blend in a gallon of hot water, preparing enough tea to douche with or use as a sitz bath for urethritis in men. Use once or twice a day.

Formulas for Behçet's Disease

Behçet's disease involves systemic vasculitis that causes blood vessel inflammation throughout the body; there are no satisfactory pharmacotherapies. Multiple organ systems are affected, with lesions in many tissues causing oral, joint, skin, eye, genital, digestive, nervous, and vascular symptoms. While the cause of Behçet's disease is unknown, it is commonly proposed to be an autoimmune disorder with underlying genetic and environmental contributors, and the HLA-B51 gene may predispose. Corticosteroids can reduce joint pain and cutaneous lesions and are used to suppress acute exacerbations; however, the long-term use of corticosteroids is undesirable because of serious consequences including diabetes, immune suppression, and osteoporosis. Behçet's disease may also affect blood vessels of the gastrointestinal tract, central nervous system, heart, and other organs. Genital, conjunctival, uveitis, and oral mucosal lesions are particularly common. Behçet's disease most commonly affects men more often than women, particularly people in their twenties and thirties, and often gradually resolves over the ensuing decades. Aphthous ulcer-like oral lesions are common,

usually followed by the emergence of conjunctival, genital, and skin lesions, like reactive arthritis. The ocular lesions of Behçet's can be severe and extensive and may even lead to blindness, making aggressive immune suppression sometimes necessary. The skin lesions are variable, including vesicles, papules, and pustules, and may be chronic, while the arthritic symptoms are often mild and do not lead to deformity. Around a quarter of those afflicted may also develop serious vascular and neurologic lesions that can be fatal or lead to aneurysm and thrombosis, heart and pulmonary damage.

Colchicine from *Colchicum autumnale* has been reported since the 1970s to be useful in treating Behçet's disease. Colchicine may be used daily for several years to support recovery from mucosal, cutaneous, pleuropericardial, and abdominal complications of this disease. The skin lesions of Behçet's disease are similar to erythema nodosum, a skin condition discussed in chapter 5, Volume 1.

Purge the Heart Decoction

This formula is based on the traditional Chinese Gancao Xiexin decoction and the more modern version, Banxia

Colchicum—An Herbal Immune Suppressant?

Colchicum autumnale, the autumn crocus, is an ancient herbal medicine, mentioned in the famous *Ebers Papyrus*, an Egyptian *materia medica* dated to circa 1500 BCE. Colchicine is derived from the bulb-like corms of *Colchicum*. Despite its long history of medicinal use, colchicine was only recently approved by the FDA as a recognized drug, prescribed primarily for acute gout flares. Research in the 1950s and 1960s revealed the colchicine alkaloid to affect microtubules of the cytoskeleton, affecting mitosis, vesicle trafficking, and cellular structure and dynamics. By binding to tubulin units, colchicine can affect cellular growth and replication in a toxic manner yet in rapidly dividing, excessively proliferating, and hyperreactive cells, colchicine has proven useful without doing significant harm to the body; it has been proven safe for children, and even for pregnant and nursing women.[181]

Colchicine may act as a natural immunosuppressant and halt acute destruction of the connective tissue in RA and other collagen vascular diseases. Colchicine quickly reduces white blood cell hyperreactivity, particularly halting neutrophilic activation and thereby reducing the release of pro-inflammatory cytokines and decreasing platelet and mast cell activation. A typical colchicine dosage is 0.5 to 1 milligram at a time with daily dose never to exceed 4.8 milligrams per day. One clinical trial evaluating colchicine for gout patients found a loading dose of 1 milligram followed by 0.5 milligrams every 1 to 2 hours

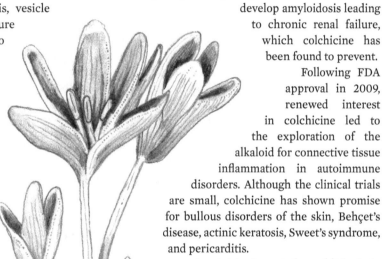

Colchicum autumnale,
autumn crocus

thereafter was effective in reducing acute pain but that gastrointestinal upset was common.

Colchicine has also been a standard treatment for familial Mediterranean fever (FMF) since the 1970s, when several clinical trials found that the daily use of colchicine reduced flares of joint pain and fever. Several long-term studies followed such patients for up to 15 years and reported favorable outcomes and safety with continued use. Furthermore, many inadequately controlled FMF patients develop amyloidosis leading to chronic renal failure, which colchicine has been found to prevent.

Following FDA approval in 2009, renewed interest in colchicine led to the exploration of the alkaloid for connective tissue inflammation in autoimmune disorders. Although the clinical trials are small, colchicine has shown promise for bullous disorders of the skin, Behçet's disease, actinic keratosis, Sweet's syndrome, and pericarditis.

An intriguing study published in 2015 examined rates of malignancy in 24,000 male gout patients in Taiwan who had either "ever-used" or "never-used" colchicine. The study reported a significantly lower incidence of all-cause cancers in the colchicine ever-users when compared with the never-users, especially significant for prostate and colorectal cancers.[182] These finding are now the impetus to further investigate colchicine for chemoprevention, or possibly an adjuvant chemotherapy. The research is on isolated colchicine, which is a prescription-only drug that can be prescribed from pharmacies. Whole-plant *Colchicum* tincture is difficult to find and may have to be made for oneself.

Xiexin Tang. Both of those formulas are known as Purge the Heart Decoction and are reported to be helpful for refractory cases of Behçet's disease,[183] providing relief for recalcitrant conditions such as persistent allergy disorders, recurrent aphthous ulcers, and relapsing ulcerative colitis.[184] Variations of this formula may include *Lonicera* when there are red pruritic skin lesions. *Smilax* and *Scrophularia* may be included for chronic symptoms as suggested by folklore.

Glycyrrhiza uralensis	15 g (½ oz)
Pinellia ternata	9 g (⅓ oz)
Scutellaria baicalensis	9 g (⅓ oz)
Ziziphus jujuba	9 g (⅓ oz)
Coptis chinensis	6 g (⅕ oz)
Zingiber officinale	6 g (⅕ oz)

Decoct 1 teaspoon of the combined herbs per cup of hot water for 20 to 30 minutes, then remove from heat and let cool for another 15 minutes. Strain the brew and drink 3 cups (720 ml) or more per day. Concentrated tea granules of this formula may be available, and can be stirred into a mug of hot water at a ratio of 1 teaspoon granules to 1 cup (240 ml) water.

Tincture for Behçet's

Because infections and viruses may trigger Behçet's disease, immune support and strong antimicrobials such as *Echinacea, Curcuma, Phytolacca, Origanum* (oregano oil), *Hypericum*, or *Astragalus* are appropriate at the onset. Immunomodulators and neurologic and vascular protectants are also logical, and include *Hypericum, Crataegus, Centella*, and *Salvia miltiorrhiza*, as well as "blood movers" like *Angelica* and *Ginkgo*. See chapter 4 of this volume for eyewash formulas that can help with Behçet's symptoms.

Hypericum perforatum	20 ml
Phytolacca decandra	10 ml
Salvia miltiorrhiza	10 ml
Centella asiatica	10 ml
Astragalus membranaceus	10 ml

Take ½ to 1 teaspoon of the combined tincture as often as hourly at initial onset. Reduce to 5 to 6 times a day after the first week to help prevent further inflammation and chronic recurrence of symptoms.

Colchicine for Behçet's

Colchicine is best known as a gout medication, but is also used to treat other autoimmune indications, as described in "*Colchicum*—An Herbal Immune Suppressant?" on page 73. Colchicine is a prescription drug available from pharmacies. Some herbalists may prefer the entire *Colchicum* herb, but it is difficult to find and considered a prescription-only botanical.

Colchicine	0.5 mg

Take one colchicine tablet, 3 times per day for a period of 1 month, or occasionally longer, such as when acute flares persist or symptoms recur as soon as medication is stopped.

Formulas for Henoch-Schönlein Purpura

Henoch-Schönlein purpura (HSP), also known as immunoglobulin A vasculitis, is the most common form of childhood vasculitis. While it can occur in children of any age, there are peaks around 4 to 6 years of age, and 90 percent of cases occur in those under 10 years of age. The primary presenting symptoms are skin lesions concomitant with gastrointestinal, musculoskeletal, and renal involvement. Vasculitis in the kidneys may be serious and lead to nephritic and nephrotic syndromes and may be the underlying cause of a very small percentage of all childhood end-stage kidney disease.[185] Renal biopsy is recommended when significant proteinuria occurs, and the typical therapy is immunosuppressive corticosteroids. The etiology is unknown but involves abnormal glycosylation of immunoglobulin A triggering humoral autoimmune response and the formation of circulating immune complexes that are deposited in the blood vessels, leading to systemic vasculitis.

Because the immunoglobulins are initially released via mucosal secretions, and the initial onset is more common in winter months, an infectious trigger is proposed. Possible infectious triggers for HSP have been reported, including common viruses such as respiratory syncytial virus, influenza, and norovirus, and some cases have been associated with streptococcal infections, dental caries, periodontitis, rhinosinusitis, tonsillitis, and otitis media. Other researchers have noted vaccinations to be an inciting trigger, although this is controversial.

The condition is more common in Asians, possibly due to certain human leukocyte antigen (HLA) alleles in this population, while other HLA types may offer protective effects against this type of vasculitis.

The skin lesions are most common on the lower limbs and buttocks and often have a bruise-like appearance, hence the name purpura, and are related to vascular damage. Skin biopsy of these lesions shows granulocytes in small arterioles or venules. During the acute presentation, up to 90 percent of patients will have joint pain, especially in the lower limbs, with some correlation to the location of the skin rash. Gastrointestinal symptoms accompany in almost 75 percent of HSP cases presenting as colicky abdominal pain, and intestinal bleeds may occur in the form of gross or occult blood in the stool, or hematemesis; both occasionally can be life threatening. Renal involvement is seen in roughly 50 percent of the cases, presenting as microscopic hematuria that can progress to gross hematuria and proteinuria. Testicular inflammation occurs in less than 20 percent of male patients and when the only presenting symptom may at first be thought to be acute testicular torsion.

Because at least half of HSP cases spontaneously improve, pharmaceutical interventions are minimal, except for severe presentation where immunosuppressants may be prescribed. A current standard of care is to offer ACE inhibitors or angiotensin receptor blocker for patients with persisting proteinuria. Alternative medical practitioners and herbalists, however, offer vascular protectants such as high-flavonoid herbs; immune modulators such as *Panax*, *Astragalus*, or *Ganoderma*; and vulnerary compresses of *Aloe*, *Calendula*, or *Hypericum* for the skin lesions. *Salvia miltiorrhiza* may be helpful for renal involvement due its vascular protective and renoprotective effects. (See chapter 3 of Volume 1 for detailed research on this.) Herbal formulas targeted at eradicating oral infections or strep throat may be indicated in some patients.

Topical Compress for HSP

Topical applications of soothing vulnerary herbs may be useful to speed the healing of skin lesions associated with HSP. This formula employs the standard wound-healing herb, *Calendula*, and combines it with *Hypericum*, emphasized in folklore for bruising and injuries to the nerves and vasculature. *Aloe* is also used for its anodyne, cooling, and wound-healing properties.

Calendula officinalis	4 ounces (120 g)
Hypericum perforatum	8 ounces (240 g)
Aloe vera gel	¼ cup (60 ml) per compress

Combine the dry *Calendula* and *Hypericum*. Steep 3 heaping tablespoons in 3 cups (720 ml) of boiling hot water for 20 minutes, then strain. Add the *Aloe* gel to 1 cup of the tea blend. Use the combined solution to saturate a fabric, such as gauze, a small wash rag, or pieces of flannel folded to the most appropriate size. Apply the compress topically, leaving in place for 20 to 30 minutes and repeating every several hours. Continue frequent topical application for several days, reducing frequency of use as symptoms improve over a week or two.

Decoction for HSP with Hematuria and Proteinuria

When the kidneys are affected by HSP, leading to blood and/or protein in the urine, herbs such as *Salvia miltiorrhiza*, thought to protect the renal vasculature, may be employed. *Ganoderma* and *Astragalus* may be used as all-purpose immunomodulators. *Equisetum* is emphasized in folklore for hematuria. *Glycyrrhiza* is used here for both its immunomodulating properties and its flavor.

Salvia miltiorrhiza
Ganoderma lucidum
Astragalus membranaceus
Glycyrrhiza glabra

Combine in equal parts; for example, 4 ounces (120 g) of each herb yields a pound of the formula. Gently simmer 2 heaping tablespoons of the combined herbs in 6 cups (1,440 ml) of water for 20 minutes, then let stand for 20 minutes more. Strain and drink the entire amount each day, reducing as urinary function improves.

Formulas for Lupus

Systemic lupus erythematosus (SLE) is an intractable, multisystemic, and relapsing autoimmune disease typified by the presence of autoantibodies and the deposition of immune complexes in multiple organs, including the skin, kidneys, heart, and joints. It can lead to inflammation and tissue damage, disfigurement, and occasionally mortality. The pathogenesis of SLE may involve genetic, environmental, and hormonal factors. SLE affects roughly 5 million people worldwide. SLE is characterized by distinct immunologic abnormalities, including antinuclear, anti-chromatin, anticytoplasmic, and anti-phospholipid antibodies. The lymphoid tissue becomes hyperplastic in SLE, B cells are hyperreactive, and organ function is impaired due to inflammation and deposition of immune complexes. Reduced organ function further contributes to the toxic and inflammatory burden in the body. Anti-chromatin autoantibodies form complexes that bind to the glomerular basement membrane in the kidney and lead to glomerulonephritis, the most serious manifestation of SLE.[186] The condition affects primarily women, roughly 100 to 400 per 100,000 of the worldwide female population.

The long-standing therapies have been corticosteroids, cyclophosphamide, and hydroxychloroquine, with only limited efficacy and all with potentially severe side effects. The development of newer immunosuppressive drugs has improved overall survival and offered a new tool to limit severe organ damage, but many patients still experience pain, fatigue, and quality of life issues. The drug belimumab is a monoclonal antibody that can control SLE, but not cure it. Bioactive peptides are a new type of immunosuppressive agent for possible therapy in SLE. For example, snake venom from *Naja naja*, the Indian cobra, is reported to activate the immune system in a manner that protects the skin and kidney in SLE, reducing concentrations of circulating globulins, anti-DNA antibody, and inflammatory cytokines IL-6 and TNF-α.[187]

There are very few studies on single herbs with benefits to lupus patients, but there are studies testing TCM formulas in animal models of lupus, and a scant number of human trials, with the most robustly studied such formulas below. Berberine-containing herbs such as *Coptis chinensis* and *Phellodendron amurense* are often featured in such formulas and credited with renoprotective effects. Baicalin, a flavonoid glycoside isolated from *Scutellaria baicalensis*, has broad anti-inflammatory and immunomodulatory activities. It has shown protective effects against oxidative damage in a variety of tissues including the kidneys.

Tincture for Lupus Skin Lesions

Lupus symptoms may include persistent scaly papules and plaques on the skin; small, shallow ulcers in the oral mucosa that may heal poorly and result in permanent scarring; and lesions on the scalp that may lead to permanent hair loss. Connective tissue and skin support may help prevent permanent atrophy and scarring. Herbal formulas for lupus vary, and can be targeted to address organ inflammation, joint pain, fever, or anemia. This formula is broadly aimed at supporting the skin and connective tissues while reducing inflammation and optimizing immunomodulation. Topical versions of this formula may also be prepared and applied to facial cartilage during acute episodes to limit tissue destruction.

Phytolacca oil, castor oil, and *Hypericum* oil can be used independently or as a base in which to mix other herbs or essential oils.

Curcuma longa	15 ml
Hypericum perforatum	10 ml
Glycyrrhiza glabra	10 ml
Ganoderma lucidum	10 ml
Tanacetum parthenium	10 ml
Phytolacca americana	5 ml

Take 1 to 2 teaspoons of the combined tincture 3 to 6 times daily for several months to several years depending on the severity of acute flares and overall course of the disease.

Tincture for Systemic Lupus with Arthralgia

As with all collagen vascular diseases, lupus can involve the joints, leading to joint pain and gradual destruction. *Ganoderma* has been shown to reduce joint inflammation in animal models of systemic lupus erythematosus.[188] *Boswellia* and *Curcuma* are both leading herbs to consider for all types of arthritis.

Ganoderma lucidum
Boswellia serrata
Curcuma longa

Combine in equal parts and take ½ teaspoon, 4 times per day. Continue for at least 3 months, evaluating results at that time, continuing long term if effective in halting or reducing joint inflammation.

Chamomile-Lemongrass Tea for Lupus

Apigenin, a flavonoid found in *Matricaria chamomilla* and other plants, is a well-known anti-inflammatory agent. It may also reduce sensitization reactions, markedly suppress T helper cell production of antinuclear autoantibodies,[190] downregulate cyclooxygenase enzymes, and induce apoptosis of inappropriately activated T cells and B cells. Citral, a lemony-smelling essential oil found in *Melissa* and *Cymbopogon*, may improve lupus-related nephritis and attenuate proteinuria via reducing lipopolysaccharide accumulation in the renal cells.[191]

Cymbopogon citratus	8 ounces (240 g)
Matricaria chamomilla	8 ounces (240 g)

Combine the dry herbs and store in a large glass jar in a dark location. Prepare a daily tea by steeping 3 tablespoons in 3 to 4 cups (720 to 960 ml) of boiling water (to taste) and allow to steep in a covered pan for 20 minutes. Strain and drink 3 cups (720 ml) or more each day.

Blood-Builder Tonic for Lupus

Anemia and long-standing fatigue are common in lupus patients. This formula contains blood-building herbs and adrenal tonics to help improve energy and vitality.

Panax ginseng
Poria cocos
Astragalus membranaceus
Lycium barbarum
Reynoutria multiflora
Ganoderma lucidum
Glycyrrhiza uralensis

Combine the herbs in equal parts. Gently simmer 2 heaping tablespoons in 6 cups (1,440 ml) of water for 20 minutes, then let stand for 20 minutes more. Strain and drink the entire amount each day, continuing for at least 3 months. If effective, the support may be maintained at a lesser dose of using just 4 or 5 days a week and taking a week or two off a few times a year.

Tincture for Lupus-Induced Glomerulonephritis

The formation of autoantibodies against chromatin is the main feature of SLE-related renal disease. Formation of anti-chromatin autoantibodies accelerates cell death and impairs phagocytosis of damaged cells; they also form complexes that bind to the glomerular basement membrane in the kidney and lead to glomerulonephritis, the most serious manifestation of SLE. The saikosaponins in

IV Artesunate for Lupus Nephritis

Artesunate is a semisynthetic derivative of artemisinin, a compound found in *Artemisia annua* that is used to treat malaria and some types of cancer. Artesunate is shown to inhibit TNF-induced production of pro-inflammatory cytokines, and animal models of lupus show increased survival and reduced progression of nephritis when the compound is administered. Artesunate may reduce the serum levels of ANA and anti-dsDNA antibody titer and improve proteinuria and serum creatinine levels.[189] Some clinics offer IV artesunate for cancer patients; such services may also benefit SLE patients. Oral artemisinin is available, but has not been researched for its impacts on lupus to the extent that IV administration has been.

Bupleurum may modulate immune-driven inflammation in SLE, while berberine-containing herbs may reduce proliferative glomerulonephritis.[192]

Angelica sinensis	20 ml
Coptis chinensis	10 ml
Scutellaria baicalensis	10 ml
Bupleurum falcatum	10 ml
Salvia miltiorrhiza	10 ml

Take 1 to 2 dropperfuls of the combined tincture 2 to 4 times per day. Clinicians may wish to start in the lower dosage range for a week, to ensure the formula is well-tolerated, and then ramp up to the higher end of the dosage range, aiming to be aggressive for this serious pathology.

Tea for Lupus-Induced Glomerulonephritis

Salvia miltiorrhiza improves circulation in the kidneys and protects against oxidative stress and fibrotic changes. *Angelica* and *Astragalus* are a circulation-enhancing duo in TCM, with immunomodulatory effects as well. Patients in the early stages of renal failure should be encouraged to drink as much of this tea as possible, as daily consumption may slow the progression of renal disease.

Salvia miltiorrhiza
Ganoderma lucidum
Angelica sinensis
Astragalus membranaceus
Rehmannia glutinosa
Paeonia lactiflora
Glycyrrhiza glabra

Combine in equal parts—such as 2 or 3 ounces (60 to 90 g) of each herb—and store in a large glass jar in a dark location. Simmer 1 teaspoon per cup of water for 20 minutes. Remove from the heat and let stand for 10 to 15 minutes or more, then strain. Larger batches may be prepared to last for several days; the tea can be stored in the refrigerator, then reheated or consumed chilled as preferred. Aim to drink 3 cups (720 ml) or more a day for several months, reducing to 3 cups (720 ml) a day 3 or 4 days per week thereafter, long term.

Formula to Clear Heat Toxins

Lupus may present as either an excess or deficiency state, according to TCM philosophy. This formula uses heat-clearing herbs for excess presentations. A heat toxin syndrome is characterized by red skin rashes, severe joint pain, oral ulcers, high fever, dark-colored urine, crimson tongue with yellow coating, and rapid, sunken pulse.

Glycyrrhiza uralensis
Panax ginseng
Reynoutria multiflora

Astragalus membranaceus
Ganoderma lucidum

Combine in equal parts and take 1 to 2 dropperfuls, 3 times each day. The dose and frequency may be reduced as symptoms improve.

Zi Shen Qing for Lupus

Zi Shen Qing has been used for deficiency of chi and yin pattern symptoms in TCM, to tonify chi, enrich yin, remove internal heat, and clear toxins. Clinical studies show the formula to reduce symptoms and flares in SLE patients, reduce dependence on corticosteroids,[193] and cool blood heat to improve inflammation.[194]

Astragalus membranaceus
Rehmannia glutinosa
Cornus mas
Paeonia lactiflora
Oldenlandia diffusa (also known as *Hedyotis diffusa*)

This formula can be found commercially prepared, in which case 10 g of the granules may be dissolved in 200 ml of hot water. This full amount should be prepared and consumed twice daily for 12 weeks. Alternately, the crude herbs can be combined in equal parts and decocted, although some of the ingredients may be difficult to source in the United States. Gently simmer 1 heaping tablespoon of the blend in 3 cups of hot water for ½ hour, then strain. Drink the entire amount each day.

Formulas for Inflammatory Bowel Disease

Inflammatory bowel disease (IBD) etiology is multifactorial with autoinflammation and autoimmune reactivity contributory. Immunosuppressants are the standard pharmaceutical approach, although there are also newer drugs. Because various inflammatory mechanisms are known to be active, herbs that inhibit leukotrienes, stabilize mast cells and platelets, and reduce NF-κB are also appropriate.[195] Cyclooxygenase is often elevated in allergic and inflammatory conditions and both *Curcuma* and *Gardenia* are noted to reduce elevated cyclooxygenase. Palmitoylethanolamide (PEA) is an amide alkaloid that occurs naturally in a variety of common foods including soybeans, egg yolks, and peanuts. PEA is available in a concentrated form from compounding pharmacists and may activate the glia of enteric cells in a manner that improves mucosal integrity.

One area of current research into IBD is the fact that low-grade parasitic infections and exposure to mild pathogens helps to suppress the overactive T cells in many autoimmune diseases. Some roundworms and flatworms (helminths) reside harmlessly in the gut and their presence may balance resting immune reactivity; their presence may support optimal immune modulation. (Schistosomal infections can damage the liver and bladder and increase cancer risk, however, and hookworms can impair iron and nutritional status.) Increasing research suggests that some of the more benign worms may be possible therapies for autoimmune disorders. Whipworms, for example *Trichuris* species, may thrive in the healthy gut without damaging the mucosa, invading the organs, or entering the bloodstream. Worms

Molecular Research on Herbs for IBD

Molecular studies and animal research support folkloric usage of herbs for IBD. Following is a brief summary of readily available herbal options.

HERB	MODE OF ACTION
Allium sativum	Protects against shifts in intestinal microbiome.
Aloe vera	Anthraquinones such as aloe emodin, aloesin, and aloin decrease myeloperoxidase, TNF-α, and IL-1β activity.
Andrographis paniculata	Andrographolide reduces elevated cytokines.
Angelica sinensis	Suppresses platelet activation, moderates the injury of endothelial cells, and improves microtransmission in IBD patients.
Artemisia absinthium	Protects against shifts in the microbiome.
Boswellia spp.	Boswellic acid reduces elevated NF-κB and pro-inflammatory cytokines, including interleukins.
Cannabis sativa	May decrease nausea, pain, and diarrhea in Crohn's disease patients.
Curcuma longa	Decreases the pro-inflammatory cytokines such as TNF-α, IFN-γ, IL-1β, and IL-12.
Gardenia jasminoides	Glycoproteins reduce nitric oxide, cyclooxygenase, and NF-κB.
Nicotiana tabacum	Nicotine may moderate leukotrienes.
Camellia sinensis	Flavones are broadly antioxidant and anti-inflammatory compounds and shown to help support the integrity of the colonic architecture.
Tripterygium wilfordii	Possible immunosuppressive action, but not readily available in United States.

activate regulatory T cells and reduce excessive immune responses, such as aberrant autoimmune reactivity. The whipworm *Trichuris suis* is currently in trials for treating Crohn's disease and ulcerative colitis. Although the results of some such clinical trials for Crohn's patients have been disappointing, some promising data propels researchers to perfect the technique.[196] For more information on ulcerative colitis and Crohn's disease, refer to chapter 2, Volume 1.

A number of plant alkaloids have been explored as possible therapies for IBD. Alkaloids are among the most pharmacologic, and sometimes outright toxic, constituent groups found in plants and do not typically have nourishing effects on tissues. Therefore, plants notably high in alkaloids are not chosen as nutritive base herbs in formulas and should be combined with gentle immunomodulators such as *Astragalus* or *Ganoderma*. Some alkaloidal plants could be prepared into teas in combination with flavonoids herbs such as *Lycium barbarum*, *Crataegus*, or *Calendula*, or blended with demulcents such

as *Avena*, *Glycyrrhiza*, or *Ulmus* to create gentle, well-tolerated formulas that can be used for many months.

Quinolizidine alkaloids may offer immunomodulating activities. They include sophocarpine and sophoridine from *Sophora* species traditionally used for rheumatism and inflammation. N-methylcytisine, an alkaloid isolated from the seeds of *Laburnum anagyroides*, *Sophora alopecuroides*, and *Actaea racemosa*, is shown to inhibit pro-inflammatory cytokines through downregulating NF-κB activation.

Isoquinoline alkaloids include berberine, found in *Berberis* species, *Coptis chinensis*, and *Phellodendron chinense*. It may help to protect the intestinal barriers through immunoregulation and positive effects on gut microbiota. Similar isoquinoline alkaloids offer numerous physiologic benefits. Palmatine found in *Phellodendron amurense*, *C. chinensis*, and *Corydalis yanhusuo* supports healing and integrity of the intestinal mucosa, and sanguinarine found in *Sanguinaria canadensis* and *Corydalis impatiens* may attenuate mucosal lesions.

Boldine in *Lindera aggregata* and tetrandrine and neferine found in *Stephania tetrandra* and *Nelumbo nucifera*, respectively, also all show positive effect on colitis.

Indole alkaloids have immunomodulating effects. Many indole compounds affect central and peripheral nervous system regulation. Isatin found in *Isatis tinctoria* reduces inflammation and limits mucosal injury, and indirubin, a bisindole alkaloid also obtained from *I. tinctoria*, may have immunosuppressive effects to limit autoimmune-driven inflammation of the colon.

Tryptanthrin from *Persicaria tinctoria* attenuates mucosal inflammatory injury.

Tincture for Ulcerative Colitis and Crohn's Disease

This formula combines antimicrobial agents *Artemisia* and *Berberis*, which can gently help intestinal dysbiosis as well as offer broad anti-inflammatory activity. *Boswellia* is better known for musculoskeletal inflammation but guggulsterones and boswellic acids have also been shown to downregulate inflammatory mediators in intestinal diseases as well. *Angelica* and *Zingiber* are vascular and tissue anti-inflammatories, and the duo are found to protect against tissue injury in animal models of inflammatory bowel disease.[197]

Artemisia absinthium
Boswellia serrata
Berberis aquifolium
Angelica sinensis
Zingiber officinalis

Combine in equal parts and take 1 to 2 dropperfuls, 2 to 6 times daily, depending of the severity of the symptoms for 3 to 6 months and possibly longer pending progress and the individual's protocol.

YunNan BaiYao for IBD

YunNan BaiYao, also called Yunnan Baiyao in the past, is a Chinese patent formula considered a first-aid staple because of its efficacy in controlling hemorrhage both internally and topically. YunNan BaiYao was introduced to the Western world when the US military learned that North Vietnamese soldiers carried the herbal blend during the Vietnam War as a life-saving treatment for acute blood loss. YunNan BaiYao has shown efficacy against intestinal hemorrhage in animal models of severe colitis.[198] The precise recipe is a closely guarded secret, maintained by certain families as their livelihood and handed down through the generations. The following

formula is a list of what ingredients have been determined over the last century, but the exact ratios and the manner of preparing the blend is unknown. The formula is listed here for interest, obtained from an FDA document involved with approving the sale of the formula, but it is recommended to obtain an authentic product directly from one of the licensed manufacturers such as the Yunnan Baiyao Group of Yunnan, China. The herbs in the formula vary slightly with different manufacturers but all have similar features in common.

Panax pseudoginseng
Dryobalanops sumatrensis
Ajuga forrestii
Inula cappa
Lycopodium complanatum
Dioscorea oppositifolia or related species
Alpinia officinarum
Erodium stephanianum
Cinnamomum camphora
Mentha canadensis

Practitioners may obtain the formula from a reputable vendor and follow the label instructions, typically a capsule at a time, 3 or 4 times daily.

Encapsulations for Ulcerative Colitis

Because ulcerative colitis, a form of IBD, can be difficult to cure, especially without steroids and other drugs, it is often necessary to create an aggressive protocol involving teas, tinctures, dietary changes, and encapsulated supplements. The following are those agents with at least some clinical research showing efficacy for IBD. See also information on these individual herbs in the "Specific Indications" section of this chapter beginning on page 85.

Cannabidiol (CBD) oil drops or capsules
Green tea catechins
Curcuma longa
Mentha piperita
Allium sativum
Andrographis paniculata

Choose one or numerous types of encapsulated herbs to use in tandem with an herbal tea and tincture. A single encapsulated herb might be taken as much as 3 or 4 capsules at a time, 3 or 4 times a day in an aggressive dosing strategy. Other approaches would be to consume 1 or 2 capsules of various kinds 2 or 3 times a day. Management of ulcerative colitis often necessitates long-term therapy of 6 months to several years or more.

Soothing Tea for IBD

It can be helpful to keep a constant trickle of soothing and healing herbs going through the intestines during acute IBD flares. Chamomile is gentle and multipurposed, *Mentha* helps allay acute cramps and distension in the intestines, and *Ulmus* can help heal ulcers and intestinal mucosal lesions. *Geranium maculatum* or *Agrimonia* could be added for bleeding, *Centella asiatica* may support healing, and a myriad of other herbs can be considered to make this basic tea more specific for individual presentations.

Matricaria chamomilla
Mentha piperita
Ulmus fulva
Hamamelis virginiana
Glycyrrhiza glabra

Combine in equal parts—such as 3 to 4 ounces (90 to 120 g) of each herb—and mix. Steep 1 tablespoon of the blend per cup of boiling water for 10 minutes, then strain. Drink 3 cups (720 ml) or more per day.

Green Drink for IBD

A drink such as this one, consumed once or twice daily, may speed the healing of intestinal lesions and help control symptoms. This drink can also provide nutrition during fasting or an elimination diet. Barley green powder is a possible substitute for barley green juice, but fresh juice is desirable. Barley grass can be grown at home and prepared into a juice using a wheat grass juicer. Some health food stores carry live barley grass for juicing, or green juices that contain barley grass. The Soothing Tea for IBD can be substituted for the plain chamomile if desired. No sweetener is included here. If necessary, *Stevia* powder or fresh fruit, such as blueberries, may be added.

2 cups (480 ml) *Matricaria chamomilla* tea
1 ounce (30 g) *Aloe vera* gel
1 tablespoon glutamine powder
1 ounce (30 g) barley green juice

Prepare the chamomile tea by steeping 2 tablespoons dried leaves in 2 cups (480 ml) of boiling water for 10 minutes, then straining. Combine the chamomile tea, hot or cold as desired, with the *Aloe* gel, glutamine powder, and barley green juice in a blender or immersion mixer, blending until homogenized. Drink at least one such blend each day, at least 5 days per week, for 2 months and evaluate for efficacy. If desired, this drink can be consumed twice or even 3 times per day as part of an elimination diet or modified fast.

Formulas for Multiple Sclerosis

Multiple sclerosis (MS) is a chronic, progressive T-cell-mediated autoimmune disease of the central nervous system (CNS). The pathogenesis of MS involves the activation of T helper cells that infiltrate the CNS via a disrupted blood-brain barrier, leading to the release of inflammatory mediators and microglial activation. As a result, the inflammatory process further activates the immune system and ultimately damages neurons and myelin. The condition is typically characterized by remissions and relapses related to demyelination and axonal loss in the CNS. MS can lead to paralysis and significant neurological disabilities where demyelination of the nerves causes musculoskeletal weakness, numbness and tingling, visual disturbance, bladder and bowel dysfunction, and general motor and mental dysfunction. Fatigue frequently accompanies.

Although the precise cause is not known, MS is presumed to be an autoimmune disease, and exposure to a variety of toxins may initiate the immune reactivity. For example, organic solvents[199] and exposure to allergens, *Candida* overgrowth, and intestinal irritants may trigger immune reactivity that contributes to MS. Dairy has been identified as a possible trigger, as has exposure to mold. Environmental and chemical antigens may also be involved. All possible antigens should be avoided as a foundational approach to treating MS.

Oligodendrocytes are cells of the nervous tissue capable of producing myelin to repair neuronal axons when injured. Any method of supporting oligodendrocytes is therefore important to treating MS. Research targeting specific signaling pathways, stem cell therapy, suppressing molecules involved with the inflammatory damage to myelinated axons, and attempting to convert glial cells into oligodendrocyte precursors are the foci of current therapy and drug development. As is typical for all autoimmune diseases, immunosuppressive drugs are the primary therapy used to treat acute flares or maintain remission and include glucocorticoids and possibly other

Botanical Therapies for MS

A number of medicinal plants have been reported useful for MS due to antioxidant, immunomodulating, and neuroprotective effects.[200] Neurotrophins support the survival, maintenance, and regeneration of neurons in the CNS and include neuronal growth factor (NGF), brain-derived neurotrophic factor (BDNF), and other neurotrophins. These are among the all-purpose brain herbs reported to help slow the progression of dementia and possibly treat other neurodegenerative disorders.

Ammi visnaga. Khellinone is a possible MS therapy shown to reduce T-cell-mediated autoimmune inflammation.[201]

Andrographis paniculata. Human clinical trials have shown *Andrographis* to improve fatigue in MS patients.[202]

Boswellia species. One clinical trial on MS patients showed *Boswellia* to significantly improve visuospatial memory.

Curcuma longa. Curcumin may decrease CNS inflammation and demyelination via modulating pro-inflammatory cytokines. Animal studies have suggested an ability to support remyelination.

Eucommia ulmoides. Iridoid glycosides including geniposidic acid in *Eucommia* may be neuroprotective.

Gastrodia elata. Widely used in TCM for stroke recovery, brain injury, and migraines; current research suggests neuroprotective effects that may benefit MS patients.

Ginkgo biloba. Ginkgolides exert neuroprotective activity via antioxidative effects. Clinical trials have shown ginkgo to improve fatigue in MS patients.

Huperzia serrata. Huperzine alkaloids may protect against cognitive deficits in MS and upregulate nerve growth factor.

Magnolia officinalis. Honokiol and magnolol may inhibit neuronal cell death.

Nigella sativa. May offer neuroprotective effects and support remyelination.

Oenothera biennis. Evening primrose oil is one of the most abundant sources of γ-linolenic acid or GLA. Clinical trials show evening primrose oil to improve manual dexterity and to reduce the frequency and severity of relapses in MS patients.

Panax ginseng. Ginseng may inhibit the proliferation of T cells and production of inflammatory cytokines. *Panax* ginsenosides protect dopaminergic neurons via promotion of neurotrophic factors.[203]

Vitis vinifera. Resveratrol is able to cross the blood-brain barrier and offer neuroprotective effects and inhibit microglial activation. Resveratrol may enhance circulation in the brain and protect the vasculature from damage and is reported to increase hippocampal NGF. The use of resveratrol capsules can be a useful addition to MS protocols.

Zingiber officinale. Gingerols and the derivative shogaols can modulate the expression of interleukins and increase the levels of NGF in the hippocampus, offering neuroprotective effects.

immunomodulating pharmaceuticals such as interferon beta (IFN-β). IFN-β and other immunosuppressants are expensive drugs, and the use of IFN-β increases the risk of stroke and can cause headache, migraine, and depression, emphasizing the need for alternative therapies.

Dietary and natural medicine approaches to MS include avoiding poor quality or allergenic fats. Strict avoidance of dairy fats and hydrogenated oils such as margarine and shortening has been recommended by some since the 1950s. Supplementing beneficial essential fatty acids, such as eating fish as often as possible, and supplementing cod liver oil is an approach that is most effective when adopted as a long-term lifestyle. Other nutritional supplements to consider include vitamin D because the greater the exposure to sunlight, the lower the incidence of MS. Vitamin B_{12} is crucial to the synthesis of myelin, and the amino acid L-carnitine may improve fatigue by supporting the transport of fatty acids into cells and enhance basic metabolic function. Hormonal balance may be important, as some women experience sustained improvements of MS during pregnancy, and DHEA is shown to be slightly lower in MS patients compared to controls.

Berberine Capsules for MS

Berberine, an isoquinoline alkaloid found in *Coptis chinensis*, *Berberis aquifolium*, and *Phellodendron amurense*, has demonstrated numerous neuroprotective effects including neurotrophin-mediated neuroprotection.

Berberine capsules   500 mg

Take 1 or 2 capsules, 1 to 3 times a day as part of a larger MS protocol. A standard dose can range from 500 to 2,000 milligrams per day, while larger dosages of 3,000 milligrams per day may be useful during acute exacerbations of MS symptoms.

Tincture for MS

This formula offers *Withania* as an adaptogen, *Zingiber* and *Boswellia* as specific musculoskeletal anti-inflammatories, and *Andrographis* and *Curcuma* as immunomodulating agents. *Perilla* is a rich source of the antioxidant rosmarinic acid. *Curcuma*, *Zingiber*, and *Boswellia* are multipurpose musculoskeletal anti-inflammatories, and *Withania* can help blunt stress responses and offer neuroprotective effects.

Withania somnifera
Zingiber officinale
Boswellia serrata or *papyrifera*
Andrographis paniculata
Curcuma longa
Angelica sinensis
Perilla frutescens

Take 1 to 2 dropperfuls of the combined tincture 3 to 6 times a day for many months and possibly many years, taking occasional week-long breaks with long-term use.

Cannabis for MS Symptoms

Numerous studies show cannabinoids to reduce muscle stiffness, bladder disturbance, spasms, neuropathic pain, and sleep disorders in MS patients. The combination of tetrahydrocannabinol (THC) with CBD appears to be the most effective in treating spasticity, and simply smoking marijuana is shown to improve muscle stiffness,

Encapsulation Protocol for MS

Because MS is a seriously debilitating disease with no known cure, aggressive protocols are often appropriate. These herbs and concentrated nutrients are all available as tablets or capsules. Several may be chosen to complement the tea and tincture in this section for a powerful combination. Take 1 or 2 capsules, 3 or more times a day.

Andrographis paniculata	*Ginkgo biloba*
Nigella sativa	Resveratrol
Curcuma longa	Evening primrose oil
Boswellia serrata	Rosmarinic acid

tremors, pain, and emotional disturbances. Synthetic CBD products are also being explored. Oral sprays and oil capsules are all readily available. In my apothecary, we use full-spectrum oils, meaning that they contain terpenes and cannabigerol (CBG) in addition to CBD.

CBD oil to deliver 2 to 5 mg CBD per dose

Take 2 to 4 times daily as part of a regular maintenance therapy for MS symptoms, increasing at times of exacerbations and decreasing or taking a break from the medicine during periods of sustained remission.

Tea for MS

The herbs in this formula contain powerful antioxidants, many with demonstrated neuroprotective effects. These include rosmarinic acid in *Rosmarinus*, apigenin in *Matricaria*, and catechins in *Camellia*. Rosmarinic acid has been found to reduce antibody-driven inflammation of neural tissue.[204]

Hypericum perforatum
Matricaria chamomilla
Camellia sinensis
Glycyrrhiza glabra
Rosmarinus officinalis

Combine in equal parts and steep 1 tablespoon of the blend per 1 cup (240 ml) of hot water for 10 to 15 minutes, strain and drink 3 cups (720 ml) or more per day, long term.

Formulas for Miscellaneous Autoimmune Issues

CREST is an acronym for calcinosis, Raynaud's syndrome, esophageal dysfunction, sclerodactyly, and telangiectasia. These conditions involve diffuse fibrotic, degenerative changes in the connective tissues of the skin, joints, organs, intestines, and esophagus. The cause of CREST syndrome is unknown but since it overlaps with scleroderma and involves auto-inflammatory destruction of connective tissue, it is usually classified as an autoimmune disorder. There are no pharmaceuticals that have shown to be effective in halting the course of the syndrome other than steroids and immunosuppressive agents, thus herbal approaches would be of great value.

Herbs that protect connective tissue can be valuable in treating these conditions. Depending on the severity and presentation, demulcent teas may be used to assist swallowing in atrophic esophagitis, *Crataegus* and high doses of flavonoids may help protect the heart for fibrotic changes, and *Salvia miltiorrhiza* can help protect the kidneys and slow the development of renal insufficiency. *Ginkgo*, *Allium*, and *Zingiber* may serve as general vascular anti-inflammatories and support peripheral circulation for Raynaud's symptoms. *Curcuma*, *Centella*, *Calendula*, and *Equisetum* may be all purpose in tinctures, capsules, and teas for helping to protect the skin and connective tissue throughout the body.

Herbs Highest in Essential Fatty Acids

The seeds of the following plants are rich in EFAs. Although consuming the whole nuts and seeds is nourishing and has health benefits, it is difficult to obtain therapeutic dosages of EFAs through food alone, and many physicians prescribe these oils in daily tablespoon dosages to help treat allergies and offer neuroprotection. If using the oils, they should not be used for frying or brought to high temperatures, as they are fragile and heat-labile. They can easily be used for making salad dressings and sauces, added by the spoonful to smoothies and blender drinks, and worked into cooked foods once the cooking is done.

Hippophae rhamnoides	*Ribes* spp.
Linum usitatissimum	*Rosa canina*
Oenothera biennis	

Tincture for CREST Syndrome

This formula for CREST includes *Salvia miltiorrhiza* to help protect against vascular fibrosis, and *Curcuma* and *Centella* to protect the skin and connective tissue. *Calendula* and *Equisetum* might also be used for the latter purpose.

Centella asiatica
Salvia miltiorrhiza
Silybum marianum
Curcuma longa

Combine in equal parts and take 1 to 2 teaspoons, 3 to 6 times per day.

Tincure for CREST Syndrome with Raynaud's

Ginkgo, *Allium*, and *Zingiber* may serve as general vascular anti-inflammatories and support peripheral circulation for Raynaud's symptoms. *Crataegus* works as a nutritive base for this formula to help protect the heart and vasculature in general.

Crataegus oxyacantha	20 ml
Ginkgo biloba	20 ml
Allium sativum	10 ml
Zingiber officinale	10 ml

Combine the herbs in a 2-ounce (60 ml) bottle and take 1 dropperful, 4 times daily, long term.

Supportive Tincture for Iritis

A number of autoimmune diseases affect the iris, and may even destroy vision unless controlled. Steroidal eye drops are often necessary to suppress the inflammatory process and protect vision; this tincture will not replace such medications but may be complementary and enhance the efficacy. *Ginkgo* may promote circulation to the eyes and offer antioxidant and anti-inflammatory effects. *Curcuma* is a broad-acting anti-inflammatory and antioxidant herb, and is also an excellent liver-supportive herb. In TCM, the liver is linked with the eyes, and it is therefore valuable to consider *Curcuma* or *Silybum* for all eye complaints. *Vaccinium* berries are classic for supporting eyesight and improving microcirculation in the eyes. Other flavonoid-rich berries should be included liberally in the diet.

Vaccinium myrtillus	1 ounce (30 ml)
Ginkgo biloba	½ ounce (15 ml)
Curcuma longa	½ ounce (15 ml)

Take 1 teaspoon of the combined tincture 3 or more times daily.

Specific Indications:
Herbs for Allergic and Autoimmune Conditions

This review includes herbs used folklorically for allergies, atopic skin and airway disorders, arthritic disorders, and what was once referred to as rheumatism. In many cases these folkloric herbs have been found to exert numerous anti-inflammatory and immunomodulating activities such that precise molecular mechanisms of action can now be described.

Allium cepa • Onion

Allium cepa can be used topically for burns, to promote healthy wound healing, and to reduce keloid formation. *A. cepa* is also used as a homeopathic for allergic burning and watering of the nose and eyes, chronic upper respiratory mucus, and chronic neuralgic pains. It is a leading homeopathic for allergic rhinitis. *A. cepa* is high in quercetin, and onion syrups have been traditional for treating lung congestion, cough, and asthma, and for use in topical poultices. *A. cepa* may reduce cytokine release in the lungs in a manner that helps to treat allergic airway disorders.[205]

Allium sativum • Garlic

Allium is stimulating to mucous membranes and will thin respiratory, nasal, and digestive secretions. It can be difficult for those with weak digestive systems to digest and tolerate garlic, so larger doses may require a corrigent in formulas. *Allium* is best as a synergist or specific in allergy formulas for those with cold, damp, stuck symptoms or constitutions and abundant thick mucus with a tendency to marked congestion that leads to secondary bacterial bronchitis and sinusitis. Garlic also exhibits immunomodulatory activity, tempering allergic reactions and lymphocyte proliferation.[206] Researchers are proposing that sulfur, and particularly sulfated polysaccharides, in the body may play a role in mast cell activity and histamine release, and inadequate sulfation can contribute to allergic conditions. *Allium* is notably high in sulfur compounds, which may be important for urothelium integrity and regulation of mast cell histamine responses in bladder and other allergic disorders.

Ammi visnaga • Khella

This Apiaceae family plant has been the inspiration for a number of drugs, including cromolyn sodium (sodium cromoglycate), a derivative of khellin, and it is shown to inhibit the release of mediators. *Ammi* has been prepared into nonsteroidal bronchodilating inhalers since the 1960s (see Cromolyn Sodium Inhaler for Allergic Airways on page 46). *Ammi* may be used orally to reduce allergic and inflammatory phenomena specifically for eczema and concomitant asthma, and for inflammation and reactivity in the lungs and urinary system. *A. visnaga* compounds khellinone and its many variants are natural potassium channel blockers, and also noted to suppress human T-cell proliferation.[207] One group of researchers has reported khellinone to be a possible therapy for MS, targeting excessive T-cell-mediated autoimmune inflammation.[208]

Andrographis paniculata • King of Bitters

Andrographis is a foundation herb bitter liver tonic in Ayurvedic medicine for infections and as an antitoxin. *Andographis* is traditionally used for autoimmune disorders including RA and allergies such as chronic rhinosinusitis with nasal polyps or for when allergies, hives, or migraines follow bowel upset, liver burden, and general toxicity states. Andrographolide is an active component shown to modulate altered immune responses and attenuate excessive or inappropriate macrophage activation[209] and may reduce elevated cytokines and inflammatory mediators in colitis,[210] lung injury, and endometriosis.[211]

Angelica sinensis • Dong Quai

Angelica sinensis is commonly used in TCM to treat allergic disorders including allergic skin and blood reactivity, allergic airway diseases, and allergic reactivity in the urinary system. *Angelica* roots have "blood moving" effects that make it useful for clotting disorders, blood vessel inflammation,[212] and chronic urticaria. As a blood and yin tonic, *Angelica* has been emphasized for women with infertility, and to improve poor circulation or congestion in the pelvis and vital organs and extremities, including cold hands and feet. *Angelica*'s constituents ferulic acid and ligustilide inhibit vasoconstriction induced by cold temperatures, explaining its traditional use to treat Raynaud's syndrome and frostbite.[213]

Apis mellifica • Honey Bee

Apis is not an herb but an insect, the honey bee. It has a long-standing history in Eclectic and homeopathic medicine for treating sensations that are itching, burning, and stinging, which are common in allergic inflammation. Material doses of mother tincture are suggested for edema and renal engorgement and 3x dilutions are suggested for bladder irritation.

Bee venom, also referred to as apitoxin, has been explored as a medical therapy, referred to as apitherapy, for various inflammatory diseases including musculoskeletal pain, and for some skin lesions to prevent scarring. Apitherapy may take the form of administering live stings, injecting of venom into arthritic joints or inflammatory skin lesions, or using venom acupuncture to relieve pain and to treat chronic inflammatory diseases such as RA and MS. Apitox is an FDA-approved subcutaneous injectable purified bee venom product for relieving pain and swelling associated with RA, tendinitis, bursitis, and MS.

Apium graveolens • Celery

Apium is nourishing and nontoxic, suitable for long-term use and as a base herb in allergy formulas. Apium is specific to reduce allergic reactivity in cases of dermatitis, urticaria, and ulcerative lesions with profuse discharges. Apium seeds, roots, and juiced stalks have all been used medicinally. Celery juice and seed extracts including teas and tinctures are used as nerve, bronchial, and urinary relaxants. Folkloric herbals suggest Apium to be specific for allergic and inflammatory reactivity of the bladder, ureters, and bronchi associated with tightness and spasm. Apium is also said to be specific for itchy, blotchy skin lesions and a sensation of creeping or crawling on the skin. Essential oil of celery seed may be used with carrier oils and in topical compresses for wheezing, bladder cramps, and renal colic and applied to the affected area and covered with heat. Modern research has suggested Apium to reduce inflammation in animal models of RA.[214]

Astragalus mebranaceus • Milk Vetch

Astragalus is used in TCM for chi (qi) deficiencies and blood deficiencies, to support vital energy and improve stagnant blood flow, and to build the blood and act as a yin tonic. Astragalus may be used orally to reduce allergic skin conditions including severe allergic and anaphylactic reactivity following exposure to toxins, allergens, and chemicals, as well as to reduce conditions resulting from severe drug reactivity. Astragalus may also help those with chronic infections by offering immune support to reduce infection susceptibility, as well as by limiting inflammatory cascades at the onset of an infection and reducing excessive allergic reactivity

for those prone to wheezing, hives, or angioedema when suffering from simple colds. For allergic and autoimmune disorders, Astragalus is often combined with Angelica, a duo confirmed to improve the cardiovascular function by stimulating nitric oxide production in endothelial cells. Animal models of chronic asthma have shown A. membranaceus to reduce airway reactivity and reduce hypersecretion of mucus.[215] Animal models of dermatitis have also shown Astragalus to reduce swelling and normalize T-cell responses following exposure to known allergens and skin sensitizers.[216]

Avena sativa • Oats

Avena is most used for allergic conditions as an herbal bath, plaster, or poultice to allay itching. Avena baths and poultices are indicated for all sorts of itching, including dermatitis, chicken pox, mosquito bites, pityriasis, and other pruritic skin conditions. Oatmeal baths have moisturizing, cleansing, antioxidative, and anti-inflammatory properties, and clinical research has shown that Avena has a very low allergenic potential, making it safe for atopic individuals, even in the midst of an acute episode of allergic dermatitis.[217] Use as a full-strength bath for dermatitis, itchy dry skin, and nervous skin complaints.

Boswellia serrata • Frankincense

Many Boswellia species are referred to as frankincense, a common name said to have derived from a French term meaning "pure incense," due to the long history of collecting the hardened tree sap for burning and ceremonial purposes. The oleoresin is referred to as olibanum and is collected from incision of the tree trunk and used to prepare various topical and oral medicines, especially to treat chronic inflammatory disorders including asthma, arthritis, chronic bowel diseases, and cancer. Boswellic acids, a group of terpenoids isolated from oleo-gum-resins are credited with numerous immunomodulating and anti-inflammatory actions including lipoxygenase inhibition.[218] Due to ever-increasing popularity, take care to use Boswellia only from ethically harvested and sustainable sources.

Camellia sinensis • Green Tea

As an all-purpose systemic antioxidant, anti-inflammatory and antiallergy herb, Camellia may help allergic disorders when consumed liberally as a daily tea. Camellia may also be used topically to reduce allergic skin symptoms and may improve the absorption of other compounds in topical formulas such as compresses, lavages, and sitz baths. The well-known antioxidant activity of *Camellia* offers neuroprotective activity, credited largely to epigallocatechin gallate (EGCG), which is one of the most important active compounds of green tea leaves. Researchers have suggested that green tea be included in protocols for MS because of its protective anti-inflammatory effects on the CNS and because regulation of energy expenditure may improve MS-related fatigue.

Cannabis sativa • Cannabis

Cannabis research and product development are booming at the time of this writing, with pain management being one of the primary indications. Arthritic and rheumatic pain and fibromyalgia are among the main types of chronic pain that may respond to CBD products. CBD at a dose of 10 to 25 milligrams per kilogram orally inhibits the release of inflammatory mediators in animal models of arthritis.[219]

Celastrus aculeatus • Celastrus

In TCM, the roots, stem, and leaves of *Celastrus aculeatus* have been used for centuries to treat RA, osteoarthritis, lower back pain, and autoimmune disease. *Celastrus* contains celastrol, a terpenoid compound shown to inhibit interleukins and other inflammatory mediators. Research suggests that celastrol limits the progression of arthritis in part by balancing the ratios of T helper and T regulatory cells in the target organ to facilitate immune regulation.[220] In addition, celastrol may influence T-cell activation and cellular migration into the joints.[221] Animal models of RA show *Celastrus* to help protect against bone damage[222] and to prevent the progression of joint disease in a manner similar to the immunosuppressive drug methotrexate.[223] *Celastrus* is being explored as an immunosuppressant alternative or as an adjuvant to immunosuppressants from methotrexate to corticosteroids to NSAIDs.[224]

Centella asiatica • Gotu Kola

Centella asiatica has been used in many traditional systems, including Ayurvedic, African, and Chinese medicines, in the management of many chronic conditions. The leaves have been juiced to consume as a beverage, and extracts have been prepared into creams for cosmetics, wounds, and skin diseases. *Centella* is a very nourishing herb upon which an allergy or autoimmune formula might rest, especially when skin or mucous membrane lesions and ulcers are present. *Centella* is also specific for allergic reactivity with a tendency to keloid formation in the skin, fleshy growths with chronic eye allergies, nasal polyps with chronic hay fever, scarring in the bladder with interstitial cystitis, auto-inflammatory and allergic inflammation in the joints and connective tissue, fibrosis, and scleroderma.

Cinnamomum cassia • Chinese cinnamon

Also known as Chinese cassia, *Cinnamomum cassia* is a tropical evergreen tree commonly used to treat dyspepsia, gastritis, blood circulation disturbances, and inflammatory diseases in TCM. The leathery leaves, buds, and bark are all used medicinally and contain cinnamaldehyde, cinnamic acid, and coumarin; the bark is the primary medicine in common usage. *C. cassia* has potent anti-inflammatory and antimicrobial activity. *C. cassia* is noted to significantly decrease joint swelling and elevated IL-1β and TNF-α levels in animal models of RA[225] and to reduce intestinal inflammation in animal models of colitis.[226]

Cinnamomum verum • True Cinnamon

Cinnamomum verum is cultivated in Sri Lanka, Myanmar, and coastal areas of southern India, where the tree bark has long been a traditional medicine for treating abdominal and pelvic pain as well as pain due to rheumatism, neuralgia, wounds, and toothache. Cinnamon bark is one of the oldest herbal medicines mentioned in traditional texts for inflammation and as an analgesic and antipyretic against colds, fever, headache, myalgia, and arthralgia. Animal research has shown *C. verum* to reduce C-reactive protein when elevated, to inhibit leukocyte migration and prostaglandin synthesis, and to exert antiarthritic activity and significantly reduce elevated serum TNF-α concentration without causing gastric ulcerogenicity in animal models of RA.[227] This species is also known as *C. zeylanicum* (Saigon cinnamon).

Cnidium monnieri • Snow Parsley

Dried fruits (seeds) of *Cnidium monnieri* have been used as traditional remedies for skin disease, women's health, and stasis of the blood, as well as in formulas to reduce allergic tendencies and atopic conditions.[228] *Cnidium* is often mentioned in folkloric texts and formulas for its effects on dermatitis and itchy skin lesions and can be included in herbal formulas as an immunomodulatory adjuvant to treat allergic inflammation mediated by type 2 T helper cells. Osthol is a well-studied coumarin in *Cnidium* shown to have numerous anti-inflammatory and antiallergy mechanisms, including decreasing NF-κB activation, and the release of TNF-α, nitric oxide, and cyclooxygenase.[229]

Colchicum autumnale • Autumn Crocus

The bulbs of the autumn "crocus" are the source of the colchicine alkaloid, which has been used for over a thousand years as a treatment for gout, a use that continues to the present. Colchicine is also an FDA-approved drug for familial Mediterranean fever and the associated complication of amyloidosis. The 2009 FDA approval of colchicine led to new research, and large studies on its long-term use by gout patients have demonstrated additional benefits of colchicine in oncology, immunology, cardiology, and dermatology. Colchicine is also showing promise for some autoimmune disorders, including bullous disorders of the skin, vasculitis, and aphthous stomatitis.[230] See also "*Colchicum*—An Herbal Immune Stimulant?" on page 73.

Coleus forskohlii • Coleus

Coleus is a mint-family plant native to China; due to its striking leaf colors and patterns, coleus has been cultivated around the world. The leaves have been used as a traditional herbal medicine, and the constituent forskolin caught the attention of researchers when it was shown to promote cAMP, an important conductor of cell membrane responsivity. Forskolin is now well established as a research tool in cell biology. Among its many effects, forskolin may attenuate inflammatory diseases by affecting white blood cell reactivity and reducing their release of pro-inflammatory cytokines, interleukins, and TNF.[231] *Coleus* may be used as a synergist in allergy, psoriasis, and chronic dermatitis formulas especially where there is atony, sluggishness of general metabolic functions, and excessive inflammatory response.

Commiphora mukul • Guggul

Guggul may be included in allergy and autoimmune formulas when concomitant with metabolic dysfunction, including elevated lipids, diabetes and hyperglycemia, or hypothyroidism. Guggul is specifically indicated in folkloric traditions for slow metabolism, obesity, and a sense of weight or dragging in the pelvis. Guggul is also an antiseptic and a mucous membrane stimulant and can be included in formulas as a specific for excessive discharges.

Commiphora myrrha • Myrrh

Commiphora myrrha is a hot antimicrobial herb with particular affinity to the mouth, gums, and oral and pharyngeal mucosa. The resin is collected from mature tree trunks and powdered or otherwise processed into tincture, oils, ointments, and medicines. Use this herb as a specific or synergist in cases of mucosal lesions in the mouth, including infectious or inflammatory causes such as lichen planus and oral manifestations of autoimmune or other systemic illnesses.

Cordyceps species • Caterpillar Fungus

Cordyceps sinensis is an unusual fungus that grows on decaying insects and is specific for immune deficiency and serious infections with difficult recovery. Because the habitat is small and the naturally occurring plant material is sparse, authentic *C. sinensis* may sell for $40,000 per pound. As market demand has grown, a cultured type of *Cordyceps* has been developed. It is referred to as *C. militaris* and is used in most commercial products, even many of those labeled as *C. sinensis*. Various *Cordyceps* species have been used in the Himalayas and Tibetan and Chinese herbal medicine to replenish the kidneys and soothe the lungs, mostly by the aristocracy due to the rarity of the plant, also known as "winter worm," as it is searched for on snow-covered mountains. The caterpillar fungus has also been used in traditional medicine to treat numerous diseases and has been reported to possess antitumor and immunomodulatory activities. Modern research reports immunomodulating, anticancer, and antimutagenic effects. *Cordyceps* may be used as a preventive when allergic reactivity produces excessive mucus that is a breeding ground for microbes, and airway reactivity progresses to serious bronchitis and pneumonia. *Cordyceps* is shown to have renoprotective activity,[232] so might be included in formulas for lupus nephritis or any autoimmune disease involving the kidneys.

Cornus mas • Cornelian Cherry

The *Cornus* genus contains over 65 species, with only two, *C. mas* and *C. officinalis*, being traditional medicines of ancient Europe and southwest Asia. *C. mas*, Cornelian cherry, is also known as European cornel, and the edible fruits contain iridoids, anthocyanins, phenolic acids, and flavonoids.[233] *C. officinalis*, Asiatic dogwood, has also been used in TCM for more than 2,000 years for its ability to replenish the liver and kidney.[234] It is also called cornel dogwood or Japanese cornel. *C. officinalis* is used to treat allergic asthma and chronic inflammatory disease of the airways characterized by wheezing, coughing, breathlessness, and airway inflammation.

Crataegus species • Hawthorn

Hawthorn berry and hawthorn leaf and flower may be used as trophorestorative ingredients in formulas for vasculitis and collagen vascular disease. Also use *Craetaegus* in formulas for bruising and skin trauma, to improve microcirculation in the skin, for easy contusions with slow recovery, and for vascular inflammation and allergic reactivity including rosacea, hives, telangiectasias, and phlebitis. Animal studies have shown hawthorn to modulate inflammatory cascades in the vasculature in a manner that limits injury and supports tissue repair.[235] Older herbals mention *Crataegus* as specific for excessive perspiration. *C. laevigata*, *C. oxyacantha*, and *C. monogyna* are all used medicinally.

Curcuma longa • Turmeric

Curcuma is an all-purpose systemic antioxidant, anti-allergy, anti-inflammatory, alterative, and antimicrobial herb whose bright yellow roots are used to prepare oils, creams, and tinctures and are powdered for encapsulation. *Curcuma* is useful internally to promote skin repair, and prevent keloid formation. *Curcuma* is an antimicrobial strong enough to treat staph and other infections and may reduce excessive dermal proliferation in cases of keloids, psoriasis, and allergic inflammation due to the presence of pathogens, fungus, microbes, or allergens. *Curcuma* is also an excellent hepatic support herb and general alterative to assist in hormone and toxin clearance via the liver, which may also in turn reduce allergic reactivity. *Curcuma* may improve intestinal dysbiosis and propensity to fungal and other infections that may trigger allergic hyperreactivity and is specific when liver or biliary issues underlie allergic conditions.

Eleutherococcus senticosus • Siberian Ginseng

Native to Northern Asia, *Eleutherococcus* roots are considered a ginseng-like medicine of Siberia and are used as an adaptogenic herb for allergic and autoimmune conditions. *Eleutherococcus* is indicated for individuals with poor immune response and a tendency to infection, as well as excessive immune response and autoimmune skin conditions, as is the case for many adaptogens. *Eleutherococcus* may prevent joint destruction and delay the progression and severity of arthritis.[236] *Eleutherococcus* is often included in herbal formulas for atopic disorders that follow viral infections such as chronic rhinosinusitis in which mast cells become activated and T helper cells are recruited to release cytokines. *E. senticosus* may significantly inhibit mast cell–mediated anaphylaxis.[237] *E. senticosus* is also known as *Acanthopanax senticosus*.

Ephedra sinica • Ma Huang

Ephedra is a folkloric medicine for airway reactivity, asthma, influenza, fever, bronchial asthma, hay fever, hives, and eczema.[238] The photosynthetic stems have no true leaves and have been used for thousands of years in TCM. When used as a tea, *Ephedra* is mildly stimulating due to caffeine-like methylxanthines, most notably ephedrine. This herb may become very stimulating when concentrated in pills and taken in large dosages. It can occasionally promote jitters and heart palpitations in those sensitive to caffeine, especially when used in large, repetitive dosages. The related species *E. nevadensis*, Mormon tea, has less ephedrine and is therefore a weaker medicine. *Ephedra* is available mainly as a tea at the present time due to FDA sanctions forbidding concentrated products, though many herbalists make their own tinctures.

Eugenia caryophyllata • Cloves

Eugenia may be used in allergic disorders as an essential oil applied topically to allay neuralgic pains or reduce itching in chronic eczema. The isolated terpene eugenol has shown antiasthmatic effect via broad anti-inflammatory effects, inhibiting pro-inflammatory mediators such as cyclooxygenase, lipoxygenase, and leukotrienes.[239] Clove essential oil may reduce the occurrence of angioedema or other inflammatory disorders of the skin in those with sun sensitivity. It should be diluted with a nourishing carrier oil such as jojoba, argan, or another quality skin oil and topically applied, or dotted on a moist compress to dilute and prevent potential skin irritation. Avoid applying near the eyes or on broken skin. For oral use, *Eugenia* is best combined with soothing anti-inflammatory base herbs.

Euphrasia officinalis • Eyebright

Euphrasia is an annual plant found throughout Europe, Asia, and North America whose small leaves and young sprouts are used as a medicine for eye complaints and allergic disorders of the ears, nose, and throat. *Euphrasia* is specific for the itchy watery eyes of hay fever and has been traditionally used in eye drops and eye lotions to treat conjunctivitis, ophthalmia, sties, and ocular allergies. *Euphrasia* can also help decongest the ears and sinuses in cases of hay fever, respiratory allergies, and otitis. The dry herb can easily be prepared into compresses for eyes or skin, or even prepared into eye drops, and ear and sinus lavage blends. See also the *Euphrasia* entry in chapter 4 on page 196.

Foeniculum vulgare • Fennel

Foeniculum seed medicines can be used for numerous types of allergies. It is safe, high in minerals, gentle in its actions, and has a very pleasant flavor, making it particularly useful in formulas for children. *Foeniculum* is also specific for bowel reactivity and for IBS with gas, pain, and cramping. It is used orally to reduce allergic tendencies in cases of chronic dermatitis. *Foeniculum* seeds or seed powder infusions are also a classic ingredient in eyewash formulas for conjunctivitis and allergic eye symptoms.

Fucus vesiculosus • Bladderwrack

Fucus is a common seaweed that has many medicinal applications due to its antioxidant and anti-inflammatory actions, ranging from the treatment of cellulite to allergic and inflammatory disorders including RA, asthma, atherosclerosis, diabetes, psoriasis and other skin diseases, and cancer. *Fucus* is indicated for dermatitis with dry skin associated with hypothyroidism or for patients with slow metabolic rates and endocrine disruption that contributes to allergic phenomena. *Fucus* is also specific for allergic disorders concomitant with obesity, goiter, exophthalmia, constipation, and flatulence. *Fucus* species are edible with high nutritional value and are notably high in minerals, especially iodine, and contain fucoidans shown to protect the liver from cytokine-induced fibrotic inflammation of tissues. *Fucus* may therefore be considered for autoimmune diseases associated with liver disease,[240] ulcerative colitis, and Crohn's disease.[241]

Ganoderma lucidum • Reishi

Reishi is a woody mushroom traditionally used as an immunomodulating herb[242] and its ß-glucan polysaccharides are shown to protect cellular DNA[243] and normalize skewed T-helper-cell ratios in atopic disorders,[244] reactive airway diseases,[245] and inflammatory bowel diseases.[246] These polysaccharides are also shown to suppress autoimmune diabetes in animals.[247] *Ganoderma* may improve oral ulcers in lupus or Behçet's disease patients.[248] *Ganoderma* is indicated for those with low-grade allergic and infectious conditions that linger, or for allergic congestion that slowly turns into secondary bacterial infections. Injections of purified *Ganoderma* glycopeptides have been developed in China and are used widely to treat immune disorders,[249] progressive muscular dystrophy, myotonic dystrophy, dizziness, and autonomic nerve functional disturbance caused by vestibular dysfunction. Such injections are also used as an adjuvant therapy for cancer and hepatitis.

Gastrodia elata • Tian Ma

Gastrodia is a saprophytic orchid and the dried rhizome has been used in TCM for thousands of years for headaches, convulsions, vertigo, and hypertension, as well as in the treatment of neurodegenerative disorders such as MS. (Gastrodin is an active component credited with neuroprotective effects. *G. elata* may optimize neuronal regeneration by supporting the protein metabolism in neurons, inhibiting stress-related proteins, and mobilizing neuroprotective genes in a manner the promotes neurosynaptic plasticity.[250]

Ginkgo biloba • Maidenhair Tree

The leaves and fruits of the *Ginkgo* tree have been used medicinally for thousands of years for vascular and allergic disorders including asthma, hives, migraines, chronic chemical sensitivity, and contact dermatitis. *Ginkgo* is highly specific for skin lesions due to vascular insufficiency, such as stasis ulcers, or for use in asthma or dermatitis formulas when patients also have hives, high blood pressure, clotting tendencies, or other vascular conditions. *Ginkgo* may inhibit platelet-activating factor and thereby reduce the release of histamine in allergic disorders and of other inflammatory mediators in autoimmune disorders.

Glycyrrhiza glabra • Licorice

The sweet-tasting roots of *Glycyrrhiza* are used as a systemic all-purpose anti-inflammatory and antiallergy agent, with particular affinity for mucous membrane inflammation and ulceration. The oral lesions of lichen planus, lupus, and erythema multiforme, as well as aphthous ulcers and canker sores may respond to licorice teas and solid extracts. *Glycyrrhiza* is also indicated in formulas to help wean patients whose eczema, asthma, and other allergic phenomena have been managed with steroids, to upregulate and normalize adrenal function or cortisol regulation. Glycyrrhizin is a steroidal saponin present in licorice root and contributes to its sweet flavor and has been shown to boost the effects of cortisone in RA patients when used in tandem. Glycyrrhizin is metabolized into glycyrrhetinic acid in the gut, and both compounds are credited with antiallergenic and anti-inflammatory activity.[251]

Grifola frondosa • Maitake

Grifola is a medicinal and edible mushroom sometimes referred to as hen of the woods. It is known to be an immunomodulator and antiviral. *Grifola* helps treat many types of allergies and infections due to immuno-modulatory effects, such as enhancing NK-cell response, macrophage and cytotoxic T-cell activity, and antibody response.[252] Include *Grifola* in formulas for treating colitis[253] and RA and to prevent immunosuppression from chemotherapy.[254] *Grifola* also helps clear mercury from the tissues, which may reduce toxin-triggered autoimmune disease and protect the bone, liver, and kidneys.[255] *Grifola* contains ß-glucan, an immune polysaccharide also found in other medicinal mushrooms and shown to have immunomodulating activity.[256]

Grindelia squarrosa • Gumweed

Grindelia is a resinous herb with stimulating, expectorating, and warming qualities used for respiratory disorders and topically as an antidote to the itchy rashes of poison oak and poison ivy and other forms of dermatitis. *Grindelia* is also specific topically for burns and blisters and for herpetic, vesicular, and papular eruptions with itching and burning sensations. Folkloric herbals report *Grindelia* to be specifically indicated for chronic indolent skin ulcers surrounded by purplish discoloration.

Hispidulin is a methoxyflavone produced by *Grindelia* shown to inhibit platelet aggregation 100 times more potently than theophylline,[257] explaining *Grindelia*'s traditional use as an antiasthmatic herb. Hispidulin is credited with antispasmodic effects on the airways via calcium channel activation. Hispidulin exerts significant antiallergy effects by reducing mast cell activation, as well as inhibiting the release of histamine, prostaglandins, and leukotrienes.[258]

Hamamelis virginiana • Witch Hazel

North American First Peoples used boiled witch hazel leaf and bark preparations, and commercial witch hazel products have been produced in the United States since at least the mid 1800s. *Hamamelis* may check itching discharges in cases of allergic eruptions or atopic dermatitis and can be a stand-alone topical preparation for allergic skin conditions that are moist and suppurative. *Hamamelis* can also be taken internally for vascular inflammation and bleeding from the bowels or urinary lesions. *Hamamelis* tannins have significant antiviral[259] and antiallergy[260] actions. *Hamamelis* shampoos and head washes have been shown to improve dandruff and seborrhea of the scalp.[261]

Hippophae rhamnoides • Sea Buckthorn

The genus name *Hippophae* is from ancient Greek; *hippos* means horse and *phaos* means shiny. The Greeks fed sea buckthorn fruits to their horses to make their coats healthy and shinier. *Hippophae* fruits can be eaten or processed into a red-orange oil for both topical and oral use to treat allergic disorders in the skin. *Hippophae* promotes wound healing and limits scars and keloids via stimulating dermal fibroblasts to synthesize collagen.[262] Due to these mechanisms, *Hippophae* might be included as a medicinal food in protocols for scleroderma and other autoimmune diseases affecting the dermis.

Hypericum perforatum • St. Johnswort

Hypericum flower buds are rich in the bright red flavonoid hypericin. *Hypericum* may be included in formulas for allergic and inflammatory skin conditions associated with vascular inflammation, such as rosacea and telangiectasia. Folkloric literature reports *Hypericum* to be specific for eczema of the hands and face with intense itching, and for sensitive ulcers and sores in the mouth. *Hypericum* might also be considered in formulas for skin eruptions due to underlying viral infections. *Hypericum* has a calming effect on the nerves and can be included in formulas to help manage pain and is mentioned in the older literature for shooting and tingling sensations, as well as intense itching and hypersensitivity of the skin and skin lesions. *Hypericum* in teas may reduce bladder irritability and hypersensitivity of cystic mucosa.

Linum usitatissimum • Flaxseed

Flax has been extolled as a medicinal food as long ago as 650 BCE, when Hippocrates advocated flax for the relief of abdominal pains. Flaxseed oil is a rich source of EFAs, including linoleic acid and α-linolenic acid, which regulate prostaglandin synthesis. Flaxseed oil is specifically indicated for dry, lichenified skin, but is useful in all chronic allergic conditions.[263] One study on elderly patients found that after 1 month of consuming flaxseed oil, levels of pro-inflammatory lipid-based oxylipins could be reduced to levels usually seen only in younger people.[264] Flaxseed oils may also reduce inflammation in the brain to benefit cognitive function in MS patients.

Lobelia inflata • Indian Tobacco, Pukeweed

Young flower buds of *Lobelia* are a traditional expectorant and lung herb. *Lobelia* is most indicated in lower respiratory airway reactivity, where it has bronchodilating action due to direct adrenergic effects on bronchial smooth muscle. It may be an effective tool for reducing reliance on steroidal inhalers. *Lobelia* makes a good synergistic "kicker" herb in oral formulas or in topical compresses for acute asthmatic wheezing.

Matricaria chamomilla • Chamomile

Matricaria flowers make a gentle anti-inflammatory to use as a compress over small skin eruptions, or as a bath to immerse babies suffering from diaper dermatitis. *Matricaria* is very effective orally for intestinal upset with cramping and bloating, often providing relief after a single cup or two of herbal tea. *Matricaria* is specific for diarrhea, loose stools, nausea, and vomiting, whether due to a microbial or allergic trigger. Chamomile medications may inhibit histamine-, cyclooxygenase-, and lipoxygenase-driven inflammatory and allergic processes. Topical compresses or use of the essential oil may reduce acute itching in atopic dermatitis.[265]

Melaleuca alternifolia • Tea Tree

Tea tree oil is most discussed for its significant antimicrobial effects and activity against skin fungus, lice, scabies, and mites. Dozens of published research papers detail specific antimicrobial mechanisms for treating skin disorders, periodontitis, and scalp disorders. However, tea tree oil also has anti-inflammatory activity, reducing a number of inflammatory processes and exerting an overall immunomodulating effect[266] and inhibiting histamine release.[267]

Mentha piperita • Peppermint

The essential oil of *Mentha* is an important herbal tool to allay itching when topically applied, including chronic pruritus of renal, hepatic, and diabetic origins,[268] and the generalized pruritus referred to as pruritus gravidarum that occurs in 1 to 8 percent of pregnant women.[269] The cooling sensation and antipruritic effects on the skin are due to the ability of menthol, one of the volatile compounds in peppermint essential oil, to activate a specific transient receptor potential (TRP) receptor, the TRPM8 receptor and its minty-cool ion channels.[270]

Nigella sativa • Black Cumin

Records from ancient Egyptian and Greek physicians note *Nigella* seeds and the seed oil as treatment for headache, epilepsy, and neurologic disorders,[271] as well as other conditions. The seed oil contains thymoquinone, which has an inhibitory effect on the production of inflammatory mediators.[272] Thymoquinone may exert neuroprotective effects for those with MS and prevent the peroxidation of phospholipid liposomes in the brain.[273] *N. sativa* oil may also improve skin lesions of vitiligo.[274] Other common names are black seed and love-in-a-mist.

Oenothera biennis • Evening Primrose

Oenothera biennis seeds are processed to obtain EFAs credited with numerous anti-inflammatory and immunomodulating effects. Evening primrose oil is one of the best known, readily available sources of γ-linolenic acid (GLA). Many people who suffer from atopic dermatitis have been shown to have deficient levels of Δ-6-desaturase, the enzyme responsible for converting linoleic acid into GLA,[275] making evening primrose oil more effective than other plant oils for these individuals.

Panax ginseng • Ginseng

Ginseng roots are one of the most well-known and robustly researched immunomodulating herbs. *Panax* can be included in formulas for atopic tendencies, as well as for individuals with poor immune response and tendency to infection or excessive immune response and autoimmune skin conditions. *Panax* is appropriate to include in formulae when weaning patients from steroids, or when fatigue and chronic infections accompany any allergic phenomena. *Panax* ginsenosides may offer neuroprotection in formulas for Parkinson's disease, Alzheimer's disease, Huntington's disease, and MS. Clinical trials have indicated ginseng to improve fatigue in MS patients and reduce brain inflammation.

Perilla frutescens • Perilla

Perilla frutescens has many common names, including zisu in China and shiso in Japan. It has been recorded in Chinese medical classics since around 500 CE, as a treatment for allergies. *Perilla* has a long tradition of use as a medicinal food, serving as an antidote for allergic reactivity to fish and crab ingestion. Use *Perilla* in teas, tinctures, or extracts in formulas for bronchial asthma, allergic rhinitis, and atopic dermatitis. *Perilla* seeds are also used medicinally to treat allergies, and seed preparations have been injected into acupuncture points to reduce asthmatic inflammation.

Petasites hybridus • Butterbur

Petasites roots are an excellent allergy and inflammation herb especially appropriate as a lead ingredient in asthma, allergy, cough, and migraine formulas due to broad anti-inflammatory and antihistamine effects. *Petasites* has compared favorably to antihistamines in the treatment of allergic rhinitis.[276] Look for pyrrolizidine alkaloid-free products, or limit use to short term to prevent possible hepatotoxicity. See also *"Petasites* for Inflammation" on page 41.

Phyllanthus amarus • Stone Breaker

Both the roots and the aerial parts of many species in the large genus *Phyllanthus* are used medicinally, especially for renal and hepatic disease. *P. amarus* is credited with many anti-inflammatory and immunomodulating activities, and *Phyllanthus* may be included as a mild immunosuppressive agent for patients with autoimmune-driven destruction of the kidneys and as a long-term therapy aimed at preventing renal failure.

Picrorhiza kurroa • Kutki

Picrorhiza is native to the Himalayas, and the rhizomes have been used traditionally in Ayurvedic medicine to treat many inflammatory disorders and as an all-purpose allergy herb appropriate for acute itching in the skin and dermatitis. Modern research has revealed direct antihistamine effects. A concentrated mixture of two *Picrorhiza* iridoid glycosides, picroside and kutkoside, may reduce inflammation in colitis.[277] The iridoid glycosides are also shown to have significant hepatoprotective effects and may also protect against drug-induced immune suppression.[278]

Rehmannia glutinosa • Chinese Foxglove

Rehmannia roots have been emphasized in TCM for yin deficiency and used for a variety of allergic, inflammatory, and autoimmune conditions including asthma, urticaria, eczema, RA, and chronic renal disease. *Rehmannia* offers hepatoprotective, renoprotective, and neuroprotective effects and may limit autoimmune and allergic hyperreactivity. This herb has been included in TCM formulas for autoimmune disease such as Sjögren's syndrome[279] and used topically on allergic dermatitis.[280] Catalpol, an iridoid glycoside in *Rehmannia*, exerts a neuroprotective effect in animal models of MS.[281]

Salvia miltiorrhiza • Dan Shen

The red roots of *Salvia miltiorrhiza* are a valuable vascular herb, also known as red sage. *S. miltiorrhiza* reduces inflammation via a variety of mechanisms, particularly those involving blood cells and endothelial adhesion. *S. miltiorrhiza* reduces endothelial and microcirculatory injury and inhibits the release of cytokines from mast cells.[282] It may be included in formulas for hives, migraines, allergies, and vasculitis. *S. miltiorrhiza* may help protect blood vessels from fibrosis in cases of collagen vascular disease.

Sanguinaria canadensis • Bloodroot

Sanguinaria roots exude a bright orange-red, potentially caustic resin and may be used medicinally in small cautious dosages orally for thrush, and for oral lichen planus and oral erythema multiforme. **Caution:** *Sanguinaria* tinctures and extracts should be properly diluted with more soothing herbs or a sufficient quantity of water to prevent irritation when used orally, and even topically they can be harsh and even ulcerating. *Sanguinaria* is often highly diluted to prepare homeopathic medications, and use of the crude herb should be limited to skilled herbalists. *Sanguinaria* may be an effective synergist in formulas for vasomotor symptoms, for burning sensations in the palms and soles of the feet, and for red blotchy eruptions that burn and itch. *Sanguinaria* is also a possible antidote for irritation resulting from poison oak and poison ivy exposure.

Scutellaria baicalensis • Scute

Mature roots of *Scutellaria baicalensis* are used medicinally as a broad-acting anti-inflammatory agent. *S. baicalensis* inhibits the release of histamine from basophils and mast cells and is appropriate as a lead herb internally to reduce allergic response and atopic skin conditions. Flavonoids in the roots modulate prostaglandin via cyclooxygenase inhibition,[283] explaining traditional use in treating asthma, allergy, and dermatitis. Baicalein is one such flavonoid with numerous anti-inflammatory and antiallergic activities,[284] reducing mast cell degranulation[285] and inhibiting the release of eotaxin from fibroblasts.[286] Wogonin is another *Scutellaria* flavonoid shown to reduce mite-induced inflammatory cytokine release and contributing to the antiallergy effects of the herb.[287] *S. baicalensis* has been extensively used in TCM for thousands of years, with the common name of huang qin, where it contributes cooling and drying anti-inflammatory effects to formulas.

Smilax ornata • Sarsaparilla

Smilax may be included in formulas for boils, eczema, and skin eruptions related to hot weather. *Smilax* is a traditional remedy for seborrhea on the scalp and behind the ears and for deep fissures of the hands and feet that are worse in spring and summer. *Smilax* is an aromatic alterative agent with hormonal and adrenal supportive properties and is specific for fatigue, muscle weakness, and poor digestion. *Smilax* has slightly warming properties.

Stillingia sylvatica • Queen's Root

Stillingia is specific for chronic dermatitis of the hands and fingers. Chronic skin ulcers, chronic skin complaints associated with swollen glands, and "scrofula" (an old medical term used to refer to congested tissues prone to infection) are indications for including *Stillingia* in formulas. *Stillingia* is also indicated where skin eruptions are accompanied by liver congestion with jaundice and constipation.

Tanacetum parthenium • Feverfew

Tanacetum is an excellent antiallergy herb and can be a foundational or lead ingredient in formulas for hives, itching in the skin, dermatitis, migraine, hay fever, and febrile allergic phenomena associated with underlying blood reactivity and auto-inflammation. *Tanacetum* may reduce allergic reactivity, drug side effects, and vascular inflammation. *Tanacetum* is not terribly nourishing, but neither is it dangerous for aggressive dosing or long-term use. Chewing fresh leaves, however, is occasionally reported to irritate the mouth, a side effect not reported with teas, tinctures, or encapsulations.

Urtica dioica, V. urens • Nettle

Urtica is a highly nourishing herb with broad utility in herbal medicine. In terms of allergic conditions, *Urtica* is specific for hives and for itchy, blotchy skin lesions as well as genital itching, pruritus ani, pruritus vulvae, and itching of the scrotum. *Urtica* is also indicated for profuse discharges from mucous membranes. *Urtica* may also be used for angioedema, both acute and chronically recurring. *Urtica* is used as a urinary tonic and for prostatic enlargement and urinary inflammation. *Urtica* is extremely safe, tonifying and appropriate in teas, tinctures, or encapsulations, for short- or long-term use as a base, synergist, or specific herb, as most appropriate.

Vaccinium myrtillus • Blueberry

Due to *Vaccinium*'s stabilizing and anti-inflammatory effects on the vasculature, products that contain *Vaccinium* leaves and berries can be useful in cases of hives, blood vessel wall reactivity, and easy bruising and fragile skin in the elderly. This species is also called bilberry.

Zingiber officinale • Ginger

Zingiber officinale is native to India and commonly grown in Asia, tropical Africa, and Latin America. Modern research on *Zingiber* has identified a broad array of anti-inflammatory mechanisms. *Zingiber* is appropriate for inclusion in small amounts in formulas for allergic disorders of the digestive organs and is specifically indicated for those with cold and stagnant constitutions. Gingerols occur in fresh ginger roots, and their derivatives, the shogaols, are formed upon drying. Both types of compounds exhibit significant anti-inflammatory effects, inhibiting the expression of inducible nitric oxide synthase and cyclooxygenase in macrophages.

— CHAPTER THREE —

Creating Herbal Formulas for Musculoskeletal Conditions

Musculoskeletal conditions affect almost everyone at some point in life and treating muscle and joint aches and pains is very common in a general family practice. Trauma-related ailments, autoimmune disorders affecting the connective tissue, and disorders of bone growth are common presentations. Chronic muscle stiffness that is not due to underlying arthritis and autoimmune disease commonly occurs due to a general state of inflammation in the body. Easy fatigue and muscle weakness are also common complaints that lack an exact diagnostic category in Western medicine and often occur due to hormonal, adrenal, and thyroid imbalances, or simply due to a general deficiency state with lowered vitality. Food allergies, leaky gut, and toxicity states in the body can also contribute to inflammatory burden throughout the body and contribute to stiff muscles, joint aches, and chronic muscle pain. In such situations that defy a traditional diagnosis, dietary, lifestyle, nutritional, and herbal approaches have a lot to offer.

All connective tissue is synthesized from glycosaminoglycans, commonly referred to as GAGs. Herbal and nutritional supplements that help the body to produce these important GAG molecules are appropriate therapies for all manner of connective tissue disorders. Bones, joints, and tendons are supported by glycosaminoglycan production. The living bony matrix includes these substances, along with other proteins and trace minerals, grouped together in living crystals with a complex and intelligent architecture that is constantly undergoing both breakdown and synthesis. Healing broken bones, improving bone density, and repairing connective tissue often involves supporting vital processes in the body—circulation, digestion, liver function, kidney function—in addition to simply offering calcium, vitamin D, minerals, and other nutrients.

The Bowels, Liver, and Musculoskeletal System

There is a phrase in TCM that the "liver rules the joints and tendons," which embraces the age-old belief that connective tissue health is impacted by liver function. Although foreign to Western medicine, the concept is quite logical in that, if the liver fails to remove metabolic waste products and other impurities from the bloodstream, their presence in systemic circulation can contribute to inflammatory processes, particularly of the musculoskeletal system. Intestinal dysbiosis, a condition in which pathogenic bacteria replace the beneficial intestinal microbes, can contribute to systemic disease including musculoskeletal inflammation. Leaky gut syndrome, in which various intestinal stresses impair cell adhesion and damage the integrity of the mucous membrane's barrier, allows inappropriate absorption through or between the epithelial cells, and may contribute to allergies, autoimmune disorders, chronic inflammation, and other problems, including effects on the musculoskeletal system.

The delicate intestinal microenvironment controls what is absorbed into and what is excluded from the body. When toxins are excessive, or when the intestinal control over what is absorbed is deranged, intestinal mucosal cells are more easily inflamed and infected. The intestines may become more permeable and allow toxic, improperly digested substances to be absorbed into the internal tissues.[1] Aspirin and NSAIDs increase intestinal permeability in an undesirable way.[2] Ironically, NSAIDs are used to treat arthritis, bowel diseases, migraines, and other conditions that are typically caused by bowel toxemia, yet increase intestinal permeability, and a vicious cycle for pharmaceutical use may be established.[3] Increased absorption of antigenic substances may also trigger autoimmune and autoinflammatory diseases,[4] including rheumatoid arthritis.[5] Therefore, many formulas in this chapter, aimed at long-term treatment of chronic musculoskeletal inflammation, include herbs capable of improving liver function, intestinal health, gut barriers, and digestion.

Formulas for Musculoskeletal Trauma

Strains, sprains, and bone injuries can all be treated using many of the same herbs. Among the best herbs emphasized in historical literature for acute trauma, such as falls, motor vehicle accidents, and sports injuries, are *Arnica montana*, *Symphytum officinale*, and *Hypericum perforatum*, each of which has its own specific indications. *Symphytum* contains allantoin, an agent shown to speed cell proliferation, and is a folkloric classic for broken bones, as well as soft tissue injury. Several human clinical studies have shown topically applied *Symphytum* ointment to match the efficacy of pharmaceutical anti-inflammatories for the management of pain of a sprained ankle.[6] Bromelain, an anti-inflammatory from pineapples, is also very valuable to treat pain and speed healing of musculoskeletal injuries (see "Bromelain for Musculoskeletal Pain" on page 108).

In addition to the three all-purpose trauma herbs, herbs that are tonic for connective tissue may be indicated in formulas for musculoskeletal trauma. When there are broken bones, or when a laxity of joint, tendon, or connective tissue integrity predisposes to repetitive musculoskeletal injuries and impaired healing, such tonic herbs can be included in herbal formulas. *Centella asiatica*, *Aloe vera*, *Medicago sativa*, *Curcuma longa*, and *Equisetum* species are among the high-mineral herbs that can strengthen connective tissue and promote healing of fractures and osseous injuries. *Centella* has been used folklorically for skin lesions, trauma, and fractures, and studies show positive effects on wound healing.[7] *Curcuma*,[8] and its flavonoid curcumin, support healing of severe bruising.[9] Studies have described a variety of molecular effects such as enhancement of collagen synthesis, uptake of proline, and reduction of inflammatory mediators. Many herbs historically reported to be connective tissue tonics may have their effect by supporting GAG molecules in the tissues. Modern research has shown that *Curcuma*'s effects on GAGs, hydroxyapatite, proline, collagen, and fibroblasts, as well as many other mechanisms can affect wound healing and support bone, joint, tendon, ligament, and muscle regeneration. Since tendons and ligaments have no direct blood supply as muscles and bones do, injuries to these tissues tend to have a much longer healing time. Immobilization such as splinting, wrapping, and slinging are helpful to limit motion, prevent further injury and inflammation, and promote repair.

Liniment for Musculoskeletal Trauma

This formula incorporates the three main traditional herbs for sprains—*Arnica*, *Hypericum*, and *Symphytum*—and essential oils to offer immediate pain-numbing effects. The castor oil may help drive the herbs deeper into tissues.

Symphytum officinale oil	30 ml
Arnica montana oil	30 ml

Hypericum perforatum oil	30 ml
Castor oil	15 ml
Gaultheria procumbens essential oil	5 ml
Syzygium aromaticum essential oil	5 ml
Mentha spp. essential oil	5 ml

Combine the ingredients in a 4-ounce (30 ml) bottle and shake well. Apply the oil by hand or with a compress hourly, in between ice pack applications, on the first day of the injury. After 24 hours, begin to cover with heat such as a heating pad or hot moist pack after applying the oil, or apply immediately after an Epsom salt soak. Apply the oil frequently through the day.

Tincture for Acute Trauma

After severe trauma from events such as motor vehicle accidents and bad falls, tinctures and teas that are taken internally may speed healing, reduce scarring, and provide pain relief. This formula may be especially appropriate in cases of severe bruising. *Curcuma* is an all-purpose anti-inflammatory, *Boswellia* is specific for musculoskeletal pain, and *Symphytum* and *Hypericum* are specific for trauma. *Salix* contains salicylates for further anti-inflammatory and pain-relieving effects.

Curcuma longa	15 ml
Hypericum perforatum	15 ml
Symphytum officinale	10 ml
Salix alba	10 ml
Boswellia serrata	10 ml

Take every 15 to 60 minutes, reducing as symptoms improve. Supplements that can also help in cases of severe bruising include bromelain (500 to 1,000 mg, 3 to 6 times daily on an empty stomach) and vitamin C (1 g at each meal). Bioflavonoids also help, such as those from berries, isolated resveratrol, and concentrated anthocyanidins.

Witch Hazel Compress for Acute Trauma

Hamamelis is another option for treating trauma, and this remedy is inexpensive and easy to prepare. *Hamamelis* has astringent effects on tissue, helping to reduce swelling and inflammation of skin and blood vessels.

Hamamelis virginiana tea	1 cup (240 ml), prepared

Prepare the *Hamamelis* tea by gently simmering 2 tablespoons dry bark in 2 cups (480 ml) of water for 10 minutes, let stand 10 minutes more, and then strain. Saturate a gauze pad or soft cloth and apply topically for 15 to 20 minutes at a time. Refresh throughout the day every few hours.

Antispasmodics versus Anti-Inflammatories

Antispasmodic herbs should be added to formulas when trauma or subluxations induce muscle spasms and uncomfortable tightness in the tissues. With its reliable antispasmodic effects on the muscles, *Piper methysticum* (kava) is one of the most powerful, indeed drug-like, herbs. *Passiflora incarnata*, *Piscidia piscipula*, *Valeriana sitchensis*, *Viburnum prunifolium*, and *Lobelia inflata* are other very helpful antispasmodic herbs, especially for tetanic spasms of skeletal muscles. Anti-inflammatories with a particular affinity for the muscles, joints, and connective tissues include *Boswellia serrata*, *Hypericum perforatum*, *Yucca filamentosa*, *Actaea racemosa*, and bromelain (from pineapples). These herbs may be less effective for acute spasm but may reduce chronic inflammation more powerfully over time than antispasmodic herbs do. Either or both antispasmodic and anti-inflammatory herbs can be blended into formulas for pain, recovery from injuries, fibromyalgia, arthritis, and autoimmune disease.

Musculoskeletal and Connective Tissue Anti-Inflammatories

The following herbs are among the best musculoskeletal anti-inflammatories and may be included in teas, tinctures, and topical applications aimed at treating muscle and bone pain. See the "Specific Indications" section on page 131 for more information about each.

Arnica montana	*Ananas comosus*
Equisetum spp.	*Hamamelis virginiana*
Hypericum perforatum	*Actaea racemosa*
Curcuma longa	

Acute Treatments for Sprains and Traumatic Injuries

Immediate icing, homeopathic remedies, *Arnica montana*, *Symphytum officinale*, and *Hypericum perforatum* can be extremely helpful for traumatic injuries, especially when implemented promptly. Epsom salt soaks in the bathtub are also helpful following trauma. Topical application of *Hippophae rhamnoides* (sea buckthorn) oil is also noted to enhance the healing of tendons in cases of strains.[10] Baths of *Equisetum* are a classic herbal treatment for musculoskeletal pain, and one nearly forgotten at present. Essential oils from mint to camphor to cayenne extracts provide anodyne effects, and cannabidiol (CBD) ointments are increasingly popular for reducing musculoskeletal pain.

Facial Oil for Black Eyes and Bruising

The same herbs effective for sprained ankles and joint trauma can be used to treat facial bruising. However, clove and mint oils, which are used in the Liniment for Musculoskeletal Trauma, may sting the eyes and be harsh for the face. *Helichrysum* and carrot seed oils are alternatives, used in high-end cosmetics to prevent wrinkling. This light oil for facial trauma can double as a beauty product.

The specialty oils are readily available from aromatherapy and essential oil companies. Use only pure, natural essential oils (i.e., never perfumed or synthetic oils).

Hypericum perforatum oil	25 ml
Sea buckthorn (*Hippophae rhamnoides*) oil	25 ml
Carrot seed (*Daucus carota*) oil	5 ml
Helichrysum italicum essential oil	5 ml

Combine the oils and mix together. Apply to the face, with a cotton ball or fingertip, every several hours throughout the day.

Tea for Strains, Sprains, Fractures, and Trauma

This tea is not a powerful anodyne on its own, but provides bone and connective tissue healing support. Combine this tea with a topical agent and possibly an anti-inflammatory tincture for a more aggressive protocol that better addresses acute pain.

Centella asiatica
Symphytum officinale
Hypericum perforatum
Medicago sativa
Pueraria montana var. *lobata* (finely cut or powdered)
Mentha piperita
Glycyrrhiza glabra

Combine the dry herbs in equal parts, adding more licorice or mint for flavor if desired. Combine 6 tablespoons of the tea blend in 6 cups (1,440 ml) of water, along with the juice of 1 lemon for optimal mineral extraction, then let soak for 4 to 5 hours, or overnight. Bring the tea to a gentle simmer for 1 to 2 minutes, then turn off the heat, cover the pan, and steep for 15 minutes. Strain and drink 3 to 6 cups (720 to 1,440 ml) per day, until fully healed.

Formulas for Connective Tissue Weakness

There is not a formally recognized diagnostic category for "connective tissue weakness," but the astute practitioner will notice when patients' injuries heal poorly or scar excessively, when strains and sprain injuries are chronic, or when a person has unusually fragile fingernails and poor connective tissue integrity. In such cases, rule out any malnutrition and malabsorption issues, and otherwise treat with herbs that help build connective tissue. These herbs are also appropriate for nonunion fractures, or indeed any fractures and bone trauma.

The quality of the nails, hair, and skin may provide clues as to the health of connective tissue in the body. Individuals with dry, peeling, and cracking fingernails; dry, frizzy hair; and poor-healing wounds may have general nutritional deficiencies, hypothyroidism, or poor connective tissue integrity. Some autoimmune diseases have characteristic dermatologic changes such as the

Chemical Exposure and Connective Tissue Disease

Exposure to chemical solvents may trigger connective tissue disease, including scleroderma, systemic lupus erythematosus, and rheumatoid arthritis. A small number of epidemiologic studies have shown weak, but statistically significant, associations between solvent exposure and connective tissue disease including scleroderma.[11] Both genetic and epigenetic factors contribute to the development of these autoimmune diseases, and exposure to chemicals, particularly organic solvents, may trigger autoimmune reactivity in those with genetic predisposition. Those with a family history of lupus, rheumatoid arthritis, or other autoimmune disease should avoid exposure to organic solvents at all costs.[12] Scleroderma[13] and lupus,[14] in particular, have been associated with exposure to chemical organic solvents. Hair dyes are also associated with an increased risk of developing lupus,[15] as are certain pharmaceutical drugs, including procainamide and hydralazine.[16]

Organic solvents of concern include dry cleaning fluids (tetrachloroethylene), paint thinners (toluene, turpentine), nail polish removers and glue solvents (acetone, methyl or ethyl acetate), spot removers (hexane, petroleum ether), and other common paint removers, industrial detergents, and synthetic perfumes. Biomarkers of exposure to relevant common solvents are available in the form of urinary metabolites; however, all must be measured during exposure or within 24 hours of the end of exposure in order to be meaningful. See chapter 2 of this volume for further information on the effects of chemicals on the body.

red cheeks of lupus, referred to as a malar rash, the thin shiny fingers of scleroderma, or the obvious psoriatic lesions of psoriatic arthritis. Connective tissue tonic herbs are appropriate to include in the treatment programs for all of these conditions. Loss of muscle strength and muscle mass in the finger pads or palms is seen with muscular dystrophy, and using herbs that build muscle and tissue are appropriate. Deformation of the fingers toward the ulnar bone, known as ulnar deviation, is seen with rheumatoid arthritis and mixed connective tissue disorders, and connective tissue tonics are also appropriate for all types of arthritides.

Capsules for Bone and Connective Tissue Support

I've been making these mineral capsules for many years to help provide organic mineral complexes to patients with bone and connective tissue fragility. Ingesting whole herbs in food, mineral ash, or capsule form provides more minerals than either teas or tinctures. Use the following high-mineral herbs long term for osteoporosis, or for a few months after a fracture. A formula such as this may also help with arthritis, poor healing, connective tissue weakness, and fragile hair or fingernails.

Equisetum hyemale powder
Centella asiatica powder
Medicago sativa powder

Mix the freshly ground powders in equal parts and encapsulate in size 00 capsules. Take 2 capsules with each meal.

Tea for Broken Bones and Fragile Nails

All plants contain minerals, but these herbs are particularly mineral-rich and specifically indicated for fractures, sprains, arthritis, and connective tissue fragility. Extended soaking during preparation and the addition of lemon juice or vinegar enhances mineral extraction.

Symphytum officinale leaf	1 ounce (30 g)
Equisetum arvense leaf	1 ounce (30 g)
Urtica urens leaf	1 ounce (30 g)
Mentha piperita leaf	1 ounce (30 g)

Mix the herbs together, then combine 1 tablespoon of the herb mixture per 1 cup (240 ml) of cold water, adding the juice of 1 lemon for optimal mineral extraction. Soak for 4 to 5 hours, or overnight. After the soak, bring the mixture to a low simmer for several minutes. Then remove from heat, cover the pan, and allow the brew to

High-Mineral Herbs

These herbs are nutritive, high in minerals, nontoxic, and safe to use long term for bone disease, connective tissue weakness, or long-standing malnutrition and debility. Magnesium has been shown to be the most easily extracted mineral, with just over 55 percent of all of the magnesium in plant material able to be extracted into a tea. Those minerals and trace minerals that are moderately extractable—from 55 percent down to 20 percent—include sodium, phosphorus, boron, zinc, and copper. Elements that are poorly extractable—less than 20 percent—include aluminum, iron, manganese, barium, calcium, and strontium.[17] To optimize extraction, soak the fresh or dry plant material in a bit of citrus juice or vinegar for a few hours or overnight.

Alcea rosea	*Pueraria* spp.
Avena sativa[18]	*Palmaria palmata*
Centella asiatica	*Rubus idaeus*
Equisetum spp.	*Symphytum officinale*
Foeniculum vulgare	*Taraxacum officinale*
Fucus spp.	*Camellia sinensis*
Lepidium meyenii	*Ulmus fulva*
Medicago sativa	*Urtica* spp.[19]
Mentha spp.	

steep for 10 to 15 minutes. Strain and drink 3 cups (720 ml) or more per day. This tea can also be simmered for a longer period of time or heated in a crock pot for an entire day, yielding greater mineral extraction. However, the volatile oils and flavor will dissipate, resulting in a dark, earthy brew. This brew can be combined with bone broths and consumed long term for serious fractures and bone injury.

High-Mineral Vinegar

Vinegar extracts minerals from herbs more readily than does alcohol or water. Additionally, the use of vinegar on or in foods promotes digestion and assimilation of minerals. This recipe can be provided to patients so they can prepare culinary and medicinal vinegars at home, or you can make the vinegar and dispense the finished product to patients who need mineral support. Organic balsamic, wine, or apple cider vinegar may be chosen depending on price, availability, and taste preferences. The minerals in organic, plant-based molecules build bone density, help heal musculoskeletal injuries, and support healthy skin, hair, and nails.

Vinegar (balsamic, apple cider, wine, etc.)	4 cups or enough to fully saturate and cover the herbal powders
Medicago sativa powder	½ cup (25 g)
Symphytum officinale powder	½ cup (25 g)
Urtica urens powder	½ cup (25 g)
Fucus spp. powder	½ cup (25 g)

Combine the powdered herbs in the vinegar and macerate for at least 2 months, shaking daily. Fresh fruits or culinary spices may also be added to the recipe as desired. Because *Fucus* has a salty and fishy flavor, it may be omitted to create preferred flavors as desired. Strain or simply decant the finished vinegar and store in a glass bottle. Use in marinades, salad dressings, and drizzle onto steamed vegetables and other foods.

Formulas for Skeletal Fractures and Bone Injuries

High-mineral herbs provide the nutrients the body needs to produce new bone tissue, and many of these herbs are also anti-inflammatory to the musculoskeletal system, making them useful for bone fractures and trauma. Phytoestrogens, such as the isoflavones found in soy and other legumes, have been noted to promote healing of fractures,[20] as well as improve bone density in cases of osteoporosis. Phytoestrogens found in *Actaea racemosa*,[21] *Medicago sativa*, *Pueraria* species, *Glycyrrhiza glabra*, and many legumes bind estrogen receptors and elicit weak hormonal effects, which is the proposed mechanism of bone regeneration. The common isoflavone genistein enhances bone integrity, as evidenced by increased mineral deposition and bone density.[22] Isoflavone-containing herbs can be used in teas, tinctures, and in some cases foods, to support the healing of skeletal fractures. In addition to isoflavones, *Angelica sinensis* promotes the growth of bone cells[23] and *Lepidium meyenii* (maca root) has demonstrated acceleration of healing broken bones in animals.[24] *Lepidium* is available

Equisetum for Joints and Bones

The spindly photosynthetic stems of *Equisetum* are notably high in silica, and consumption may lead to increased absorption of calcium and stimulate collagen synthesis. In addition to its robust mineral content, *Equisetum* contains phytosterols, triterpenes, and phenolic compounds that may contribute to osteogenic effects. *Equisetum*'s secondary metabolites, such as quercetin, kaempferol, luteolin, apigenin, oleanolic acid, betulinic acid, and ursolic acid, have an anabolic activity on osteoblasts. *Equisetum* consumption in a variety of forms may decrease the number of osteoclastic cells, increase the activity of osteoblasts, and increase the synthesis of GAGs. Use *Equisetum* in teas, vinegars, and capsules as a nutrient to help rebuild connective tissue, bones, and mucous membranes and in cases of malnutrition and malabsorption. Consumption of *Equisetum* has been found to improve bone density in animal studies,[25] and one study showed *Equisetum* to be as effective as raloxifene in preventing bone loss in an animal model of osteoporosis.[26] Other species of *Equisetum* are used in identical manners, including *E. hyemale* and *E. giganteum*, a South American species referred to as cola de caballo. *E. giganteum* is noted to have anti-inflammatory effects in animal models of rheumatoid arthritis (RA), protecting against T-cell driven osseous destruction.[27] *Equisetum* may benefit RA by modulating TNF-α and limiting excessive proliferation of B and T lymphocytes. Kynurenic acid is one *Equisetum* constituent shown to inhibit the proliferation of synoviocyte and exert analgesic effects.

Equisetum arvense, horsetail

as powder and is palatable enough to use a tablespoon or two in blender drinks.

Tincture to Promote Healing of Fractures

When there is concern over nonunion fractures, or teas are impractical, the use of an herbal tincture may be complementary. Tinctures do not provide as many minerals as teas, but may still promote healing. A tincture can complement a tea to create a more aggressive treatment than a tea alone.

Symphytum officinale	20 ml
Equisetum arvense	10 ml
Medicago sativa	10 ml
Angelica sinensis	10 ml
Hypericum perforatum	10 ml
Ruta graveolens	4 ml

Take 1 teaspoon of the combined tincture hourly, reducing over time as symptoms improve.

Tincture for Fractures in the Elderly

This is a formula for both acute use and as part of longer-term herbal approaches for treating osteoporosis, as discussed in the section "Formulas for Osteoporosis" on page 124. *Actaea* is included for pain relief, as well as for its isoflavone content. *Angelica* and *Ginkgo* are circulatory-enhancing herbs and may promote healing in the elderly by helping to deliver blood and oxygen to the bones. *Symphytum* is a specific herb for all types of fractures and musculoskeletal trauma. Complement with one of the mineral-rich teas described in this chapter and with convalescent and preventive approaches to continue building bone strength and supporting circulation.

Actaea racemosa	15 ml
Angelica sinensis	15 ml
Symphytum officinale	15 ml
Ginkgo biloba	15 ml

Take 1 teaspoon of the combined tincture 4 times a day for prevention, or 6 to 8 times daily for acute issues.

Formulas for Bursitis and Tendinitis

Bursitis can be acute or chronic and is most common in the shoulders, followed by the elbows, knees, ankles, and pelvic girdle. Both tendinitis and bursitis are related to overuse and repetitive motion pain, and it is essential to rest the joint to encourage recovery. Bursitis may result from trauma or the presence of a pathogen or irritant, such as uric acid, in the tissues, where the affected bursa is a form of gout.

The affected joint should be rested, and the use of a sling, splint, or bandage can reduce the temptation to use the joint, even minimally. However, total immobilization of severely inflamed joints can result in adhesions and contractures, especially in the shoulder with the case of adhesive capsulitis, more commonly known as frozen shoulder. For this reason, it is advised to immobilize the joint as much as possible, but to remove the splinting 2 or 3 times a day, to gently move the joint following a heat pack application and/or an herbal liniment massage when possible. When bursitis or tendinitis are chronic due to repeated trauma or exertional strain, individual bursa may become calcified or form adhesions, resulting in reduced range of motion.

The main herbal treatments for tendinitis are anti-inflammatory agents with an affinity for the joints, such as *Boswellia serrata*, *Curcuma longa*, and *Harpagophytum procumbens*. Moistening agents and demulcents may also improve pain, burning, crepitus, and restricted motion by supporting GAG molecules. *Althaea officinalis*, *Asparagus racemosus*, and *Symphytum officinale* are among such useful demulcent herbs, supporting lubrication of the joint by synovial fluid, and enhancing connective tissue repair. For infectious bursitis, *Echinacea purpurea*, *Phytolacca americana*, or other specific antimicrobials can be combined with the anti-inflammatories. If gout symptoms or a joint aspirate demonstrates uric acid crystals, the use of renal depurants such as *Urtica urens* and *Equisetum* species would be appropriate additions to formulas (see "Renal Depurants for Gout" on page 123). Epsom salt soaks and hydrotherapy treatments are complementary topical treatments, and castor oil, wintergreen liniments, and cayenne ointments can be implemented for pain relief and further healing support. Bromelain supplements are also helpful for treating tendinitis and bursitis.

Homemade Heat Pack

If you have a microwave oven, you can make an excellent hot wet pack for home therapies. Moist heat penetrates more deeply than dry heat, such as a heating pad. Fold a large bath towel into a 1-foot square. Get the towel soggy, but not dripping wet. Place the wet towel in a large ziplock bag. Leave the bag open and place in the microwave, heating on high for 10 to 15 minutes. The bagged towel will become so hot it must be handled with oven mitts or corn tongs. After removing the bagged towel from the microwave, seal the bag and wrap the bag in 1 or 2 dry towels. The pack should feel quite hot but not painful on the skin. Adjust the number of outer wrappings to achieve the desired heat. The plastic bag must have no holes because water leaking through the plastic could be hot enough to scald the skin. This hot pack will stay steamy hot for at least half an hour, and warm for a full hour. Dry towels should never be heated in a microwave oven, only well-moistened ones.

Tincture for Acute Tendinitis and Bursitis

This formula is for acute tendinitis pain, reducing inflammation and discomfort as quickly as possible. Because tendinitis can become chronic, patients need to be coached to avoid repetitive motions and thereby the risk of flaring up the joint again. *Harpagophytum* and *Boswellia* are both specifically indicated for musculoskeletal pain and inflammation. *Curcuma* and *Glycyrrhiza* are general systemic anti-inflammatories, and *Centella* promotes repair and regeneration of tissues.

Curcuma longa	12 ml
Glycyrrhiza glabra	12 ml
Harpagophytum procumbens	12 ml
Centella asiatica	12 ml
Boswellia serrata	12 ml

Combine the tinctures in a 2-ounce (60 ml) bottle. Take 1 dropperful of the combined tincture every hour, for 24 to 48 hours. Reduce to 4 or 5 times per day thereafter, if needed. For those with chronic bursitis or tendinitis, refill the bottle as needed to continue at a lower dose and follow with a connective tissue restorative tea. Bromelain, hydrotherapy, and Epsom salt soaks, as well as one of the topical applications described in this section, can complement this tincture and speed resolution.

Topical Liniment for Bursitis and Tendinitis

A liniment combines oils and tinctures and must be shaken before each use, as the ingredients are immiscible. The massaging action with which liniments are applied is part of the medicine, stimulating circulation and driving the medicine into the joint. This formula is an example of a "counterirritant," as *Zanthoxylum* and *Capsicum* are mildly irritating and stimulating. The

Arnica and castor oil base helps drive the anti-inflammatory and pain-relieving oils into the tissue.

Arnica montana oil	30 ml
Castor oil	30 ml
Capsicum annuum	30 ml
Zanthoxylum clava-herculis	15 ml
Cinnamomum camphora essential oil	5 ml
Lavandula angustifolia essential oil	5 ml
Pinus sylvestris essential oil	5 ml

Combine the ingredients in a 4-ounce (120 ml) bottle and shake well, as the essential oils will separate from the tinctures. Apply an additional small amount of pure castor oil to the skin over the joint or problem area, then massage ¼ to 1 teaspoon of liniment into the skin. Use a bit more liniment to saturate a gauze pad or thin cloth, then place over the affected area and cover with heat.

Therapies for Contractures

Contractures are fibroproliferative disorders in the musculoskeletal system, such as Dupuytren's contracture, that may progress rapidly to a disabling contracture of the fingers or may emerge as hard nodules and ropy cords on the tendons of the hands that persist for decades without leading to contracture. The tendency to Dupuytren's and other contracture is in part hereditary but may be aggravated by smoking, heavy alcohol use, hyperlipidemia, and diabetes. Peyronie's disease is a fibroproliferative disorder of the penis in which collagenous plaques develop within the tunica albuginea. There is weak evidence that Peyronie's onset may follow the use of β-adrenergic blockers or urethral instrumentation, or it may follow plantar fascial contractures, tympanosclerosis, radical prostatectomy, and gout.[28] Third-degree burns may also result in contractures, especially in areas with many tendons, such as the knee and the neck. Morphea is an unusual local proliferation of scar tissue that is poorly understood, sometimes categorized as a localized form of scleroderma.[29] Plantar fasciitis is a condition of inflammation of the tendons and connective tissue in the plantar surface of the foot but is less prone to the formation of contractures. In all types of contractures, there is an excessive proliferation of type III collagen related to an upregulation of transforming growth factor β and basic fibroblast growth factor.[30]

The main therapy is surgical removal of the excessive scar tissue, but the condition may recur. Another leading therapy involves the injection of collagenase enzymes into the lesions with the intent to soften and dissolve the proliferative tissue. However, a side effect may be tendon rupture, and this therapy is usually reserved for advanced disease. Several collagenase creams are available by prescription, but they are quite expensive. Steroids have not been shown to be highly effective, but some case studies report success, such as for superficial morphea with occlusive dressings. Para-aminobenzoate is noted to decrease plaque size in Peyronie's disease and may prevent disease progression, but this medication is also expensive and may cause gastric upset. Pentoxifylline, a nonspecific phosphodiesterase inhibitor that has anti-inflammatory and antifibrogenic properties, has been shown to be helpful in Peyronie's disease, raising the question of whether natural cAMP inhibitors may be helpful. Calcineurin inhibitors such as topical tacrolimus 0.1 percent ointment are reported to have high efficacy for localized scleroderma but commonly have side effects such as pruritus, burning, or erythema in the application site. Imiquimod is a topical immune response modifier that may inhibit the production of collagen and has been reported helpful for scleroderma patients. Topical nitroglycerin has been found helpful for Raynaud's disease, scleroderma, and other connective tissue

Connective Tissue Nutrients

Glucosamine sulfate is a carbohydrate and amino acid–based molecule that occurs naturally in animals as a component of the proteoglycans in cartilage. Glucosamine contributes to the formation of GAGs, a type of proteoglycan having an ability to absorb water, become compressed without damage by losing water, and act as a resilient shock absorber in cartilage and connective tissue. Studies have shown glucosamine supplementation to help rebuild cartilage.[31]

Commercially available glucosamine is derived from the exoskeletons of shellfish such as crabs. Chondroitin is similar to glucosamine but of animal origin, such as from bovine sources. There are claims that glucosamine is better absorbed than chondroitin, but there has been little published investigation. Glucosamine does appear well absorbed in the intestines[32] and is observed to concentrate mostly in the cartilage, with lesser amounts in the liver.[33]

Glutamine is a simple amino acid of low molecular weight that is used by the body to synthesize endogenous GAGs. Glutamine is used to form the amino portion of the proteoglycans. The synthesis requires the enzyme glucosamine synthetase. This enzyme appears widely distributed in the tissues and forms glucosamine as well as hyaluronic acid.[34] L-glutamine supplementation has also been shown to elevate human growth hormone[35] and supplement regeneration of GAG-based tissues.

disorders and may prevent recurrence of contractures following Dupuytren's surgery.[36]

There is very little research on herbal or any alternative therapies for fibroproliferative disorders. *Curcuma longa* is noted to reduce excessive scar tissue in the skin and can do no harm. Oral or topical vitamin E may decrease pain when present. Topical vitamin D may be helpful and has been shown to reduce erythema, dyspigmentation, telangiectasia, and induration in morphea cases, and would do no harm applied to contractures. Other proteolytic enzymes such as bromelain may be worth a try. Castor oil may be a useful base for placing vitamin E, *Curcuma* oil, or other options. Dimethylsulfoxide (DMSO) can be used both orally and topically for arthritic pain and may be considered for contracture; however, be aware that some people are allergic to sulfur-containing substances. Traditional herbal connective tissue tonics such as *Hypericum perforatum*, *Centella asiatica*, or *Equisetum* species seem logical but there is no published clinical research on these precise disorders. *Aloe vera* and α-lipoic acid have been shown to help deter fibrosis due to radiation and may be considered topically or internally in patients with contractures and fibrotic pathologies of the musculoskeletal system.[37] Colchicine from *Colchicum autumnale* has been shown to halt other acute inflammatory disorders from gout to autoimmune disorders (see "Colchicum—An Herbal Immune Suppressant?" on page 73). Doses of 1 mg per day are commonly used in other disorders and *Colchicum* tincture may be considered in formulas for contracture. The topical application of colchicine has been shown to reduce cellular proliferation when applied topically to actinic keratosis.[38] One percent colchicine gel is commercially available.

Supplements for Contracture

These nutritional and herbal supplements can be taken orally to help reduce inflammatory reactivity in connective tissue and to support repair and regeneration. Clinicians may search out combination products that blend several of these ingredients to reduce the number of capsules or pills to be consumed.

Vitamin A 10,000 IU per day
Vitamin E 800 IU per day
Vitamin D 5,000 IU per day
Bromelain
Curcuma longa capsules
α-lipoic acid
Aloe vera
Glucosamine
Methylsulfonylmethane (MSM)
Dimethylsulfoxide (DMSO)

Choose 2 or 3 of the above supplements and consume orally at the dosage recommended on the product label for 3 to 6 months. If contracture softens during this time, continue to take the supplements long term. If not, choose others from the list and evaluate for efficacy after another 3 to 6 months.

Tincture for Dupuytren's Contracture

The herbs in this formula may offer antifibrotic and anti-inflammatory effects and may be most effective when taken promptly as a contracture begins to form. This tincture is also appropriate for fasciitis or burn recovery. **Caution:** *Colchicum* has potential toxicity and use is limited to knowledgeable clinicians.

Curcuma longa	25 ml
Hypericum perforatum	25 ml
Colchicum autumnale	10 ml

Take 1 dropperful of the combined tincture 3 times per day for at least several months.

Topical Oil for Contracture

Commercial colchicine and retinoid gels are used in this skin oil and may require a prescription to obtain. The ingredients in this formula may be combined in various ways, and precise amounts of ingredients can be amended as desired.

Castor oil	1 tablespoon
1 percent colchicine gel	1 tablespoon
2 percent retinoid gel or cream	1 tablespoon
Hypericum perforatum oil	1 tablespoon
Vitamin E oil	2 teaspoons
Vitamin D oil	1 teaspoon

Place all ingredients in a small bowl and blend by whisking with a fork. The resulting 5-tablespoon quantity can be stored in a small glass jar and massaged into the affected area 3 or more times daily.

Formulas for Myalgia, Muscle Pain, Tension, Cramps, and Tics

These conditions are grouped together because they will be improved by more or less the same herbs and by similar formulas. Herbal antispasmodics will relax muscle spasms due to trauma, stress, or issues with skeletal alignment. Anyone with acute muscle tension or cramping should be given calcium, magnesium, and potassium in frequently repeated dosages. Low mineral status may induce charley horses, and these minerals can help treat all types of spasms. If stress and emotional difficulties are thought to blame, the use of muscle-relaxing nervine herbs such as *Viburnum opulus* or *V. prunifolium*, *Actaea racemosa*, *Valeriana* species, *Passiflora incarnata*, and *Piper methysticum* may be therapeutic.

Some cases of chronic muscle pain and stiffness defy diagnosis but may respond to general dietary, nutritional, and liver-supportive measures (as discussed in The Bowels, Liver, and Musculoskeletal System on page 95). If the liver and intestines are not clearing the bloodstream of metabolic wastes, their presence in muscle tissues can trigger low-grade, persistent muscle and joint pain. Alterative herbs such as *Achillea millefolium*, *Arctium lappa*, *Berberis aquifolium*, *Rumex crispus*, *Silybum marianum*, and *Phytolacca americana* might be included in muscle pain formulas or given to those with chronic cramps if there is any indication of liver congestion, toxemia, intestinal dysbiosis, or leaky gut.

If connective tissue weakness is involved with repetitive sports injuries and strains, connective tissue tonics such as *Urtica urens*, *Equisetum* species, *Apium graveolens*, and *Centella asiatica* can be included in formulas. These high-mineral herbs also act as diuretics and often relieve myalgia by helping the kidneys to eliminate metabolic wastes such as uric, lactic, and oxalic acids via the urine.

Another contributor to consider for chronic myalgia is perimenopausal, hormonally induced muscle stiffness. In these cases, myalgia often responds to hormonally supportive herbs, as well as the isoflavone-containing legumes such as *Pueraria montana* var. *lobata* and *Glycyrrhiza glabra*.

Food allergies and liver inflammation can cause chronic myalgia. The nightshade family is particularly noted to trigger myalgia, but any food may be a cause. Dairy, wheat, and alcohol are all common culprits, and myalgia is a side effect of many pharmaceuticals.

Tincture for Muscle Tics and Twitching

Tics and twitches can result from mineral deficiencies, as well as from muscle fatigue and anxiety states. These herbs are specific for muscle cramps and twitches and do double duty as anxiolytics. *Veratrum* is described in the folkloric literature as being effective for tremors, convulsions, and states of excess mania, but can be used only in small doses.

Herbal Actions for Myalgia

Herbs having these actions are discussed throughout this volume and may be specifically selected to create herbal formulas for musculoskeletal issues.

CATEGORY OF HERBS	CONDITIONS
Muscle antispasmodics	Tightness, cramps, charley horses, spastic muscles, tics, twitches
Muscle anti-inflammatories	Chronic stiffness and pain concomitant with a general inflammatory state
Liver herbs and alteratives	Musculoskeletal pain due to poor digestion, oxidative stress, toxicity
Intestinal and leaky gut restoratives	Muscle pain in tandem with digestive issues
High-mineral renal depurants	To improve the assimilation of minerals via nutrition and renal function
Connective tissue restoratives	Chronic or familial tendency to weak joints and poor healing
Hormone, reproductive, adrenal, or thyroid herbs	Muscle weakness and sarcopenia
Antiallergy herbs	Muscle pain following exposure to various allergens (in atopic individuals)

Caution: *Veratrum* can produce toxicity, so do not alter this formula or apply more than specified below.

Valeriana officinalis	18 ml
Scutellaria lateriflora	18 ml
Actaea racemosa	18 ml
Veratrum viride	6 ml

Take ½ to 1 teaspoon of the combined tincture 3 to 6 times daily. Complement this tincture with calcium, magnesium, and other mineral supplementation if appropriate.

Tincture for Acute Muscle Cramps

Acute back spasms and torticollis will respond to musculoskeletal antispasmodics such as *Viburnum* and *Piper methysticum*. The spasms themselves can quickly initiate an inflammatory cascade. Anti-inflammatories such as *Salix* and *Harpagophytum* may shorten the duration of the episode, if taken immediately. *P. methysticum* may also be effective on its own.

Viburnum prunifolium
Salix alba
Piper methysticum
Harpagophytum procumbens

Combine in equal parts, then take ½ teaspoon every 10 to 30 minutes, reducing as acute cramping subsides. Complement with a calcium or magnesium supplement, a soak in hot bathwater, and heat packs applied to the affected area.

Tincture for Myalgia with Stress

This formula is more appropriate for muscle tension associated with anxiety, while the Tincture for Acute Muscle Cramps is specific for acute inflammatory muscle spasms. This formula contains a greater proportion of general anxiolytics, while the other has a greater anti-inflammatory effect. The logic is that severely spastic muscles require strong-acting antispasmodics and anti-inflammatories, while general tension due to stress and anxiety requires treating the mental/emotional realm.

Piper methysticum
Valeriana officinalis
Hypericum perforatum
Actaea racemosa

Combine in equal parts and take ½ to 1 teaspoon, 3 to 4 times a day for chronic, low-grade myalgia. This formula combines well with a nervine tea and a regular exercise program.

Tea for Stress-Related Muscle Tension

Simply stopping to brew and drink a tea benefits nervous tension. I often prescribe teas for any medical complaint with stress as an underlying cause, for the self-care moment the ritual can offer. *Passiflora* and *Scutellaria* are general nervines, while *Avena* is a nourishing mineral tonic with trophorestorative effects on the nervous system. *Glycyrrhiza* supports the adrenal glands and their involvement in stress responses, plus it sweetens up the tea.

Glycyrrhiza glabra	2 ounces (60 g)
Avena sativa	2 ounces (60 g)
Passiflora incarnata	1 ounce (30 g)
Scutellaria lateriflora	1 ounce (30 g)
Equisetum arvense	1 ounce (30 g)

Steep 1 tablespoon of the combined herbs per 1 cup (240 ml) hot water for 10 minutes. Strain and drink 3 or more cups a day for 5 to 10 days, reducing as symptoms improve.

Tincture for Myalgia with Toxicity

Some cases of chronic muscle stiffness may relate to the health of the liver or bowel, as metabolic wastes or toxins can trigger chronic inflammation. Consider this formula when symptoms of liver congestion or toxicity suggest the need for alteratives. Patients with frequent illness, right upper quadrant pain, skin blemishes, and bowel symptoms—rather than emotional tension or anxiety—may respond better to a formula such as this for chronic myalgia.

Silybum marianum
Curcuma longa
Berberis aquifolium
Glycyrrhiza glabra

Combine in equal parts and take 1 to 2 dropperfuls at a time, 5 to 6 times a day for 5 to 10 days, reducing as symptoms improve.

Tea for Myalgia with Weakness

Some people are easily injured as a result of even moderate activities and remain stiff and sore for days after exerting themselves. While the liver may sometimes play a role, in other cases the connective tissue may be weak and in need of nourishment. *Equisetum* and *Urtica* are mineral-rich connective tissue trophorestoratives that help tone connective tissue and support the muscles.

Glycyrrhiza glabra	3 ounces (90 g)
Avena sativa straw	1 ounce (30 g)
Equisetum arvense	1 ounce (30 g)
Urtica urens	1 ounce (30 g)
Centella asiatica	1 ounce (30 g)
Medicago sativa	1 ounce (30 g)

Steep 1 tablespoon of the combined herbs per 1 cup (240 ml) of hot water for 10 minutes, then strain. Drink 3 or more cups per day for several months.

Piper methysticum for Muscle Spasms

Piper methysticum roots are one of our most powerful, fast-acting musculoskeletal antispasmodic herbs, helpful for acute muscle spasms due to trauma and stress, as well as underlying inflammatory disorders in which muscle and joint pain are chronic. Kava root tinctures, teas, and encapsulations often soothe the discomfort of muscle stiffness and tightness when taken daily; they can alleviate acute spastic muscles when taken every half hour for a short period of time. Kava roots are known to have an anesthetic action on the oral mucosa, and the lactones (or pyrones) have been shown to be responsible.[39] It is likely that *kavain*, one such kava lactone, works in a manner similar to cocaine. Kava may have some activity at GABA receptors,[40] as well as glycine receptors,[41] both provoking a sedative, muscle-relaxing effect. Kava has been noted to prevent strychnine-induced convulsions,[42] indicating the chemical constituents may have some effects on indolamines or tryptophan and glycine metabolism. Kava has known hepatotoxicity when used long term; this is most commonly seen when used in tandem with alcohol, NSAIDS, or other hepatotoxins. Limit kava use to one month's duration or less, and avoid use in alcoholics or in patients with preexisting liver disease.

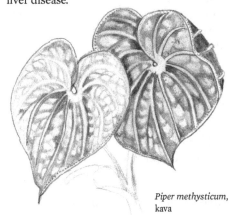

Piper methysticum, kava

Myalgia, Muscle Pain, Tension, Cramps, and Tics

Bromelain for Musculoskeletal Pain

Bromelain is a collection of easily absorbed[43] proteolytic enzymes extracted from pineapples, *Ananas comosus*, and is a folkloric remedy to enhance digestion of meat or heavier meals, as well as for parasites, obesity, and musculoskeletal pain. Bromelain may improve the pain of strains, sprains, phlebitis, osteoarthritis,[44] tendinitis,[45] and pain associated with autoimmune disorders. In India bromelain is traditionally combined with turmeric, and studies suggest the duo to be an effective anti-inflammatory for degenerative joint diseases[46] and to reduce reliance on pharmaceutical and OTC pain medications. A number of human clinical trials have shown bromelain to improve knee pain,[47] and several randomized clinical trials suggest improvements in osteoarthritis pain.[48]

Bromelain is also valuable to reduce inflammation and risk of clot formation and speed healing for patients undergoing surgery. Bromelain speeds postoperative recovery, reduces adhesions,[49] and reduces swelling and the need for additional analgesia.[50] A meta-analysis suggested bromelain to improve acute thrombophlebitis by beneficial effects on platelet aggregation and arterial walls.[51] Numerous wound-healing mechanisms are identified, including beneficial effects on cytokines, platelet aggregation, and fibrin formation.[52]

Even though bromelain reduces excessive and inappropriate clotting, it doesn't appear to interfere with appropriate clotting, and is believed to be safe for surgical patients prior to and after surgical procedures. One human clinical trial showed that bromelain had no increased risk of hemorrhage in knee surgery patients, even when used concomitantly with heparin.[53] The typical dose of bromelain is one or two 500 mg pills at a time, taken 2 to 3 times per day between meals for arthritic pain, and as often as every few hours for acute postsurgical pain or to help resolve a thrombus.

Tincture for Myalgia with Infections

Viral infections, chronic fatigue syndrome, hepatitis C, and other chronic infections may induce muscle stiffness and myalgia. *Eupatorium* is specific for influenza- and virus-induced myalgia. *Hypericum* and *Glycyrrhiza* provide additional antiviral and anti-inflammatory action. Some patients with chronic muscle stiffness following mononucleosis or Epstein-Barr viral infections may respond to this formula as well.

Eupatorium perfoliatum	1 ounce (30 ml)
Hypericum perforatum	½ ounce (15 ml)
Glycyrrhiza glabra	½ ounce (15 ml)

Take ½ to 1 teaspoon of the combined tinctures 5 to 6 times daily in cases of acute influenza, or 3 to 4 times a day in cases of chronic myalgia, reducing as symptoms improve.

Tincture for Myalgia in Elderly

Many patients become stiff as they age, especially morning stiffness, and lose general flexibility. Herbs that stimulate circulation, cellular function, and general repair and regeneration may be more effective than simple anti-inflammatories in treating myalgia in the elderly with cold constitutions. This formula uses *Panax* as a general warming and chi tonic herb, while *Zanthoxylum* and *Actaea* offer musculoskeletal anti-inflammatory effects. *Medicago* provides bone-building minerals and phytosterols, while *Zingiber* and *Coleus* act as warming and stimulating synergists. *Coleus* acts at the cellular level, helping individual muscle cells' basic metabolic functioning.

Panax ginseng	15 ml
Medicago sativa	12 ml
Zanthoxylum clava-herculis	12 ml
Actaea racemosa	12 ml
Zingiber officinale	7 ml
Coleus forskohlii	6 ml

Take ½ to 1 teaspoon of the combined tincture 3 to 6 times daily. Be sure to stretch and engage in gentle exercise, sufficient to warm the muscles, each day.

Tincture for Myalgia with Menopause

Menopausal hormonal decline is associated with muscle stiffness, especially in the back and spine. When the onset of chronic muscle stiffness occurs with menopause, this formula is frequently helpful. *Actaea* is specific for both menopause and myalgia, and *Hypericum* can reduce spinal irritation and act as a nerve-supportive agent. *Pueraria* supports bone density and connective tissue via hormonal mechanisms, to round out the formula.

Actaea racemosa	1 ounce (30 ml)
Pueraria montana var. *lobata*	½ ounce (15 ml)
Hypericum perforatum	½ ounce (15 ml)

Take 1 teaspoon of the combined tincture 3 or more times a day for several months reducing as symptoms improve. For best results, women should also adopt a regular exercise and stretching program, which might include yoga, walking, or swimming.

Liniment for Muscle Spasms and Myalgia

Acute muscle spasms are highly debilitating, and topical pain-relieving agents may help speed resolution and relieve suffering. This formula is similar to the basic Liniment for Musculoskeletal Trauma on page 96; both contain the all-purpose anti-inflammatories *Hypericum* and *Arnica*. *Lobelia* is often overlooked for its antispasmodic effects, and is added here to give the formula greater specificity for muscle spasms. The essential oils provide immediate comforting effects.

Hypericum perforatum oil	60 ml
Lobelia inflata tincture	15 ml
Arnica montana tincture	15 ml
Cinnamomum camphora essential oil	10 ml
Syzygium aromaticum essential oil	10 ml
Mentha piperita essential oil	10 ml

Combine the oils and tinctures in a 4-ounce (120 ml) bottle. Shake well to avoid separation. Apply the liniment to the painful area with a massaging action, then cover with heat. Leave in place for 15 to 20 minutes. Avoid getting the product near the eyes due to its relatively large quantity of essential oils.

Tincture for Leg Pain with Edema

Some "muscle" pain may actually be caused by circulatory insufficiency, as is the case with intermittent

claudication. This formula combines the circulatory-enhancing herbs *Ginkgo* and *Angelica* with *Aesculus*, an excellent choice for vascular congestion. *Apocynum*, which contains cardiac glycosides, may be added when congestive heart failure underlies circulatory insufficiency. *Apocynum* is a prescription-only herb and may be omitted from the formula.

Ginkgo biloba
Angelica sinensis
Aesculus hippocastanum
Apocynum cannabinum

Combine in equal parts and take ½ to 1 teaspoon, 3 to 5 times daily, long term.

Tincture for Leg Pain with Weakness

Occasionally, I have seen patients with chi deficiency appear weak and frail and suffer from various aches and pains that seem to be related to their depleted, deficient state. Using chi tonics such as *Panax* may improve the strength of the body and muscles. *Capsicum* is used here as a synergist to enhance digestion and stimulate circulation, adding a fiery spark to the formula. Using maca powder (*Lepidium meyenii*) in separate blended drinks and smoothies may offer further muscle-enhancing effects.

Panax ginseng	30 ml
Smilax ornata	15 ml
Rhodiola rosea	10 ml
Capsicum annuum	5 ml

Combine in a bottle and shake well. Take ½ to 1 teaspoon, 3 to 4 times daily for 6 months to a year.

Musculoskeletal Antispasmodics

These herbs are among the most potent musculoskeletal antispasmodics to use in tinctures, teas, and topical applications for acute muscle spasms or chronic tightness and stiffness.

Piper methysticum	*Viburnum opulus*
Valeriana sitchensis	*Lobelia inflata*
Dioscorea villosa	*Piscidia piscipula*

Formulas for Fibromyalgia

Fibromyalgia is a particularly challenging type of chronic muscle pain that is not well understood. There are few to no effective pharmaceuticals, making alternative treatments particularly important.[54] Fatigue, headaches, insomnia, stress symptoms or anxiety, and bowel symptoms typically accompany. Some investigators also report an association with depressive disorders,[55] and others with chronic fatigue syndrome symptoms.[56] Due to the association of fibromyalgia with insomnia, anxiety, and depression, many researchers are exploring a possible central nervous system component in the processing of pain signals that creates hypersensitivity to pain.[57] Abnormally elevated potassium levels may induce heightened nervous and electrical sensitivity and be involved with chronic pain phenomena.[58]

Due to variable presentations of fibromyalgia, formulas for fibromyalgia are best tailored to the individual, although several herbs might serve as useful base herbs. Neuroendocrine- and neurotransmitter-regulating herbs can be helpful in formulas for fibromyalgia. I often select *Piper* and *Actaea* as the foundation of formulas for fibromyalgia and add other herbs, such as anxiolytics and adaptogens, on a case-by-case basis. *Piper methysticum* inhibits sodium and calcium channels on muscle cell membranes; it may also relax muscle spasms and alleviate muscle pain. *Actaea* was used extensively by physicians in the early 1900s for rheumatic muscle pain, and a condition referred to as "neurasthenia," which had a symptom list nearly identical to that of fibromyalgia.

Tincture for Fibromyalgia

The combination of *Actaea* and *Piper methysticum* is the most effective I have tried for quick relief of musculoskeletal pain from fibromyalgia; both herbs are specific for muscle tension and inflammation. However, long-term management requires adrenal support, provided by the *Withania*. Some patients may benefit from additional nervines, digestive support, antiallergy herbs, or connective tissue herbs.

Actaea racemosa	1 ounce (30 ml)
Piper methysticum	½ ounce (15 ml)
Withania somnifera	½ ounce (15 ml)

Take 1 to 2 teaspoons, 3 to 5 times daily for several months, reducing as symptoms improve.

Tea for Fibromyalgia

This basic formula combines the adaptogen *Glycyrrhiza*; the connective tissue tonic *Equisetum*; the nerve anti-inflammatory, mood-enhancing *Hypericum*; and the general muscle anti-inflammatory *Zingiber*. This formula can be tailored to suit the diverse needs of individual patients.

Glycyrrhiza glabra	2 ounces (60 g)
Equisetum arvense	2 ounces (60 g)
Hypericum perforatum	2 ounces (60 g)
Zingiber officinale	1 ounce (30 g)

Gently simmer 1 heaping teaspoon of the combined herbs per cup (240 ml) of water for 10 minutes, then strain. Drink 3 or more cups per day for several months, and as desired thereafter long term.

Formulas for Muscle Weakness and Poor Stamina

Loss of muscle mass and strength is normal as a gradual progression to old age, but when sudden, unusual, or severe, may be a symptom of cancer and consumptive illness or other serious muscle-wasting disease. Botanical medicines may prevent or retard the loss of muscle mass and muscle strength in such pathologies and possibly even retard normal aging. Animal studies have noted *Ginkgo* and *Panax*, for example, to delay the progress of sarcopenia[59] in some disease states. Adrenal herbs help treat stress, fatigue, and the age-related decline in strength and stamina and may be the basis of many muscle-supportive formulas.

Some phytosterols have anabolic effects and are possibly useful in building and strengthening muscles.[60] These adaptogenic steroids and saponins may support adrenal DHEA and support complex hormonal balances, in a manner that improves muscle tone and athletic endurance. Adaptogenic herbs such as *Rhodiola rosea*, *Panax ginseng*, *Glycyrrhiza glabra*, *Smilax ornata*, and *Lepidium meyenii* may support adrenal activity and optimize androgen levels in the body. *Rhodiola* may improve strength in exhaustive disorders.[61] *Serenoa repens* may promote testosterone levels in both men and women when abnormally low. The seeds of *Trigonella foenum-graecum* (fenugreek) have been shown to improve exercise endurance in animals[62] and may act via the adrenal–pituitary axis, and possibly via enhancing glucose metabolism.

In other cases, poor circulation to the muscles and vital organs—or poor lung, kidney, and liver function—may contribute to muscle weakness and lack of stamina. Those with heart disease would of course be best treated with cardiac- and circulatory-support herbs. Animal studies have shown *Silybum marianum* to improve the ability of muscle cells to resist oxidative stress,[63] and thus might be considered appropriate in formulas for athletes, or for muscle wasting due to circulatory insufficiency.

Tincture for Muscle Weakness in the Elderly

The elderly, and those devitalized by a long illness, will benefit from a course of adrenal support. This formula is based on the adrenal- and vitality-supportive herbs *Panax*, *Lepidium*, and *Smilax*. *Coleus* may stimulate general cellular metabolic function, and *Zingiber* acts as a synergist to enhance absorption and assimilation of the lead herbs.

Panax ginseng	20 ml
Lepidium meyenii	20 ml
Coleus forskohlii	10 ml
Smilax ornata	5 ml
Zingiber officinale	5 ml

Take 1 to 2 dropperfuls at a time, 2 to 3 times a day long term, reducing dose over time.

Schisandra Simple for Sarcopenia

Sarcopenia is a progressive loss of muscle strength and mass due to aging and some diseases. The condition causes the downregulation of protein synthesis in muscle cells. Those affected lose muscle tone and sometimes suffer dramatic decreases in muscle mass, leading to frailty and an emaciated appearance. This simple uses a tincture, but *Schisandra* is tasty enough to use in teas, or its powdered berry in foods.

Schisandra chinensis

Take 1 to 2 dropperfuls, 3 to 4 times a day as part of a broader protocol or dietary program featuring high-protein intake.

Adrenal Support for Muscle Weakness

The following plants have been emphasized in folklore to help maintain muscle mass, strength, and endurance with age and to preserve muscle mass in various disease states.

Panax ginseng	*Schisandra chinensis*
Eleutherococcus senticosus	*Avena sativa*
Withania somnifera	Common nuts, seeds, and oils
Smilax ornata	

Tea for Muscle Weakness Due to Disease

This tea, aimed at muscle support for those with chronic disease, includes the better-tasting herbs that provide hormone and adrenal support. It covers the bases of the major organ systems and metabolic functions to support general health, forming a tea flavorful enough to consume regularly and long term. *Astragalus* supports the immune system, *Glycyrrhiza* is an adrenal tonic, *Avena* is a nerve trophorestorative, *Trigonella* supports healthy blood sugar and hormonal balance, *Apium* is a mineral tonic, and *Zingiber* is warming and enhances circulation.

Avena sativa (whole oat groats)
Astragalus membranaceus finely cut
Glycyrrhiza glabra cut or shredded
Trigonella foenum-graecum seeds
Apium graveolens seeds
Zingiber officinale root, chopped

Combine the dried herbs in equal parts. Gently simmer 1 heaping teaspoon per cup (240 ml) of hot water for 15 minutes in a covered pan. Let stand, covered, for 15 minutes more, then strain. Drink as much as possible, using long term.

Ergogenic Tincture for Athletes

Athletes, especially high-level competitors, are frequently interested in legal alternatives to harmful anabolic steroids to enhance muscle mass, strength, and endurance. Such herbs may be referred to as ergogenic, enhancing energy efficiency within muscle cells, reducing oxidative stress, staving off energy depletion and exhaustion, and shortening recovery time. Caffeine has ergogenic effects, and consuming caffeine-containing beverages prior to exercise is noted to improve both endurance and anaerobic performance associated with

Schisandra for Muscle Strength

Schisandra chinensis is a vining plant that bears flavorful fruits, sometimes known as the five flavor fruit, used traditionally as a longevity tonic in Asia and Russia. The clusters of red berries may be dried for tea or ground in powder for food supplements or to prepare into a variety of medicines. Also known as magnolia vine, *Schisandra* is indicated for chronic muscle weakness and fatigue. The foods and teas are reported to stave off hunger during exertion and to maintain strength and vitality into old age. The most-studied chemical constituents are the *Schisandra* lignans, sesquiterpenes that include several distinctive cycloartanes such as the schinchinenins and schinchinenlactones, and common antioxidant polyphenols.[64] *Schisandra* has shown a chondroprotective effect in animal models of osteoarthritis in which the

Schisandra chinensis,
five flavor fruit

protective effects were associated with a reduction in inflammatory cytokines and TNF-α and an increase in cartilage matrix proteins and collagen. *Schisandra* fruits have also been traditionally used for rheumatoid and degenerative arthritis in traditional Korean medicines. Several of the lignans, including schisandrin and gomisin A, have been shown to possess a stimulating activity of osteoblastic proliferation, limiting osteoclast differentiation, and to also protect against osteoporosis due to activity at estrogen receptors.[65] Dietary supplementation with schisandrin B was shown to ameliorate age-related impairment of mitochondrial antioxidant functions in various tissues, and rats with accelerated aging fed a diet rich in *Schisandra* lignans showed a normalization of all biomarkers used to monitor the aging process.[66]

increased serum catecholamine levels. Coffee, green tea, and yerba mate are the healthiest options, while sugar- and chemical-laden sports drinks should be avoided. *Panax, Rhodiola, Cordyceps,* and *Tribulus* are among the herbs with ergogenic effects.[67] Be sure to use quality products; there have been incidences of manufacturers placing pharmaceutical steroids in their products without reporting this on the products' labels.

Tribulus terrestris
Cordyceps sinensis
Panax ginseng
Rhodiola rosea

Combine in equal parts. Take 2 dropperfuls, 3 to 4 times daily for several months. If effective, the tincture can be consumed at a smaller dosage—1 dropperful, 2 or 3 times daily—for up to 6 months or even intermittently for a few years. When using herbs long term such as for years on end, it is recommended to take breaks of a few weeks to a month several times per year.

Electrolyte Drink for Athletes

When people exercise and sweat in hot climates, they may lose large amounts of electrolytes. It is important to replenish sodium, potassium, magnesium, and calcium. Endurance athletes may use a commercial product that contains these minerals. Coconut water contains many natural electrolytes and can be a stand-alone sports drink in a pinch. Green, leafy herbs such as *Equisetum, Medicago, Rubus, Urtica,* and *Taraxacum* are excellent sources of the above minerals. This formula uses *Urtica*, but this can be swapped out or combined with the others according to preference. *Camellia*, or green tea, is a caffeine source used here for its ergogenic effects. Adding a pinch of salt or mineral and trace mineral drops to the finished tea can further boost the sodium, chloride, and other electrolytes.

Coconut water	1 cup (240 ml)
Fruit juice, optional	½ to 1 cup (120 to 240 ml)
Camellia sinensis, dried	4 tablespoons
Urtica urens, dried	4 tablespoons
K/Ca/Mg powder	1 teaspoon
Himalayan salt	¼ teaspoon

Prepare the *Camellia* and *Urtica* into a tea by steeping the combined herbs in 8 cups (1,920 ml) of boiling water. Add the salt and mineral powder to the steeping water at this time. Let steep until cool, then strain into a pitcher or large canning jar and refrigerate. To prepare the sports drink, combine roughly 3 cups (720 ml) of the tea blend with the coconut water and fruit juice, if desired.

Fenugreek Capsules for Endurance Athletes

Fenugreek is a folkloric herb used for diabetes and obesity due to its metabolism-enhancing effects. Modern research has shown fenugreek seeds to share steroidal saponins in common with *Dioscorea* (wild yam), a well-known adaptogen and adrenal tonic. Fenugreek seed powder has positive effects on glucose metabolism, fatty acid utilization, and endurance. Human studies have shown 300 mg of fenugreek powder twice a day to have beneficial effects on body fat, free testosterone levels, and serum creatinine in male subjects undergoing resistance training.

Trigonella foenum-graecum   300 mg capsules

Take one capsule twice a day.

Tincture for Muscle Pain with Exertion

Cramps in the calves with walking is a common presentation of peripheral artery disease. The condition tends to become progressively worse, with no effective cure and few therapies. Patients should stop smoking immediately and work with a cardiologist or other specialist to improve circulatory health. Reducing oxidative stress in the blood vessels and improving lipid levels may help. *Ginkgo biloba* flavonoids can help improve muscle perfusion, but it might take many months of regular consumption to note any improvement. Patented *Ginkgo* products such as Tanakan or Tebonin are commonly prescribed in Europe for this purpose. *Ginkgo* and *Rhodiola* form a common combination that may improve endurance. For more on peripheral vascular insufficiency, consult chapter 2, Volume 2.

Ginkgo biloba
Rhodiola rosea

Combine in equal parts and take 1 to 2 dropperfuls, 3 or more times a day, long term. If, after 3 to 6 months of continuous use, improvements are seen, it may be possible to maintain the results with a smaller dosage.

Formulas for Arthritis

Arthritis simply means inflammation of a joint, and herbal treatments vary according to the specific type and cause, such as osteoarthritis, RA, and the many types of autoimmune arthritises, gouty arthritis, and infectious arthritis. In general, all arthritis formulas might include osteoprotective, bone-building, and connective-tissue-supportive herbs. Antimicrobial botanicals can be added for infectious types of arthritis, and renal depurant diuretics are helpful in gouty arthritis. Autoimmune arthritic disorders such as RA or ankylosing spondylitis are very difficult to cure with herbs or with any medical therapy. Significant and lasting remission is achieved by only 5 to 10 percent of treated patients.[68] See chapter 2 of this volume for more on RA.

Periodontal disease commonly occurs in tandem with cardiovascular diseases, diabetes, and RA and is worthy of addressing in the management of any chronic inflammatory condition of the musculoskeletal system. Pathogenic bacteria such as *Porphyromonas gingivalis* that can occur in the mouth have also been found in the synovial fluid from RA patients and is believed to activate inflammatory cascades. Therefore, agents noted to reduce *P. gingivalis* in the mouth may be a target in treating arthritis as well as periodontal disease. (See "Formulas for Dental and Gingival Issues" on page 174.)

There is a great deal of research on the mechanisms of action of various anti-inflammatory herbs that can help with pain management and protecting joints from destruction. Such herbs would be appropriate for both osteoarthritis and RA. Herbal anti-inflammatories with an affinity for the musculoskeletal system include *Curcuma longa, Actaea racemosa, Boswellia serrata, Zingiber officinale, Equisetum* species, and *Yucca schidigera*, as well as bromelain (extracted from pineapples). The salicylate-containing herbs, including *Salix* species, *Populus* species, and *Filipendula ulmaria*, have been used folklorically for arthritic pain. *Actaea* also contains salicylates and may reduce immune-triggered release of histamine from mast cells[69] and reduce destruction of collagen by abnormally elevated enzymes.[70] Iridoid glycosides in *Harpagophytum procumbens* have significant musculoskeletal anti-inflammatory effects[71] and offer pain relief via opiate pathways,[72] and clinical studies show benefit to arthritis pain.[73] *Boswellia* may reduce hyperimmune reactivity[74] and modulate inflammatory mediators;[75] limited clinical studies suggest efficacy in osteoarthritic patients.[76] *Curcuma* may also benefit autoimmune disorders and is shown to reduce elevated cytokines in rheumatoid and psoriatic arthritis patients.[77] *Zingiber*'s anti-inflammatory mechanisms include reduction of lipoxygenase and cyclooxygenase enzymes[78] and it may improve pain, stiffness, mobility, and swelling.[79] *Piper methysticum* may not only reduce musculoskeletal pain in arthritis, it may also deter oral biofilms.[80]

Diuretics may also be helpful in formulas for rheumatic swelling of the joints, supporting the ability of the kidneys to eliminate inflammatory substances by providing minerals that enhance the excretion of wastes. Many herbalists contend that diuretics do not just temporarily reduce swelling but provide significant organic minerals that act as connective tissue restoratives. *Equisetum, Medicago sativa, Urtica urens, Ulmus, Petroselinum crispum,* and *Apium graveolens* are all diuretic and tonifying to connective tissues. *Apium* is high in glutamine, which promotes GAG formation. *Urtica* may reduce inflammatory mediators released by articular chondrocytes and synovial macrophages, and one clinical trial reported reduced osteoarthritic pain in one week's time.[81] *Medicago* improves bone mineral density in poultry.[82] *Ulmus* is also a great source of organic mineral complexes, with an affinity for GAG molecules. *Ulmus* is used in China for RA and is noted to increase alkaline phosphatase and prolyl-hydroxylase enzymes involved with osteoblastic and bone-building activity.[83]

Many alternative practitioners will investigate possible contribution of oral and intestinal health to musculoskeletal pathophysiology via the "gut–joint axis." Although not curative, searching for the optimal diet and supporting digestive and eliminative function will often help reduce pain and slow the progression of the disease. Various restrictive diets—from gluten-free, to vegetarian to chemical-free—have been shown to reduce symptoms in some specific populations.[84] Alterative herbs and attention to the health of the digestive

Arthritis Formulas in a Nutshell

The following therapeutic options are guidelines for choosing herbs to create formulas for arthritis. The guidelines help create herbal formulas specific for an individual, rather than for the diagnosis.

1. Choose one or more anti-inflammatories with an affinity for the connective tissue.
2. If an autoimmune type of arthritis, include immunomodulating ingredients and search for all possible triggers, including toxins, leaky gut, or oral dyshiosis, that may be initiating immune reactivity.
3. Choose one or more high-mineral herbs to help repair and regenerate the joints and connective tissue. If gout or poor renal function is suspected, emphasize diuretics and agents that enhance circulation to the kidneys.
4. Choose one or more therapies aimed at bowel health, liver support, and detoxification, especially if indicated by other systemic symptoms.
5. Include bioflavonoid-rich herbs to help protect collagen and the vasculature when autoimmune vasculitis may be contributing to joint pain, inflammation, and destruction.

Offer one or more formulas and topical agents aimed at pain relief to help manage symptoms while the primary alterative, immunomodulating, and connective tissue restoratives have time to begin working.

system can improve nutrient assimilation, and assisting the body to eliminate toxins may reduce the underlying trigger of autoimmune phenomena. Consider *Taraxacum officinale*, *Curcuma*, *Berberis aquifolium*, *Arctium lappa*, or other gentle tonic herbs as base ingredients in formulas for autoimmune arthritis and toxin-induced musculoskeletal pain. *Taraxacum* may help RA patients, when used long term, to improve mobility and reduce the progression of arthritic symptoms.[85]

Herbs that nourish and support connective tissues are another important category to include in herbal formulas for arthritis. Since damage to joints involves inflammation and destruction of various connective tissues, plants that offer specific nutrients to chondrocytes and bone are of value in arthritis. *Equisetum*, *Medicago*, *Urtica*, *Avena sativa*, *Astragalus membranaceus*, *Pueraria montana* var. *lobata*, and/or *Centella asiatica* offer such nutrients and support repair and regeneration of connective tissues, cartilage, and bone. Large and long-term dosage of these herbs is most effective, keeping in mind that capsules and long-macerated teas will deliver more minerals than tinctures.

While osteoarthritis is managed with osteoprotective and bone-building support, RA requires vigilant efforts to temper immune hyperreactivity by all means possible to protect against joint damage. Collagen vascular disorders and mixed connective tissue disorders frequently involve joint pain, if not serious joint destruction. Ankylosing spondylitis, Sjögren's syndrome, relapsing polychondritis, scleroderma, psoriatic arthritis, RA, lupus, CREST syndrome, dermatomyositis, polymyositis, reactive arthritis (Reiter's syndrome), and other such diseases are very difficult to cure. All involve hyperimmune reactivity, in which the body attacks connective tissue and inflames the blood vessels and joints. Similar to their effect in treatment of allergies, herbs such as *Tanacetum*, *Curcuma*, *Zingiber*, *Urtica*, or *Glycyrrhiza* may normalize and optimize excessive and inappropriate immune response. Vasculitis is a common denominator of all the collagen vascular diseases, therefore herbs that protect collagen and vasculature are important to saving the joints and blood vessels from fibrotic demise. The colorful procyanidin- and anthocyanidin-containing herbs, such as *Crataegus* species, *Vaccinium myrtillus*, *Hippophae rhamnoides*, and *Hypericum perforatum* are among the herbs useful to strengthen vascular connective tissues and should be used in food-like ways to complement herbal formulas. Some researchers have found green tea to help protect the salivary glands from damage, and to prevent the dry mouth that occurs as Sjögren's disease progresses.[86] Another powerful flavonoid extracted from pine bark is Pycnogenol, which is shown to have powerful anti-inflammatory effects and a protectant effect on blood vessels. Lupus patients show improved immune modulation following the use of Pycnogenol.[87]

Rheumatoid Arthritis (RA)

The mechanism of autoimmune-driven joint pain and destruction are discussed in more detail in chapter 2 of this volume. RA is the archetypal autoimmune type and is a systemic, debilitating, chronic inflammatory disorder that affects roughly 1 percent of the world population. While the precise pathogenesis is likely multifactorial, most research suggests that a combination of environmental and genetic factors may contribute to the development of RA. RA and other autoimmune disorders have been associated with specific human leukocyte antigen (HLA) alleles. Joint pathology progresses as genetic tendencies to glycosylate various proteins irregularly follows exposure to pathogens, antigens, and environmental triggers. Diagnostic tests for RA include confirming elevated levels of rheumatoid factors, anticyclic citrullinated peptide autoantibodies, and antimannose binding lectin autoantibodies. The primary allopathic therapies for RA are steroidal immunosuppressant such as sulfasalazine. In the modern era many rheumatologists begin therapy for RA with the combination of hydroxychloroquine, sulfasalazine, and methotrexate. The newer disease-modifying antirheumatic drugs include anti-TNF-α therapy, sometimes referred to as *biologics* because they target specific molecules rather than suppress the entire immune system.

Tincture for Osteoarthritic Pain

Curcuma, *Harpagophytum*, and *Filipendula* provide an anti-inflammatory base for this formula. *Apium* offers minerals, helps the kidneys eliminate waste, and prevents the primary herbs from irritating the stomach with its carminative action. *Zingiber* has anti-inflammatory effects and enhances the absorption and action of the lead herbs. This formula is best combined with encapsulated nutritional supplements—such as calcium, magnesium, and vitamin D—and other bone-building approaches.

Curcuma longa	15 ml
Harpagophytum procumbens	15 ml
Filipendula ulmaria	15 ml

| *Apium graveolens* | 10 ml |
| *Zingiber officinale* | 5 ml |

Take ½ teaspoon of the combined tincture 4 or more times daily, continuing for several months before evaluating the results. If helpful, the formula can be used long term.

Protective Duo for Joint Inflammation

Yucca and *Ulmus* are recommended for long-term use to protect the joints and support connective tissue repair.

Yucca schidigera powder
Ulmus fulva powder

Combine in equal parts. This bland-tasting powder can be added to smoothies or oatmeal.

Mineral Tonic Tea for General Arthritis

This tea includes several gentle, safe, and nourishing high-mineral herbs. They have healing effects on bone and connective tissue and are best used long term, if not for life. *Glycyrrhiza* offers additional anti-inflammatory activity and lends a sweet flavor to the brew. This formula is just an example—any or all of the following herbs may be selected, and flavoring ingredients such as mint or licorice can be increased or decreased according to personal preference.

Avena sativa
Equisetum spp.
Urtica spp.
Centella asiatica
Mentha piperita
Rubus idaeus
Ulmus fulva
Medicago sativa
Glycyrrhiza glabra

Combine equal parts of the dry herbs, or mix to taste. To optimize mineral extraction, soak the herbs in a saucepan for 4 or 5 hours or overnight using 1 cup (240 ml) of water for each tablespoon of herb blend and adding 1 teaspoon or more of fresh lemon juice or vinegar for every cup of water. Bring the mixture to the lowest possible simmer for 5 to 10 minutes, then let stand for 15 to 20 minutes more, covered, and then strain. Drink 3 to 6 cups (720 to 1,440 ml) per day. With long-term use as a mineral tonic it is recommended to take a few days to a week off every few months.

Nightshades and Joint Pain

Some individuals are sensitive to Solanaceae-family foods and find that regular consumption of potato, tomato, eggplant, tobacco, and peppers causes joint inflammation and pain.[88] These plants contain cholinesterase-inhibiting glycoalkaloids and steroids that in some people can induce inflammatory reactions that worsen with large and repetitive consumption. It is always a worthwhile experiment to try a trial elimination of all such foods for a period of at least 1 month to see whether chronic musculoskeletal pain improves with abstinence. Possible offending compounds include capsaicin in peppers, which paradoxically is often used topically to alleviate pain due to its warming sensation achieved via activity on TRP channels in pain fibers. Other inflammatory molecules in Solanaceae-family plants include nicotine in tobacco, solanine in potato and eggplant, and tomatine in tomatoes. Many Solanaceae family plants also contain vitamin D_3, otherwise quite rare in plants, and excessive daily consumption may cause calcium deposition in the joints and muscles and promote inflammatory reactivity. Susceptible people may be affected by immunoglobulin-mediated hypersensitivity reactions following the ingestion of nightshades. Nightshade sensitivity involves initiation of inflammatory cascades that promotes joint pain and may contribute to the pathophysiology of osteoarthritis, RA, and atopic dermatitis. Those with the HLA-DR$_4$ antigen found on B lymphocytes and macrophages may develop RA from the ingestion of Solanaceae-family plants.

Ayurvedic Tincture for Osteoarthritis

Boswellia and *Curcuma* are a common duo for reducing pain and inflammation in arthritic joints. Clinical trials have shown the combination to compare favorably with cyclooxygenase-inhibiting drugs. This formula blends the classic duo with other widely studied traditional antiarthritic herbs. Variations on this formula may include taking individual capsules of ginger, turmeric, and *Boswellia*, which are usually readily available in the commercial marketplace, and formulating a complementary tincture of the remaining herbs.

Zingiber officinale
Tinospora cordifolia
Phyllanthus emblica
Uncaria tomentosa
Boswellia serrata
Curcuma longa

Combine in equal parts. Take 1 to 2 dropperfuls at a time, 3 to 6 times daily, reducing as pain improves. Continue a maintenance dose long term.

Juice Cleanse for Arthritic Flares

A juice such as this can be useful as part of a cleansing and elimination diet, where all common allergens, sugar, grains, nightshades, processed foods, and food chemicals are avoided for 2 months. It is beneficial to start such a cleanse with a period of fasting, or with a vegetable-only diet for several days. This juice could serve as a breakfast, followed by a raw salad for lunch and steamed vegetables or a soup for dinner. This formula can also be used in place of ginger, turmeric, and bromelain capsules, as part of a long-term protocol aiming to use food as medicine.

6 stalks celery
3 cups (600 g) raw pineapple
2-inch piece fresh ginger
1-inch piece fresh turmeric

Combine all ingredients in a high-speed blender to pulverize, or process in a juice extractor. Drink once or twice per day while avoiding all known or suspected food allergens.

Tincture for RA

In contrast to the formulas for osteoarthritis in this section, this formula features agents noted to have immunomodulating activity for autoimmune-mediated joint inflammation. *Harpagophytum* and *Boswellia*

Topical Considerations for Arthritis

These agents are time-honored topical approaches to alleviating musculoskeletal pain in patients with arthritis and various types of joint pain.

Epsom salts can be added to hot water for a whole-body bathtub soak; repeat on a daily basis.

Liniments and sore muscle rubs can be gently worked into affected joints. Ginger, mustard, and horseradish are all rubefacient and stimulating to circulation, and like cayenne, numbing to pain as well. Wintergreen (*Gaultheria procumbens*) contains salicylates and is temporarily pain relieving, and the essential oil can be added to other massage oils.

Cayenne and capsaicin ointments are effective for temporary pain relief.

Essential oils may be both pain relieving and anti-inflammatory. Essential oils of black pepper, cloves, ginger, lavender, myrrh, mint, rosemary, pine, balsam fir, and wintergreen are among the most widely used aromatic pain-relieving agents. Many are too harsh to use directly on the skin and should be diluted with water or other fixed oil. Such essential oils may also be placed on a hot wet rag or towel and applied topically. Essential oils may also be added to liniments.

Dimethylsulfoxide (DMSO) is a sulfur-containing compound that reduces arthritic pain and inflammation and, like methylsulfonylmethane (MSM) and glucosamine sulfate, may have rejuvenating effects on connective tissue.

are specific for musculoskeletal inflammation, while *Curcuma* and *Tanacetum* are broad-acting anti-inflammatories. This formula contains no nourishing ingredients or trophorestoratives, so is best combined with mineral tonics and bone-building nutrients and teas, as mentioned throughout this chapter.

Curcuma longa	15 ml
Harpagophytum procumbens	15 ml
Tanacetum parthenium	15 ml
Boswellia serrata	15 ml

Take 1 to 2 teaspoons of the combined tincture 3 to 6 times daily, reducing as symptoms improve.

Tincture for RA in the Hands

This formula is based on the herbs recommended in the Eclectic literature for RA, specifically in the hands. It is included here as an example of emphasizing alteratives (*Berberis*, *Curcuma*), lymph herbs (*Phytolacca*), and liver herbs (*Curcuma*) in combination to treat autoimmune arthritis. *Caulophyllum* is recommended in the folkloric literature for heavy, tense, aching musculoskeletal pain, including discomfort in the hands.

Berberis aquifolium
Curcuma longa
Caulophyllum thalictroides
Phytolacca americana

Combine in equal parts, and take 1 to 2 teaspoons, 3 to 6 times daily for several months, continuing long term if effective, reducing to a maintenance dose over time.

Boswellia serrata for Musculoskeletal Pain

Boswellia serrata is a tree of arid locales native to the Middle East, North Africa, and India. Injuring the bark yields an oleoresin known as salai guggul that is used as a traditional medicine for arthritis and musculoskeletal pain, including in formulas for bursitis, tendinitis, osteoarthritis, RA, and trauma. The resin dries into a solid rocky mass that is powdered and dissolved in alcohol, glycerin, oil, honey, and other liquids for oral consumption used as an anti-inflammatory and arthritis medicine. Modern research has investigated the boswellic acids in the resin for over 50 years, and commercial popularity is placing the plant in peril if not thoughtfully grown in a sustainable fashion. *Boswellia* is shown to improve pain in randomized clinical trials of osteoarthritis patients,[89] and improve knee pain, knee flexion, and walking distance, compared to placebo.[90] One study dosing 250 milligrams of *Boswellia* reported arthritic pain to improve as early as 7 days after beginning therapy.[91] Molecular investigations demonstrate *Boswellia* to improve inflammatory mediators in a manner that protects cartilage from enzymatic and inflammatory destruction.[92] The boswellic acids work by many of the same mechanisms as NSAIDs,[93] including inhibition of lipoxygenase.[94] However, while pharmaceuticals can impair GAG synthesis and can induce digestive ulcers and accelerate joint damage, boswellic acids significantly reduce GAG degradation.[95]

The dosage of *Boswellia* for arthritis ranges from 200 to 400 milligrams, 3 to 4 times a day, or 1,000 to 1,500 milligrams in total each day. Isolated and concentrated individual boswellic acids may be dosed at 100 milligrams at a time, several times per day.

Boswellia serrata, Indian frankincense

Tincture for RA Flare-Up

Like the Tincture for RA in the Hands, this formula is based on historical literature wherein lymph and immunomodulating herbs were emphasized for RA more so than anodynes and anti-inflammatories. *Atropa* is used as a synergist when joints are red, swollen, hot, and throbbing. Consume this tincture in a small amount of water to avoid possible irritating effects of *Phytolacca* with this aggressive dosage. This formula can be complemented with topical anodynes, hydrotherapy, and bone nutrients to prevent the joint destruction and fibrosis that may result from acute episodes. **Caution:** *A. belladonna* has potential toxicity and should only be used by licensed clinicians and skilled herbalists.

Phytolacca americana	28 ml
Stellaria media	28 ml
Atropa belladonna	4 ml

Mix ¼ to ½ teaspoon of the combined tinctures in a sip of water, and take every 1 to 3 hours, decreasing as the flare-up abates.

Topical Oil Blend for Arthritic Pain

Arthritic pain can sometimes necessitate long-term use of potentially harmful pharmaceutical anodynes; anything that reduces the need for those drugs is very valuable. This is an example of a formula that may offer pain relief for both RA and osteoarthritis. This formula may be altered to incorporate a person's favorite essential oils, or for best economy. The castor oil base helps draw the oils deeper into the tissue, and the *Hypericum* fixed oil offers connective tissue anti-inflammatory effects. *Cryptocarya aganthophylla* (clove nutmeg) is also known as *Ravensara aromatica*.

Castor oil	1½ ounces (45 ml)
Hypericum perforatum oil	1 ounce (30 ml)
Zingiber officinale essential oil	¼ ounce (8 ml)
Cryptocarya aganthophylla essential oil	¼ ounce (8 ml)
Rosmarinus officinalis essential oil	¼ ounce (8 ml)
Juniperus communis essential oil	¼ ounce (8 ml)
Apium graveolens essential oil	¼ ounce (8 ml)
Syzygium aromaticum essential oil	⅛ ounce (4 ml)
Commiphora myrrha essential oil	⅛ ounce (4 ml)
Cinnamomum camphora essential oil	⅛ ounce (4 ml)

Combine ingredients in a 4-ounce (120 ml) bottle. Rub ¼ teaspoon of the oil gently over painful joints, and cover with heat to enhance the effects.

Roses and Resveratrol for Osteoarthritis

Rosehip powder from the seeds and husks of *Rosa canina* fruits has been used extensively in traditional medicine as a source of vitamin C, and for immunomodulating effects. Modern research has shown rosehips to have a wide variety of anti-inflammatory and antioxidative properties. A meta-analysis of randomized, controlled trials using *Rosa canina* (dog rose) powder for symptomatic treatment of osteoarthritis showed statistically significant reduction in the use of analgesics in the rosehip users compared to those receiving a placebo.[96]

Resveratrol, a polyphenolic phytoalexin in grapes, blueberries, cranberries, red wine, cocoa powder, and peanuts, has numerous anti-inflammatory, immunomodulatory, and antioxidative properties including protective effects on chondrocytes.[97] (See chapter 2 of Volume 2 for more information on resveratrol.)

Rosa canina, dog rose

Arthritis

Sesame Seed Oil for Arthritis

Oil extracted from sesame seeds has long been used in both oral and topical pain relief formulas in Asia and the Middle East. Sesame oil may also reduce complement system activation. Human clinical studies on patients with osteoarthritis of the knee have suggested that oral consumption of sesame oil may improve the outcome of other therapies, as compared with the outcomes of those therapies alone. Use sesame seed oil in smoothies and salad dressings, or in specific medicinal oil blends, as shown below. *Nigella* oil has a strong thyme-like flavor and may be omitted from blends when this flavor is not suitable for the intended use.

Sesamum indicum oil	3 ounces (90 ml)
Nigella sativa oil	½ ounce (15 ml)
Hypericum perforatum oil	½ ounce (15 ml)

Use a few drops on steamed vegetables and foods at the time of serving or use in teaspoon quantities when making salad dressings and sauces.

Encapsulated Herbs for RA

These herbal supplements are readily available and can be used alone, in tandem, or in combination with the herbal tinctures and teas in this section. They may be used to reduce pain, reduce dependence on pharmaceutical anti-inflammatories and immunosuppressants, and protect the joints from permanent damage.

Boswellia serrata capsules
Bromelain capsules
Curcuma longa capsules

Take 2 capsules of each item, 1 to 3 times daily.

Tea for Autoimmune Joint Degeneration

This tea combines herbs noted for healing broken bones and sprained ankles and strengthening weak fingernails and connective tissue. This formula is not highly immunomodulating or antiallergy, and therefore works best as one component of a larger treatment protocol.

Centella asiatica
Equisetum arvense
Glycyrrhiza glabra
Urtica urens

Combine in equal parts, storing extra in a glass jar or plastic bag. Place 3 to 4 tablespoons of the herb blend in a medium saucepan, then cover with a small amount of boiling water and several tablespoons of lemon juice to optimize mineral

extraction. Let stand overnight, or for at least 4 to 5 hours. After macerating, add 3 to 4 more cups (720 to 960 ml) of water and bring to a brief simmer for 10 minutes. Strain and consume at least 3 cups (720 ml) per day, long term.

Huangqi Guizhi Wuwutang Decoction

One modern clinical trial suggested that this classic Chinese herbal formula was helpful in reducing symptoms in RA patients.[98]

Ziziphus jujuba	16 ounces (480 g)
Astragalus membranaceus	8 ounces (240 g)
Cinnamomum cassia twig (Gui Zhi)	8 ounces (240 g)
Paeonia lactiflora	8 ounces (240 g)
Glycyrrhiza uralensis (Gan Cao)	8 ounces (240 g)

Combine the dried herbs to yield a 3-pound (1.4 kg) mixture, which can be stored in a glass jar in a dark, cool location. Prepare the tea by gently simmering 1 teaspoon per cup (240 ml) of water for 10 minutes. Drink 3 or more cups per day. If effective, the formula can be continued long term at a maintenance dose of several cups per day 4 or 5 days a week.

Tincture for Psoriatic Arthritis

Autoimmune conditions like psoriatic arthritis are particularly difficult to treat. This formula utilizes alterative and liver-supportive *Arctium* and *Curcuma* as the base herbs. *Phytolacca* offers lymphatic and immunomodulating activity, and *Tanacetum* is aimed at optimizing the hyperreactive immune component of psoriatic arthritis.

Arctium lappa
Curcuma longa
Phytolacca americana
Tanacetum parthenium

Combine in equal parts and take 1 teaspoon, 3 to 6 times daily.

Tincture for Psoriatic Arthritis

The Eclectic literature of the 1800s mentions the herbs in this formula specifically for psoriasis of the palms and soles of the feet. *Berberis* and *Iris* support liver function, an action important in the treatment of all sorts of skin lesions.

Berberis aquifolium	20 ml
Centella asiatica	20 ml
Stillingia sylvatica	10 ml
Iris versicolor	10 ml

Take 1 teaspoon of the combined tincture, 3 to 6 times daily, long term.

Curcuma longa for Musculoskeletal Pain

Curcuma longa (turmeric) is a common culinary spice used widely in herbal medicine, including for biliary disorders, rheumatism, cancer, obesity, diabetes, nerve pain, kidney diseases, vascular diseases, infections, skin inflammation, and digestive disorders. Turmeric has been extensively researched for systemic immune and antioxidant effects,[99] with many thousands of scientific and clinical studies conducted.[100] The bright yellow flavonoid curcumin, and related compounds referred to as curcuminoids, are credited with numerous medicinal effects, but since the compounds are poorly absorbed by the body, taking the powder with black pepper,[101] fat, and phospholipids is recommended to enhance bioavailability.

Turmeric may help reduce connective tissue inflammation, provide pain relief, and reduce disease progression in osteoarthritis and autoimmune disease, and is shown to halt joint degeneration in animal models of arthritis.[102] *Curcuma* protects against overgrowth of synovial fibroblasts in RA[103] and inhibits the breakdown of cartilage cells and joint destruction via numerous effects on inflammatory mediators.[104]

Among the mechanisms whereby *Curcuma* supports joint regeneration are inhibition of osteoclasts[105] and extracellular matrix degradation,[106] along with limitation of connective tissue demise via progressive fibrosis in autoimmune disease.[107] Due to many positive effects on the cytoskeleton and inhibition of autoantibody binding to connective tissues, curcumin may help treat autoimmune diseases, including Sjögren's syndrome and systemic lupus erythematosus.[108] Human clinical studies on arthritis patients have shown curcuminoids to reduce inflammatory cytokines in a manner that can match or exceed the efficacy of indomethacin[109] and other NSAIDs,[110] and without any significant side effects.[111]

Curcuma longa, turmeric

Anti-Inflammatory Tincture for Connective Tissue

While the Tea for Autoimmune Joint Degeneration provides optimal mineral absorption, this tincture features the less-tasty, systemic anti-inflammatory herbs. *Curcuma* is liver-supportive and an all-purpose anti-inflammatory. *Symphytum* is included for its reputation as a bone- and joint-healing herb. *Harpagophytum* and *Boswellia* are specific for arthritis and joint pain.

Curcuma longa	30 ml
Symphytum officinale	15 ml
Harpagophytum procumbens	15 ml
Boswellia serrata	4 ml

Take 1 to 2 teaspoons of the combined tinctures 3 to 6 times daily.

Tincture for Arthritis in Allergic Constitutions

This formula and the Tincture for Arthritis in Weak Constitutions demonstrate how to tailor autoimmune disease formulas for those with signs of excessive reactivity throughout the body versus those with signs of devitalization and immune depletion. This formula addresses the former, using herbs known to be effective for hay fever, hives, and allergies. It may improve arthritic inflammation associated with excess immune-reactive states in the blood.

Tanacetum parthenium
Ginkgo biloba
Curcuma longa
Scutellaria baicalensis
Angelica sinensis

Combine in equal parts and take 1 to 2 teaspoons, 3 to 6 times daily, long term.

Tincture for Arthritis in Weak Constitutions

This formula is specific for weak, frequently ill, or otherwise devitalized patients, in contrast with the Tincture for Arthritis in Allergic Constitutions. It features herbs classically known to support appropriate immune responses in those who are fatigued, cold, stressed, toxic, and/or frequently ill with slow recovery periods.

Panax ginseng
Zanthoxylum clava-herculis

Smilax ornata
Zingiber officinale
Astragalus membranaceus

Combine in equal parts and take 1 to 2 teaspoons, 3 to 6 times daily, long term.

Tincture for Autoimmune Disorders with Toxicity

As emphasized in the introductory discussion of arthritic disorders, the health of the gastrointestinal tract can have a huge impact on, or even initiate, autoimmune disorders. This formula features alterative and liver-supportive herbs aimed at improving gut health and promoting the removal of wastes and toxins from the body.

Echinacea angustifolia
Achillea millefolium
Silybum marianum
Andrographis paniculata
Schisandra chinensis

Combine in equal parts and take 1 to 2 teaspoons, 3 to 6 times daily, long term, reducing to a lower maintenance dose over time.

Formulas for Gout

Gout is typically characterized by acute arthritic attacks that come on suddenly and are often so painful as to severely restrict mobility. Because the joint pain is due to uric acid accumulation, both dietary changes and support of renal elimination of wastes are needed for repeated episodes, which can lead to permanent damage of both the joints and the kidneys in severe cases. Gout is increasing in prevalence in tandem with the metabolic syndrome and diabetes epidemics, and obesity is associated with both conditions. In addition to being acutely painful, gout is associated with an increased risk of cardiovascular disease and heart failure,[112] attesting to its association with metabolic stress in the body.

Gout is more common in men than women, and the incidence in women has been shown to increase postmenopausally. However, the use of hormone replacement therapy may reduce the risk of gout in women.[113] The consumption of beverages high in fructose is shown to increase the occurrence of gout in women.[114] Coffee consumption may improve renal excretion of wastes,

and one large epidemiologic study showed coffee consumption to reduce the occurrence of gout.[115] Vitamin C supplementation may prevent flare-ups in recurrent episodes of gout.[116] Gout attacks are frequently ascribed to triggering events such as high-purine food intake. Investigations have shown that acute consumption of purines induces a fivefold increase in the risk of an episode in those with recurrent gout. The intake of animal purines is more problematic than plant purines. Those with gout should avoid purine-rich foods, especially those of animal origin.[117] A period of fasting, eating very lightly, or adopting a liquid diet is also highly recommended to reduce the renal burden of eliminating wastes.

Allopurinol has been the primary drug to reduce the incidence of chronic gout from the 1960s through the turn of the century, and only in the following decade did any new pharmaceuticals become available, including those aimed at reducing elevated pro-inflammatory interleukins.[118] In acute situations, pain-relieving anti-inflammatory herbs are appropriate.

Acute episodes typically respond quickly and dramatically to colchicine from *Colchicum autumnale*; however, this medication is a potent immunosuppressant that halts white blood cell–driven inflammatory process and does nothing to correct the underlying ills. Worse, it can have toxic effects if not used within a narrow therapeutic window. The use of the whole *Colchicum* plant may be a safer alternative than synthetic colchicine but it too can hardly be considered holistic or metabolically corrective. Furthermore, whole plant *Colchicum* tincture is hard to find on the market. For these reasons, prescription colchicine might be considered for short-term pain relief in acute episodes while offering other herbs and lifestyle approaches to help prevent chronic episodes of gout.

Chronic gouty arthritis might be better addressed with diet, mineral supplements, and herbs that help the kidneys to excrete wastes, an action referred to as a renal depurant (see "Renal Depurants for Gout" on page 123). For acute episodes, renal depurants might be combined with general anti-inflammatories. Agents known to enhance blood flow to the kidneys, such as *Angelica sinensis* and *Scutellaria baicalensis*, are used in China for gout and other types of arthritis, and research has shown these herbs to reduce uric acid–induced inflammation.[119] *Withania somnifera* has been shown to reduce uric acid–induced inflammation in animal models of gout via a variety of mechanisms.[120] *Phyllanthus emblica* is a traditional renal supportive herb, and animal studies show the plant to help reduce serum uric acid.[121]

Nettle Truffles

Bland high-mineral herbs can be prepared into herbal "bonbons" by preparing a dough with nut butter, coconut oil and flakes, honey, and dried fruit. In this formula, cashew butter may be substituted with peanut or almond butter; honey with molasses, carob syrup, maple syrup, or agave nectar; and raisins with any finely chopped dried fruit. Nettle is the herb incorporated here, but oatstraw and horsetail powders are possible substitutes.

1 cup (50 g) cashew butter
½ cup (25 g) honey
½ cup (25 g) chopped raisins
⅛ cup (6 g) coconut oil
1½ cups (360 grams approximate) nettle powder
coconut flakes or sesame seeds, to coat

Combine the nut butter, honey, raisins, and coconut oil, blending with a fork to create a sticky dough. Mix in a few tablespoons of the nettle powder at a time to create

Renal Depurants for Gout

Diuretic herbs such as *Urtica* (nettle), *Apium graveolens* (celery seed), *Equisetum* species (horsetail), and *Galium* (cleavers) may help the kidneys eliminate wastes and acids from the body. *Urtica dioica* and other species of stinging nettle have been used historically as renal depurants in treating rheumatism and gout. A renal depurant is different than a simple diuretic. Simple diuretics stimulate the removal of water from the body, while renal depurants stimulate the removal of solid wastes from the body. *Urtica* and other high-mineral herbs may enhance the removal of uric acid, crystals, and accumulated wastes. *Urtica* is very nourishing and mineral-rich and is best consumed as a tea on a regular basis for those suffering from chronic episodes of gout. The other diuretics mentioned are also very high in minerals and all may be philosophically considered renal depurants.

a drier dough that can be worked with the hands. The honey and nut butter can be adjusted for desired taste and consistency. Roll the dough into small balls of roughly 1-inch diameter. Roll the balls in finely flaked coconut or sesame seeds, and store in an airtight container. Consume 2 to 4 bonbons per day for a supply of high-mineral herbs and a healthy alternative to other sweets and snacks.

Tincture for Acute Gout

Angelica is chosen in this formula because it may enhance circulation to the kidneys and promote renal excretion of uric acid. *Scutellaria baicalensis* and *Colchicum* are anti-inflammatory and help allay the acute pain of gouty arthritis. *Phytolacca* may help lymphatic clearance of urates, and *Urtica* acts as a renal depurant, helping the kidneys clear metabolic wastes including uric acid. Any of the high-mineral connective tissue teas mentioned in this chapter can complement this tincture. Patients should also be assisted in establishing a low-protein, high-vegetable diet with abundant water intake. This tincture can be continued long term to help prevent recurrences following an acute episode of gout. **Caution:** *Colchicum* is a potentially toxic plant and should be limited to skilled herbalists and licensed clinicians only.

Angelica sinensis	15 ml
Scutellaria baicalensis	15 ml
Urtica urens	15 ml
Phytolacca americana	13 ml
Colchicum autumnale	2 ml (optional)

Take 1 dropperful as often as every 30 minutes for a half a day, reducing to hourly, then every 2 or 3 hours as symptoms improve over the course of 3 or 4 days.

Topical Herbal Paste for Gout

Colchicine from *Colchicum autumnale* is a classic medicine for gout, but it may have harmful side effects when taken orally, long term or improperly dosed. Whole *Colchicum* tincture may also help when applied topically but may be difficult to source.[122]

Colchicum autumnale tincture	5 ml
Capsicum annuum oil	5 ml
Cinnamomum camphora essential oil	5 ml
Zingiber officinale powder	1 to 2 tablespoons

Place the *Zingiber* powder in a small bowl and add the liquids, whisking briefly with a fork. Apply gently to the affected area and leave in place for ½ hour, then rinse or shower off. Repeat 3 or more times a day for several days discontinuing once the flare-up has entirely resolved.

Tea for Chronic Gout

Agents that help remove wastes from interstitial fluid, and especially those that assist the kidneys to clear uric and metabolic wastes from the blood, may deter the recurrence of gout. The kidneys need a supply of minerals in order to carry out their function of excreting wastes. *Apium*, *Urtica*, and *Equisetum* are mineral-rich diuretics, and *Glycyrrhiza* enhances the flavor of the tea. If the kidneys are poorly functioning or circulation is deficient from diabetes or other disease, including a circulatory tonic such as *Salvia miltiorrhiza* or *Angelica sinensis* may add a renal perfusion-enhancing dimension to the formula.

Apium graveolens seeds, crushed (or powder where available)	1 ounce (30 g)
Urtica urens leaf	1 ounce (30 g)
Equisetum arvense	1 ounce (30 g)
Glycyrrhiza glabra	1 ounce (30 g)

Add 1 tablespoon of the combined herbs per 1 cup (240 ml) hot water, and steep for 10 minutes. Strain and drink 3 cups (720 ml) or more per day, long term.

Formulas for Osteoporosis

Osteoporosis, the most common metabolic bone disease in the United States, is defined as a general decrease in the formation of new bone relative to the rate of bone resorption. Although the bony matrix may remain intact, the density of the trabecular bone is greatly diminished. Cortical bone is affected to a lesser degree. Cortical bone is considered non–weight bearing, with trabecular bone being weight bearing and having the ability to absorb impact. The loss of bone density in trabecular bone mostly affects the vertebral bodies and results in spontaneous or trauma-induced compression fractures of the spine. Vertebral compression fractures may be asymptomatic and evidenced only with radiology. Loss of cortical bone most severely affects the metacarpals, femoral head and neck, and distal radius. Heavy metal exposure throughout life damages bone as the individual metal atoms become incorporated in the crystalline structure of the bone matrix. Avoid environmental exposure and contaminated food, because lead, mercury, cadmium, and other heavy metals are very difficult to excrete once established.

The disease is a pathology of older adults, especially females, that results when the skeletal mass begins to decline as part of the aging process. The most rapid time of demineralization occurs at natural or surgical menopause, or upon withdrawal from pharmaceutical estrogens. It is estimated that 10 to 40 percent of women in the United States will experience spontaneous fractures by age 70. Although such fractures are rarely fatal, they are associated with a poor prognosis. Around 30 percent of people will experience life-threatening disease, hospitalization, or death within a year, and 50 to 60 percent will be admitted to convalescent homes or require living assistance. Osteoporosis may progress silently without outward signs or symptoms except backache and a loss of stature. Nocturnal leg

Mushrooms and Seaweed as Mineral Sources

Cultivated mushrooms, including *Agaricus* species, *Pleurotus* species, and *Lentinula edodes,* are good sources of minerals. Growing the mushrooms on special media may increase mineral content even further and decrease the risk of heavy metal and radionuclide contamination. Efforts are underway to amend soil and growing medium with selenium to create selenium-rich mushrooms.[123] Chanterelles, for example, have been found to accumulate a wide range of desirable minerals while excluding aluminum, lead, and other toxins.[124] Other edible mushrooms offer a similar array of minerals and trace minerals. Those with bone thinning or degeneration should be encouraged to include mushrooms in the diet wherever possible. Seaweeds are also excellent sources of minerals and trace minerals, including iodine, copper, selenium, and molybdenum.[125]

Seaweed can be used fresh, dried, or powdered in cooking, macerated in vinegar to use in beverages and dressings, or steeped in broths for use in soups.

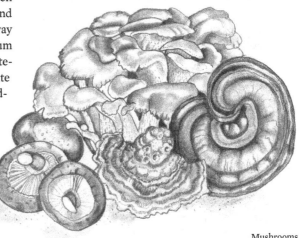

Mushrooms

cramps, abnormal or fragile hair and fingernails, or periodontal disease may hint at poor mineral balance in general in the body.

The bisphosphonate drugs commonly offered to treat osteoporosis do halt the breakdown of bone, but also halt the regenerative, remodeling processes involved with renewing the living crystals of the trabecular bone. These drugs are increasingly being recognized as causing necrosis of the jawbone. Animal studies have revealed that bisphosphonate drugs may cause trabecular irregularities and imperfect mineral and collagen deposition,[126] and most naturopathic physicians do not recommend them. The greatest amount of herbal research regarding osteoporosis has been on the ability of isoflavones to improve bone density. Isoflavones are phytosterol compounds found in many legumes such as *Medicago sativa,* *Glycine max* (soy), and *Pueraria* species. *Lepidium meyenii* has also been shown to help build bone and improve bone density in situations of declining estrogen.[127] Agents that protect the bone marrow and bone cells may also help preserve the regenerative capacity of the bone. Kaempferol is a common dietary and herbal

flavonoid shown to have osteoprotective effects.[128] Improving blood flow through the trabecular bone may be therapeutic to bone density, as may controlling blood sugar and avoiding smoking or other agents that damage vasculature. *Ginkgo biloba,* *Vaccinium myrtillus,* and other herbs that protect the microvasculature and support osteogenesis can be utilized. Animal studies show *Ginkgo* to help protect against steroid-induced osteonecrosis,[129] and the herb may protect mitochondria from heavy metal toxicity and reduce oxidative stress in the bone.[130] Asiaticoside found in *Centella asiatica* improves microcirculation and is a traditional remedy for broken bones; asiaticoside may be included in formulas for osteoporosis. Curcumin, found in *Curcuma longa,* protects the kidneys from heavy metals and may be an osteoprotective herb in formulas for osteoporosis. Additional alternative therapies focus on prevention, exercise, and diet—in particular paying attention to adequate vitamin D and minerals. Kaempferol and related flavonoids may protect against injury from bone-damaging steroids and other drugs, but are not shown to be capable of reversing bone loss in advanced osteoporosis.

Legumes and Isoflavones for Bone Density

Edible legumes, also known as pulses, and leguminous herbs including *Astragalus membranaceus*, *Pueraria* species, *Medicago sativa*, and *Glycyrrhiza glabra* are studied for numerous hormonal and metabolic effects due to their isoflavones, a group of phytoestrogens known to bind to estrogen receptors and to act as agonists or partial agonists, as well as to exhibit antagonist actions. Soy isoflavones are among the most studied. Isoflavones in legumes are metabolized in the intestines and absorbed into systemic circulation, and daidzein from soy is also metabolized into equol in some people. Intestinal metabolism of isoflavones is highly variable, with only one-third of people able to produce the desirable equol, due to differences in intestinal microbes, diet, and history of antibiotic use.[131] Isoflavones bind to α- and β-estrogen receptors on bones and exert a bone-building effect. Isoflavones may also support osteoblasts, induce apoptosis in osteoclasts, and exert an overall anabolic effect on bone.[132] *Medicago*, *Glycyrrhiza*, and *Pueraria* have all been researched to have positive effects on bone density yet they do not promote hormonal cancer or vascular disease, as synthetic pharmaceutical hormone replacement therapy is noted to do. *Actaea* species are a nonlegume source of isoflavones, and may improve osteoporosis, as animal studies have shown *Actaea cimicifuga* (also known as *Cimicifuga foetida*) and *Actaea heracleifolia* to inhibit bone resorption,[133] and have bone-building effect in ovariectomized rats.[134]

Edible legumes

Osteoporosis Risk Factors

Lack of weight bearing exercise, sedentary
   habits; immobilization, invalidism

Smoking

Consuming more than 2 ounces of alcohol
   per day

Coffee consumption of over 5 cups
   (1,200 ml) per day

High-protein diet, especially animal proteins

High-sodium intake

Nulliparity

Because teas and capsules deliver more minerals than tinctures, these types of herbal preparations may be best for treating osteoporosis. Herbal support for this condition must be used long term, and DEXA scans performed every few years can monitor progress. Complementary supplements may include boron, fluoride, strontium, vitamin K_2, and vitamin D. Collagen, glucosamine, hydroxyproline, and other supplements are also appropriate, and may be complemented by the vascular tonics in this chapter.

In TCM philosophy, the health of the bones is related to the health of the kidneys, which maintain mineral status in the body and eliminate heavy metals harmful to the bones. The kidneys are said to "govern the bones" and are foundational to producing the "sea of marrow," which includes the bone marrow and solid substance of the bones. Many traditional renal tonics double as bone marrow tonics, assimilating nutrients and enhancing their delivery into the trabecular bone to be incorporated into the living crystalline structure. Indeed, the kidneys convert vitamin D into calcitriol, its most active form, and both the kidneys and the bones maintain calcium and phosphorus homeostasis in the body. Many kidney diseases are associated with bone damage or an increased risk of osteoporosis. For example, diabetes mellitus, an epidemic in the modern era, is said (in TCM) to involve deficient kidney essence and, as the disease progresses, to involve nephropathy. Calcium–phosphate metabolism is altered and leads to excessive calcification of the vasculature and inadequate mineralization of the bones. Endocrinologists may use the term bone–kidney axis to refer to how calcitriol and parathyroid hormone

(PTH) provide tight control of plasma calcium levels by regulating intestinal and renal uptake balanced against skeletal release of calcium and phosphate. Vitamin D_3 alone is an inadequate therapy for treating bone disease, especially when renal function is poor. Physicians may offer calcitriol and parathyroid hormone in such cases, and herbalists offer herbs that improve microcirculation to both the kidneys and the bones and protect both tissues from oxidative stress and destruction. Research has also shown the fibroblast growth factors play important roles in regulating bone growth and are secreted by osteoblasts and osteocytes, which in turn help to modulate homeostasis of calcitriol and phosphate via the kidneys.[135] Many herbs that are renoprotective may also have osteoprotective effects, and herbs that enhance renal microcirculation are likely to improve trabecular circulation as well. Examples include *Pueraria*, *Salvia miltiorrhiza*, *Ginkgo*, and high-flavonoid herbs.

Decoction for Kidney Essence Deficiency

Shen-An is a TCM formula to treat kidney essence deficiency, and is used to help the kidneys retain vital substances and excrete metabolic wastes and toxins. This formula is included here because these herbs are also noted to support osteogenesis via a number of mechanisms, including antioxidant and hormonal supportive effects. *Epimedium*, which contains icariin, is a kidney yang tonic with osteoprotective effects. The whole herb may exert broad hormonal actions as well. *Astragalus* is a nutritive and immunomodulating tonic whose saponins, such as the astragalosides, are shown to have broad anti-inflammatory actions. Chinese rhubarb, *Rheum officinale*, has numerous antifibrotic effects and is most noted for its ability to help clear inflammatory toxins from the body.

Epimedium brevicornu	8 ounces (225 g)
Astragalus membranaceus	8 ounces (225 g)
Rheum officinale	4 ounces (112 g)

Combine the dry herbs and store in glass jars in a cool, dark cupboard. Gently simmer 1 teaspoon of the herb blend per 1 cup (240 ml) of water for 10 minutes, then strain. Drink 3 cups (720 ml) or more per day, long term. Additional herbs, such as licorice, cinnamon, or ginger, may be added for flavor.

High-Mineral Capsules for Osteoporosis

Herbs that have been shown to improve osteoporosis in the medical literature include *Glycine max* (soy),

Phytic Acid, Phytase Enzymes, and Mineral Absorption

Phytic acid, also known as phytate or inositol hexaphosphate, is considered to be an antinutritional factor that influences the bioavailability of essential minerals by forming complexes with them and converting them into insoluble salts. Phytates occur in plants and potently inhibit the absorption of iron, zinc, and calcium. Phytates are the highest in grains (100 milligrams per 100 grams) and legumes (approximately 600 milligrams per 100 grams). Starchy tubers contain less than 20 milligrams per 100 grams.[136] Phytates are present in soy milk and can interfere with absorbing the minerals in the milk or in foods consumed at the same time.[137] A high-fiber diet, while desirable, will inevitably increase phytate consumption.[138] High-fiber foods supply minerals, but may actually result in a net mineral loss due to the fact they also are high in phytates.[139] One group of researchers concluded, however, that significant amounts of wheat bran (16 grams) did not impair mineral absorption except for zinc.[140] Although phytates interfere with mineral absorption, they also offer benefits, including reduced tendency to renal calcification, hypolipidemia, and possible anticancer effects.[141]

Each phytate molecule has multiple binding sites for calcium ions, and forms insoluble tricalcium and tetracalcium phytate salts.[142] Phytic acid can impair mineral absorption—particularly of iron by as much as 48 percent, zinc by 62 percent, and copper by 31 percent—and this, over time, will also reduce blood levels of important minerals. Iron differs from other minerals in that there is no physiologic mechanism for excretion,[143] so dietary intake and absorption determine iron levels in the body. Phytic acid also decreases crypt depth of intestinal villi in the jejunum in animal studies, and reduces sodium-dependent glucose transporter gene expression in the duodenum.

Phytase enzymes may help reverse these effects.[144] Fermented foods may help inactivate phytates by introducing phytase-producing bacteria into the body. Pulque is a traditional Mexican beverage produced by fermenting *Agave* sap, or aguamiel, with microorganisms present in the environment including several lactic acid bacteria and yeasts such as *Saccharomyces cerevisiae*. Moderate pulque intake appears to increase the bioavailability of iron and zinc bound by phytate in corn.[145]

Actaea, *Pueraria*, and *Medicago*. In the folkloric tradition, *Equisetum* and *Centella* are used to support general connective tissue regeneration. I make these capsules for my patients and my apothecary because nothing similar exists on the commercial marketplace. I purchase organic freshly ground herb powders and have them professionally encapsulated. I simply call them "mineral caps," and consuming these herbs as powders delivers more minerals than would tinctures or teas.

Medicago sativa
Equisetum arvense
Centella asiatica

Obtain powdered herbs and encapsulate professionally or in small batches by hand using an encapsulating stand or simply tapping them full. Empty gelatin and vegan gel capsules can be obtained from most herb wholesalers.

Take 2 or 3 capsules of each herb, each day—6 to 9 capsules per day.

Tea for Osteoporosis

While coffee may promote the loss of calcium and minerals when consumed in excess, herbal teas can provide important minerals. For the most effective approach, this tea should be combined with herbal capsules and other supportive therapies.

Medicago sativa
Pueraria candollei var. *mirifica*,
　Pueraria montana var. *lobata*
Mentha piperita
Rubus idaeus
Avena sativa
Trifolium pratense

Equisetum arvense
Centella asiatica
Glycyrrhiza glabra

Combine in equal parts, using more mint and licorice for flavor if desired. Bring 1 tablespoon of the herb blend per 1 cup (240 ml) of water to a gentle simmer for 1 minute, then cover the pan and steep 10 to 15 minutes more. Strain and drink freely, as much as desired, consuming 2 or 3 cups (480 or 720 ml) throughout the day, 4 to 5 days per week, long term.

Beetroot Juice for Bone Health

Beets increase nitric oxide levels and are shown to improve general circulation. Beet juice consumption has also been found to increase comfortable walking distance in those with peripheral artery disease. Beet consumption may protect the kidneys and the sensitive hematopoietic cells of the bone marrow from nephrotoxins.[147] Beets contain many powerful antioxidants, including betalains, which have been shown to have a protective effect on osteoblasts and to upregulate osteogenic gene expression via a number of signaling pathways.[148] Beet juice, kvass, and powder are all readily available.

Beet juice or powder

If using powder, stir 1 teaspoon to 1 tablespoon into 2 cups (480 ml) liquid or blender beverages and smoothies

Digestive Support for Osteoporosis

Digestive ailments may impair the absorption of calcium,[146] as well as many vitamins and minerals. Optimize digestion with bitter herbs and alterative agents and if necessary, supplement with digestive enzymes and hydrochloric acid. The use of *Rumex* tea, lemon juice, or herbal vinegars in water before meals is a basic start. The long-term use of antacids and the popular proton pump inhibitors (PPIs) are increasingly being realized to interfere with the absorption of vital nutrients and contribute to osteoporosis. Get off of such medications as fast as possible and work on improving digestion.

to taste. Drink 1 to 3 cups (240 to 720 ml) per day, long term. Drink 1 to 2 cups (240 to 480 ml) prior to workouts or exercise. For those with peripheral artery disease, drink 1 to 2 cups (240 to 480 ml) before simple walking as exercise.

Formulas for Paget's Disease (Osteitis Deformans)

Paget's disease, also referred to as osteitis deformans, is a chronic skeletal disease in which regions of the bone become hyperactive and result in deformed and enlarged areas of poor quality, weakened bone. Paget's disease involves an excessive activity of osteoclasts. As the condition progresses, the deformities can lead to crippling and abnormal gait. When the condition occurs in the skull, the head may become deformed and cause damage to the bony ossicles of the ears, which can result in deafness. The condition may be painless at onset but progress to involve moderate aching and then severe pain. The pain is typically worse at night, and fractures can occur spontaneously. Herbal pain management strategies may be helpful, with formulas similar to those offered in the section of this chapter on arthritis. Fluoride supplementation may help improve the synthesis of

new bone, and one study noted fluoride to help relieve bone pain and decrease spontaneous fractures.[149] *Glycyrrhiza glabra* is one of the richest known plant sources of organic fluoride. Herbs that promote basic osteoclastic activity and reduce aberrant inflammation and immune reactivity are logical to support bone cell regeneration, reduce inflammation, and help mitigate pain, but the existing herbal research is mostly molecular, and there have not yet been clinical trials using herbal medicines to treat Paget's disease. Salmon calcitonin at a dose of 100 IU subcutaneously may heal osteolytic lesions and be pain relieving when given daily; treatment may be followed by a maintenance dose of 50 to 100 IUs three times weekly. Bisphosphonate pharmaceuticals may be a more powerful option for Paget's disease, but they are controversial due to numerous side effects, some quite serious, and most naturopathic physicians avoid them or consider them a last resort.

Tincture for Osteitis Deformans

Pueraria is included here to support both the bones and blood vessels. *Petasites* and *Actaea* are included to reduce inflammation, allay pain, and further support bone growth and density.

Pueraria candollei var. *mirifica*
Petasites hybridus
Actaea racemosa

Combine in equal parts, and take 1 to 2 teaspoons, 3 to 6 times per day, long term.

Tea for Osteitis Deformans

This is a supportive tea, combining herbs that are general anti-inflammatories and offer many organic minerals— including fluoride and trace minerals—to assist normal cell turnover in the bones. *Apium* is anti-inflammatory and contains an impressive amount of minerals. *Medicago* provides isoflavones and trace amounts of vitamins D and K. *Ulmus* and *Althaea* provide GAG-like molecules that may support bone synthesis. *Glycyrrhiza* contains safe amounts of fluoride, as well as isoflavones. It also contributes to the tea's flavor.

Centella asiatica
Medicago sativa
Apium graveolens
Glycyrrhiza glabra
Ulmus fulva
Althaea officinalis

Combine in equal parts or to taste. For best results, soak the *Ulmus* and *Althaea* separately in cold water overnight. Add the remaining herbs to the macerate and steep 1 tablespoon of the mixture per 1 cup (240 ml) of hot water for 10 minutes. Drink a minimum of 3 cups (720 ml) per day, long term.

Tincture for Paget's Pain Management

For cases of osteitis deformans where there is significant pain, the following herbs may be helpful. This is an example of a formula based on folkloric specific indications, where no modern research is available to guide herb selection. *Eupatorium* and *Ruta* have a folkloric reputation of improving a deep aching sensation in the bones, but have been little-studied in modern times. *Petasites* helps many types of difficult and nociceptive pains, and both *Salix* and *Stillingia* are noted in the historical literature for bone pain and inflammation.

Eupatorium perfoliatum
Salix alba
Petasites hybridus
Stillingia sylvatica
Ruta graveolens

Combine in equal parts. Take 1 to 2 teaspoons, 3 times per day to hourly, as needed.

Therapies for Herniated Intervertebral Discs and Sciatica

Degeneration of intervertebral discs is a major cause of chronic low back pain. The treatment of herniated discs can be extremely conservative (such as bed rest) or may be aggressive (such as spinal fusion) depending on the severity of the herniation, the degree of pain and disability, and the health and condition of the vertebral column in general. Surgical repair techniques may preserve spinal motion, but fusion results in permanent loss of motion in the spinal column. Newer techniques aimed at regenerating the discs utilize injections of hyaluronic acid,[150] immature chondrocytes,[151] and stem cells[152] show great promise, as does gene therapy. Many patients who undergo surgical therapies may still experience chronic low back pain within a few months to a few years after the procedure,[153] as the surgery itself may accelerate further degeneration.[154] Therefore, herbal therapies may be a wise complement for anyone undergoing surgical interventions. Aim to optimize health, improve metabolic function, support digestion, and reduce oxidative stress as much as possible prior to surgical interventions to reap the best possible outcome.

Intervertebral discs are avascular and heal slowly as sulfated proteoglycans diffuse from blood vessels at the disc margin into the surrounding matrix and reach chondrocytes to regenerate cartilage. Glucosamine sulfate and related compounds are helpful in treating arthritis and may also help repair damaged intervertebral discs, yet carry none of the side effects or risks associated with the use of steroids or NSAIDs.[155] There is little evidenced-based herbal therapy for repairing intervertebral discs, but therapies aimed at regenerating bone and connective tissue are logical, in the absence of supportive research. *Hypericum perforatum* is specific for various types of nerve pain[156] and may be included in formulas to promote healing.

Tincture for Leg Pain with Sciatica

Sciatic pain can be excruciating and keep people immobilized for days. When sciatic pain is due to a severely herniated disc, surgical intervention may be required to prevent muscle atrophy and permanent loss of function. This formula can provide comfort for those recuperating in bed or awaiting or recovering from surgery. Topical applications and Epsom salt soaks can complement this tincture. All three herbs in this formula are specifically recommended in traditional literature for neuritis. Minute doses of 1 to 2 milliliters of *Conium maculatum*, *Aconitum napellus*, or *Gelsemium sempervirens* may be added to the tincture to deaden nerve pain. **Caution:** All three of these herbs are by prescription and for use by experienced clinicians only.

Hypericum perforatum	30 ml
Zanthoxylum clava-herculis	15 ml
Piper methysticum	15 ml

Take ½ to 1 teaspoon of the combined tincture every 30 to 60 minutes, reducing as symptoms improve.

Hypericum Simple for Sciatica

Sciatic nerve injury can result in debilitating neuropathic pain when excessive generation of reactive oxygen species perpetuates inflammation and pain, which is slow to subside. *Hypericum* may help speed resolution, with specific antioxidant effects on neurons and the dorsal root ganglion involved with sciatic pain. Excessive release of calcium ions can result from inflammation, mechanical injury, and ischemia and contribute to neuropathic pain. *Hypericum* may limit the release of calcium, and deter inflammatory cascades and painful activation of TRP channels on the sciatic nerve.[157] Both topical *Hypericum* oils or liniments and oral *Hypericum* tincture or encapsulation may be helpful, and can be complemented by Epsom salt soaks and bed rest.

Hypericum perforatum tincture
Hypericum perforatum oil

Take 1 to 2 dropperfuls of *Hypericum* tincture as often as hourly, reducing gradually over several days' time and continuing at a lower dose of 1 dropperful, 3 or 4 times daily for 1 month. The oral tincture may be complemented with *Hypericum* oil. Rub into the affected area hourly, covering with heat. Reduce the frequency of application as pain subsides. Because *Hypericum* is photosensitizing at the high dosage used here, avoid direct sunlight. Gentle walking and stretching following hot Epsom salt baths is encouraged, but otherwise the body should be rested to encourage healing and repair.

Specific Indications: Herbs for Musculoskeletal Conditions

Drawing from both folkloric traditions and modern molecular and clinical research, the following descriptions feature herbs that offer anti-inflammatory effects for joint pain and trauma, trophorestorative effects on connective tissue, and immunomodulating effects for autoimmune disorders. Rounding out this section are listings of complementary herbs that offer alterative and depurant actions to reduce inflammatory triggers as well as high-mineral plants to support bone density.

Achillea millefolium • Yarrow

Young leaves and flowers of *Achillea* have alterative, cholagogue, and antimicrobial properties, and may be included in formulas for musculoskeletal complaints of an infectious origin such as osteomyelitis or infectious

arthritis, and in other formulas as an alterative and cholagogue adjuvant. *Achillea* is one of the ingredients in the homeopathically prepared Traumeel ointment for musculoskeletal trauma, and is a folkloric remedy for cuts, to control acute bleeding following trauma, and to treat skin and other infections.

Aconitum napellus • Aconite

Caution: *Aconitum* is a very powerful herb with toxic potential to be used by experienced clinicians only. *Aconitum* roots have a paralyzing effect on nerves but may sometimes be included in formulas to help address acute pain and neuralgia. Folkloric literature emphasizes *Aconitum* in the initial stages of infectious and inflammatory processes when there is a sense of urgency or fear, restlessness, increased blood flow, and hyperemia in the affected part, such as a bursa or joint, and increased sensory nerve irritation such as with gout. Aconite can be applied topically for these same symptoms, but bear in mind that the alkaloids are readily absorbed, and can suppress the heart and other nerve centers as readily as they can inhibit local sensory nerves. Arthritic pain, myalgia, gout, and neuralgic pain may all respond to small doses of *Aconitum*.

Actaea racemosa • Black Cohosh

Actaea roots are prepared into tinctures and encapsulated products specifically indicated for restlessness and cramping pain, aching stiffness, and for jerking and choreic motions in the muscles. *Actaea* can be included in formulas for a sensation of heaviness, aching, and tightness, for fibromyalgia, and for muscle stiffness and joint pain associated with menopause or osteoporosis. This herb is also known as *Cimicifuga racemosa*.

Aesculus hippocastanum • Horse Chestnut

Aesculus seeds are processed into tinctures to include in formulas for osteomyelitis associated with circulatory insufficiency and vague backaches due to circulatory congestion or associated with lower bowel symptoms and hemorrhoids. By relieving portal congestion, *Aesculus* may also improve aching pain in the legs due to venous stasis and varicose veins. *Aesculus* is specifically indicated for portal congestion with tenderness in the right upper quadrant. Aesculus seeds are also processed in oil to produces salves and leg rubs for varicosities.

Aloe vera • Aloe

The mucilaginous pulp of the leaves of various succulent *Aloe* species may be used internally to support repair and regeneration of epithelial and connective tissues. *Aloe* is also high in allantoin, noted to enhance wound healing by speeding cell proliferation. Since *Aloe* has been noted to increase GAGs and hyaluronic acid formation, it might be useful for all manner of tissue, muscle, and bone laxity, from osteoporosis and osteogenesis imperfecta to dermatomyositis and polymyositis. *A. barbadensis* is a related species that may be used interchangeably. *Aloe* polysaccharides are being used as biomaterials to bioengineer human tissues including bone and connective tissues. Acemannan is one of the most studied polysaccharides in *Aloe vera* and is approved by the US FDA for the treatment of wounds and alveolar osteitis.[158] The use of *A. vera* powder on

tooth sockets following dental extractions is shown to prevent bone loss and support healing.

Ananas comosus • Pineapple

Pineapple fruits contain a mixture of sulfur-containing proteolytic enzymes known as bromelain. Taking bromelain supplements has an anti-inflammatory effect, as does consumption of whole pineapple, and these can be offered for arthritis, traumatic injuries, and inflammation in musculoskeletal tissues. Bromelain favors the production of the prostaglandin E series and has been shown to be fibrinolytic. Bromelain has been found to prevent the breakdown of curcuminoids, key constituents in *Curcuma*, and many traditional formulas use the two in combination. Bromelain is considered safe pre- and postsurgery and may be considered for patients undergoing hip or knee replacement or musculoskeletal surgery to reduce swelling, bruising, and pain and speed healing. Bromelain may be included in pain-management protocols for osteoarthritis[159] and may help reduce the need for pharmaceutical analgesics.

Andrographis paniculata • King of Bitters

The bitter-tasting leaves of *Andrographis* are a long-standing folkloric remedy for chronic inflammatory states including degenerative musculoskeletal disease related to aging and autoimmune-driven arthritic pain such as RA. The labdane diterpenoid andrographolide is one of the most widely studied constituents and shown to be rapidly absorbed and nontoxic and to have a variety of well-documented anti-inflammatory effects, including the ability to reduce inflammatory mediators commonly elevated in RA.[160] *Andrographis* may serve as a complementary herbal ingredient in formulas for musculoskeletal pain and stiffness due to underlying infectious, inflammatory, or toxicity states.

Angelica sinensis • Dong Quai

Angelica is specific for muscle discomfort due to vascular congestion, pelvic stagnation, menstrual cramps, or allergies. *Angelica* species may be included in formulas for musculoskeletal complaints associated with allergic phenomena and vascular hyperreactivity. *Angelica* is traditionally regarded as a blood mover in TCM, and enhancing circulation to the bones may be one mechanism whereby *Angelica* improves bone density, healing capacity, and inflammatory processes in the bone. *A. sinensis* polysaccharides may protect cartilage from anti-inflammatory damage and improve new GAG synthesis in chondrocytes.[161] *Angelica* may be included in formulas for osteoporosis and to support healing following joint replacement surgery. *Angelica* supports restoration of trabecular and cortical bone in animal models of osteoporosis and osteoarthritis. Animal studies have shown *Angelica* to improve bone density in ovariectomized rats equally to estradiol.[162] The molecular constituent ligustilide is credited with anti-inflammatory and protective effects on bone cells.[163]

Apium graveolens • Celery

Use *Apium* seed preparations to reduce allergic reactivity in cases of allergy-related muscle pain and for anodyne effects for general arthritic pain. *Apium* is also high in minerals and may both provide nutrients to build the bone and connective tissue, and assist the kidneys in eliminating wastes in cases of gouty arthritis. *Apium* seeds may be included in formulas for osteoporosis and connective tissue fragility as nutritional agents, and in tincture or tea formulas for gout to assist in the elimination of uric acid. For the best mineral content, consuming *Apium* powder, thorough decoctions of seeds, or fresh celery juices are superior to tincture.

Arctium lappa • Burdock

Include *Arctium* roots in formulas for autoimmune disorders to help the liver eliminate potentially pro-inflammatory substances. *Arctium* may be used for rashes and eruptions on the extremities, and limb pain that radiates downward to the hands and feet, and fingers and toes. Also consider *Arctium* for those with poor digestion, to improve the absorption and utilization of nutrients and minerals, and to support repair and regeneration of bones and connective tissues. *A. lappa* constituents inhibit several inflammation processes, such as inducible nitric oxide synthase (iNOS), pro-inflammatory cytokines TNF, and IL-6 expression, helping to explain the inclusion of the alterative herb in traditional formulas for arthritis and musculoskeletal pain.[164]

Arnica montana • Leopard's Bane

The young leaves and flowers of *Arnica* have traditionally been used to prepare oils and salves for topical use. Diluted tinctures and homeopathics are used orally for bruising, musculoskeletal trauma, and soft tissue injury, as well as complaints related to muscle strains and overuse.[165] *Arnica* is not recommended for topical use over abrasions, lacerations, or open skin wounds, and is more specific for bruising, strains, sprains, swellings, and injuries where the dermis is intact. *Arnica* ointments may be included in protocols for the management of RA and connective tissue disorders. Animal models of collagen-induced arthritis show *Arnica* to protect against joint deformation associated with a decrease in nitric oxide, TNF-α, and IL-1β, IL-6, and IL-12. *Arnica* application is also noted to decrease anti-type II collagen antibodies and improve antioxidant status in the underlying tissues. One human investigation compared *Arnica* and ibuprofen for pain relief in patients with osteoarthritis of the hands and reported equal efficacy.

Atropa belladonna • Deadly Nightshade

Caution: *Atropa belladonna* is a potentially toxic botanical for use by experienced herbalists only and used in small drop to single milliliter doses at any one time. Belladonna is used homeopathically, or in small doses botanically, for skin that is dry, hot, and bright red, such as with streptococcal infections and erysipelas. Belladonna is specific for sudden onset of high fever with marked vascular congestion, with restlessness and throbbing or burning sensations. Belladonna is also specifically indicated for hyperesthesia over injured tissue, and for musculoskeletal complaints associated with shooting pains, twitching, and tremors, and for musculoskeletal spasms and tics. *A. belladonna* is also indicated for acute gout or acute flares of RA, where the joints are swollen, red, and hot, or for acute cellulitis when there is red streaking up the arm or legs. Also consider *A. belladonna* for musculoskeletal complaints associated with fever and heat; infection and delirium; vascular engorgement; and restlessness, heat, and burning sensations.

Avena sativa • Oats

The ripe groats (hulled kernels) of *Avena* are indicated for nervous exhaustion and debility following chronic disease and digestive derangements and also for mental, emotional, and physical weakness. The dried leaves of oats are referred to as oat "straw." The young, juicy seeds referred to as "milky" oats are highly nutritious and are emphasized for nervous exhaustion. Oat straw tea may be used to provide nutrients and minerals; it is best prepared with long maceration and gentle simmering with a bit of vinegar or lemon juice added in the water. Consider *Avena* in its various forms for those with muscle weakness, loss of muscle mass, and laxity of joints and tissues, related to poor nutritional states.

Berberis aquifolium • Oregon Grape

Include *Berberis* root bark in formulas for rheumatic pains, especially in shoulders, arms, and hands, and for swelling of the finger joints. *Berberis* is helpful as an adjuvant support herb in musculoskeletal formulas for liver congestion and slow digestion, intestinal dysbiosis, skin eruptions due to poor liver and digestive health, coated tongue, and dyspepsia. Berberine in *Berberis* and its relatives may also improve actual osteoarthritis via anti-inflammatory effects and attenuating interleukins in a manner that offers a protective effect on chondrocytes. Animal studies show berberine to increase levels of proteoglycans in cartilage matrix and the thickness of articular cartilage.[166] *B. aquifolium* is also known as *Mahonia aquifolium*.

Boswellia serrata • Indian Frankincense

Due to the popularity of the gum-resin obtained from incising the trunks and main branches of *Boswellia* trees, the species is becoming threatened. Increased cultivation and conservation is needed. *Boswellia* sap is commonly prepared into proprietary encapsulations aimed at treating arthritis. Boswellic acids are shown to inhibit lipoxygenase, a key enzyme for synthesizing pro-inflammatory leukotrienes. They also inhibit leukocyte elastase, an enzyme that can hydrolyze collagen and elastin of the extracellular matrix and play a role in the progression of osteoarthritis.[167] Numerous other anti-inflammatory effects are identified, and various combinations of boswellic acids have demonstrated antiarthritic effects. *B. sacra* and *B. frereana* are also used medicinally. For more information, see "*Boswellia serrata* for Musculoskeletal Pain" on page 118.

Calendula officinalis • Pot Marigold

Unopened *Calendula* flower buds or petals pulled from young flowers are a traditional remedy for wound healing and often prepared into oils and salves. Include *Calendula* in musculoskeletal formulas when there are chronic infections and poor healing. *Calendula* may help muscles heal from trauma or surgical procedures when taken internally. *Calendula* may improve microcirculation in the dermis when used topically and can be considered to slow the progression of collagen vascular diseases and other connective tissue diseases. It can also be considered to prevent fibrotic degeneration in autoimmune disorders or in musculoskeletal complaints associated with poor circulation and a tendency to cellulitis or infections.

Capsicum annuum • Cayenne Pepper

The ripe fruits of various species of *Capsicum* are used to extract capsaicin, a resin with warming anodyne effects on musculoskeletal pain with topical application. *Capsicum* tincture, glycerin, or syrup may also be included in formulas for infections, including streptococcal infections, especially when associated with chilliness and mucous congestion. *Capsicum* tincture may be diluted with glycerin and taken orally to promote the healing of extensive ecchymosis. *Capsicum* is specific for leg pain and sciatica that is worse bending or coughing. See also chapter 4 of Volume 4 for details on the extensive research on capsaicin that explains cayenne's traditional use in the treatment of neuralgic, traumatic, and musculoskeletal and arthritic pain.

Caulophyllum thalictroides • Blue Cohosh

The roots of *Caulophyllum* entered the herbal *materia medica* from North American native traditional medicine, in which it was used for menopausal symptoms and to treat musculoskeletal inflammation and rheumatic pain. Folkloric writings emphasize *Caulophyllum* for tight, tense, drawing pains and for stiffness in the small joints of the fingers, toes, wrists, and ankles. *Caulophyllum* has not yet been extensively researched but extracts are shown to reduce lipopolysaccharides, nitric oxide, TNF-α, IL-1β, and IL-6, all of which are typically elevated in RA and general inflammatory processes.[168] Triterpene saponins, including caulosaponin, are credited with immunomodulating and anti-inflammatory effects. **Caution:** *Caulophyllum* alkaloids may disrupt mitochondrial respiration cycles and this has caused the herb to fall out of favor among midwives, as its use to induce labor has been associated with perinatal stroke, acute myocardial infarction, congestive heart failure, multiple organ injury, and neonatal shock.[169]

Centella asiatica • Gotu Kola

Centella grows in tropical swampy areas and in some regions the young leaves are juiced or eaten as vegetables, and the whole plants are prepared into oral and topical medicines. *Centella* promotes healing of various types of wounds and traumatic injuries; it contains triterpenoid saponins including asiaticoside, madecassoside, asiatic acid, and madecassic acid that are credited with vulnerary effects, reducing excessive fibrosis and protecting collagen in arthritic disorders. Following injury, consider *Centella* to improve connective tissue integrity and repair, and *Centella* may be included as a powder in a smoothie or other food blend to use long term for those with arthritis and osteoporosis to help support osseous regeneration. Asiatic acid may support the differentiation of mesenchymal stromal cells in bone marrow into osteoblasts and improve osteogenesis.[170]

Cinnamomum verum • Cinnamon

Cinnamon bark is an ancient herbal medicine used extensively in Ayurvedic medicine and TCM for pain and inflammation. Polyphenols in the bark of cinnamon trunk and branches have many anti-inflammatory actions and the bark may be included in formulas for arthritis, metabolic disorders with muscle stiffness, and chronic pain and inflammatory disorders.[171] Cinnamon also has immunomodulatory effects in arthritic disorders. Consider cinnamon in formulas for osteomyelitis secondary to diabetes and circulatory insufficiency. *Cinnamomum sieboldii* is another Asian species used as an anti-inflammatory in Japan. *C. zeylanicum* is a synonym for *C. verum*.

Cnicus benedictus • Blessed Thistle

Cnicus is an alterative herb in the same family as *Silybum marianum* (milk thistle), and it may be used for musculoskeletal complaints related to poor digestive and hepatic function. *Cnicus* has not been extensively studied, but is known to contain alkaloids, mucilage, polyacetylene, triterpenoids, lignans, flavonoids, tannins, phytosterines, and volatile oils, as well as benedictin and cnicin. Cnicin is well known as a galactogogue and has other medicinal properties.

Cnidium monnieri • Snow Parsley

Cnidium medications are used orally to reduce allergic tendencies in hyperreactive inflammatory conditions, and a variety of anti-inflammatory effects may have a protective effect on bone. *C. monnieri*, also called she chuang zi, is commonly used in TCM to improve bone strength. Coumarins in *Cnidium* species are credited with anti-inflammatory and hormone-modulating effects and may support bone health by promoting osteoblast proliferation. One specific coumarin, osthol, is shown to promote osteoblast proliferation and inhibit bone resorption by decreasing the formation, differentiation, and activity

of osteoclasts.[172] Imperatorin and bergapten are other coumarins that have shown similar bone-building effects, inhibiting osteoclastic resorption and promoting osteoblastic proliferation.

Colchicum autumnale • Autumn Crocus

Colchicine, an alkaloid in *Colchicum*, is used pharmaceutically to halt white blood cell responses in cases of acute gout. *Colchicum* is rarely used botanically, but may be considered in homeopathic or very diluted preparations for acute muscle, joint, and bone pain. **Caution:** The plant has potential toxicity, so it is used in specially prepared medications by skilled clinicians and by prescription only. *Colchicum* is specific for pains that are excruciating to the touch, for tearing stinging pains, for the sensation of pins and needles, and for musculoskeletal pain that is worse in the evening and with warm temperature. The topical application of colchicine may be considered following knee surgery to prevent adhesions.[173] It can also be used in foot soaks for gout or liniments for contractures. *Colchicum* is discussed in greater detail in chapter 4, Volume 4.

Coleus forskohlii • Coleus

Coleus is a cellular and metabolic stimulant used in traditional medicine for muscle weakness, atony, sluggish metabolism, and inflammation. Forskolin, a diterpene from *Coleus* roots, has direct effects on adenylate cyclase, an enzyme that activates cyclic adenosine monophosphate, or cyclic AMP (cAMP). This enzyme is sometimes referred to as the "second messenger" because it transmits signals from the cell membrane inward. Fat and muscle cells are especially dependent on cAMP, and forskolin may enhance energy utilization by muscle cells and support weight loss, metabolism, and improved muscle function in deficiency states. *Coleus* may be used to support dietary efforts and exercise in patients with diabetes, hormonal insufficiency, and metabolic insufficiency.

Commiphora mukul • Guggul

The small species of *Commiphora* trees are a source of steroid-rich gum resin, known in Ayurvedic medicine as guggul. The plant sterols in the resin are referred to as guggulsterones and have been used for centuries in India to treat arthritis. *C. mukul* is specifically indicated for elevated lipids due to diabetes, hypothyroidism, slow metabolism, and a sense of weight or dragging in the pelvis. Consider *C. mukul* in musculoskeletal formulas where muscle weakness or discomfort are associated with metabolic syndrome or metabolic disorders. Guggulsterones are shown to have anti-inflammatory effects via inhibition of NF-κB signaling and may be included in arthritis formulas.[174]

Conium maculatum • Poison Hemlock

Conium is used in very small drop dosages to allay intolerable neuralgic pain. The topical use of *Conium* may allay the pain of cancerous growths and ulcers or help relieve dental pain. The alkaloid coniine has analgesic and anti-inflammatory affects.[175] **Caution:** *Conium* is a potentially toxic botanical for use by experienced clinicians only. See chapter 4 of Volume 4 for more information on molecular research about *Conium*.

Cordyceps sinensis • Caterpillar fungus

Cordyceps is an entomopathogenic fungus native to high mountainous regions of Asia. Due to its relative rarity and expense, cell cultures of *C. sinensis* have been developed to synthesize the molecular constituents. *Cordyceps* is credited with ergogenic activity and has been used by endurance athletes to improve performance and endurance. *Cordyceps* is noted to increase hemoglobin levels and reduce muscle fatigue.

Crataegus species • Hawthorn

Both the ripe berries and the young flower buds and leaves of *Crataegus* are used dry in teas and to prepare a variety of medicinal extracts. Include *Crataegus* as a support herb in formulas for bruising and musculoskeletal trauma and to improve microcirculation in the tissues. *Crataegus* is specifically indicated for easy contusions with slow recovery and for connective tissue disorders accompanied by vascular inflammation and allergic reactivity, including rosacea, hives, telangiectasias, and phlebitis.

Curcuma longa • Turmeric

The bright yellow roots of *Curcuma* are used as a culinary spice and medicine. *Curcuma* may reduce inflammatory damage of the joints in osteoarthritis and autoimmune disease. *Curcuma* may reduce excessive collagen and fibrin deposition in inflammatory disorders of the muscles, blood vessels, and connective tissue. *Curcuma* is useful in cases of traumatic injury with extensive bruising and muscle inflammation. *Curcuma* also promotes detoxification and metabolic functions of the liver, helping to clear hormones, metabolic wastes,

and toxins, which may improve chronic muscle stiffness and vague myalgia that defies a typical diagnosis.

Dioscorea villosa • Wild Yam

The edible tubers of hundreds of species of wild yam have been used medicinally for centuries. While *Dioscorea* is not specific for musculoskeletal disorders, it may support stamina and muscle mass and prevent sarcopenia of disease and adrenopause when used as an adrenal and hormonal tonic. *Dioscorea* is specific for weakness in the back and pain with bending over, as well as for sciatica and shooting pains, aching and stiffness in the joints, cramps, and prickling pains in the fingers and toes.

Echinacea species • Coneflower

The roots of *Echinacea angustifolia*, *E. purpurea*, and *E. pallida* excel in protecting against tissue decay and destruction, preventing the dissemination of infection through cellular and lymphatic channels, and deterring the breakdown of hyaluronic acid. *Echinacea* may help protect connective tissues from destruction in cases of infectious, toxic, and inflammatory processes. Although sometimes said to be an immunostimulant, *Echinacea* is more accurately termed an *immune modulator* and is not contraindicated for individuals suffering from various autoimmune disorders.

Among the several thousand published papers on *Echinacea*, many researchers report the plant to have dual activities: an ability to both stimulate and suppress the production of the cytokine TNF-α depending on the background physiologic situation, and widely varying (even polar) manners of regulating cytokines via various chemical constituents.[176]

Eleutherococcus senticosus • Siberian Ginseng

The roots of *Eleutherococcus* have immunomodulating, adrenal-supportive, and hormone-regulating effects. Include Siberian ginseng, also known as eleuthero, in muscle pain formulas where underlying allergies or atopic tendencies are present, as well as for individuals with poor immune response and tendency to infection, or excessive immune response and autoimmune skin conditions. Most molecular research on Siberian ginseng has focused on the eleutherosides, a group of steroidal saponins similar in structure to the ginsenosides found in *Panax ginseng*. These steroidal compounds have been widely studied for enhancing sports performance, possibly due to an ability to reduce oxidative stress and prevent fatigue in working muscles. Siberian ginseng may also have testosterone-enhancing effects via adrenal-supportive actions, without being an actual anabolic steroid.

Epimedium brevicornu • Yin Yang Huo

Epimedium brevicornu has been used in TCM for thousands of years for sexual function and reproductive health as well as for osteoporosis and bone disease. The phytoestrogens in the roots may inhibit postmenopausal bone loss without having a proliferative effect on the uterus. *Epimedium* has also been used for male impotence and gonadal support. Among the many phytoestrogenic compounds are icariin, epimedins, and related compounds showing osteogenic and antiosteoclastogenic effects. Additional anti-inflammatory effects may also protect the bone from damage in cases of autoimmune-driven destruction of bone cells. Furthermore, *Epimedium* may promote bone regeneration through hormonal and anti-inflammatory pathways.

Equisetum arvense • Horsetail

The spindly photosynthetic stems of *Equisetum* are notably high in silica, and consumption may lead to increased absorption of calcium and stimulate collagen synthesis. In addition to its robust mineral content, *Equisetum* contains phytosterols, triterpenes, and phenolic compounds that may all contribute to osteogenic effects. *Equisetum*'s secondary metabolites, such as quercetin, kaempferol, luteolin, apigenin, oleanolic acid, betulinic acid, and ursolic acid, have all shown to have an anabolic activity on osteoblasts.

Eschscholzia californica • California Poppy

The roots of the California poppy were used by West Coast Native American populations as a general anodyne agent, for muscle spasms and trauma-related pain. Use *Eschscholzia* for acute muscle spasms and for chronic tension associated with stress and anxiety. *Eschscholzia* alkaloids bind to GABA receptors, producing a relaxing effect on the muscles, as well as an anxiolytic effect.[177]

Eupatorium perfoliatum • Boneset

Eupatorium perfoliatum is indicated for muscle complaints associated with viral and bacterial infections. *Eupatorium* is specific for muscle aching associated with the flu and for thirst, chilliness, and nausea associated with the flu. Tinctures are generally used and specific for deep aching in the bones and back, gouty inflammations, and myalgia associated with febrile illnesses. *Eupatorium* is recommended for weakness and worn-out presentations in aging alcoholics with sluggishness in all organs and organ functions.

Filipendula ulmaria • Meadowsweet

Young leaves and flowers of *Filipendula* are a traditional anti-inflammatory and febrifuge. *Filipendula* is indicated for muscle and joint pains associated with inflammation of the intestinal mucous membranes, where the stools are loose and watery. *Filipendula* contains salicylates, which can help thwart acute inflammation at the onset of strains and acute muscle spasms.

Gaultheria procumbens • Wintergreen

Wintergreen leaves are most commonly used to extract methyl salicylate, a component of the plant's essential oil. Both the extract and the essential oil are often used in topical anodyne creams and ointments. Whole plant extracts are also used for the anti-inflammatory actions of the flavonoids, pentacyclic triterpenes, and sterols. Wintergreen is specific for rheumatism, sciatica, and neuralgia. The essential oil is especially appropriate to include in liniments and topical formulas for musculoskeletal pain. Although *Gaultheria* is commonly regarded to have more anodyne than trophorestorative effects, research suggests topical use may offer more than simple palliation in musculoskeletal formulas. Wintergreen may limit lipoxygenase and hyaluronidase activity and thereby help limit inflammatory damage in connective tissues.[178] Additional species may be used in the same way, including *G. fragrantissima*, *G. nummularoides*, *G. paniculata*, and *G. subcorymbosa*.

Gelsemium sempervirens • Yellow Jessamine

Caution: *Gelsemium* has significant toxicity due to high alkaloid content and is used in small amounts only in oral medication. Even topical application warrants caution as the alkaloids can be systemically absorbed through the skin. It is also prepared into a homeopathic medicine. *Gelsemium* has been used traditionally to treat pain, but due to its toxicity, it is usually reserved for intractable pain such as severe neuralgic and musculoskeletal pain, migraines, and cancer (such as when tumors cause nociceptive pain or metastatic bone pain). The vine bark contains gelsemine and over 120 other alkaloids. *Gelsemium* has an antinociceptive effect that does not appear to lessen over time, as is the case with the rapid onset tolerance to opioids.[179] *G. sempervirens* occurs in

North America and a related species, *G. elegans*, occurs in China and East Asia.

Ginkgo biloba • Maidenhair Tree

The leaves of the *Ginkgo* tree are most commonly known for the circulatory effects that may extend to enhancing perfusion to the bone. *Ginkgo* is shown to have positive effects on bone mineral density and bone microstructure and may significantly reverse bone loss in various research models.[180] Consider *Ginkgo* for poor circulation to the bone and for bone infections and diseases associated with circulatory insufficiency related to long-term smoking, diabetes, chronic renal failure, or other disease. Ginkgo may help protect cellular mitochondria from various toxins and have renoprotective effects in cases of kidney disease and help to slow the resulting progression of bone loss.

Glycyrrhiza glabra • Licorice

Include *Glycyrrhiza* root in musculoskeletal formulas where underlying allergies or atopic tendencies are present, as well as for individuals with poor immune response and tendency to infection or excessive immune response and autoimmune conditions in the connective tissue. *Glycyrrhiza* may also benefit muscle weakness associated with long-term stress and adrenal fatigue.

Grifola frondosa • Maitake Mushroom

The *Grifola* mushroom has immunomodulating and antiviral properties and may be a supportive herb for musculoskeletal complaints associated with immune disorders or triggered by an underlying viral infection. Like other adaptogens, *Grifola* can also support convalescence and recovery following surgeries or debilitating illnesses.

Hamamelis virginiana • Witch Hazel

Drawing from Native American traditions, bark from the young twigs has been used to steam-distill commercial witch hazel extracts since the 1800s. *Hamamelis* has anti-inflammatory effects on the skin and mucous membranes and is also specific for skin trauma and bruising. Use *Hamamelis* topically to control bleeding in wounds, or prepare compresses from strong teas to use over strains, sprains, and especially contusions. *Hamamelis* tincture taken internally may improve pelvic congestion, and thereby circulation in the lower limbs, and help remedy a sense of heaviness and aching in the legs. Include *Hamamelis* in formulas where muscle pain and swelling are related to vascular congestion. *Hamamelis* is specifically indicated for tired feelings in arms and legs, soreness in the muscles and joints, chilliness in the back and hips that radiates down the legs, neuralgia, and sciatica in the legs.

Harpagophytum procumbens • Devil's Claw

Harpagophytum is a traditional medicine from Africa, where the tree bark is used as an all-purpose musculoskeletal anti-inflammatory. *Harpagophytum* is presently used for arthritis, autoimmune-related joint pain, bursitis, and tendonitis. It is useful in both general and specific musculoskeletal pain formulas. Modern research has suggested that regular use may slow the progression of degenerative diseases of the musculoskeletal system via chondroprotective activity. One of the molecular constituents, harpagoside, is noted to reduce the inflammatory mediators TNF-α, inducible nitric oxide synthase (iNOS), and NF-κB and to improve the ratios and activity of various interleukins. Harpagoside is an iridoid glycoside; along with harpagide and procumbide, harpagoside exerts anodyne and antiarthritic effects.

Hedeoma pulegioides • Pennyroyal

The aromatic leaves of pennyroyal are used for tea, to distill essential oils, and to process into herbal tinctures. *Hedeoma* is specific for pain in the thumb joints and for pain, coldness, twitching, and soreness in the muscles. *Hedeoma* is also specific for soreness and a sprained sensation in the Achilles tendons.

Hypericum perforatum • St. Johnswort

The young flower buds are prepared into topical oils used to prevent stretch marks and to promote healing following trauma to the skin and joints. *Hypericum* is an excellent connective tissue tonic for vascular fragility, easy bruising, and chronic hemorrhoids; it is specifically indicated for injury to highly innervated areas. *Hypericum* may be used for musculoskeletal pain associated with underlying viral infections. *Hypericum* may improve osseous healing in bone graft surgery, increasing collagen deposition and decreasing fibroblast migration and deposition, speeding healing time.[181] Hyperforin activates TRP channels and stabilizes neuronal membranes, limiting inflammatory processes.

Lomatium dissectum • Biscuitroot

The roots of *Lomatium* were widely used by North American native peoples for respiratory symptoms, and this Apiaceae family plant is presently regarded primarily as an antiviral herb used for respiratory infections and influenza. *Lomatium* may be included in oral formulations for myalgia due to underlying viral infections and for general muscle pain accompanying infectious illnesses, viral bronchitis, colds, and flu. Be aware that in rare cases the use of *Lomatium* can induce a maculopapular cutaneous rash that resolves over several weeks' time.

Medicago sativa • Alfalfa

The young leaves of alfalfa are dried for teas or fresh processed into medicinal extracts. Use *Medicago* as a mineral tonic in cases of poorly healing wounds, weak fingernails and connective tissue, and osteopenia and osteoporosis. Vinegar macerations, also known as acetracts, are useful for extracting minerals from *Medicago*, and such acetracts may be used in beverages, marinades, and salad dressings. Vitamins D and K are rare in plants but do occur in *Medicago*; these vitamins, along with the plant's robust mineral content, make *Medicago* highly nourishing to the bones. Like many legumes, *Medicago* is also high in hormonally active isoflavones, which may improve bone density due to estrogen agonism or other hormonal affects.

Melaleuca alternifolia • Tea Tree

The aromatic leaves of tea tree are used to steam distill essential oils that are widely used as an antimicrobial agent. *Melaleuca* may be considered in topical applications for osteomyelitis and infectious arthritis and bursitis, especially when combined with castor oil that may help drive the essential oil into the tissues. One published case study describes the satisfactory treatment of antibiotic-resistant osteomyelitis using a combination of essential oils that included *Melaleuca* injected into the bone through the skin.[182]

Melissa officinalis • Lemon Balm

Lemon balm leaves are beloved for their lemony aroma and widely used to prepare herbal extracts. *Melissa* may be included in formulas for muscle pain due to stress and emotional upset or to viral infections. Drinking lemon balm tea before bedtime may improve sleep in those with stress-related muscle tension.

Panax ginseng • Asian ginseng

Panax is helpful in musculoskeletal formulas for fatigue and poor stamina, as well as in a variety of situations in which underlying allergies or atopic tendencies contribute to pain or inflammation in the connective tissue. *Panax* is also appropriate for individuals with poor immune response and tendency to infection, or excessive immune response and autoimmune skin conditions. *Panax* is specifically indicated for a sensation of swelling in the hands, where the skin feels tight and impedes clenching the fingers. Consider *Panax* for stiffness in the back with a feeling of heaviness, fatigue, and weakness. *Panax* may be a supportive herb in formulas to treat issues related to stress and overwork, and for musculoskeletal complaints associated with fatigue. *Panax* is also an option for sarcopenia and to improve stamina and exercise tolerance. One clinical trial noted *Panax* to improve energy in those with idiopathic fatigue concomitant with a reduction in reactive oxygen species,[183] suggesting an improvement in inflammatory burden.

Phytolacca americana • Pokeweed

Phytolacca roots are a traditional lymphagogue herb, and extracts are appropriate to include in formulas for chronic infections and for muscle and connective tissue complaints that may originate from viral or other systemic pathogens. Historic literature mentions *Phytolacca* to be specific for shooting pains in the shoulders, with stiffness and difficulty raising the arms, and for rheumatic pains that are worse in the morning. *Phytolacca* is also said to be specific for changing and shifting pains that are shooting and lancinating in nature, and for neuralgia in the feet and toes. Due to the herb's affinity for tissue congestion, *Phytolacca* is also specific for swelling in the feet and lower legs, with aching in the heels and relief from elevating the legs. Historical literature describes an affinity for "scrofulous constitutions," and also recommends *Phytolacca* to stimulate the muscles and circulation in children with rickets and difficulty walking.

Pinus pinaster • Maritime Pine

The bark of the maritime pine is a rich source of potently antioxidant polyphenolic compounds, reportedly 50 to 100 times more potent than vitamin E. These compounds are commonly concentrated and sold under the tradename Pycnogenol. Pycnogenol is used for various inflammatory disorders and often included in protocols to support cardiovascular health. Pycnogenol may support bone health by enhancing microcirculation, increasing capillary permeability, and significantly reducing oxidative stress. Research suggests that Pycnogenol active metabolites can be found in the synovial fluid of arthritis patients, where they exert an anti-inflammatory effect lasting roughly 14 hours.[184] Clinical trials show Pycnogenol to inhibit the release of cartilage-destructing proteases and pain-producing cyclooxygenases and that regular use may reduce reliance on more-harmful pharmaceutical anodynes, reducing joint stiffness and improving mobility.

Piper methysticum • Kava Kava

The large roots of the "intoxicating pepper," as *Piper methysticum* is sometimes called, have been used to create psychoactive beverages in the Fiji Islands. Kava is a strong and fast-acting antispasmodic indicated for acute musculoskeletal pain such as torticollis or lumbar spasm. *P. methysticum* may be included in formulas for acute, and occasionally chronic, musculoskeletal pain, including fibromyalgia, and rheumatic complaints characterized by tight, spastic muscles. One of kava's constituents, kavain, deters periodontitis induced by *Porphyromonas gingivalis*, a bacterium known to act like an antigen that increases vascular reactivity and increases the risk of RA.[185]

Pueraria montana var. lobata • Kudzu

The Chinese herb *Pueraria montana* var. *lobata* is widely used in TCM, where it goes by the name gegen. *Pueraria* is a classic renal and hormonal tonic, used in the daily diet, in teas, and in targeted formulas. *Pueraria* is a legume and like many legumes contains hormonally active isoflavones. Puerarin is one well-studied isoflavone shown to support bone mineral density and prevent loss of trabecular bone structure in animal models of menopause. Puerarin increases the proliferation of osteoblasts, stimulates bone growth, and inhibits the formation of osteoclasts. Puerarin protects bones from breakdown and destruction by drugs and toxins. These mechanisms explain *Pueraria*'s historic use to treat menopause, osteoporosis, renal disease, and other bone loss diseases. For more details on *Pueraria*, see chapter 4 of Volume 3.

Rheum officinale • Chinese Rhubarb

Rheum roots are traditionally used in Asia for constipation and to enhance intestinal motility. Include *Rheum* in formulas for musculoskeletal pain where malabsorption and poor digestion underlie. *Rheum* is specifically indicated for a sour smell to the body and for diarrhea. It is also indicated for those who are hungry but easily become overfull, for colicky pain about the umbilicus, and for sour-smelling stool passed with cramping and straining.

Rhodiola rosea • Arctic Rose

The astringent roots of small, creeping, herbaceous *Rhodiola* have been used as an energy tonic and an adaptogen-like herb. *Rhodiola* may also help prevent altitude sickness, improve stamina and endurance at high altitudes, and support athletic endurance and performance in general in cases of muscle weakness and fatigue. Salidrosides in *Rhodiola* are shown to promote synthesis and activity of the neurotransmitters norepinephrine, serotonin, dopamine, and acetylcholine. They are also thought to explain *Rhodiola*'s ability to increase attention and memory, as well as to enhance muscle endurance and physical performance in athletes. Roughly 200 milligrams per day is the dosage shown to enhance athletic endurance, but may also be considered as a therapy for those with muscle weakness and poor stamina.

Rumex crispus • Curly Dock

Rumex roots are a traditional alterative agent in Western herbalism and may be added to support elimination of wastes and support the assimilation of nutrients. Include *Rumex* in formulas for joint pain and pathology related to poor digestion, liver congestion, and toxemia. *Rumex* is specifically indicated for vague joint pain and chronic stiffness secondary to digestive insufficiency, hypochlorhydria, malabsorption, and constipation. *Rumex* is also indicated when nausea and anorexia, flatulence and abdominal pain, a sore and coated tongue, heartburn, hiccups, or chronic gastritis accompany chronic musculoskeletal pain and stiffness.

Salix alba • White Willow

Bark from the young branches of many *Salix* species may be used to prepare salicylate-rich medication used for inflammation, fever, and pain. *Salix* tinctures and teas offer palliative relief of joint pain, traumatic pain, and arthritis with acute short-term usage. *Salix* reduces inflammation via a number of mechanisms, and one study showed a chondroprotective effect via limiting NF-κB activation induced by IL-1β.[186] More than an "herbal aspirin," *Salix* may be used long term in formulas for arthritis, tendinitis, or other chronic musculoskeletal issues due to its many beneficial anti-inflammatory actions.

Sanguinaria canadensis • Bloodroot

Caution: The bright orange resin in the roots of *Sanguinaria* can be irritating to the skin and mucous membranes, and when concentrated, can be outright caustic. Therefore, *Sanguinaria* is considered to be a prescription-only botanical for use by skilled herbalists. However, small amounts of *Sanguinaria* tincture are sometimes diluted with other herbs in tincture formulas and dosed in small quantities, which avoids oral or gastric side effects. According to folklore, *Sanguinaria* may be used in small amounts to treat rheumatic pains, especially those that are burning in quality; for shoulder, neck, and hip joint pain; and for burning sensations in the palms and soles of the feet.

Schisandra chinensis • Magnolia Vine

Schisandra fruits are recommended in TCM for diseases of the gastrointestinal tract, respiratory failure, and cardiovascular diseases as well as for states of body fatigue and weakness, excessive sweating, and insomnia. The herb is also native to parts of Russia, where it has been described as a tonic used to reduce hunger and fatigue and to enhance vitality and delay the aging process. *Schisandra* helps the liver clear hormones, drugs, and poisons from the blood, and has a warming quality for chronic musculoskeletal and arthritic complaints that feel cold, or are worse with cold.

Scutellaria baicalensis • Scute

The leaves of this Asian species of *Scutellaria* have an anti-inflammatory action of use in treating musculoskeletal inflammation and pain. *Scutellaria* may reduce excessive fibroblast activity and fibrin deposition common in collagen vascular diseases and mixed connective tissue inflammatory disorders. Scute is used internally to reduce allergic and hypersensitivity reactions in the muscles and joints and may improve autoimmune inflammation. *Scutellaria*'s flavonoids are shown to have chondroprotective effects by reducing the activity of pro-inflammatory interleukins.[187] In the Middle East, *S. baicalensis* is often combined with *Acacia catechu* to treat joint pain,[188] and such proprietary blends are shown to strongly inhibit cyclooxygenase and lipoxygenase pro-inflammatory enzyme pathways, making them useful to treat joint pain and osteoarthritis.[189]

Scutellaria lateriflora • Skullcap

Include the North American species of *Scutellaria* in musculoskeletal formulas where nervousness or fear underlies or contributes to the symptoms. Skullcap is specific for tics, twitches, and restless legs and may be included in musculoskeletal pain formulas where restlessness, anxiety, and insomnia accompany.

Sesamum indicum • Sesame Seed

The seeds of *Sesamum* are pressed to yield a light oil used in cooking, cosmetics, and medicines. Sesame seed oil contains the lignans sesamin and sesamolin and the phenolic compound sesamol. Sesamol is shown to limit joint inflammation, cartilage degradation, and periarticular bone resorption in an animal model of adjuvant-induced arthritis. Sesamin is also shown to retard inflammatory degradation of GAGs and collagen in the joints. Sesame oil may be used as a base oil to make arthritis liniments and also may be consumed orally.

Spilanthes acmella • Paracress

Spilanthes flowers have a numbing effect in the mouth when chewed and have been traditional medicines for toothache and to treat oral pain (it is also commonly called toothache plant). *Spilanthes* is also prepared into pastes, liniments, and salves for topical use on headaches and joint pain, and it may be included in formulas for Lyme disease, muscle pain, and achiness related to infections. Alkylamides, such as spilanthol in the aerial parts, are responsible for the topical anesthetic effects and are shown to act on cannabinoid type 2 receptors as well as to exert immunomodulatory effects.[190]

Stellaria media • Chickweed

Stellaria is specific for sharp, transient, ever-changing rheumatic pains that are associated with systemic stasis, congestion, and sluggishness. *Stellaria* is indicated for joint pains that are worse each morning, and for stiffness and darting pains that are sore to the touch and worse with motion. *Stellaria* is also specific for sharp pain in the small of the back, over the lower ribs, and in the gluteal muscles with pain radiating down the legs. Older textbooks mention *Stellaria* for chronic pain in the calf muscles, bruised sensations, and chronic tendinitis.

Stillingia sylvatica • Queen's Root

Stillingia roots are prepared into medicines that are specific for sensation of deep aching in the bones, especially when associated with chronic skin ulcers and chronic skin complaints, as well as liver congestion with jaundice and constipation. No significant molecular or clinical research has yet been done on the plant.

Symphytum officinale • Comfrey

Symphytum leaf and root preparations are among our best remedies for traumatic injuries to the bones, tendons, and joints and may promote healing in nonunion fractures and improve bone density in cases of osteoporosis. *Symphytum* has been used for fractures, sprains, rheumatism, and thrombophlebitis both topically and internally, and it may improve chronic pain and promote healing. *Symphytum* contains pyrrolizidine alkaloids known to cause veno-occlusive liver disease in cattle who graze on large amounts and may cause liver damage to an unborn fetus when consumed during pregnancy. When using *Symphytum* short term to heal a fracture, monitor liver enzymes to ensure safety. There are no such concerns with topical use of comfrey oils, creams, or gels.

Tanacetum parthenium • Feverfew

The young leaves of feverfew are prepared into hay fever, migraine, and antiallergy medications. *Tanacetum* reduces allergic activity in the blood and blood vessels and may be included in formulas for musculoskeletal complaints where allergic reactivity underlies, such as extreme muscle soreness and swelling following a bee sting or immunization. Consider *Tanacetum* in formulas for autoimmune conditions to reduce vasculitis and inflammatory processes. *Tanacetum* may also reduce drug side effects that cause muscle pain or swelling.

Taraxacum officinale • Dandelion

Taraxacum leaves act as a general diuretic and mineral tonic for long-term use in osteoporosis. They may be eaten as bitter greens, and vinegars prepared from the leaves may provide minerals and nutrients in cases of poor nutrition or assimilation. *Taraxacum* roots are especially useful for complaints related to liver congestion, constipation, or poor nutritional status.

Tribulus terrestris • Puncture Vine

Tribulus has been used as a traditional genitourinary tonic and also by athletes as a cardiovascular and endurance tonic for enhancing performance. *Tribulus* contains steroidal compounds including diosgenin and protodioscin and may support testosterone and luteinizing hormone production as well as muscle growth. For research on *Tribulus* as a sexual and libido tonic, refer to Volume 3.

Trifolium pratense • Red Clover

Trifolium is traditionally considered to be a "blood mover" and general nutritional herb. Modern research shows hormonal effects, and *Trifolium* may be used in formulas for postmenopausal bone density support. *Trifolium* contains isoflavones with phytoestrogenic activities shown to support bone growth.[191] Teas, tinctures, and encapsulated isoflavone concentrates may improve connective tissue laxity, osteoporosis, skin complaints, and weak nails. *Trifolium* isoflavones, including daidzein, genistein, and formononetin, may reduce postmenopausal bone loss by inhibiting bone resorption and the proliferation and differentiation of osteoclasts.

Ulmus rubra • Slippery Elm

The bark of slippery elm trees is dried for use in teas or processed into a variety of medicinal extracts. *Ulmus* is a nutritive and demulcent herb to include in musculoskeletal formulas, and to soothe the oral and other mucous membranes in cases of Sjögren's syndrome, lupus, and other collagen vascular diseases. Some studies suggest that the nutritive properties may improve bone density.

Urtica species • Nettle

The leaves of *Urtica dioica, U. urens,* and other nettle species are indicated for connective tissue weakness, osteoporosis, loss of bone mass, and arthritis. *Urtica* is also very valuable to include in formulas for gout to facilitate the excretion of uric acid. Consumption of nettle as food, in teas, and in powders, tincture, and other medications are a traditional means of treating osteoarthritis and may help patients reduce their dose of anodyne pharmaceuticals and avoid the prolonged use of NSAIDs. Purposeful stinging of arthritic joints with fresh nettle leaves, sometime referred to as urtication, is also a long-standing traditional approach to alleviating joint pain. Nettles may also exert numerous anti-inflammatory effects that may reduce autoimmune phenomena and benefit RA.[192] *Urtica* extracts can inhibit lipopolysaccharide production and resultant cytokine release from cells, thereby offering broad immunomodulating effects.[193]

Vaccinium myrtillus • Blueberry

Vaccinium fruits are high in antioxidant anthocyanidins well known to offer protective effects on the vasculature, but also noted to prevent oxidative stress in the bone marrow mesenchymal and exert antiosteoclastogenic properties.[194] *Vaccinium* may also be included in formulas for collagen vascular conditions, to help mitigate vasculitis, and to prevent loss of vessel integrity and function in muscle and connective tissue. *Vaccinium* leaves and berries are available in dried form for teas, in tinctures, and in concentrated products, often standardized to their anthocyanin content. All such products can be used for bruising and skin trauma and to improve microcirculation in the skin.

Valeriana officinalis • Valerian

Valerian is specific for muscle pain and for tension associated with underlying stress. Include *Valeriana* as a supportive herb in formulas for rheumatic muscle pain and sciatica, and possibly a lead herb in formulas for acute muscle cramps and jumping and twitching of the muscles. *Valeriana* is specifically indicated for tension in the body and for mental turmoil preventing sleep. When nervousness, hypersensitivity, obsessive thinking, symptoms from worry, and stress headaches accompany musculoskeletal pain, *Valeriana* may be a useful adjunct herb.

Veratrum viride • False Hellebore

Veratrum is a potentially cardiotoxic plant but the roots may be prepared into medicines used to allay neuralgia and nerve hyperexcitability. *Veratrum* is specific for acute and severe musculoskeletal pain, twitching, convulsions, and lancinating pain, especially when associated with pronounced fatigue and prostration and tissue congestion. **Caution:** Due to its great toxic potential, *Veratrum* is for use by skilled clinicians in small dosages only. The plant is also well-known to be teratogenic and must be avoided during pregnancy.

Verbena hastata • Vervain

The aerial parts of *Verbena* have been traditionally used as a gentle, broad-acting anti-inflammatory. Include *Verbena* in musculoskeletal formulas where nervousness underlies pain, tension, or spasm. *Verbena* can help reduce pain in bruising and promote the absorption of extravasated blood.

Viburnum opulus, V. prunifolium • Crampbark

Viburnum is a folkloric classic herb for musculoskeletal tightness and pain. *Viburnum* roots are useful in cases of spinal stiffness, neck pain, and low back pain, especially when accompanied by a sensation of weakness and heaviness.

Withania somnifera • Ashwagandha

Ashwagandha roots are the primary plant part used as a traditional adaptogen and nervous restorative agent, although the leaves, seeds, and berries are occasionally used as well. Include *Withania* in musculoskeletal formulas when adrenal insufficiency contributes to muscle tension and lack of stamina. *Withania* may help protect hyaluronic acid from inflammatory destruction in autoimmune disorders and aging in general. *Withania* is specific in cases of fatigue and weakness with stress and insomnia.

Yucca schidigera • Yucca

Yucca species are native to the arid regions of the Americas and Caribbean, where they have been used in folk medicine as antiarthritic medicines. *Yucca* is a traditional musculoskeletal pain remedy with demonstrated anti-inflammatory action.[195] *Yucca* contains steroidal saponins credited with anti-inflammatory activity, and the plant is high in polyphenolic compounds, including resveratrol. Yucca saponins include the spirostanol saponins, also known as spirostanosides, shown to have antiproliferative, anti-inflammatory, and antihyperuricemic activities.[196] *Yucca* inhibits NF-κB with numerous immunomodulating benefits, but otherwise has not been rigorously studied.[197]

Zanthoxylum clava-herculis • Southern Prickly Ash

Zanthoxylum may be used topically in liniments for relief of arthritic and musculoskeletal pain. *Zanthoxylum* is indicated for neuralgia, sciatica, numbness, motor weakness, and paralysis. Tinctures or teas may stimulate saliva flow in cases of dry mouth in Sjögren's syndrome. *Zanthoxylum* is a warming, stimulating remedy that brings heat to the stomach, increasing function. *Zanthoxylum* increases circulation and secretions in cases of digestive debility and insufficiency and is best in those with cold constitutions, weakness, lethargy, and poor circulation.

Zingiber officinale • Ginger

Zingiber is an all-purpose anti-inflammatory for arthritis, joint pain, and musculoskeletal inflammation.

Dosages of about 3 or 4 grams of ginger powder taken daily appear most effective, and even cooked ginger may be helpful.[198] *Zingiber* may be added to musculoskeletal formulas for various pathologies for those with cold constitutions, the elderly, or others with poor circulation and digestive insufficiency. *Zingiber* is used in many formulas for deficiency, coldness, stasis, and for musculoskeletal complaints related to digestive insufficiency. Ginger juice, powder, and extracts inhibit cyclooxygenase and lipoxygenase enzymes[199] and may help reduce inflammation and preserve the integrity of cartilage and connective tissue during aging and for osteoarthritis.[200]

— CHAPTER FOUR —

Creating Herbal Formulas for Eye, Ear, Nose, Mouth, and Throat Conditions

Conditions of the eyes, ears, nose, and throat (EENT) and oral cavity are among the most common infections and allergic conditions presenting in general practices, and nearly all people experience one or more such conditions in their lifetimes. While colds and minor sore throats are rarely life-threatening and do not need extensive therapy in many situations, in other cases they do warrant treatment due to long recovery times or the tendency to develop into more difficult or serious infections. In some cases ear infections, throat infections, and other EENT conditions become chronic, and vigilant physicians may want to address any underlying contributors such as systemic allergies, bowel toxicity, or high sugar intake and diabetes. Herbal formulas for infectious and inflammatory conditions often include alteratives, liver herbs, and antiallergy and immunomodulating herbs to address particular contributions to chronic issues. Such attention to underlying contributors and aiming to reduce chronicity is rarely attempted in allopathic medicine. In fact, the repetitive use of antibiotics and steroidal remedies without addressing underlying causes in cases of chronic EENT infections and inflammatory conditions may actually contribute to chronicity.

For example, in the case of chronic otitis media, excessive dairy intake or sensitivity to dairy may congest the oral mucous membranes and block the Eustachian tubes. The repetitive use of antibiotics may promote more virulent and drug-resistant strains of antibiotics to evolve, while the use of herbs that address underlying allergy or decongest tissues and Eustachian blockage may be superior to antimicrobial therapy alone. In other cases, blood sugar may be elevated or the liver and bowels may not be eliminating wastes efficiently, and waste materials left in the tissues and bloodstream create a hospitable environment for bacteria. In these cases, the inclusion of diabetic protocols, or the inclusion of herbs to support digestive health, may reduce the chronic tendency to otitis or strep throat.

This chapter illustrates repeatedly how herbalists can fine-tune formulas to not simply offer an herbal alternative to antibiotics, but rather offer a sophisticated combination of herbs and dietary support to address underlying allergic phenomenon, underlying mucous membrane congestion, and underlying immune deficiencies. This fine-tuning also serves to help prevent recurrence as each individual case requires.

As you review the formulas and other information in this chapter, keep in mind that some EENT infections and allergic conditions are also discussed in chapter 2 of this volume.

The Significance of Dental and Oral Health

Dental and oral health are a reflection of health of the entire body. In particular, the state of health in the lower bowels may be mirrored in the oral cavity. TCM and other medical systems of the world examine the tongue and mouth for clues as to the energetic and organ system health of the body. A coating on the tongue can be a sign of dampness or toxemia in the body, while dryness may reflect dryness or heat in the entire body, and severe dryness may be a symptom of Sjögren's syndrome. Excessive salivation can occur with heavy metal toxicity or, rarely, allergic phenomena. Enlarged tonsils may be a sign of infection, lymph congestion, or a lymphatic constitution. Pallor of the gums can be indicative of anemia, and a blue or purplish discoloration may indicate vascular congestion. Grayish papules surrounded by a ring of red inflammation in the mouth may signify measles and are referred to as Koplik's spots. Halitosis can indicate a shift in desired probiotic microflora or poor dental health, or it may be a reflection of dysbiosis and toxemia in the lower gastrointestinal system. Oral lesions occur with viral infections, autoimmune deficiencies, and nutritional deficiencies such as vitamins C or B. For all these valuable windows into the health of the entire body, most physicians examine the oral cavity as part of a routine health assessment.

Herbal treatments for oral lesions might be aimed at improving immune function in the case of chronic aphthous ulcers, or at providing antiviral effects in the case of acute herpes. Other formulas might be aimed at improving nutritional status and connective tissue integrity to treat bleeding gums or gingivitis. Chronic gingivitis and gum disease are also associated with heart disease, both involving loss of the integrity of the connective tissues. In some situations, pain-relieving ingredients are warranted to treat oral lesions. Sinus pain may sometimes refer to the teeth and oral cavity. Cellulitis in the oral cavity may follow dental extractions and often responds well to herbal remedies. Parotid gland enlargement is palpable and may be associated with liver disease, ductal stones, or autoimmune disorders.

Herbal medicine for dental procedures might include typical trauma herbs such as *Centella asiatica*, *Calendula officinalis*, *Arnica montana*, and *Hypericum perforatum* in mouthwashes and compresses. For infectious problems in the mouth, *Commiphora myrrha*, *Sanguinaria canadensis*, *Coptis chinensis*, or *Berberis aquifolium* are all good choices. For pain, irritation, or ulceration in the mouth, *Glycyrrhiza glabra*, *Ulmus fulva*, *Symphytum officinale*, *Echinacea purpurea*, and *Aloe vera* may promote healing. In some people, dental procedures can initiate a herpes flare, and *Hypericum*, *Melissa officinalis*, *Glycyrrhiza*, and lysine supplementation may be used preventively. In those with rheumatic heart disease or prolapsed heart valves, antibiotics are routinely given to prevent any potentially serious infection that may follow routine dental work, even teeth cleaning. I have used herbal antimicrobial mouthwashes in such cases, before and after dental work, for people who do not tolerate antibiotics.

Formulas for Eye Conditions

Serious ocular trauma and vision-threatening conditions are best treated by ophthalmologists and are not covered in this chapter, even though many herbal treatments may speed recovery and address underlying conditions. Acute eye pain, sudden visual disturbances, and eye trauma should be immediately referred to eye specialists.

The eyes are an important site for evaluating the health of the entire body and might offer clues to systemic pathologies or constitutional tendencies. The eyes are the only place where blood vessels of the body can be directly viewed (with the aid of an ophthalmoscope) and may provide clues to the vascular health of the entire body. The appearance of the conjunctival mucous membranes is another opportunity to assess the constitution. For example, pigmentation and opacities suggest fluid stasis or toxin accumulation. Pale lids suggest anemia, and abnormal pupillary light reflexes suggest imbalances in parasympathetic and sympathetic tone. In TCM, the eyes are said to be related to the liver, an example being jaundice seen with hepatitis, and vague chronic eye complaints may benefit from alterative and liver herbs.

General herbal considerations for eye complaints include antimicrobial herbs used topically for conjunctivitis, many examples of which are given in the formulas in this section. When eye symptoms are chronic and don't fit any particular diagnosis, many patients may respond to *Silybum marianum*, *Taraxacum officinale*, *Chelidonium majus*, *Berberis aquifolium*, *Arctium lappa*, or *Cynara scolymus* in the formula. Vascular protectants are indicated to treat and prevent retinal damage related to diabetes. Circulatory protective herbs include *Ginkgo biloba*, *Crataegus* species, *Vaccinium myrtillus*, and *Sambucus nigra*.

Severe infections of the eye, such as orbital cellulitis, require expert evaluation and likely antibiotic therapy. Hospital records suggest that ethmoid sinusitis may predispose and that the most commonly involved organisms include *Streptococcus pneumoniae*, *Moraxella catarrhalis*, *Haemophilus* species, and *Staphylococcus aureus*.[1] Fungal infections of the orbit can also occur and may be overlooked or confused for other pathologies.[2] While I have treated orbital cellulitis without antibiotics or steroids, conventional treatment is typically penicillin and its relatives, and in many cases this may be best given the serious risk of loss of vision. Otherwise, heat packs and aggressive doses of *Echinacea*, *Hypericum*, *Phytolacca americana*, *Origanum vulgare*, *Curcuma longa*, *Andrographis paniculata*, and other antimicrobials may be orally dosed every 15 to 30 minutes while keeping a close watch. The development of severe neck pain, worsening eye pain, ptosis, and/or photophobia may signal that the meninges are becoming inflamed, and pharmaceutical antibiotics, and in some cases hospitalization, may be appropriate. Around 20 percent of acute cases may require surgical interventions when rapid suppuration within tissues necessitates incision and draining of the orbit and IV antibiotics.

Pink eye is a common condition that often readily responds to herbal therapies. Pink eye is the most common form of conjunctivitis, occurring frequently in school age children. While pink eye may spread readily through a family, the condition is typically self-limiting. Pink eye is typically viral or bacterial and often occurs alone without other upper respiratory infection (URI) symptoms,[3] but bacterial conjunctivitis may associate with otitis in children.[4] Conjunctivitis may also follow irritation from smoke or chlorine fumes, wind, dust,

---

Herbal Specifics for Eye Complaints

These herbs and homeopathic remedies are all indicated for specific eye issues, from trauma to infections to allergic reactivity. Please see the "Specific Indications" section on page 192 for more information on the unique symptoms for which these agents may be useful in oral and eyewash formulas.

Achillea millefolium	*Hydrastis canadensis*
Aloe vera	*Hypericum perforatum*
Apis mellifica	*Iris versicolor*
Arnica montana	*Matricaria chamomilla*
Berberis aquifolium	*Panax ginseng*
Calendula officinalis	*Phytolacca americana*
Cantharis vesicatoria	*Salvia officinalis*
Centella asiatica	*Silybum marianum*
Chelidonium majus	*Symphytum officinale*
Coptis chinensis	*Tanacetum parthenium*
Euphrasia officinalis	*Vaccinium myrtillus*
Hamamelis virginiana	

---

or bright sun. Enterovirus conjunctival infections with conjunctival hemorrhage have occurred in epidemics in Africa and Asia, and trachoma due to *Chlamydia trachomatis* (found in the water of tropical locales) is a leading cause of blindness. Conjunctivitis in a newborn (ophthalmia neonatorum) is worrisome, and chlamydia, herpes, gonorrhea, or other pathogens must be ruled out with bacterial and viral cultures. Most neonates receive prophylactic eye drops as a matter of course. Silver nitrate eye drops are used prophylactically, and gonorrhea or other causes of neonatal conjunctivitis is treated with gentamicin or neomycin; however, breastmilk, and especially colostrum, instilled in newborns' eyes may also be effective.[5] *Echinacea*, *Allium sativum*, and *Thymus vulgaris* might be included in oral herbal formulas when symptoms or systemic bacterial infections are present, and all types of conjunctivitis can be treated with eyewashes of *Calendula*, *Berberis*, or *Euphrasia officinalis*, which can make a good base for an all-purpose eyewash formula.

Tincture for Orbital Cellulitis

Orbital cellulitis is a serious acute inflammation of the cells surrounding the orbit, including those of the mucosal and lymphatic tissues, the immediately adjacent nasal and sinus passages, and the connective tissue and vascular connections. Orbital cellulitis may follow trauma or occur when dental or respiratory infections enter the deeper tissues. When severe, it may cause fever, malaise, nausea, and pain in conjunction with eye movement. Since the condition can progress to inflame the optic nerves, possibly leading to blindness or affecting the meninges, orbital cellulitis must be treated aggressively. The following herbal formula may complement antibiotics in acute cases. *Echinacea* protects hyaluronic acid, which prevents the breakdown of the structure of the connective tissue and prevents infection and allergic inflammation from disseminating into deeper tissues. *Hypericum* has an affinity for nerve inflammation and may help protect the optic nerves.

Echinacea angustifolia	30 ml
Hypericum perforatum	10 ml
Origanum vulgare	10 ml
Phytolacca americana	10 ml

Take 1 teaspoon of the combined tincture every ½ hour, reducing as symptoms improve. Consider prescribing antibiotics if there are no improvements within 4 hours.

All-Purpose Eye Drop Formula

Bacterial conjunctivitis usually involves thick, yellow-to-greenish, purulent discharges that may crust and glue the eyelashes together, while viral conjunctivitis is more common[6] and may involve profuse watery discharges. Both bacterial and viral conjunctivitis can be treated similarly with herbs that have broad antimicrobial effects, while allergic and inflammatory types of conjunctivitis are better addressed with soothing, anti-inflammatory, demulcent herbs, specifically *Euphrasia*.[7] *Berberis* (and *Hydrastis*, *Coptis*, and *Phellodendron*) contains isoquinoline compounds that have broad antimicrobial effects, especially effective for purulent infections in mucous membranes. *Calendula* has healing and soothing properties, while *Euphrasia* is highly specific for allergic and viral eye complaints. This formula is "all-purpose" because it addresses bacteria, viruses, allergies, and general irritation.

Berberis aquifolium bark	1 ounce (30 g)
Calendula officinalis flowers	½ ounce (15 g)
Euphrasia officinalis leaves	½ ounce (15 g)

Add 1 heaping tablespoon of the combined herbal mixture to 1 cup (240 ml) of boiling water and steep for 15 minutes. Strain once through a wire mesh or another strainer, then strain a second time through a coffee filter to remove all remaining particulates. Cool to body temperature or cooler, then apply using an eyecup or an eye dropper. Apply the eye drops hourly, or every ½ hour when possible. Discard any remaining liquid at the end of the day, making a fresh mixture each morning. Continue as needed, reducing frequency as symptoms improve. Systemic immune support or antiallergy therapies, such as the Tincture for Viral Conjunctivitis and Tincture for Hay Fever–Related Conjunctivitis remedies below, may complement eye drops when systemic symptoms such as acute viral infections or hay fever necessitate.

Tincture for Viral Conjunctivitis

When conjunctivitis is accompanied by sore throat, aching muscles, and runny nose, systemic antiviral herbs are appropriate. The following herbs are antiviral and particularly active against respiratory and influenza viruses.

Euphrasia officinalis	15 ml
Sambucus nigra	15 ml
Lomatium dissectum	15 ml
Glycyrrhiza glabra	15 ml

Take 1 dropperful of the combined tincture hourly, reducing as symptoms improve.

Tincture for Hay Fever–Related Conjunctivitis

Seasonal allergies are easy to diagnose, and can be treated with herbs affecting mast cells, histamine, and allergic activity. In some cases, it is appropriate to use an antiallergy protocol for several months prior to peak pollen or allergy months; see chapter 2 of this volume for formulas for this purpose. Botanical agents noted to stabilize mast cells are also appropriate for allergic conjunctivitis, and include the high-quercetin[8] herbs *Ginkgo*, *Allium cepa* and *sativum*, and *Curcuma*. In addition to quercetin, many other flavonoids may reduce allergic phenomena—in part through stabilizing effects on mast cells—and may be taken orally or used in eyewashes.

Euphrasia officinalis	20 ml
Tanacetum parthenium	15 ml
Astragalus membranaceus	15 ml
Curcuma longa	10 ml

Take 1 dropperful of the combined tincture as often as every ½ hour for acute symptoms, reducing as symptoms improve.

Tea for Chronic Conjunctivitis

Chronic conjunctivitis usually occurs in a person who has allergies and is related to daily exposure to the allergen—mold, cleaning products, chlorinated pool water, and so forth. In such cases, antiallergy protocols can be implemented (see chapter 2 of this volume), although patients should do all they can to eliminate the offending exposures and help calm the hyperreactivity. Antiallergy botanicals include *Astragalus, Petasites, Angelica, Ginkgo,*[9] *Scutellaria baicalensis, Tanacetum,* and *Euphrasia.* These are useful in reducing systemic allergic reactivity. When allergies do not appear to be the underlying cause of chronic conjunctivitis, consider several months of liver and digestive support, especially if constipation, elevated liver enzymes, gallbladder insufficiency, or digestive complaints are present. The herbs in this formula have antiallergy and anti-inflammatory effects and can help reduce histamine and mast cell activation when consumed regularly for many months.

Glycyrrhiza glabra	3 ounces (90 g)
Euphrasia officinalis	1 ounce (30 g)
Ephedra sinica	1 ounce (30 g)
Urtica dioica	1 ounce (30 g)
Ginkgo biloba	1 ounce (30 g)
Camellia sinensis	1 ounce (30 g)

Combine the herbs and prepare the tea using 1 tablespoon of the mixture per cup of hot water. Steep for 10 minutes and then strain. Drink 3 cups (720 ml) or more daily for many months to reduce allergic reactivity. The tea may be prepared without *Ephedra* when it is unavailable.

Tincture for Chronic Eye Irritation

Liver congestion and toxicity are often associated with vague eye symptoms, such as frequent irritation. *Curcuma* is included in this formula not only for its antioxidant properties, but also for its liver-regenerative properties. *Ginkgo* improves general circulation to the eyes, while *Vaccinium* supports microcirculation of the retina. *Euphrasia* is indicated for allergic irritation of the eyes. This formula is therefore a kitchen-sink formula for eye irritation, and may be amended given individual circumstances.

Curcuma longa
Euphrasia officinalis
Vaccinium myrtillus
Ginkgo biloba

Combine in equal parts. Take 1 to 2 dropperfuls, 3 or more times daily.

Herbal Specifics for Eye Symptoms

The herbs listed here are specifically indicated for common types of eye irritation. They can be employed as simples or included in formulas to treat conjunctivitis, hay fever, and dry eyes.

SYMPTOM	RECOMMENDED HERBS
Profusely watering eyes	*Euphrasia officinalis* *Salvia officinalis* *Tanacetum parthenium*
Dry, easily irritated eyes	*Althaea officinalis* *Calendula officinalis* *Euphrasia officinalis* *Iris versicolor*
Year-round eye allergies	*Angelica sinensis* *Astragalus membranaceus* *Euphrasia officinalis* *Ginkgo biloba* *Petasites hybridus* *Tanacetum parthenium*

Basic Eye Drops for Dry Eyes

This eye drop formula is to be prepared in small amounts for use over 48 hours, at which time the remainder should be discarded to prevent microbial contamination. Tamarind polysaccharides are similar to mucin secreted from the corneal goblet cells, and research suggests the plant's mucoadhesive properties are useful in eye drop formulas, helping other ingredients and molecular compounds adhere to mucosal surfaces and enhance the therapeutic effects.[10] Be sure to use quality, authentic rosewater because some commercial products employ synthetic rose perfumes.

Rose water	Roughly 1.5 ounces (45 ml)
Aloe vera gel	1 tablespoon
Pilocarpus jaborandi	tincture 1 teaspoon
Tamarind paste	½ teaspoon

Combine the *Aloe* gel, *Pilocarpus* tincture, and tamarind paste in a 2-ounce (60 ml) bottle and fill the remaining space with rose water. Shake vigorously until homogenized. Gently instill 5 dropperfuls at a time into each eye, repeating as often as hourly and reducing frequency as symptoms improve.

Options for Treating Dry Eyes

Dry eyes or keratitis sicca is most common in adult women. There are no lasting or curative remedies other than pilocarpine, frequent use of palliative eye drops, and a variety of mechanical approaches, such as special eyeglasses to help hold in humidity. Sjögren's syndrome, scleroderma, or other autoimmune disease may also cause dry eyes and lead to sclerotic lesions. More extensive and deeper cases of scleritis can damage the sclera to the point of rupture and must be treated aggressively by an eye specialist as soon as the condition emerges.

Pilocarpine. Pilocarpine is a secretory stimulant used to promote lacrimation.

Essential fatty acids. Omega-3 fatty acids are noted to improve dry eyes.[11]

Vitamin A. Retinoic acid, a metabolite of vitamin A, may improve chronic dry eyes.[12]

Vitamin D. Many people are found to be low in vitamin D when tested; supplementation may improve chronic dry eye syndrome.[13]

Hormones. Low androgen levels may contribute to chronic dry eye syndrome in elderly men,[14] and chronic dry eye syndrome may emerge postmenopausally in women.[15] The lacrimal gland possesses androgen receptors,[16] and possibly other hormone receptors, and hormones may help maintain ocular mucosal surfaces.

Calendula. *Calendula* may be used locally to soothe dry eyes and possibly enhance conjunctival circulation. *Calendula* polysaccharides may adhere to mucous membranes,[17] providing temporary moistening effects.

Centella. *Centella asiatica* may help prevent excessive deposition of fibrotic tissue in cases of autoimmune connective tissue inflammation.[18] Thus, it may help with treatment of sclerosis of the eye.

Tincture for Conjunctivitis with Discharges

These herbs are chosen for their drying effects, being specifically indicated for those who experience chronic conjunctivitis with profuse eye-watering and allergic reactivity.

Salvia officinalis
Euphrasia officinalis
Tanacetum parthenium

Combine in equal parts. Take 1 to 2 dropperfuls as often as hourly for acute inflammation, or 2 to 3 times daily as part of a larger protocol for chronic allergic reactivity.

Tincture for Conjunctivitis with Dry Eyes

Chronic conjunctivitis can also result in dry eyes. These herbs are chosen for their moistening and secretion-stimulating effects, being specifically indicated for burning sensations.

Althaea officinalis	20 ml
Euphrasia officinalis	20 ml
Calendula officinalis	15 ml
Iris versicolor	5 ml

Take 1 to 2 dropperfuls of the combined tincture 3 or more times a day. For acute eye irritation, the formula may be taken every 1 to 15 minutes for several hours, reducing to hourly as symptoms improve.

Tincture for Herpes-Induced Ophthalmia

The global incidence of herpes simplex–related keratitis is roughly 1.5 million cases per year, including 40,000 new cases involving significant visual impairment or blindness each year.[19] Both herpes simplex and herpes zoster infections can affect the eyes, causing pain and suffering. When severe, they can lead to uveitis and loss of vision, necessitating aggressive treatment and possible referral to an eye specialist. Steroids are sometimes prescribed for severe inflammation and cases of uveitis; however, steroids suppress the immunologic response, thereby allowing the herpes virus to proliferate. Gene therapy is also an option in treating herpes keratitis. Whether keratitis is being treated with or without pharmaceutical medications, antiviral and antiherpetic herbs are appropriate. These include *Glycyrrhiza*,[20] *Hypericum*,[21] and *Melissa*.[22] Nutritional agents are also indicated, including vitamin E, to help prevent sequelae of herpes infections; lysine has additional antiviral effects. Immune support may also be accomplished through the use of *Panax*, *Echinacea*, *Eleutherococcus*, and *Ganoderma* as synergists in formulas. *Echinacea* has shown activity against the herpes simplex virus.[23] See chapter 2 of this volume for more information and research on herbs for treating herpetic infections.

Glycyrrhiza glabra	15 ml
Hypericum perforatum	15 ml
Melissa officinalis	15 ml
Echinacea angustifolia	15 ml

Take 1 teaspoon of the combined tincture every several hours, reducing as symptoms improve. Honey prepared in eye drops may also be complementary to other conventional therapies in the treatment of keratitis.[24]

Tincture for Corneal Ulcers

Infections caused by *Staphylococcus*, *Streptococcus*, herpes, *Chlamydia trachomatis*, *Neisseria gonorrhoeae*, and various fungi can all result in ulcerative lesions of the cornea, as can extremely poor nutrition. *Pseudomonas* infection is associated with the use of contact lenses and may also lead to severe keratitis, which could progress to ulceration. Glycyrrhizin from *Glycyrrhiza* may protect against *P. aeruginosa*–induced keratitis.[25] Endemic soil fungi in tropical areas may put agricultural workers at risk for fungal infections of the eye in some regions.[26] UV light therapy may provide urgently needed antimicrobial effects, and is emerging as a possible therapy with riboflavin used as a photosensitizer.[27] Infectious keratitis can break down corneal connective tissue and lead to ulceration as bacterial endotoxins and enzymes dissolve cellular collagen fibers. *Echinacea* inhibits the breakdown of hyaluronic acid and is good to include in ulcer formulas no matter the underlying cause.

Glycyrrhiza glabra
Echinacea angustifolia

Combine in equal parts and take as often as hourly for corneal ulcers, reducing as symptoms improve. This tincture can be complemented with the Supportive Tea for Corneal Ulcers and one of the eye drop or eyewash formulas in this section.

Pilocarpus for Dry Eyes

Pilocarpus jaborandi is a prescription-only botanical agent used for Sjögren's patients or other patients with dry eyes and mouth. Pilocarpine, an active alkaloid in *Pilocarpus*, promotes tears and saliva due to cholinergic effects,[28] and is orally dosed at 5 milligrams three or four times daily,[29] or may be instilled directly in the eyes. Herbalists usually prefer to use the whole plant, as opposed to the isolated alkaloid, and research suggests the number of the mucus-secreting cells in the ocular epithelia may actually increase under the influence of *Pilocarpus*, and that the whole plant offers more than just temporary tear production.[30] The plant and isolated pilocarpine are limited to prescription status because the powerful cholinergic effects can induce neural excitation and seizures if dosed inappropriately. *Pilocarpus* may also help reduce intraocular pressure in urgent situations, or as part of a long-term therapy. Eye drops of 0.05 percent pilocarpine are instilled in the eyes for acute glaucoma episodes.

Pilocarpus jaborandi, Jaborandi

Supportive Tea for Corneal Ulcers

Urtica, *Glycyrrhiza*, *Equisetum*, *Centella*, and *Symphytum* are herbs noted to help repair ulcers and connective tissues and offer nutritional benefits. For active infections, antimicrobial formulas such as the Tincture for Herpes-Induced Ophthalmia are necessary alongside a nutritive tea. This supportive tea is not a substitute for antimicrobial treatment prescribed by an eye expert, but may complement antibiotic and other therapies.

Glycyrrhiza glabra	5 ounces (150 g)
Vaccinium myrtillus leaves	3 ounces (90 g)
Urtica dioica	2 ounces (60 g)
Equisetum arvense	2 ounces (60 g)
Centella asiatica	2 ounces (60 g)
Symphytum officinale	2 ounces (60 g)

Mix the herbs to prepare a pound of dry tea. Gently simmer 1 to 2 teaspoons per 1 cup (240 ml) of water for several minutes. Cover and let stand for 15 to 20 minutes, then strain. Drink as frequently as possible, such as 1 cup (240 ml) each waking hour for several days, reducing to 3 to 6 cups (720 to 1,440 ml) per day as symptoms improve. When strained well and cooled to body temperature, this tea could also be used as an eyewash.

Formulas for Cataracts

Although cataracts can occur congenitally, opacity of the ocular lens most commonly develops over a lifetime, related to the cumulative effects of long-term inflammatory damage. Antioxidant agents that decrease free radicals, such as vitamin C and α-lipoic acid or *Camellia sinensis* (green tea),[31] may retard the progression of cataracts. Plant flavonoids absorb UV light and can help protect the tissues from sun.

Eye Drops for Cataracts

One herbal option for dissolving cataracts is *Jacobaea maritima* (also known as *Cineraria maritima*),[32] the fresh juice of which is instilled in the eyes. Though little research is available, many clinicians, including myself, have anecdotally reported good results. The first several doses are painless, but the eye drops become increasingly irritating with continued use. I instruct patients to skip dosing for several days once irritation occurs. After that, they can develop an intermittent dosage schedule by trial and error. Due to its causticity, *Jacobaea* is available by prescription only. Commercial products are hard to come by, and eye drops sometimes need to be made by hand.

Jacobaea maritima eye drops

Instill 1 drop in each eye once or twice a day until results are achieved.

Tincture for Cataract Prevention in the Elderly

This cataract-prevention formula and the Tincture for Cataract Prevention in Diabetes both feature *Vaccinium* and *Ginkgo* to support circulation and offer antioxidant and UV radiation protection. Early animal studies suggest that *Ocimum tenuiflorum* may prevent lens opacities from oxidative stress,[33] and *Ginkgo biloba* may also protect the eye from UV radiation due to superoxide dismutase and other antioxidant constituents. This general formula combines these herbs with *Curcuma* and *Centella*, agents noted to reduce fibrosis and scarring, as a further measure to prevent or slow cataract development.

Vaccinium myrtillus	15 ml
Ginkgo biloba	15 ml
Curcuma longa	15 ml
Centella asiatica	15 ml

Take ½ to 2 teaspoons of the combined tincture 3 times per day, long term.

Tincture for Cataract Prevention in Diabetes

Vaccinium may prevent cataract progression with long-term use.[34] Because diabetes is also a risk factor in cataract development, herbs such as *Vaccinium* that protect the vasculature from hyperglycemia and hyperlipidemia are recommended for diabetic patients for lifelong use. A small amount of research suggests that the glucose-regulating plants *Gymnema*,[35] *Pterocarpus marsupium*, and *Trigonella foenum-graecum*[36] may also contribute to cataract prevention.

Vaccinium myrtillus	15 ml
Ginkgo biloba	15 ml
Gymnema sylvestre	15 ml
Trigonella foenum-graecum	15 ml

Take ½ to 2 teaspoons of the combined tincture 3 times per day, long term.

Formulas for Blepharitis, Styes, and Lacrimal Gland Strictures

Inflammation of the eyelids, known as blepharitis, may have infectious or allergic origins. In some individuals, blepharitis can be a localized type of eczema or seborrheic dermatitis characterized by red, thickened eyelids with a tendency to scales, crusts, and small ulcers on the eyelid margins; the condition may involve secondary bacterial infections of the meibomian glands. Patients with seborrheic blepharitis may also have similar lesions in the eyebrows or nasolabial folds or frank eczema in the usual places. Blepharitis can be painless but usually involves itching and burning sensations of the eyelids. Chronic blepharitis may result in loss of the eyelashes.

Although both hordeolums and chalazions are commonly referred to as styes, hordeolums are infectious in origin, while chalazions are a proliferation of fibrous tissue. *Staphylococcus* is the most common underlying pathogen.[37] Styes tend to be recurrent and emerge rapidly and painfully from Zeis or Moll glands of the external lid margin, or from meibomian glands at the inner lid margins. Chalazions emerge slowly and painlessly or may result from an incompletely healed hordeolum. Additionally, sebaceous cysts may occur at the lid margin or inner canthi, sometimes related to clogged sebaceous glands that may result from the use of heavy eye creams. Since the causes of hordeolums and chalazions are not entirely known, it is unknown how to prevent recurrent episodes. If liver or gallbladder issues are present, alteratives and *Silybum*, *Curcuma*, and *Chelidonium* may resolve chronicity. Hot compresses are the main treatment both for pain and to speed resolution. Homeopathic Pulsatilla and Silica, as well as other homeopathic remedies are sometimes effective and, as they are inexpensive and safe, are certainly worth a try.

Congenital strictures of the lacrimal duct are commonly noted in infants within the first weeks of life, evidenced by constant tearing, usually unilaterally. In other cases, the tear duct can become blocked later in life following head trauma. The duct may be opened surgically but requires general anesthesia, making effective alternatives highly desirable. Simple warm compresses may open the affected duct in some cases, and parents can be encouraged to gently massage the inner canthi with warm water at every diaper change. Homeopathic Pulsatilla or Silica has proved effective in opening up the lacrimal duct in some cases. The secondary bacterial infection is typically minimal, and antibiotics would not open the duct and only risk development of more virulent strains. If the infection threatens to worsen and affect the lacrimal sac, eyewashes such as those in "Formulas for Eye Conditions" section beginning on page 146 may be prepared. Infections of the lacrimal sac are referred to as dacryocystitis and may rarely develop into abscesses and even fistulas, which then would require antibiotics, and even surgical incision and drainage.

Tincture for Chronic Styes

For patients with chronic styes, *Curcuma* may provide immune support and quell inflammation, and *Euphrasia* may reduce hyperreactivity in ocular mucous membranes. *Silybum* supports liver function, and *Ginkgo* may support circulation to the head while also offering antiallergy effects. Taking this formula internally would not be a very powerful therapy for acute styes or to allay pain but might be employed for many months to reduce chronicity.

Curcuma longa
Euphrasia officinalis
Silybum marianum
Ginkgo biloba

Combine in equal parts and take ½ to 1 teaspoon 3 to 4 times daily for at least 6 months or until condition improves.

Tincture for Chronic Blepharitis

This formula reduces allergic processes that may contribute to eyelid inflammation. For best results, combine with the use of a topical compress and avoid inhalation of fumes and contact with chemicals, which may aggravate allergies. Simple hot compresses to the eyes are helpful,[38] though compresses prepared from hot herbal teas such as *Euphrasia*, *Calendula*, or *Berberis aquifolium* are even better. *Euphrasia* has an affinity for eye inflammation according to folkloric literature, and modern research shows numerous anti-inflammatory and immunomodulating activities.[39]

Euphrasia officinalis	15 ml
Tanacetum parthenium	15 ml
Calendula officinalis	15 ml
Curcuma longa	15 ml

Take ½ to 2 teaspoons of the combined tincture 3 times per day long term.

Eyewash for Pterygiums

Pterygiums and other fleshy growths are most commonly due to repetitive exposure to strong sun and wind. Eye protection may help prevent these conditions, and herbs such as *Centella*, *Calendula*, or *Curcuma* may encourage resolution.

Centella asiatica

Steep 1 tablespoon of the dried herb in 1 cup of hot water for 10 minutes. Strain well and allow to cool to a comfortable temperature; use an eye dropper to instill as drops or use an eyecup.

Tea Tree Oil Eyelid Wash #1

Demodex mites are a normal skin parasite of the face and eyelids, feeding on the skin, hormones, and oils in hair follicles. However, the mites may proliferate in the elderly, or in those with disturbances in tear production, hygiene, or immune function, and contribute to chronic blepharitis.[40] Although antibiotic remedies are offered allopathically to treat these mites, herbalists often recognize bacterial presence as secondary to the underlying ecosystem that supports, if not invites, the microbes, and would address such conditions directly wherever possible. Gently washing the eyelids with diluted tea tree oil can help eradicate Demodex mites.[41] Ophthalmologists may use anesthetics to help scrub the inner eyelids with a cotton swab soaked in tea tree oil, which has been shown to be helpful in such procedures. Terpinene is the substance in tea tree oil that can improve blepharitis that results from overgrowth of Demodex mites. Tea tree oil must be highly diluted prior to using it near the eyes.

Dr. Bronner's or similar simple liquid baby soap   1 drop
Melaleuca alternifolia essential oil   1 drop

Place the drops of baby soap and tea tree oil in ¼ cup of warm water or *Euphrasia* tea. Dip a cotton swab in the combined liquid. Gently scrub the undersides of the eyelids with the cotton swab, going as deep as possible

Healing Support for Ocular Injury

Due to ample immunoglobulins in tears and eye fluids and to its rich vascularity, the eye has an excellent ability to heal quickly from minor scratches and burns due to hot grease splatters, as well as even more serious traumas such as blunt blows to the eye. Foreign bodies in the eye, lacerations of the eyelid margins, or other serious traumas should be referred to an eye specialist immediately. Care must be taken to ensure sterility of all products used in the eye, especially in cases of scratches and trauma.

Homeopathic remedies. Homeopathics for eye trauma include low-potency *Symphytum*, *Arnica*, and *Natrum muriaticum*. Homeopathic *Apis* and *Cantharis* are choices for burns.

Saline rinses. For chemical burns, the eye should be copiously rinsed with saline solution, or even tap water, as available. Scratches and abrasions, although extremely uncomfortable, tend to improve in 24 hours, and demulcent herbs can be soothing when placed directly in the eye.

Demulcent herbs. Demulcents, such as thick mucilaginous preparations of *Althaea*, *Symphytum*, and *Ulmus*, may blur the vision, but will allay pain almost instantly. Patients might be encouraged to simply rest with the eyes closed, using demulcent drops frequently.

Astringent herbal washes. When there is profuse purulent mucus in the eye following injuries, astringent herbs such as *Hamamelis* and *Quercus* may be helpful. Oxyquinoline-containing disinfectant herbs *Berberis*, *Coptis*, *Hydrastis*, and *Chelidonium* are usually highly effective for mucous discharges.

Oral vulnerary herbs. *Symphytum officinale*, *Calendula officinalis*, *Hypericum perforatum*, and/or *Centella asiatica* can be used singly or combined and consumed orally to support healing of all types of wounds and physical traumas. These four herbs are among the most commonly emphasized agents to promote healing and can be taken internally for a week or more following trauma.

without causing pain. Use twice each day for at least 10 to 14 days, and then once a day thereafter for an additional 2 weeks.

Tea Tree Oil Eyelid Wash #2

This is an alternative to the Tea Tree Oil Eyelid Wash #1 formula, with the addition of glycerin and povidone-iodine.

Warm water	½ cup (8 tablespoons)
Vegetable glycerin	1 teaspoon
Baby shampoo	½ teaspoon
Povidone-iodine	5 drops
Tea tree oil	1 drop

Combine the ingredients in a dish and stir vigorously. Use a soft cloth or gauze pad soaked in the liquid to gently scrub the eyelids. Rinse thoroughly with plain water. Repeat as often as possible until the problem resolves. Be aware that iodine can permanently stain fabric.

Tincture for Black Eye

"Black eyes" are less often due to eye trauma and more frequently due to general head trauma, where microbleeds may pool in the eye socket. Because head trauma and concussions are increasingly being found to predispose cognitive impairment years later, it is important to support healing even for relatively minor head injuries. The following herbs are believed to be specific for head trauma. Bromelain, at a dosage of 500 to 1,500 milligrams 3 times a day, may speed the healing of bruises and reduce pain and swelling.

Hypericum perforatum
Centella asiatica
Panax ginseng

Combine in equal parts in a 1- to 2-ounce (30 to 60 ml) bottle. Take ½ to 1 teaspoon 5 to 6 times per day until gone. A similar tea might also be prepared and used to complement or replace the tincture.

Belladonna Eye Drops for Pain

In rare cases where demulcents and homeopathics do not effectively allay pain, *Atropa belladonna* is available by prescription in the form of eye drops, typically sold as 1 percent ophthalmic solutions in which each milliliter contains 10 milligrams of atropine sulfate (1 percent Atropine solution eye drops) but is contraindicated for those with glaucoma. A homemade version is as follows.

Atropa belladonna	10 drops
Isotonic saline solution	100 drops

Combine the tincture and the saline. Instill 1 drop in each eye every hour. Discard the eye drops at the end of the day, making fresh each day.

Formulas for Glaucoma

Glaucoma is a serious condition that can lead to visual impairment due to gradual damage from intraocular pressure. Primary glaucoma is spontaneous pressure elevation in the eye; secondary glaucoma develops when cataracts, uveitis, or other condition leads to increased ocular pressure. Increased oxidative stress in the eye may occur with a variety of conditions and may lead to inflammation and, over time, destruction of the ground tissue—the trabecular meshwork of the eye.[42] Closed-angle glaucoma can frequently develop more acutely in sudden episodes, usually unilaterally, accompanied by visual disturbance and pain, and is a medical emergency. Because some people also have nausea and vomiting, some cases are mistaken for migraines, allowing damage to vision to ensue. Pathologies of increased oxidative stress, such as diabetes, may predispose to developing glaucoma. Because steroid use in the eye can, in rare instances, initiate a sudden acute increase in intraocular pressure, people must be evaluated for undiagnosed glaucoma before such medicines are used. Due to the slow and insidious progression of the disease, primary glaucoma is often not diagnosed until it has progressed to the point of causing vision damage. This is one reason for routine eye screens, even in people with 20/20 vision who have no eye complaints. Pilocarpine is a primary treatment for primary open-angle glaucoma. Medical marijuana is also a leading medical therapy for reducing intraocular pressure and slowing the progression of the disease.

Herbal Support Tincture for Glaucoma

Because oral glycerin alone can reduce acute glaucoma attacks,[43] *Vaccinium* and *Ginkgo*[44] might be taken long term in glycerin form for prevention, maintenance, and acute episodes. These herbs are selected because of their ability to enhance circulation in the eye, their excellent safety profiles, and their broad anti-inflammatory effects. *Ginkgo* is reported to significantly increase blood flow to the ophthalmic artery,[45] and to even reverse visual damage in some glaucoma patients.[46] *Salvia miltiorrhiza*, which is well known to enhance circulation, may reduce optic nerve damage from chronic elevated intraocular pressure[47] and chronic oxidative stress in the diabetic state.[48] Fennel (*Foeniculum vulgare*), although not included in this formula, may also have an ocular hypotensive effect.[49]

Ginkgo biloba glycerite	15 ml
Vaccinium myrtillus glycerite	15 ml
Salvia miltiorrhiza	15 ml
Curcuma longa	15 ml

Take ½ to 1 teaspoon of the combined tinctures and glycerites 3 times a day as a maintenance dose long term. Sudden eye pain and visual disturbance is a medical emergency. It is appropriate to administer 1 teaspoon every 5 minutes while seeking medical attention, but urgent evaluation is required for acute eye pain or sudden visual disturbances in glaucoma patients.

Pilocarpine Eye Drops for Glaucoma

Pilocarpine, an alkaloid found in *Pilocarpus jaborandi*, is available as prescription eye drops in standard pharmacies. See also "*Pilocarpus* for Dry Eyes" on page 151.

Pilocarpine 0.5 percent eye drops

Use 2 drops in each eye once or twice a day long term, typically for life.

Cannabis and Cannabidiol (CBD) for Glaucoma

Cannabis can reduce intraocular pressure, an effect credited to the cannabinoids. Medical marijuana may be smoked to treat glaucoma where legal, while a wide variety of CBD products offer an alternative to smoking.

CBD

Take 1 milligram once or twice a day, increasing the dose as needed.

Formulas for Retinopathies

Retinopathies are various conditions that lead to inflammatory damage of the retinal blood vessels. Retinopathies most commonly accompany diabetes, arteriosclerosis, hypertension, and autoimmune disease. Diabetic retinopathy is a leading cause of visual impairment worldwide. Specific mechanisms include microvascular hemorrhage, other exudates and fibrosis, and larger retinal vein occlusion or bleeds. Exudative retinopathy involves the accumulation of blood debris and particulates accumulating in the back of the eye as retinal blood vessels are damaged. There is no pharmaceutical treatment other than addressing the underlying hypertension, elevated lipids, and blood sugar. Herbs that improve circulation to the eye, such as *Ginkgo* and *Vaccinium*, can offer vasoprotective effects. Herbs that support insulin response include *Cinnamomum verum*, *Hibiscus sabdariffa*, and *Opuntia ficus-indica*. Herbs that may lower cholesterol, such as *Commiphora mukul* and *Curcuma*, may help to slow the progression of diabetic and other types of retinopathy. (For more discussion of herbal therapies for diabetes, see chapter 3, Volume 3.) Herbs that control hypertension, such as *Allium* and *Angelica*, may also help to address underlying contributors as specifically indicated.

Herbal Supplements for Retinopathies

The following herbs are available singly. Some combination products aimed at treating diabetes or metabolic syndrome may contain all of these herbs.

Ginkgo biloba capsules
Allium sativum capsules
Commiphora mukul capsules
Curcuma longa capsules

Take 1 or 2 capsules of the chosen herb, 1 to 3 times daily as preferred considering the severity and urgency of a case, patient tolerance of consuming multiple pills each day, and cost of the supplements. Some patients may prefer to rotate through the options, taking just one or two recommended herbs at a time. For long-term management of retinopathy, patients with metabolic syndrome may add chromium, vitamin D, and inositol compounds. Patients with hypertension can also try hypertension formulas discussed in chapter 2 of Volume 2, as well as using nervine herbs for stress and beta-blockers for palliation, taking mineral tonics, and exercising. Patients with elevated cholesterol or triglycerides might take separate formulas to include *Commiphora mukul, Curcuma, Allium*, liver herbs, and lipotropic formulas.

General Retinopathy Tincture

Formulas for retinopathies include vascular protectants such as *Ginkgo biloba, Allium sativum, Angelica sinensis, Salvia miltiorrhiza*, and *Vaccinium myrtillus*. The herbs below are chosen for their ability to address elevated lipids, hypertension, and elevated glucose, as well as offer vascular protection.

Salvia miltiorrhiza
Ginkgo biloba
Curcuma longa
Commiphora mukul
Allium sativum

Combine in equal parts and take 1 to 2 dropperfuls, 3 to 6 times per day long term, depending on the severity.

Tea for Diabetic Retinopathy

Studies show the following herbs to protect against retinopathy in animal models of diabetes[50] and ischemic eye diseases.[51] *Astragalus* has also shown promise in human clinical trials for controlling hyperglycemia.[52] *Lycium* (goji berry) polysaccharides are a traditional Chinese remedy for the eyes, and contain various constituents known to protect the eyes against aging and oxidation, including zeaxanthin and other carotenoids.[53] This formula, which is based on traditional Chinese formulas, may be altered to accommodate personal taste, availability of herbs, and cost.

Angelica sinensis
Astragalus membranaceus
Panax notoginseng
Paeonia officinalis
Curcuma longa
Lycium barbarum

Combine in equal parts. Gently simmer 1 teaspoon per cup of hot water for 10 minutes. Let stand 20 minutes more, then strain. Drink 2 to 4 cups (480 to 960 ml), at least 5 days per week. Continue long term.

General Retinopathy Tea

The herbs in this tea address many underlying contributors to retinopathy, and there are similar formulas for diabetes in chapter 3 of Volume 3, and for hypertension

Beverages to Protect Retinal Vessels

Vaccinium and many berries that contain anthocyanin, resveratrol, and related red, blue, and purple pigments can be very valuable in the long-term management of retinopathies[54] and should be stressed in the diet, in teas, and in medicines. Try these berry-based beverages and other beverages:

Prickly pear juice, as available
  and affordable
Pomegranate juice
Blueberry juice
Grape juice
Hibiscus tea
Smoothies made with frozen berries
Cold water or seltzer with frozen
  berries added
Nut milk, coconut milk
Fish oil (added to smoothies and milk blends)

and hyperlipidemia in chapter 2 of Volume 2. Adjust this tea's flavor by adding more or less *Stevia* for sweetness, *Hibiscus* for tartness, or cinnamon and ginger for spiciness.

Vaccinium myrtillus leaves	3 ounces (90 g)
Crataegus spp. leaves and flowers	2 ounces (60 g)
Stevia rebaudiana leaves	2 ounces (60 g)
Crataegus spp. berries, powdered or whole	1 ounce (30 g)
Hibiscus sabdariffa calyces	1 ounce (30 g)
Sambucus nigra berries	1 ounce (30 g)
Cinnamomum verum finely chopped bark	1 ounce (30 g)
Zingiber officinale finely minced root	1 ounce (30 g)

Mix the herbs together and store in an airtight vessel. Soak 1 tablespoon per cup of water for 1 to 2 hours, or overnight. Add 3 more cups (720 ml) of water and bring to a gentle simmer for 10 minutes. Remove from heat and let sit, covered, for 15 minutes. Strain and drink as much as desired, aiming to include in the diet long term along with other beverages and options detailed in this section as part of a lifestyle approach to protecting the eyes.

Formulas for Macular Degeneration

Age-related macular degeneration is the most common cause of irreversible central blindness in the older population worldwide, affecting over 3 million people in the United States alone. In addition to atrophy of the macula, excessive vascularization in the choroidal vessels often accompanies, and is referred to as the "wet" form of macular degeneration. *Maculopathy* is a term that acknowledges a wide spectrum of age-related and inflammatory changes in the macula, the hallmark of which is the accumulation of drusen, deposits of lipids and proteins, ultimately leading to progressive and untreatable loss of vision. While there is no established medical or surgical treatment, agents that inhibit vascular endothelial growth factors (VEGFs), also called anti-VEGF drugs, are being explored. Genetic factors may also predispose,[55] including hypomethylation,[56] oxidative stress in general, and poor glycemic control, all of which accelerate macular atrophy. Therefore, general nutritional, antioxidant, and anti-inflammatory agents would be appropriate therapies. Maintaining adequate levels of omega-3 fatty acids with two servings of fish per week and following a low glycemic index diet may be particularly beneficial for those with early stage macular degeneration.[57] Carotenoids are especially valuable to prevent,[58] and possibly slow, the progression of macular degeneration.

Vaccinium Capsules for Macular Degeneration

Vaccinium capsules may complement fish oil and vitamin A supplementation to treat, or at least slow the progression of, macular degeneration. *Vaccinium myrtillus* capsules may be standardized to anthocyanidin content.

Vaccinium myrtillus capsules

Take 2 pills, 2 to 3 times daily long term.

Tincture for Macular Degeneration

Botanical agents that support circulation, protect the blood vessels, and have general anti-inflammatory and antifibrotic effects are appropriate for slowing the aging process in the eyes. Human clinical trials suggest *Ginkgo* can improve vision in people with age-related macular degeneration.[59] *Vaccinium* may be another herb specific for macular degeneration based on its historical reputation and limited animal studies.[60] It contains anthocyanin flavonoids credited with many vasoprotective actions. The flavonoid lutein, particularly, has been credited with macular supportive effects.[61] *Centella asiatica* is a traditional herb for brain injury and tissue regeneration, and can be used in tea, tincture, or encapsulation. Complement this formula with dietary and lifestyle measures wherever possible.

Vaccinium myrtillus
Salvia miltiorrhiza
Ginkgo biloba
Centella asiatica

Combine in equal parts. Take 1 to 3 dropperfuls, 3 or 4 times a day. This formula can be used long term, with occasional breaks, and can be prepared as a tea as well.

Tea for Macular Degeneration

The combination of these herbs has shown efficacy comparable to the anti-VEGF agent Avastin in human clinical trials.[62] This formula, which is based on formulas from TCM, may be altered to accommodate personal taste, availability of herbs, and cost.

Astragalus membranaceus
Angelica sinensis
Poria cocos
Panax notoginseng
Rheum palmatum
Curcuma longa

Combine the dry herbs and gently simmer 1 teaspoon per cup of hot water for 10 minutes. Let stand 20 minutes more, then strain. Drink 2 to 4 cups (480 to 960 ml) daily, at least 5 days per week. Continue long term.

Formulas for Ear Conditions

Otitis media (middle ear infection) is one of the most common ear conditions seen in general family practice. It affects people of all ages but is particularly common, and sometimes chronic, in children. Dairy or other food intolerance that leads to mucus and congestion in the upper respiratory passages may promote, but smoking in the home,[63] immune deficiencies, and allergies may also underlie. Breastfeeding infants appears to offer some protection against ear infections.[64] In many cases, otitis media follows viral or bacterial URIs, but in some people, ear infections accompany allergic congestion, and more rarely, seem to emerge out of nowhere, independent of other upper respiratory symptoms. When otitis media persists longer than 6 weeks, the tympanic membrane and ossicles can be permanently damaged, and hearing may be permanently impaired. Persistent otitis can move to the inner ear, and also spread to the mastoid bone and surrounding tissues.

Although antibiotics were once the standard of care for otitis media, they have since been realized to foster resistant bacterial strains. Furthermore, taking antibiotics does nothing to reduce any underlying propensity to develop otitis, but does impair beneficial intestinal flora. Formulas for otitis media should include both acute management options for ongoing infections to speed resolution, prevent spread, and address pain, as well as options to address underlying causes when otitis is chronic. For acute otitis infections, useful antimicrobial herbs include *Achillea millefolium, Andrographis paniculata, Curcuma longa, Baptisia tinctoria, Echinacea purpurea, Nepeta cataria, Origanum vulgare, Berberis aquifolium,* and *Allium* species and many others. *Echinacea* is actually *not* the best or most all-purpose herb for treating colds, coughs, and ear infections. Refer to the *Echinacea purpurea* entry in this chapter (page 195) to study *Echinacea*'s specific indications and to compare them with the specific recommendations for other herbs noted here. Because otitis is commonly related to upper respiratory mucus and Eustachian tube congestion, antimicrobials or support herbs with drying, expectorating, and decongesting actions are good choices for use in acute formulas. In those with hay fever, chemical sensitivity, asthma, or other allergic or atopic condition, antiallergy herbs such as *Ginkgo biloba, Camellia sinensis, Tanacetum parthenium, Petasites hybridus, Astragalus membranaceus, Scutellaria baicalensis,* and *Euphrasia officinalis* may be appropriate.

In those with chronic otitis media, alterative herbs and herbs that enhance bowel and digestive function may help cure the problem, and clues to focus on these herbs would be concomitant constipation or frequent digestive upset, liver or gallbladder symptoms, or a high toxin exposure and load in the body. Using alterative herbs and adopting a vegetable-rich, allergen-free diet, as well as supplementing with probiotics will often do wonders for such people. Other approaches to treating chronic otitis media might focus on the use of antiallergy and antioxidant herbs in those who develop otitis in association with spring pollen or smoke exposure, or in those with accompanying eczema, asthma, or other allergies. For those who experience many other infections such as frequent sore throats, coughs and colds, bladder infections, or wound infections, the use of immunomodulating herbs for many months may help to reduce chronic infections.

Formulas for otitis media can be altered to best address acute infections or to help reduce recurrences of infection. It is possible to tailor a formula for a particular presentation or constitution, or when a certain underlying condition is a contributing factor. In those with underlying allergic constitutions, secretory otitis media may occur when upper respiratory inflammation congests local tissues and impairs the drainage of the Eustachian tubes. Negative pressure is created in the middle ear, causing the tympanic membrane to appear retracted, and in many cases appears opaque or yellowish due to stagnant fluid. A fluid level or tiny air bubbles may also be visible with an otoscopic exam, and the person may report frequent cracking and popping sounds. The treatment for secretory otitis media might include formulas containing antiallergy herbs such as *Euphrasia, Astragalus, Tanacetum,* or *Petasites,* and lymphatic herbs such as *Phytolacca americana, Iris versicolor,* or *Calendula officinalis.* As with many chronic conditions, if any digestive or liver difficulties exist, the use of alterative formulas and addressing bowel health may be appropriate.

Cholesteatomas are thick fleshy growths on the tympanic membrane resulting from long-standing otitis media, as epithelial cells hypertrophy and release enzymes that further contribute to chronic inflammation. Cholesteatomas are generally well-demarcated accumulations of keratinized squamous epithelium that must be treated aggressively because they may spread to the inner ear, facial nerves, and temporal bone and cause permanent hearing impairment if allowed to persist, and even death if allowed to progress. The lesions may occur congenitally or arise in childhood or adulthood.

Tincture for Pediatric Otitis Media

Herbs available in glycerin make palatable base ingredients in liquid formulas for infants and children. Some *Echinacea* glycerites are available with zinc and vitamins C and A for additional immune support. Elderberry (*Sambucus*) syrup is also readily available and is especially indicated for viral infections where there is mucus and a cough. As children tend to produce abundant mucus, one or more drying herbs may be helpful. While *Thymus* or *Salvia* are powerful drying agents, they are strongly flavored. *Euphrasia* is moderately drying, and its mellower flavor may be more appropriate in children's formulas.

Echinacea angustifolia glycerite	20 ml
Sambucus nigra syrup	20 ml
Euphrasia officinalis tincture	20 ml
Citrus sinensis essential oil	10–20 drops

Take ½ teaspoon of the combined ingredients straight off the spoon, or mixed in a small amount of water or fruit juice. Dose 3 to 4 times a day for mild infections, and as often as hourly for acute symptoms. This formula may also be taken with a bit of fruity baby food for older babies and toddlers. Nursing infants may nurse 5 to 10 drops at a time off the mother's nipples.

Tincture for Acute Otitis Media with Discharges

Allium helps loosen stuck mucus, while *Achillea* and *Thymus* can help dry abundant mucous secretions, and all three provide antimicrobial effects. Using syrup or glycerite forms rather than tinctures will improve palatability of this formula.

Allium sativum
Achillea millefolium
Thymus vulgaris
Echinacea angustifolia

Combine tinctures, syrups, or glycerites in equal parts. Take as often as hourly, reducing as symptoms improve. *Echinacea* syrup and glycerite are commonly available. Several herbal extract companies in the United States specialize in making syrups and glycerites, and all of the herbs in this formula could be obtained in those forms

to keep on hand for when the need arises; a wise investment for clinicians who treat many children.

Tincture for Acute Otitis Media with Infection

Many "colds" start with respiratory viruses, such as simple rhinoviruses, and the mucus and inflammatory response then creates a hospitable environment for secondary bacterial infections to ensue. These herbs are chosen to address respiratory viruses that may be prominent in the initial stages of an ear infection, as well as address secondary bacterial infections typical when symptoms have persisted for a few days to a week. These herbs may be more specific for EENT infections than other antimicrobial herbs due to their affinity for respiratory mucous membranes. As noted for the Tincture for Acute Otitis Media with Discharges, substituting syrups or glycerites for tinctures will make this formula more palatable for children.

Allium sativum
Euphrasia officinalis
Thymus vulgaris
Sambucus nigra

Combine tinctures, syrups, or glycerites in equal parts. Take as often as hourly, reducing as symptoms improve.

Tincture for Chronic Otitis Media with Respiratory Allergies

While the Tincture for Acute Otitis Media with Infection contains the powerful antimicrobials needed for an active infection, this formula features antiallergy and immunosupportive agents to help reduce chronic infections due to underlying allergies.

Astragalus membranaceus	20 ml
Euphrasia officinalis	10 ml
Tanacetum parthenium	10 ml
Ginkgo biloba	10 ml
Curcuma longa	10 ml

Take ½ to 1 teaspoon of the combined tincture 2 to 4 times daily, for several months. For those with ear congestion associated with spring hay fever, begin in late winter to early spring, a full month prior to the earliest pollen.

Tincture for Chronic Otitis Media with Food Allergies

This formula features *Euphrasia*, *Tanacetum*, and *Ginkgo* to reduce respiratory allergic reactivity, combined with

Matricaria, *Achillea*, and *Curcuma* to support liver and bowel function when digestive reactivity contributes.

Achillea millefolium	15 ml
Curcuma longa	15 ml
Tanacetum parthenium	15 ml
Matricaria chamomilla	15 ml

Take ½ to 1 teaspoon of the combined tincture 3 or 4 times daily, for 3 to 4 months.

Tincture for Chronic Otitis with Toxemia

While the Tincture for Chronic Otitis Media with Food Allergies incorporates *Tanacetum* and *Matricaria* for greater antiallergy and anti-inflammatory effects, alterative herbs are better choices for chronic URIs or otitis accompanied by bowel or liver toxemia. This formula combines the alteratives *Achillea*, *Curcuma*, and *Mahonia*. *Allium* addresses possible intestinal dysbiosis, and *Astragalus* provides immune support. A separate formula would be needed to treat any new-onset episode of acute otitis.

Achillea millefolium	20 ml
Curcuma longa	10 ml
Berberis aquifolium	10 ml
Allium sativum	10 ml
Astragalus membranaceus	10 ml

Take ½ to 1 teaspoon of the combined tincture 3 or 4 times daily for 3 to 4 months.

Ear Drops for Acute Otitis Media

Ear drops are a weak solo therapy for otitis media compared to oral agents and general systemic therapies, but can be a supportive measure, and may help relieve pain. Do not, however, use ear oils or vigorous lavages when the tympanic membrane has ruptured. Instead, use gentle astringent swabs, as in the Ear Lavage Astringent Powder formula on page 161. *Verbascum* oil is a traditional remedy for ear infections, *Hypericum* provides anti-inflammatory effects, and *Allium* provides additional antimicrobial effects. Rarely, I have seen children react to *Allium* oil in the ear, developing mild papular dermatitis. Omit the *Allium* as needed.

Mullein oil (*Verbascum thapsus*)
Hypericum perforatum oil
Allium sativum oil

Combine in equal parts. Use 5 to 10 drops in affected ear. Warming the oil first may improve anodyne effects. Use a bit of cotton to seal the oil in the ear, especially when sleeping at night or when muffled hearing is not an issue.

Tea for Acute Otitis

This tea is pleasantly flavored and covers all of the bases, containing antiviral, antibacterial, anti-inflammatory, and anticatarrhal agents. The above tinctures may be more powerful in treating acute infections, but this tea will complement the tinctures and may be added to a comprehensive protocol.

Glycyrrhiza glabra shredded roots
Pimpinella anisum seeds
Inula helenium root
Thymus vulgaris leaf
Euphrasia officinalis leaf
Sambucus nigra flowers

Combine in equal parts, adding more or less licorice (*Glycyrrhiza*) to account for individual tastes. Steep 1 tablespoon per cup hot water for 10 minutes. Drink freely.

Ear Lavage Astringent Powder

Geranium and *Hamamelis* have drying properties to astringe the ear canals, accompanied here by the soothing and healing effects of the other herbs. This formula also has weak antimicrobial properties due to the *Calendula* and *Curcuma*. Alter the proportions as needed to create a formula that is more or less antimicrobial, more or less astringent, or more or less demulcent.

Hamamelis virginiana powder	1 cup (125 grams)
Centella asiatica powder	⅓ cup (40 gm)
Curcuma longa powder	⅓ cup (40 gm)
Equisetum spp. powder	⅓ cup (40 gm)
Geranium maculatum powder	¼ cup (30 gm)
Hamamelis virginiana powder	¼ cup (30 gm)
Symphytum officinale powder	¼ cup (30 gm)
Calendula officinalis powder	¼ cup (30 gm)

This is a recipe sized for office stock. Dispense to patients in a 1- to 2-ounce (30 to 60 gram) ziplock bag. Combine 1 teaspoon of the powder with ⅛ cup (30 ml) hot water. Steep and allow the powders to sink to the bottom of the container. Decant off the fluid, or simply dip a cotton swab in the fluid and use to swab the ears many times each day—the more the better.

Cholesteatoma Herbal Tincture

Due to the similarity in process between cholesteatoma and keloid formation (see chapter 5 of Volume 1), herbs that reduce tissue proliferation and inflammation, such as *Curcuma* and *Centella*, may be helpful, both locally and internally. Surgical removal is sometimes attempted when medical measures fail; however, there is an obvious risk of proliferative responses to the surgical procedure itself. Herbal and other supportive options are therefore appropriate. These herbs are chosen for their ability to promote wound healing, support optimal connective tissue regeneration, and inhibit excessive fibrosis and cellular proliferation.

Calendula officinalis
Centella asiatica
Curcuma longa
Equisetum spp.

Combine in equal parts and take 1 teaspoon, 6 to 8 times daily for 1 to 2 weeks, then evaluate progress. These same dry herbs could also be prepared as a tea by steeping 1 tablespoon in a cup of hot water and straining for use as an ear lavage or swab for localized application to inhibit cellular proliferation.

Tincture for Inner Ear Infection

Infections of the inner ear and the labyrinthine apparatus usually evolve from middle ear infections, resulting in temporary hearing loss, tinnitus, dizziness, and, in many cases, severe vertigo due to nerve inflammation. Herbal formulas for labyrinthitis might feature antimicrobial agents combined with nerve anti-inflammatory herbs such as *Hypericum*, which is specific for inflammation of highly innervated areas. *Zingiber* may improve nausea and vomiting, if present, and is both antimicrobial and anti-inflammatory. *Ginkgo* is a classic herb for dizziness, but may or may not be helpful for dizziness due to infections. Antimicrobial mucous membrane decongestants may be helpful, and include *Armoracia*, *Allium*, *Phytolacca*, *Thymus*, *Rosmarinus*, or *Salvia*. *Allium* is a mucus-thinning decongestant for cases of labyrinthitis related to ear, throat, and respiratory congestion. When infections persist or do not respond readily to treatment, *Astragalus*, *Ganoderma*, *Panax*, *Cordyceps*, or other adaptogens and immunomodulators may be offered.

Ginkgo biloba
Hypericum perforatum
Zingiber officinale
Allium sativum

Combine in equal parts. Take 1 teaspoon every hour, reducing over several days' time as symptoms improve.

Formulas for Otitis Externa

Otitis externa or "swimmer's ear" can indeed occur in swimmers who have ear canals that are frequently wet or exposed to chlorinated water, but it can also be triggered by shampoos, hair spray, and any contact allergen. Like other forms of dermatitis, otitis externa occurs most commonly in allergic individuals, and the same comprehensive protocols that might be used for eczema or hay fever are also appropriate for otitis externa. Such protocols for chronic itchy ear canals (with or without superimposed infections) might include beta carotene or vitamin A supplements, quercetin and other bioflavonoids, and essential fatty acid supplements. Some individuals may benefit from dietary changes and anti-inflammatory diets.

Otitis externa may also be infectious, with *Pseudomonas aeruginosa* and *Staphylococcus aureus* being the most common pathogens.[65] More rarely, fungal infections including *Aspergillus* and *Candida* species may cause otitis externa. Such fungal infections are sometimes said to be "malignant," meaning that they can spread rapidly and aggressively, and inappropriate treatment with antibiotics or steroidal ear drops may worsen the infection. Clinicians should be vigilant when a patient presents with headaches and systemic symptoms accompanying otalgia and ear discharge, as the consequences of failing to treat malignant otitis externa are dire.[66]

Otitis externa may begin when allergic inflammation leads to swelling and secretion, or chronic moisture from swimming or other cause provides a breeding ground for the opportunistic skin microbes. Skin desquamation and microscopic fissures that result from allergic reactivity or irritation provide a portal of entry for infecting organisms. The use of earplugs may also contribute, and some feel that regular use of cotton swabs to swab the ears can cause impaction of cerumen, which in turn retains more water in the ears, and contribute to the development of otitis externa.

I believe that *Tanacetum parthenium* and *Euphrasia officinalis* are excellent choices for treating oral otitis externa due to the reduction of allergic phenomena and histamine elevations. Other anti-inflammatory and anti-allergy options include *Curcuma longa*, *Ginkgo biloba*,

Astragalus membranaceus, and *Glycyrrhiza glabra*; these may be ingredients in oral tincture formulas or complementary teas. To help allay itching and heal the ear canals, a topical wash or ear oil should be prepared as specifically indicated. In some cases of otitis externa, the ear canals may be dry, lichenified, and itch intensely. In other cases, the canals may be pale, swollen, moist, and prone to secondary bacterial infections. Topical herbal preparations should be made either astringent or oily as specifically indicated. *Calendula officinalis* or *Centella asiatica* serve well as trophorestorative base herbs for either an astringent or a lubricating formula. *Calendula* and *Centella* could be used as powders to combine with *Hamamelis virginiana*, *Geranium maculatum*, *Quercus rubra* or *Q. robur*, or other astringents to reduce secretions and suppuration. *Calendula* oil could be used as a base to combine with *Symphytum officinale*, *Hypericum perforatum*, and *Verbascum thapsus* oils for dry presentations, or with astringent powders for moist and swollen presentations.

Ear Lavage Powder

Similar to the Drying Tincture for Otitis Externa, this formula has a gentle astringent action due to the *Geranium* and *Hamamelis* that is balanced with the demulcent and wound-healing herbs *Symphytum* and *Calendula*.

Geranium maculatum powder	¼ cup (30 gm)
Hamamelis virginiana powder	¼ cup (30 gm)
Symphytum officinale powder	¼ cup (30 gm)
Calendula officinalis powder	¼ cup (30 gm)

Mix 1 teaspoon of the combined powder into ⅛ cup (30 ml) hot water. Steep for 10 minutes and allow the powders to sink to the bottom of the container. Either decant or simply dip a cotton swab in the fluid and use to swab the ears many times each day.

Ear Oil for Otitis Externa

Prepare an oil such as this to swab the ear canals when there is dryness, flaking, crusts, and itching. Sea buckthorn oil is another option, although more expensive.

Calendula officinalis oil	⅓ ounce (10 ml)
Hypericum perforatum oil	⅓ ounce (10 ml)
Symphytum officinale oil	⅓ ounce (10 ml)

Insert a cotton swab into the combined oil, saturate, and use to swab the ear canals many times each day. An eye dropper may also be used to instill the oil into the ears. Apply at night before bed, placing a small piece of cotton ball in the outer ear to help keep the oil in place overnight.

Preventing Swimmer's Ear

Wet ear canals can trigger otitis externa in the same way that frequent dishwashing can trigger dermatitis of the hands in some atopic individuals. In people prone to otitis externa, the ear canals can be protected by a coating of *Calendula* salve, *Hippophae* (sea buckthorn), or other herbal oil, applied once or twice per day, especially just prior to showering or swimming. Similarly, drying the ear canals with cotton swabs after all types of water exposure can help prevent continual inflammation for those susceptible to chronic swimmer's ear.

Tincture to Reduce Ear Canal Inflammation

These herbs would be supportive therapy in treating allergic inflammation occurring with otitis externa by helping reduce allergic reactivity and histamine responses.

Curcuma longa
Ginkgo biloba
Tanacetum parthenium
Euphrasia officinalis

Combine in equal parts and take 1 teaspoon or more, 3 times daily for many months.

Drying Tincture for Otitis Externa

This formula has a drying and anti-inflammatory effect due to the *Hamamelis* and *Curcuma*, and a moistening effect due to the *Symphytum*. The formula can be amended to contain more *Hamamelis* or more *Symphytum* root as required. Prepared as an aqueous solution, the formula can be used to swab or lavage purulent and inflamed canals.

Curcuma longa
Hamamelis virginiana
Calendula officinalis
Symphytum officinale

Combine in roughly equal parts. Dilute 1 teaspoon with 1 to 2 tablespoons of warm water in a small container. Use to swab or lavage the ears many times per day. Prepare fresh each day.

Therapies for Ménière's Disease

Ménière's disease is a chronic condition with a prevalence of 200–500 per 100,000 and is characterized by episodic attacks of vertigo, fluctuating hearing loss, tinnitus, and aural pressure, as well as a progressive loss of audiovestibular functions. The pathophysiology involves endolymphatic hydrops with distension of the miniscule endolymphatic spaces of the inner ear by more than 200 percent.[67] The perilymphatic space within the cochlear duct and the semicircular canals is disrupted, leading to much of the symptomatology. Genetic predispositions may exist.[68]

Acupuncture may offer some benefit to Ménière's patients.[69] In addition to systemic pharmaceuticals, the use of intratympanic steroids, stem cell therapies, and intracochlear devices are being explored.[70] Betahistine, a histamine analogue, is a leading pharmaceutical for treating Ménière's disease; it acts as an agonist at the H1 and an antagonist at the H3 histamine receptors. Thus, histamine-modulating herbs such as *Euphrasia officinalis*, *Petasites hybridus*, or *Tanacetum parthenium* may be worth a try, especially when they specifically suit the case and presentation.

There is little to no published research on botanical medicines being effective in the treatment of Ménière's disease. Due to the pathophysiology of congested fluid and lymphatic hydrops characteristic of this condition, lymphagogues and blood movers such as *Galium aparine*, *Phytolacca americana*, and *Ginkgo biloba* may be included in broader protocols. *Ginkgo* is emphasized in folkloric traditions for treating tinnitus and dizziness, particularly when due to poor circulation, and neuromodulatory effects have been reported with modern research.[71] Because chronic and recurrent inflammation of aural nerves can lead to permanent hearing loss, therapies aimed at reducing neuritis may be appropriate. Agents to allay nausea such as *Zingiber* should also be included on a case by case basis. For more formulas for inflammation of the nerves that may be appropriate for Ménière's disease, see chapter 2, Volume 4.

Tincture for Ménèire's Disease

This kitchen sink formula uses *Ginkgo* to improve circulation to the ears and possibly help with tinnitus and vertigo. *Euphrasia* may improve fluid accumulation, and *Zingiber* exerts its own anti-inflammatory effects as well as treating nausea and vomiting. *Hypericum* is specific for nerve inflammation, and thus may retard hearing loss; it also offers anxiolytic effects. This formula is an example only, and should be altered to address specific underlying contributors, if any are identified.

Ginkgo biloba
Zingiber officinale
Hypericum perforatum
Euphrasia officinalis

Combine in equal parts and take 1 teaspoon, 3 times daily to prevent recurring episodes. Take 1 teaspoon hourly during acute episodes.

Formulas for Tinnitus

Ringing in the ears is a fairly common condition and may range in severity from a minor nuisance to a maddening condition that interferes with sleep and quality of life. Hypertension may cause an audible pulsation in the ears, but tinnitus usually refers to humming, buzzing, roaring, or other disturbing sound. One-sided tinnitus may be due to tumors, growths impinging the nerves, or actual nerve lesions in the affected side. Ototoxic drugs typically yield bilateral tinnitus, with simple aspirin being a common offender. Unilateral tinnitus may also occur in Ménière's disease and is suggested when vertigo accompanies acute episodes. Acute exposure to loud noise can damage the delicate hairs of the inner ear and result in temporary tinnitus, while chronic exposure to high-decibel environments can both damage hearing and lead to chronic tinnitus. Poor circulation to the ears can result in tinnitus, the occurrence of which increases with age. Fluid congestion in the ears and impacted cerumen can both result in tinnitus that resolves when the fluid clears, or the canals are cleaned. Panic attacks and near fainting spells are sometimes associated with sudden onset of short-lived tinnitus and are best addressed with stress management.

Roughly 1 in 10 adults in the United States experiences tinnitus to a minor degree, and 1 in 100 to a severe

Ginkgo for Dizziness, Vertigo, and Tinnitus

Ginkgo biloba is well studied for circulatory-enhancing effects in the body, improving perfusion to the limbs in cases of circulatory insufficiency, and enhancing coronary and cerebrovascular blood flow. *Ginkgo* has also been shown to improve circulation to the vestibular apparatus of the inner ear, decreasing blood viscosity and protecting neurons from oxidative damage.[72] In Germany, *G biloba* is an approved medication in the treatment of peripheral arterial disease, tinnitus, vertigo, and "dementia related syndrome."[73] Such *Ginkgo biloba*–based drugs are reimbursed by German health insurance entities, and the products typically contain 240 milligram of *Ginkgo* concentrates.[74] One randomized, double-blind, multicenter clinical trial on patients with moderate to severe dizziness reported that more than 70 percent of the patients responded to *Ginkgo* in the 12-week study, and that *Ginkgo* was as effective as betahistine in the treatment of vertigo.[75] Another study on patients with sensorineural hearing loss found that patients with underlying ototoxicity and hearing loss related to head injury, as well as idiopathic cases, responded best to *Ginkgo* medication, with greater improvements in audiometric findings than with conventional vasodilators. Statistically significant improvements in tinnitus were also noted.[76] Another multicenter, randomized, double-blind phase III study on patients with sudden idiopathic sensorineural hearing loss reported that the use of *Ginkgo* sped recovery and increased the likelihood of full recovery within a weeks' time when dosed at 120 milligrams, twice a day.[77] A similar study showed equal efficacy to the standard pharmaceutical therapy pentoxifylline for sudden deafness.[78]

Ginkgo biloba, ginkgo

degree. Despite the prevalence of tinnitus there are no FDA-approved medications specifically for treatment of the condition. The pharmaceutical options commonly attempted include IV lidocaine to block sodium channels in the cochlea and the CNS; benzodiazepines; antidepressants, particularly nortriptyline and amitriptyline; anticonvulsants such as carbamazepine and gabapentin; antiglutaminergic agents; and the paradoxical use of both dopaminergic and anti-dopaminergic agents.[79] Herbal medicines with similar mechanisms of action such as herbal GABA agonists (*Valeriana, Piper methysticum, Hypericum, Passiflora incarnata,* and others) may be worth exploring. There have been few research investigations, except on *Ginkgo biloba*. Puerarin, an isoflavone from the legume *Pueraria*, was found to boost the efficacy of betahistine alone in the treatment of vertigo in human clinical trials.[80]

Herbal approaches to the treatment of tinnitus therapy should be directed at the underlying cause when possible. *Ginkgo* is a fairly specific herb to act as a base for many formulas for tinnitus because it improves circulation to the ear, improves blood pressure, and has antiallergy effects. Also, in TCM *Ginkgo* is said to have an affinity for the kidney meridian, which passes through the ears. It is best to be patient with *Ginkgo* therapy and not give up on it until a full 3 or even 6 months of therapy has been employed. When circulatory insufficiency is believed to underlie tinnitus, other blood movers and warming circulatory tonics may be included in formulas. *Zingiber* is specific for nausea and dizziness related to motion sickness and might be included in formulas for tinnitus accompanied by dizziness or Ménière's disease. When nerve damage and sensorineural lesions are to blame, neural anti-inflammatories such as *Hypericum* may be included in formulas. When no specific therapy, surgical or otherwise, is possible, and when herbal and other alternative therapies are ineffective, the use of white noise generators or earphones playing soothing music may improve sleep at night and concentration during the day by masking the distracting tinnitus.

Ginkgo Simple for Chronic Tinnitus

When no underlying cause can be identified, try *Ginkgo* for tinnitus.

Ginkgo biloba

Take 1 teaspoon 4 times a day for 6 months, evaluate efficacy, and continue as indicated. Alternatively, use commercial encapsulations. Take 2 capsules, 2 or 3 times per day.

Tincture for Tinnitus from Nerve Damage

Hypericum, Panax, and *Rosmarinus* all are reported to have neuroprotective effects and may be employed in formulas for tinnitus when it is suspected to be due to sensorineural damage.

Ginkgo biloba
Hypericum perforatum
Panax ginseng
Rosmarinus officinalis

Combine in equal parts. Take 1 teaspoon, 4 times a day for 6 months, evaluate efficacy, and continue as indicated.

Tincture to Aid Recovery from Ototoxic Drugs

This formula is aimed at enhancing liver clearance of ototoxic drugs with *Curcuma, Hypericum*, and *Silybum*. *Ginkgo* and *Hypericum* may provide neuroprotective effects, speeding recovery.

Ginkgo biloba
Hypericum perforatum
Curcuma longa
Silybum marianum

Combine in equal parts. Take 1 teaspoon, 6 times a day for several weeks.

Formulas for Vertigo

Disturbed proprioceptive sense may result from abnormal fluid in the inner ear, otoliths, or semicircular canals, as well as from inner ear infections, nerve disorders, ear lesions, and tympanic membrane growths. Ménière's disease may involve acute episodes of vertigo, and ototoxic drugs and panic attack can also cause acute dizziness. The sensation of spinning is intensely debilitating and can be so severe as to provoke nausea and vomiting. The profound symptoms can necessitate people to hold on to walls and objects when attempting to walk, and to close their eyes to avoid becoming acutely nauseous. Benign paroxysmal positional vertigo is the most common type of vertigo and occurs due to poorly distributed otoliths in the inner ear due to age or "inflamaging." Benign paroxysmal vertigo is characterized by sudden spinning episodes of short duration, usually in association with a change in position such as rolling over in bed.

Treatments for vertigo should be directed at any underlying cause identified. Epley's maneuver is a physical therapy technique to help settle disturbed otoliths and is helpful for benign paroxysmal vertigo. Otherwise, herbal formulas might be aimed at improving circulation, reducing nerve inflammation, clearing fluid from the ears, or alleviating stress and panic, as may be specifically indicated for individual patients.

Tea for Vertigo in the Elderly

Puerarin may improve cerebrovascular circulation, inhibit platelet aggregation, and promote microcirculation, thus easing central vestibular vertigo. *Salvia miltiorrhiza* is noted to enhance circulation in many tissues, including the brain.[81]

Pueraria montana var. *lobata*
Angelica sinensis
Centella asiatica
Rosmarinus officinalis
Salvia miltiorrhiza

Combine in equal parts, or alter to taste, adding cinnamon or licorice for flavor, if desired. Gently simmer 1 teaspoon per 1 cup (240 ml) of water for 10 minutes,

Motion Sickness, Scopolamine, and Marijuana

Motion sickness is a complex syndrome that may be triggered not only by physical motion, such as riding a merry-go-round, but also visual and virtual sensations of motion. Although nausea and vomiting are the hallmark of motion sickness, less obvious symptoms include cold sweating, pallor, increases in salivation, drowsiness, headache, and even severe pain. Sopite syndrome is profound drowsiness and persistent fatigue as a variant of motion sickness, with repetitive yawning being an observable sign. Sopite syndrome can persist for hours or even days; when exposure is prolonged, it may last even longer and may also trigger personality changes typified by apathy.

Scopolamine, a tropane alkaloid found in Solanaceae family plants including *Datura stramonium* and *Atropa belladonna*, has long been used for motion sickness and may also improve vertigo due to various causes. Scopolamine acts as a nonselective muscarinic antagonist[82] and can be administered by transdermal, oral, subcutaneous, ophthalmic, or intravenous routes, often in the form of scopolamine hydrobromide. Scopolamine hydrobromide-loaded microparticles can be ingested without water or chewing, making them particularly useful for geriatric, pediatric, and traveling patients who suffer from motion sickness.[83] Blockade of the muscarinic receptors will allay the nausea of motion sickness fairly reliably, but anticholinergic effects can cause the side effect of dry mouth, throat, and nasal passages. Overdose of scopolamine or any of the tropane alkaloids can result in impaired speech, blurred vision, light sensitivity, constipation, and difficult urinating.

The cannabinoids are among the active components of marijuana, *Cannabis sativa*, and specific endocannabinoid pathways and cannabinoid neurotransmitter receptors have been identified in the brain. Endocannabinoid pathway activation also improves motion sickness,[84] as well as allays nausea and vomiting due to chemotherapy.[85] Agonism of cannabinoid receptors inhibits emesis[86] and may provide symptomatic relief in cases of motion sickness. Individuals with impaired endocannabinoid activity are prone to the most severe motion sickness. The cannabidiols, nonpsychotropic components in marijuana, also exert antinausea effects due to activity at serotonin receptors.[87]

then let stand 10 minutes more. Strain and drink 3 cups (720 ml) or more per day long term.

Tincture for Vertigo in the Elderly

Vertebrobasilar ischemia is an otolaryngological disease, common in elderly patients, and related to poor circulation to the head. When vertebrobasilar arteries become atherosclerotic or otherwise compromised, the resulting dizziness or vertigo is referred to as central vestibular vertigo. Elderly individuals with difficulty balancing may benefit from supporting circulation to the vestibular apparatus. Several studies have shown puerarin, an isoflavone in *Pueraria*, to boost circulation in the basilar arteries and boost the efficacy of betahistine.[88] Alkaloids from *Vinca*, the vincamines, have pronounced cerebrovasodilatory and neuroprotective activity[89] and might help alleviate this type of vertigo.

Ginkgo biloba
Vinca minor
Pueraria montana var. *lobata*
Panax ginseng

Combine in equal parts and take ½ teaspoon, 3 or more times daily for at least 6 months, watching for gradual improvement. Continue long term if the results are encouraging.

Tea for Vertigo Based on TCM

Gastrodia elata and *Ligusticum striatum* (also known as *L. chuanxiong*) are a traditional duo, long used in TCM to enhance circulation to a variety of organs and tissues, including the brain. The duo may enhance cerebrovascular circulation and offer neuroprotective effects in ischemic situations.[90]

Ginger for Motion Sickness

Car sickness and motion sickness (kinetosis) could be considered a type of vertigo, as the rapid motion affects the semicircular canals and the vestibular apparatus of the inner ear. Ginger, *Zingiber officinale*, is a long-standing remedy for motion sickness. Among the molecular constituents that give ginger its hot and spicy flavor are the gingerols and shogaols. Approximately a dozen studies report ginger medications to improve nausea and vomiting due to anesthesia,[91] chemotherapy,[92] and medication side effects,[93] and several others show efficacy in nausea of pregnancy.[94] The efficacy of ginger for vertigo and nausea due to motion sickness is less studied. A chewing gum containing

gingerols and shogaols is shown to have a protective effect against motion sickness.[95]

Zingiber officinale, ginger

Gastrodia elata
Ligusticum striatum

Combine in equal parts. Decoct 1 teaspoon per cup of hot water for 10 minutes. Aim to consume 2 to 4 cups (480 to 960 ml) a day, long term.

Tincture for Vertigo with Infection

Inner ear infections and acute labyrinthitis can provoke vertigo (see the Tincture for Inner Ear Infection on page 162). This formula is aimed at reducing nerve inflammation with *Hypericum* and treating ear congestion and infection with *Allium* and *Origanum*.

Ginkgo biloba
Hypericum perforatum

Allium sativum
Origanum vulgare

Combine in equal parts and take 1 teaspoon, 3 to 4 times daily, long term.

Tincture for Vertigo with Panic Attacks

When episodic vertigo is associated with dyspnea, tachycardia, and an urgent sense of fear, nervine herbs, adaptogens, and stress management may be helpful.

Hypericum perforatum
Withania somnifera
Eschscholzia californica
Rhodiola rosea

Combine in equal parts and take 1 teaspoon, 3 to 4 times daily, long term.

Formulas for "Colds"

Current medical thinking on "colds" and URIs revolves around pernicious "germs" as the underlying cause, but herbalists and naturopathic physicians contend that microorganisms are not truly the culprits, because "germs" can thrive only when a hospitable ecosystem is provided. Therefore, remedies for "common colds" should not be *anti*biotics and *anti*histamines but rather agents that simultaneously improve the tissue status, immune system, blood sugar, or other factors that contribute to the fertile soil for rhinoviruses and other microbes.

Herbal therapies for acute colds (characterized by nasal congestion and rhinorrhea) might include drying agents to help decrease the breeding ground in which such microbes thrive and antiallergy herbs to decrease secretions that contribute to the fertile ecosystem. Immunosupportive, lymphagogue, and alterative herbs are also indicated when colds are chronic, or tend to be extreme, persistent, and debilitating, or repeatedly progress into coughs, bronchitis, sinusitis, or otitis infections. For those with chronic and severe colds, lifestyle approaches are appropriate as individually indicated, and might include dietary changes such as elimination of dairy or sugar, adrenal support, and use of nervine herbs. Getting more rest is also therapeutic for those with stress-related immunodeficiency. The use of immunomodulating herbs for several months is useful for those who become quite ill with simple colds, have poor and lengthy recoveries, or display other sorts of recurrent infections.

Therefore, a formula that could be called a "cold remedy" would mainly be a palliative tool to shorten the duration of a cold and help alleviate any uncomfortable symptoms. For those who are rarely ill and for whom colds do not tend to move to the ears, sinuses, or chest, such simple palliatives might be all that is needed. On the other hand, for those who are febrile, miss work or school often, and have repetitive colds and other infections, a more thoughtful approach is required. This is also the case when superficial colds lead to deeper infections such as otitis, sinusitis, or bronchitis. Chapter 2 of this volume provides additional discussions on supporting the immune system and boosting microbial resistance, as well as further guidance on reducing allergic reactivity, which often yields damp mucous membranes that are hospitable to opportunistic organisms. Some people are constitutionally "damp" and may benefit from diet changes, such as avoidance of dairy products, and from the use of upper respiratory astringents and prophylactic steam inhalation or saunas during cold and flu season. The following formulas are crafted to address the many underlying situations that may predispose to URIs.

Tincture to Relieve Acute Cold Symptoms

This all-purpose formula can both treat and palliate simple colds and be all that is needed to speed recovery. This formula is just a sample, as there are many other herbs capable of doing similar things. I chose these particular herbs because of their ability to both move stuck mucus (*Allium* and *Armoracia*) and dry excessive secretions (*Thymus* and *Euphrasia*), while simultaneously supporting the immune system and providing antimicrobial activity.

Allium sativum
Thymus vulgaris
Euphrasia officinalis
Phytolacca americana
Armoracia rusticana

Combine in equal parts and take 1 teaspoon hourly, reducing as symptoms improve.

Tincture for Frequent Colds

Those with deficient constitutions and poor immunity would benefit from taking a formula such as this regularly for 3 to 6 months, especially prior to and during cold and flu season, allergy season, or during times of stress or poor sleep. Look for signs of immune deficiency, such as lingering infections of various types (e.g., sinus, tonsil, or ear), that tend to worsen and require antibiotics if left untreated, or simple colds that develop into sore throats, headaches, muscle aches, and systemic malaise.

Astragalus membranaceus	15 ml
Panax ginseng	15 ml
Allium sativum	15 ml
Curcuma longa	15 ml

Take ½ to 1 teaspoon of the combined tincture 3 to 4 times a day for many months. Increase frequency to once every 1 to 3 hours for acute-onset colds.

Tincture for Nasal Congestion Due to Allergies

While winter is the time when many patients develop nasal stuffiness, other patients may develop URIs during the spring and summer months when allergic secretions provide a hospitable ecosystem for microbes. This formula features anti-inflammatory and antiallergy herbs appropriate to use as a preventive medicine for those who develop bacterial infections secondary to allergies. Bear in mind that some patients experience allergies in wintertime rather than spring, when winter rain and dampness allow for mold proliferation. Thus, clinicians should not limit the use of an allergy protocol to hay fever season only.

Curcuma longa
Euphrasia officinalis
Thymus vulgaris
Tanacetum parthenium
Armoracia rusticana

Combine in equal parts. Take 1 teaspoon, 3 to 4 times a day for many months. Increase frequency to once every 1 to 3 hours for acute-onset colds.

URI Specifics

The following herbs are specifically indicated in various types of URI presentations.

PRESENTATION	HERBS
Cold that moves to the sinuses	Armoracia rusticana Euphrasia officinalis
Cold with thick mucus and stuffy nose	Allium spp. Armoracia rusticana Eucalyptus globulus
Cold with sneezing and watery eyes	Euphrasia officinalis
Cold that moves to the lungs	Asclepias tuberosa Lomatium dissectum Sambucus nigra Thymus vulgaris
Cold with thick throat mucus and cough	Morella cerifera Thymus vulgaris
Cold with throat pain that moves to tonsillitis	Commiphora myrrha Phytolacca americana
Cold with fever and systemic malaise	Actaea racemosa Eupatorium perfoliatum Glycyrrhiza glabra
Cold that evolves slowly from allergic congestion	Astragalus membranaceus Petasites hybridus Tanacetum parthenium
Chronic and lingering infections	Adaptogens and/or alteratives

Cold Tincture for "Damp" Constitutions

Those with chronic postnasal drip and profuse, easy respiratory mucous membrane secretions are said in energetic traditions to be "damp." Warm, astringent herbs with specificity for the respiratory passages, such as *Commiphora*, *Thymus*, and *Morella* are appropriate for such individuals. Warming, drying adaptogens include *Rhodiola* and *Phytolacca*, which have an affinity for the lymphatic tissue useful for congestion associated with chronic ear and throat infections.

Thymus vulgaris	10 ml
Morella cerifera	10 ml
Curcuma longa	10 ml

Rhodiola rosea	10 ml
Phytolacca americana	10 ml
Commiphora myrrha	10 ml

Take 1 teaspoon of the combined tincture 3 to 4 times a day for many months, increasing to every 1 to 3 hours for acute-onset colds.

Tea for Seasonal Allergies Based on TCM

This tea is based on a formula from TCM. These dry herbs are available in the United States from a variety of suppliers. Individual ingredients may be omitted when not available.

Paeonia × suffruticosa roots	3 ounces (90 g)
Mentha piperita leaves	3 ounces (90 g)
Scutellaria baicalensis roots	2 ounces (60 g)
Glycyrrhiza glabra	1 ounce (30 g)
Platycodon grandiflorus root	1 ounce (30 g)
Ziziphus jujuba fruits	1 ounce (30 g)
Coptis chinensis roots	1 ounce (30 g)
Citrus aurantium peels	1 ounce (30 g)
Lonicera japonica flowers	1 ounce (30 g)

Mix all ingredients together, except for the *Mentha* and *Lonicera*. Decoct 1 heaping cup of the blend in 12 cups (2,880 ml) of water in a crock pot set to the lowest setting. Simmer gently for 45 minutes, then turn off the heat and add the *Mentha* and *Lonicera*. Steep for 20 minutes before straining. Aim to drink 3 cups (720 ml) or more per day during allergy season, or to treat acute symptoms.

Tincture for Chest Colds

Many of the herbs in the Cold Tincture for "Damp" Constitutions have specificity for the upper respiratory mucous membranes and throat. In this tincture for a cold that has moved to the chest, *Allium* and *Lomatium* help warm and dry the pulmonary membranes and have a greater affinity for the lungs. A separate formula that addresses underlying immunodeficiency; dairy, wheat, or other intolerance; or allergic tendencies may also be appropriate when chest colds occur frequently.

Echinacea angustifolia	15 ml
Allium sativum	15 ml
Lomatium dissectum	15 ml
Thymus vulgaris	15 ml

Combine in equal parts. Take 1 teaspoon, 3 to 4 times a day for many months, increasing to every 1 to 3 hours for acute-onset colds.

Tincture for Sinus Colds

This formula is more specific for a cold that moves into the sinuses, due to the inclusion of *Euphrasia* and *Armoracia*. *Euphrasia* reduces both viral and allergic inflammation in the sinuses. *Armoracia* is a heating and expectorating agent specific to the sinuses, helping to thin congestion, reduce sinus pain, and prevent bacterial infections. *Achillea* and *Thymus* are warming and drying, and give this formula antimicrobial power.

Euphrasia officinalis	15 ml
Armoracia rusticana	15 ml
Thymus vulgaris	15 ml
Achillea millefolium	15 ml

Take 1 teaspoon of the combined tincture every 30 to 60 minutes, reducing as symptoms improve. Best used at the onset of new symptoms. Use frequently to prevent sinus infections in those with chronic tendencies or use twice a day during cold and allergy season.

Tincture for Colds with Tonsillitis

When a cold moves to the throat or ears, the result can be tonsillitis or otitis typically involving lymph congestion and swelling of the pharyngeal mucosa, impairing the drainage of the inner ear into the throat. *Phytolacca* is specific for lymphatic congestion, and *Commiphora* is highly specific for the throat and tonsils.

Euphrasia officinalis
Allium sativum
Phytolacca americana
Commiphora myrrha

Combine in equal parts. Take 1 teaspoon every 30 to 60 minutes at the onset of new symptoms, reducing as symptoms improve.

Formulas for Sinusitis

Chronic rhinosinusitis may affect over 10 percent of the general population. Bacterial, viral, and allergic phenomena may underlie sinusitis. Allergic rhinitis involves rapid onset of sneezing and itching sensations as mast cells are activated and release histamine; it quickly progresses to include obstructive symptoms and evolves into secondary bacterial infections. Ciliary impairment, which may result from long-term smoking, or chronic exposure to respiratory irritants and toxins may predispose to rhinosinusitis. AIDS patients, those taking systemic steroids long term, or those on chemotherapy (all of which compromise the immune system) may be prone to sinus infections that do not respond well to antimicrobial therapies. Less common are fungal infections of the sinuses, and when such patients are mistakenly given antibiotics or steroidal nasal inhalants, the infections worsen, and can be a grave illness in immunocompromised individuals.

The two Western herbs that are possibly the most specific for infections in the sinuses are *Armoracia rusticana*, which moves stuck mucus and reduces pressure, and *Euphrasia officinalis*, which is an antiallergy agent having an affinity for the sinus passages. These herbs can be combined with antiviral, antibacterial, and other antiallergy agents as appropriate. Steam inhalation, moist heat packs over the face, and saline or herbal nasal sprays can be complementary options. Treat any underlying immune deficiencies, or allergic hyperreactivity between acute episodes, as appropriate, with long-term immunomodulating therapies. Quercetin and/or *Hydrastis canadensis* may be formulated in nasal sprays, and neti pot lavage may help reduce chronic infections (see Herbal Sinus Lavage on page 173).

Eucalyptus globulus is a traditional herbal remedy to open the airways, including the sinuses. Cineole, a widely studied terpenoid in eucalyptus essential oil, is credited with antiseptic and decongestant properties.[96] Camphor and menthol are additional volatile oils useful to open obstructed respiratory passages; they are found in pine and mint essential oils, respectively. Bromelain, a proteolytic enzyme obtained from pineapple, has an anti-inflammatory and expectorating effect. The aromatic isothiocyanates in mustard and horseradish (*Armoracia*) have expectorating effects at small doses and are featured in many of the formulas in this chapter. *Andrographis paniculata* is traditionally used to treat upper respiratory tract infections and shown to shorten the duration of the common cold. Elderberry preparations (*Sambucus nigra*) are also traditional to treat common colds and flu and may be included in sinusitis formulas for their antiviral and mucolytic effects. *Scutellaria baicalensis* and *Eleutherococcus senticosus* may exert mucolytic and anti-inflammatory effects in the respiratory mucous membranes.[97]

Tincture for Acute Sinusitis

Due to the effects of histamine, allergic reactivity in the upper respiratory passages often results in profuse secretions that then thicken and serve as a fertile breeding ground for secondary bacterial infections. Reducing histamine and drying excessive secretions can reduce the tendency for stagnant fluid to result in a sinus infection. *Euphrasia* reduces the underlying allergic phenomena, *Armoracia* keeps mucus thin and flowing, and *Thymus* helps to dry and reduce abundant mucus, all serving to treat and prevent sinus infections. *Allium* is another possible ingredient in sinusitis formulas, as it is antibacterial, antiviral, and also helps thin and move stuck mucus. Use *Allium* as a medicinal food, or complement this tincture with garlic capsules.

Euphrasia officinalis
Armoracia rusticana
Curcuma longa
Thymus vulgaris

Combine in equal parts. Take ½ to 1 teaspoon every 30 to 60 minutes, reducing as symptoms improve.

Nasal Spray for Sinusitis

Nasal sprays are a handy method for delivering herbs directly to the nasal mucosa to help decongest tissues, allay pain, and deter infection. Appropriate spray bottles are available from pharmaceutical supply houses and specialty bottle companies. This formula contains no preservative and is intended for immediate use over a period of several days.

Sterile saline	30 ml
Quercetin powder	2 teaspoons
Mentha essential oil	1 drop

Warm the saline liquid and dissolve the quercetin powder by stirring vigorously. Transfer to a 1-ounce (30 ml) squeeze bottle with a nasal applicator and add the mint essential oil. Use in lieu of antihistamine or steroidal nasal sprays and to help wean from such products. Can also be used as an adjunct therapy to complement acute and chronic sinusitis protocols.

Steam Inhalation for Acute Sinusitis

Steam inhalation is a time-honored therapy to deter infectious microbes, to thin stuck mucus so it flows, and to decongest the sinuses and quickly reduce sinus pain and pressure. The addition of small amounts of essential oils to the pot of steaming water can enhance the effect.

Citrus aurantium essential oil	5 drops
Eucalyptus spp. essential oil	5 drops
Mentha piperita essential oil	5 drops

Bring a pan of water to a boil, then turn off heat. Drop the *Eucalyptus*, *Mentha*, and *Citrus* essential oils on top of the steaming water. Position head over the steam and inhale through the nose for 5 to 15 minutes. Keep the eyes closed, because the evaporating essential oils may sting. Hold a towel over the head to form a simple tent, which helps hold in the humidity and vapors and enhances the therapy.

Formulas for Nasal Polyps

Nasal polyps usually result from long-standing allergic inflammation of the nasal and sinus tissues causing the lamina propria in the nasal mucosa to proliferate. Chronic infections, heavy use of aspirin, and cystic fibrosis are also associated with benign polyps. Malignant polyps also occur and are usually single and unilateral, where inflammatory polyps are often multiple and bilateral.

Nasal polyps can interfere with olfaction, cause face pain and discomfort, and worsen the tendency to congestion and infection. When these symptoms are severe and chronic, polyps are sometimes surgically removed, although they tend to recur unless significant changes in underlying allergies or other cause can be accomplished. Antioxidant nutrients, essential fatty acids, and herbal medicines are among the best tools for helping to correct hypersensitivity, as with all allergies. Antiallergy herbs to include in formulas for respiratory allergies and nasal polyps include *Tanacetum parthenium*, *Euphrasia officinalis*, and *Ginkgo biloba*. Respiratory astringents as well as herbs that have been referred to folklorically as anticatarrhals would also be appropriate, especially when there is chronic mucus and nasal stuffiness. *Salvia miltiorrhiza*, *Thymus vulgaris*, and *Rosmarinus officinalis* make good drying agents for excessive mucus, and

Allium species and *Armoracia rusticana* are useful for thinning and moving stuck mucus. Both actions may be used as synergists in formulas for nasal polyps. Plants that contain berberine might be used in herbal snuffs and washes to halt the growth and recurrence of nasal polyps. There are similarities between the formulas in this section and those for other respiratory problems, but this section demonstrates how formulas might be altered to address polyps related to chronic allergies versus polyps related to chronic infections.

Tincture for Nasal Polyps Associated with Allergies

Euphrasia, *Tanacetum*, and *Ginkgo* all have antiallergy properties, while *Thymus* has drying effects on excessive mucus in the respiratory passages. Thus, this formula is aimed at those who have polyps associated with chronic allergic rhinitis.

Euphrasia officinalis
Tanacetum parthenium
Ginkgo biloba
Thymus vulgaris

Combine in equal parts. Take 1 dropperful 3 to 6 times a day, and as often as hourly when acute symptoms are present.

Tincture for Nasal Polyps with Infections

This formula has greater antimicrobial effects than the Tincture for Nasal Polyps Associated with Allergies. It is aimed at those with nasal polyps associated with frequent colds, sinusitis, or bacterial infections.

Euphrasia officinalis
Achillea millefolium
Berberis aquifolium
Armoracia rusticana
Eucalyptus globulus

Combine in equal parts. Take 3 to 6 times per day as a preventive between acute infections, or as often as hourly at the onset of new infections, reducing as symptoms and infections improve.

Simple Saline Nasal Lavage

Nasal saline irrigation thins secretions and supports flow and removal. Although one Cochrane review did not find a substantial benefit, a systematic review reported saline douching to be of benefit for adults and children with acute rhinosinusitis.[98] Many of my own patients credit the

> ### Nasal Lavage Herbal Options
>
> The following herbs, concentrates, and natural ingredients can be used in neti pots to lavage the nasal and sinus passages. Nasal lavage can be a helpful adjuvant therapy in the treatment of acute sinusitis and may also deter regrowth of polyps in those with a history of surgical intervention and chronic infections.
>
> | | |
> |---|---|
> | Saline | *Matricaria chamomilla* |
> | Xylitol | *Eucalyptus globulus* |
> | Quercetin powder | *Mentha piperita* |
> | Berberine powder | *Pinus* spp. |
> | *Curcuma longa* | *Capsicum* spp. |
> | *Hydrastis canadensis* | *Nicotiana tabacum* |

regular use of a neti pot nasal lavage to have broken the cycle of recurrent nasal polyps and nasal polyp surgeries.

Sea salt, or other noniodized salt

Dissolve ½ to 1 teaspoon of fine sea salt in 2 cups (480 ml) of water, warming to a comfortable temperature. Transfer the liquid to a neti pot or squeeze bottle and use to lavage each nostril at least once daily.

Herbal Sinus Lavage

The use of nasal rinses, such as weak saline in a neti pot, is reported to improve chronic inflammation of the nasal mucosa.[99] Herbal tinctures can be diluted with saline for such purposes. A better and less expensive option is to prepare small quantities of herbal tea for this purpose. *Hydrastis*, *Phellodendron*, *Coptis*, and *Berberis* have been frequently mentioned in the folkloric literature for nasal rinses, and even as a nasal snuff when finely powdered. They all contain the isoquinoline alkaloids berberine and hydrastine. The berberine alkaloids are antimicrobial and anticatarrhal agents, and are often very effective once the skill of nasal lavage is mastered.

Euphrasia officinalis
Achillea millefolium
Hydrastis canadensis

Combine in equal parts. Steep ½ teaspoon per ¼ cup (60 ml) hot water or saline for 10 minutes. Strain in a fine mesh strainer or coffee filter. Use with a neti pot or place in small nasal squeeze bottles to use as often as hourly and at least 3 times per day. The liquid should be made fresh each day.

Herbal Snuff Powder for Nasal Polyps

Nasal snuffs are not commonly employed in the modern era, perhaps due to the minor discomfort they cause, but they were popular for treating sinus infections and nasal polyps prior to the widespread availability of antibiotics. Quercetin powder may be obtained in its pure form, or it may be omitted from the formula.

Hydrastis canadensis powder
Quercetin powder
Calendula officinalis powder

Combine in equal parts. Snuff ¼ teaspoon or less into each nostril 3 or 4 times per day.

Therapies for Epistaxis

Although nosebleeds may seem like a trivial matter, in some cases they belie serious underlying illness, and can occasionally be so hemorrhagic as to be life-threatening. In most cases, however, occasional bouts of epistaxis, which may affect around 10 percent of people overall, are due to simple dry nasal capillaries or to uncontrolled hypertension. Kiesselbach's plexus, the network of capillaries in the upper nasal mucosa, may spontaneously crack and bleed in dry weather, or in heated indoor environments. A foreign body in the nose is a possible cause of long-lasting epistaxis in children, and excessive aspirin usage or mild scurvy are other possible causes in general. Serious diseases that underlie chronic epistaxis include leukemias, thrombocytopenias, coagulation disorders, tumors, and hypertensive crises. Profuse epistaxis can require cautery of the nasal vasculature in some severe cases.

Simple epistaxis due to dry air can be treated with the use of hydrating *Calendula officinalis* or other salve, directly in the nostrils several times a day. Complement this with increased water intake and essential fatty acid and vitamin C supplements. Antiallergy formulas may help those for whom allergic inflammation causes epistaxis, and of course, hypertension should be addressed directly.

In the event of profuse active bleeding, styptic formulas such as the Palliative Styptic Tincture for Epistaxis may control the situation, while systemic therapies for blood pressure, allergies, or severe dryness are addressed.

Palliative Styptic Tincture for Epistaxis

Some of the most reliable herbal styptics include *Cinnamomum*, *Achillea*, and *Erigeron*. The ergot fungus also has powerful styptic qualities due to the presence of the vasoconstrictive alkaloid ergotamine. One hydrotherapy technique to help drive blood out of the nose is to pinch the bridge of the nose while applying ice or cold rags to the back of the head and neck, and while soaking the feet in the hottest water comfortably tolerated. This helps to pull more blood volume out of the nasal passages and down into the lower body.

Achillea millefolium	½ ounce (15 ml)
Cinnamomum verum	¼ ounce (7.5 ml)
Erigeron canadensis	¼ ounce (7.5 ml)

Take 1 dropperful of the combined tincture by mouth every 5 to 10 minutes, while using the above hydrotherapy technique.

Formulas for Dental and Gingival Issues

Dental caries is promoted by plaque formation on the teeth resulting from undesirable microbes, such as *Streptococcus mutans*, which ferments carbohydrates and releases acidic compounds that may demineralize the teeth.[100] Optimizing microflora and limiting carbohydrates lingering in the mouth are key in reducing the development of caries. Babies and toddlers who fall asleep with breast milk or fruit juice in the mouth may experience demineralization of the teeth. Saliva serves an important role in creating

an environment for desirable microbes and buffering the teeth. Inadequate saliva, such as occurs with Sjögren's syndrome, and habitual mouth breathing can leave teeth unprotected and are associated with an increased tendency to caries and periodontal and gingival inflammation. *Sanguinaria canadensis* may deter these cariogenic bacteria and prevent their adherence to teeth without disrupting beneficial flora important to the oral cavity.[101] Propolis may also deter caries and is sometimes included in commercial toothpastes.[102] The triterpenes ceanothic acid and ceanothenic acid from *Ceanothus americanus*[103] and thymol from *Thymus vulgaris*[104] are active against *S. mutans*. In addition to antimicrobial herbal mouthwashes, some patients (such as those with osteoporosis) may benefit from mineral-rich herbs to support tooth density. There is no research on dental re-mineralization, but *Actaea racemosa*, *Pueraria montana* var. *lobata*, and soy isoflavones are estrogenic and promote bone mineral density.[105]

This section also addresses periodontal disease, a common oral disorder that affects over 50 percent of all adults. Periodontal diseases are inflammatory disorders that damage the connective tissue around teeth and can involve loss of alveolar bone that leads to tooth loss. Plaque formation is the primary etiological agent for periodontal disease initiation and progression. Shifts in the oral microbial ecosystem may contribute to the initiation of inflammatory processes and accumulation of pathogenic bacteria at the gingival margin that contribute to plaque formation. The accumulation of plaque triggers inflammatory processes including the production of exudate and leukocyte migration into the gingival crevice. Periodontitis occurs when the inflammatory state is severe or advanced enough to trigger cytokine production and the recruitment of neutrophils and other inflammatory cells, leading to permanent damage of gingivomucosal tissues. Periodontitis has been associated with increased incidences of adverse cardiovascular events, adverse pregnancy outcomes, diabetes, and rheumatoid arthritis. Pathogenic bacteria such as *Porphyromonas gingivalis* may travel through the bloodstream to other tissues and trigger inflammatory and possibly autoimmune reactivity.

The essential oils found in many aromatic herbs are noted to inhibit oral pathogens associated with plaque formation and periodontal disease. Tea tree oil (*Melaleuca alternifolia*), for example, has been shown to deter periodontal disease, and researchers have suspended the essential oil in methyl cellulose gel to help prolong the surface contact and adherence.[106] Clove (*Syzygium aromaticum*), thyme (*Thymus vulgare*), and many other essential oils are powerful antimicrobials that may deter periodontal disease, and Listerine mouthwash, when first released in 1879, employed essential oils as a daily preventive measure "to kill germs that cause bad breath." *Camellia sinensis* (green tea) is rich in flavonoids known as catechins, such as epigallocatechin gallate, reported to improve periodontal health by reducing inflammation, preventing bone resorption, and limiting the growth of certain bacteria associated with periodontal diseases.[107]

Piper methysticum (kava) is credited with anti-inflammatory and antiarthritic effects, and research shows oral rinses with the herb to deter periodontal and alveolar bone destruction in part by inhibiting *P. gingivalis* biofilms.[108]

Capsules to Support Tooth Density

While there is no research about the effect on tooth density of the herbal supplements recommended here, they are rich in minerals and recommended as inexpensive sources of absorbable organic mineral complexes.

Centella asiatica capsules
Urtica dioica capsules
Medicago sativa capsules

These capsules may be taken individually or in tandem, or in a repeating rotation cycle, using long term to support high mineral intake. Take 1 to 3 capsules, 3 times per day.

Organic Fluoride Herbal Tea

The herbs in this tea are among the highest sources of natural fluoride and may be an alternative to water fluoridation or the use of fluoride mouth rinses.

Glycyrrhiza glabra
Camellia sinensis

Combine in equal parts. Steep 1 tablespoon per cup hot water for 10 minutes, then strain. Drink 2 to 3 cups (480 to 720 ml) per day.

Xylitol to Inhibit Dental Plaque

Xylitol is a sweet-tasting sugar alcohol available as a sugar-like white powder; it is used in noncariogenic chewing gums. Xylitol has benefits over sucrose in that it can't be utilized or fermented by plaque bacteria in the oral cavity,[109] and yet, it can be absorbed and metabolized in the gut. Furthermore, xylitol deters *Streptococcus mutans* and has an inhibiting effect on dental plaque, and therefore, cavities. Xylitol chewing gums may reduce the tendency to dental caries,[110] and chewing also stimulates saliva flow, which itself helps protect against caries.

Fluoride and Tooth Decay

Fluoride binds with apatite, contributing to density of the tooth, and lack of fluoride can lead to tooth weakness. However, excessive fluoride damages the teeth and can cause systemic toxicity. Fluoridation of municipal water is highly controversial due to the narrow window of safety. Approximately 50 percent of US water districts choose not to fluoridate their water due to toxicity concerns. Natural sources of fluoride include *Glycyrrhiza glabra*, *Momordica charantia*, and *Camellia sinensis*.[111] Overall, moderate consumption of these herbs as teas may be good for the teeth and serve as a natural, nonindustrial source of fluoride. Regular consumption of beverages made from commercial concentrated powdered ice tea products may be problematic, however, because these products are so high in fluoride.[112]

Mouthwash to Reduce Plaque

Glycyrrhiza is used here for its flavor as well as for its fluoride content, and *Sanguinaria* is noted to deter dental plaque. Although extremely caustic in its fresh or concentrated form, *Sanguinaria* has been shown to be helpful in reducing plaque formation compared to placebo.[113] *Sanguinaria* must always be diluted to avoid irritation, and even ulceration, of the mouth and throat. See also "*Sanguinaria* for Oral Conditions" on page 180.

Glycyrrhiza glabra	20 ml
Sanguinaria canadensis	10 ml
Mentha piperita essential oil	30 drops
Xylitol	½ teaspoon, dissolved in water

Combine the two tinctures and the *Mentha* oil in a 1-ounce dropper bottle. Place 20 drops of the prepared blend in a half ounce of xylitol water and swish around mouth for 1 minute, twice a day, after brushing teeth.

Herbal Mouthwash for Periodontal Disease

Hypericum is emphasized in the folkloric literature for wound healing, and modern research supports its use for periodontitis and for limiting bone loss, cytokine expression, and inflammatory and infectious processes in general. *Hypericum* inhibits oral pathogens and promotes healing following dental extraction or other oral surgeries.[114] The topical use of *Hypericum* supports epithelialization and increases hydroxyproline levels in animal models of diabetes-related oral disease.[115] *Hypericum* may improve the success of bone grafts into a tooth socket. The essential oils in clove (*Syzygium*) and oregano (*Origanum*) are also noted to deter oral pathogens associated with periodontal disease, used here as whole herb tinctures rather than concentrated essential oil fractions.

Hypericum perforatum
Syzygium aromaticum
Origanum vulgare
Glycyrrhiza glabra

Combine in equal parts in a 4-ounce (120 ml) bottle. Add 1 to 2 dropperfuls to a glass full of tap water. Swish in the mouth for 1 to 2 minutes, then spit out. Use 2 or 3 times daily, after brushing teeth. Xylitol water or salt water may also be used as the vehicle, especially indicated when acute gingival infections are present.

Tincture for Acute Dental Pain

Belladonna as a homeopathic or in very small dosages of standardized tincture is specific for acute vascular congestion and throbbing pain and has an affinity for the teeth and head. *Hypericum* also has an affinity for the nerves and has anti-inflammatory effects, and clove essential oil has additional numbing effects on the teeth and gums.

Hypericum perforatum	15 ml
Piper methysticum	15 ml
Atropa belladonna	2 ml
Syzygium aromaticum essential oil	20 drops

Take the combined tincture every 15 to 60 minutes, as needed, to allay acute dental pain. Consider complementing with homeopathic *Belladonna* 200c every 30 minutes to several hours, acutely.

Anti-Inflammatory Dental Tincture

This formula for use following dental extractions or in cases of acute dental pain combines general anti-inflammatory anodynes, such as *Hypericum* and *Salix*, with the gum- and pain-numbing herbs *Piper methysticum* and clove oil.

Herbal Support for Dental Extraction and Root Canals

Acute and intolerable pain may occur when tooth infections affect the dental pulp, referred to as pulpitis; this is a dental emergency. Root canals or pulling the entire tooth are the usual solutions. NSAIDs such as naproxen, acetaminophen, or ibuprofen can provide significant pain relief, and botanical medicines may be used as additional anodynes and anti-inflammatories until the dentist can be seen. Among the herbs most specific for oral and dental pain are *Spigelia anthelmia*, *Hypericum*, and *Syzygium aromaticum* (clove). Clove is available as an essential oil for the greatest numbing effects. Clove, Mint (*Mentha piperita*), and Cinnamon (*Cinnamomum verum*) essential oils can also be pain relieving when placed on compresses or included in herbal formulas. *Atropa belladonna* is specific for throbbing, painful, hyperemic conditions, such as pulpitis, and *Aconitum napellus* and *Gelsemium sempervirens* can deaden nerve pain, but are to be used in drop dosages only because they are toxic in larger amounts. My approach is to consider incorporating powerful herbs such as these in severe pain formulas prior to resorting to steroids or prescription pain remedies. Recommending over-the-counter NSAIDS would be appropriate to help relieve pain as well. General anti-inflammatory anodyne herbs include *Piper methysticum*, *Curcuma*, *Zingiber*, *Salix*, and *Piscidia piscipula*. Bromelain, an anti-inflammatory derived from pineapple that has an affinity for vascular and connective tissue, is available in capsule form; it is complementary to pain relievers and herbs. Take 500 milligrams of bromelain every 2 hours following dental surgery to reduce pain and swelling and speed healing.

Piper methysticum	20 ml
Hypericum perforatum	15 ml
Salix alba	10 ml
Piscidia piscipula	5 ml
Commiphora myrrha	5 ml
Syzygium aromaticum essential oil	5 ml

Take ½ to 1 teaspoon of the combined tincture 4 to 6 times per day following dental extractions and surgical procedures, as an alternative to pharmaceuticals and synthetic medications. Reduce each day as symptoms improve.

Antimicrobial Mouthwash to Prevent Infection

I have used formulas such as this instead of antibiotics for high-risk heart patients following dental procedures, albeit with cautious vigilance. This formula combines anti-inflammatories for pain as well as powerful, broad-acting antimicrobial agents to speed healing and prevent infection.

Hypericum perforatum	10 ml
Glycyrrhiza glabra	10 ml
Commiphora myrrha	10 ml
Salvia officinalis	10 ml
Allium sativum	10 ml
Sanguinaria canadensis	5 ml
Syzygium aromaticum essential oil	5 ml

Combine the ingredients and swish 2 teaspoons in the mouth for 2 minutes, then swallow. Use 3 or more times each day. Begin 24 to 48 hours prior to dental procedures and continue for 3 or 4 days following. Those prone to infection may use more frequently and continue for a full week following dental extractions and procedures. This formula may also be used topically; saturate a gauze pad or dense piece of cotton and place against the tooth socket.

Formulas for the Lips, Tongue, and Mouth

Glossitis usually involves chronic burning pain in the tongue, worse with food and drink. When ulcers are present, herpes, streptococcal infection, TB, erythema multiforme, or pemphigus should be ruled out. When whitish patches occur on the tongue, *Candida*, leukoplakia, and lichen planus are differential considerations. A tongue that looks as though it has hair growing on it occurs due to an overgrowth of fungi or bacteria. This is usually seen only in immune-compromised individuals or in those who use antibiotics regularly. Burns and trauma to the tongue may cause it to swell, occasionally so severely as to block the airways, and necessitate hospitalization and intubation.

Due to many diverse causes of glossitis, patients should be evaluated for nutritional status, endocrine disorders, blood sugar imbalances, the presence of pathogens, digestive function, alcohol use, and other factors. Unusual lesions may be biopsied to support

Herbs for Tongue Symptoms

The tongue is regarded as being a reflection of health of the digestive system, and in fact the entire body. These herbs are mentioned in the folkloric and Eclectic physicians' writings as having a variety of specific tongue symptoms to include in herbal formulas as indicated.

CONDITION	RECOMMENDED HERBS	CONDITION	RECOMMENDED HERBS
Burning sensation in the mouth	*Aconitum napellus** *Baptisia tinctoria* *Hamamelis virginiana* *Lobelia inflata* *Sanguinaria canadensis***	Large, swollen or flabby tongue	*Artemisia absinthium* *Chelidonium majus* *Chionanthus viriginicus* *Echinacea angustifolia* *Hydrastis canadensis*
Oral catarrh	*Allium sativum*	Sensation of hair on tongue or in throat	*Allium sativum*
Excessive saliva	*Allium sativum* *Lobelia inflata***	"Strawberry" tongue of streptococcal infections	*Atropa belladonna***
Bitter or sour taste in mouth	*Aloe vera* *Baptisia tinctoria* *Dioscorea villosa* *Eupatorium perfoliatum* *Hydrastis canadensis* *Morella cerifera*	Blisters or indentations on sides of flabby tongue	*Hamamelis virginiana* *Hydrastis canadensis* *Thuja plicata*
Metallic taste in mouth	*Convallaria majalis* *Lobelia inflata***	Thick, slimy saliva; thick mucus coat on tongue	*Hamamelis virginiana* *Hydrastis canadensis* *Lobelia inflata*** *Morella cerifera*
Glossitis	*Artemisia absinthium*	Dry tongue; lack of saliva	*Iris versicolor* *Sanguinaria canadensis***
Sore tongue	*Convallaria majalis* *Echinacea purpurea*	Geographic tongue	*Phytolacca americana* *Taraxacum officinale*
Coated tongue	*Aconitum napellus** *Baptisia tinctoria* *Convallaria majalis* *Dioscorea villosa* *Echinacea angustifolia* *Eupatorium perfoliatum*	Fungal plaques on the tongue	*Tabebuia impetiginosa*

* **Caution:** Potentially toxic, drop or diluted doses only

** Drop or diluted doses only

diagnosis and rule out cancer when necessary. Dedicated oral hygiene, B vitamin supplementation, and mouth rinses with *Calendula* and antimicrobial agents can do no harm and will help many. Herbal approaches may address acute discomfort, but therapies that address the underlying cause are desirable and should be sought out. Herbal therapies may improve nutrition, soothe inflammation due to allergen and irritant exposure, or treat underlying infections.

Stomatitis, Glossitis, and Gingivitis

Stomatitis is a nonspecific term for inflammation of the mouth and lips. Glossitis is a nonspecific term describing inflammation of the tongue. There are several types of stomatitis, including herpetic, fungal, and malnutrition-related (particularly vitamin C or B deficiency). Glossitis may also occur as a chemotherapy side effect. Allergic reactivity and food intolerance can cause stomatitis, with citrus and tomatoes being common offenders. Autoimmune disease such as erythema multiforme, lichen planus, Behçet's disease, and lupus may involve stomatitis and oral lesions, and treatments might address microvasculature and connective tissue integrity. Immunomodulatory herbs may also be useful.

For ulcerated oral tissues and connective tissue weakness or inflammation, glutamine may be combined with herbs in a variety of mouth rinses to help regenerate mucosal tissue.[116] Molecular studies have shown *Centella* to normalize inflammatory interleukin levels. *Centella* and other mucosal and collagen-vascular protectants such as *Equisetum*, *Echinacea*, *Aloe*, and *Althaea* might be considered in formulas for autoimmune ulcers and fibrosis as seen with lupus and Sjögren's syndrome.

Glossitis Causes and Remedies

Glossitis may be due to presence of pathogens, nutritional deficiencies, keratinization issues, endocrine disturbances including diabetes, allergic phenomena, and exposure to irritants such as alcohol, tobacco, and spices. Other causes include deficiencies of the following vitamins and other substances:

B vitamins. Nutritional causes of glossitis usually involve B vitamin deficiencies,[117] and as with pellagra, often manifest in a tongue that is red at the tip and edges, or a tongue that is swollen, bright red, inflamed, and ulcerated.

Iron. Iron deficiency anemia may result in a tongue that is pale and smooth.

Vitamin E. Vitamin E deficiency may also be associated with glossitis.[118] The tongue may also become painful and irritated due to gastric reflux or inflammatory disorders of the GI, such as celiac disease.

Zinc. Glossitis may also be associated with zinc deficiency, so zinc might be supplemented, especially when other symptoms of low zinc occur such as alopecia, white spots on the nails, and immune dysfunction.[119]

Stomach acid. Atrophy of the tongue may be associated with low stomach acid and be a reflection of intestinal atrophy, most commonly seen in the elderly, and is associated with general malnutrition.[120]

Protein. Atrophic glossitis may also occur with systemic muscle atrophy and may be associated with protein calorie malnutrition.[121]

Astringent herbs. Astringent and high-flavonoid fruit rinds, teas, and medicines, such as pomegranate (*Punica granatum*)[122] and mangosteen (*Garcinia*)[123] are folkloric remedies for glossitis in tropical areas, and currants (*Ribes*) and blueberry leaves (*Vaccinium*) may also help heal oral tissues. *Salvia*, *Punica*, *Vaccinium*, and *Tabebuia* as teas or diluted tinctures make excellent mouthwash ingredients due to their astringent and antimicrobial activity.

Occasionally, postmenopausal women develop vague and chronic burning sensations of the tongue that appears normal upon examination. This may occur due to disruption of normal keratinization and may respond to hormonal support.

Sanguinaria for Oral Conditions

A number of commercial oral pastes and rinses are reported to treat gingivitis or prevent and reduce plaque formation on the teeth. These often contain sodium lauryl sulfate or chlorhexidine gluconate solutions; however, both can have irritating, allergenic, or toxic effects. Natural agents noted to reduce plaque formation on the teeth include zinc and *Sanguinaria*,[124] a highly caustic herb that when properly prepared is effective against gingivitis and helpful in reducing plaque formation.[125] *Sanguinaria* has a broad spectrum of antimicrobial activity in addition to anti-inflammatory activity. Some researchers have reported that *Sanguinaria* inhibits the adherence of bacteria to oral tissues without causing detrimental shifts in oral flora the way that pharmaceutical antibiotics may.[126] *Sanguinaria* mouthwashes have been noted to reduce bacteria such as *Fusobacterium* associated with gingivitis and plaque formation without leading to an increase in opportunistic organisms such as yeast, staph, or coliform strains.[127]

Sanguinaria may also inhibit the progression of precancerous oral lesions to cancer,[128] yet anecdotal reports have also raised concerns that *Sanguinaria* may promote leukoplakia.[129] One investigation reported that the use of *Sanguinaria*-containing oral products could not be linked to leukoplakia or other toxic side effects.[130] Their studies included human trials involving daily use of *Sanguinaria* for 6 months' time, leaving an unanswered question regarding exposure of many years' duration.

Sanguinaria canadensis, bloodroot

Aphthous ulcers can be extremely painful and take several weeks or more to heal. Severe cases may be accompanied by fever, malaise, and lymphadenopathy suggesting that systemic immune support be used in herbal formulas. For unknown reasons, females are afflicted more often than males. Aphthous ulcers are more common in adolescents and are chronic in some. The cause of aphthous ulcers is unknown but nutritional factors are believed to contribute. Mouth ulcers may also result from chemotherapy, glutamine powder, licorice solid extract, and *Aloe vera* gels are effective bases for formulas for ulcerative lesions of the mouth.

"Cold sores" are due to the herpes simplex virus and may occur only on rare occasion, with quick resolution, or they may occur frequently, with severely painful eruptions and a recovery period of many weeks. In many cases stress, a sunburn, or a lapse in the diet precede the emergence of neuralgic prodromal symptoms, followed by painful vesicles progressing to ulceration in the course of a day or two. The underlying herpes simplex virus is believed to be present for life but may be long dormant and emerge only with significant stressors. Cold sores are most common on the labial borders, but I have seen patients with chronic herpes simplex of the nasolabial crease, behind the ears, in the nostrils, on the immovable oral mucosa, and one case on the external skin of the throat. Any painful neuralgic facial lesion whose onset is vesicular, and quickly ulcerates, and forms thick, honey-colored crusts is suspect and should be swabbed for viral cultures to confirm. Fever and malaise may accompany initial outbreaks. *Ribes nigrum* (black currant) has been shown to inhibit the adherence of the herpes virus to cells and inhibit viral replication due to reduction of protein synthesis in infected cells.[131] *Astragalus* may have activity against the herpes virus as well as offer other immunosupportive activity.[132] *Glycyrrhiza* species have been shown to have activity against numerous viruses, including the herpes virus, and against vesicular stomatitis.[133] For chronic herpes outbreaks, antiviral immunosupportive herbs such as *Commiphora myrrha*, *Hypericum perforatum*, *Melissa officinalis*, *Glycyrrhiza*, and *Allium* therapies may be used between episodes to help prevent further outbreaks.

Allergic inflammation of the lips and the skin around the mouth is fairly common,[134] but oral mucosal allergies are less common. True buccal allergies are characterized by burning pain and profuse salivation. Metal and resins used in dentistry can sometimes trigger allergic stomatitis, and mercury toxicity is often associated with profuse salivation. In cases of allergic stomatitis, patients should eat a bland diet free of food allergens and avoid all food chemicals, cosmetics, hot spices, and other irritants. *Calendula* salve and *Hippophae rhamnoides* (sea buckthorn) preparations topically on the lips can be pain relieving and promote healing. When chronic, underlying atopy should be treated with antiallergy protocols (refer to chapter 2 of this volume for such protocols), such as essential fatty acids, antioxidant vitamins, and B vitamins.

Gingivitis afflicts the periodontal mucosa and can progress to periodontitis, pain, and dental issues. Both dental plaque and generally poor health can predispose, so if dental health appears otherwise good, search for underlying issues such as low-grade scurvy, malnutrition, alcoholism, diabetes, leukemia, or inadequate saliva such as with Sjögren's syndrome. Gingivitis may be acute or chronic and involve swelling, pain, and redness of the gums, and in more advanced cases, friable tissues that bleed and involve exudative secretion.

Topical for Oral Thrush in Infants

It is difficult to give infants medicine, so this formula is made as sweet as possible with a glycerite, and with flavor-masking, pain-relieving *Mentha* essential oil. Gentian violet is another classic remedy but is messy and can permanently stain fabric (see Gentian Violet for Oral Thrush on page 181).

Berberis aquifolium glycerite	25 ml
Commiphora myrrha tincture	7 ml
Mentha piperita essential oil	5 drops

Parents may apply a few drops of the combined mixture to infant's oral cavity with a fingertip, or saturate a small corner of a thin, soft cloth with a small amount of the formula. Spreading the formula on the buccal tissue rather than dropping into the mouth helps avoid dispensing the bitter formula directly onto the infant's tongue.

Oral Thrush in Adults

Oral fungal infections in adults are usually associated with underlying diabetes, hyperglycemia, immune suppression, and steroid use. Transplant recipients on lifelong steroids or those with severe intestinal dysbiosis are especially vulnerable to such infections. HIV positive patients may develop recalcitrant cases of chronic candidiasis. Formulas for stomatitis due to hyperglycemia and candidal infection should include herbs that help balance blood sugar, and dietary measures should be part of a long-term approach. Herbs such as *Cinnamomum verum*, *Trigonella foenum-graceum*, and *Gymnema slyvestre* can be used long term to reduce blood sugar, combined with immunosupportive herbs where appropriate.

Gentian Violet for Oral Thrush

Gentian violet is a traditional remedy for thrush that has been shown to have antibacterial and antifungal activity.[135] Oral candidiasis is one of the most common opportunistic infections in those with diabetes or depressed immune function, and gentian violet is shown to compare favorably to oral nystatin as a therapy.[136] Gentian violet is a chemical agent, not an herb, but is named as such due to the intense blue-purple color of some gentian flowers. One percent solutions are a long-standing remedy considered safe and gentle enough for infants and for HIV-positive individuals with fungal infections of the mouth.

Gentian violet 1 percent solution

For adults, place 1 dropperful in mouth to swish for several minutes and then expectorate. Prepare as a swab to paint the mouth of infants using a cotton swab or a soft cloth. Repeat at least 3 or 4 times a day until infections resolve.

Gargle for Stomatitis and Gingivitis

All of the herbs in this formula are antimicrobial and can be used for both bacterial and fungal stomatitis and plaque-related gingivitis. *Sanguinaria* is specifically inhibitory to plaque formation. A formula such as this might be used locally, while systemic and nutritional therapies might better address underlying diabetes, intestinal dysbiosis, or nutritional deficiencies.

Glycyrrhiza glabra solid extract	20 ml
Salvia officinalis	10 ml
Berberis aquifolium	10 ml
Sanguinaria canadensis	10 ml

Burning Tongue Syndrome

Burning mouth syndrome is diagnosed when uncomfortable tongue sensations recur daily for more than 2 hours per day, for more than 3 months, without clinically evident causative lesions. Because the condition is most common in menopausal women, hormone therapy may be helpful.[137] Alpha-lipoic acid supplementation may also be helpful and should be tried in all cases.[138] Long-standing iron deficiency can also lead to changes in the esophageal tissues and is referred to as Plummer-Vinson syndrome, and may sometimes include oral lesions in association with dysphagia, due to proliferation of inflammatory tissue in the esophagus. *Helicobacter pylori* may infect the oral cavity and is one of the routes of initial infection and may cause a burning tongue.[139] Allergic reactivity to sorbic acid, cinnamon, nicotinic acid, propylene glycol, sodium lauryl sulfate, and benzoic acid may also trigger this syndrome, and all should be eliminated.[140] Capsaicin in the form of hot pepper sauce may deplete substance P and provide palliative relief, but has also been reported to induce lasting remission in some cases.[141]

Other therapies that can be tried include Bi-Est Bio-Identical hormone supplementation, *Glycyrrhiza glabra* solid extract, *Aloe vera* gel, *Matricaria chamomilla* oil, *Linum usitatissimum* (flaxseed) oil. Any or all of the therapies above may be taken 2 or 3 times a day for several months, evaluating the results.

Allium sativum	5 ml
Mentha spp. essential oil	3 ml
Origanum vulgare essential oil	2 ml

Place 1 teaspoon of the combined ingredients in a cup of water or herbal tea and use as a gargle. The liquid is not harmful to ingest, but it is best to expectorate when using frequently to reduce the potential of *Sanguinaria* to irritate the esophagus. Repeat every 40 to 60 minutes, if possible, for several days, reducing as symptoms improve and discontinuing upon resolution of the infection.

Tea Tree and Clove Oral Spray for Thrush

Both *Melaleuca* (tea tree)[142] and *Syzygium* (clove)[143] essential oils are noted to deter oral candidiasis, and may be prepared into a mouthwash or spritzer.

Melaleuca alternifolia essential oil	100 drops
Syzygium aromaticum essential oil	100 drops

Add the essential oils to a 2-ounce (60 ml) bottle and top with water. Shake well before each use. Place a dropperful in a sip of water to further dilute it, and swish around the mouth for at least 2 minutes. Spit out after swishing. Repeat several times per day, or as often as every hour or 2. A spray nozzle can also be attached to the 2-ounce bottle, and the formula simply sprayed on the tongue and mouth throughout the day.

Licorice-Nutrient Rinse for Stomatitis

B vitamin deficiencies are especially associated with oral lesions and stomatitis. Low-grade scurvy often goes unrealized and is a possible cause of chronic stomatitis and should be suspected when the gums bleed readily. B vitamins are available in liquid forms for preparing mouthwashes, and vitamin C crystals can be added to oral solutions. Follow label recommendations on dosage because products vary in concentration. Aim to deliver at least 500 milligrams of vitamin C and the full spectrum of B vitamins per dose. Liquid B products may be labeled for a single daily dose, which can be divided into 3 portions to increase oral exposure to the nutrients over the course of the day. Quercetin offers broad anti-inflammatory and antiallergy effects. The combination nutrient quercetin ascorbate contains quercetin and vitamin C and may also be used. Glutamine is a sweet-tasting amino acid that is used to heal ulcers and promote tissue regeneration, and *Glycyrrhiza* has anti-inflammatory and antiulcerative effects and makes a sweet-tasting vehicle to deliver the other nutrients.

Glycyrrhiza glabra tea

Vitamin C crystals	½ to 1 teaspoon
Liquid B vitamin	1 teaspoon, or as per label advice
Glutamine powder	1 teaspoon
Quercetin powder	⅛ to ¼ teaspoon, as per label advice

Place the vitamin C, liquid B, and the glutamine and quercetin powders in 1 cup (240 ml) of *Glycyrrhiza* tea and stir to blend. Consume by holding each sip in the mouth for a moment and swishing to prolong oral exposure to the nutrients, and then swallow. Repeat the procedure twice during the course of the day to consume 3 such cups (720 ml) in total. Continue for several weeks to several months until the condition resolves.

Anodyne Oral Rinse

As most types of stomatitis and gingivitis are painful, anodyne herbs are usually essential aspects of formulas. For severe pain, steroidal and lidocaine mouth rinses may be offered allopathically, but these herbal anti-inflammatories and anodynes are excellent alternatives. In herpetic and fungal stomatitis, corticosteroids are contraindicated, as they suppress the immune system and may allow microbes to flourish. Natural anodynes include ice; baking soda mouth rinses; essential oils of mint (*Mentha piperita*), fennel (*Foeniculum*), and clove (*Syzygium*); purified menthol; and "coating" agents such as *Aloe* gel and *Glycyrrhiza* solid extract. *Commiphora myrrha* is another classic herb for oral lesions due to its significant antimicrobial activity,[144] and it is also anodyne via sodium channel blockade on mucosal cell membranes.[145]

Aloe vera gel	1 teaspoon
Glycyrrhiza glabra solid extract	½ teaspoon
Commiphora myrrha tincture	20 drops
Mentha piperita essential oil	2 drops
Syzygium aromaticum essential oil	2 drops

Combine all ingredients in a small dish or shot glass and blend. Place ¼ teaspoon in the mouth as often as hourly, reducing as symptoms improve.

Mouth Rinse for Aphthous Ulcers

One aim of this formula is to create a medicine thick and sticky enough to cling to the buccal mucosa to treat aphthous ulcers, also known as canker sores. Licorice solid extract has the consistency of molasses and provides an anti-inflammatory, ulcer-healing[146] base for the formula. *Aloe vera* gel and *Calendula* provide additional pain-relieving and vulnerary effects, and the peppermint oil provides flavor and anodyne effects.

Glycyrrhiza glabra solid extract if available	⅓ ounce (10 g)
Calendula officinalis succus	⅓ ounce (10 g)
Aloe vera gel	⅓ ounce (10 g)
Mentha piperita essential oil	10 drops

Combine these ingredients every day or two in a small, wide-mouthed container. Spoon 1 teaspoon directly into the mouth, or first place the liquid in a shot glass and top with water or herbal tea, as preferred. Use as a mouthwash, swallowing after rinsing. Repeat hourly throughout the day and reduce frequency of dosing on subsequent days as lesions resolve and pain abates.

> # Healing Agents for Cheilosis and Cheilitis
>
> Also referred to as angular stomatitis, cheilosis is a condition involving small longitudinal angular lesions occurring at the corners of the mouth. Cheilosis is associated with B vitamin deficiency, especially riboflavin and pyridoxine, and poor nutritional status in general. Cheilitis is an inflammatory condition of the lips that may include contact dermatitis-like lesions. Lipsticks, sodium laurel sulfate in toothpaste, cinnamon, or benzoates or other chemicals, or occasionally food additives may trigger inflammation of the lips and oral cavity. Alternately, cheilitis may involve actinic lesions from sun damage or smoking and may predispose to malignant transformation. The following therapies may be helpful.
>
> B vitamin supplementation and dietary guidance
> *Calendula* salve used topically[147]
> Oral supplements of fish oil, flaxseed oil, evening primrose oil, or other sources of essential fatty acids
> *Hippophae* oil used both orally and topically[148]

Herbal Tea for Canker Sores

Licorice provides anti-inflammatory and antiulcerative effects, while *Vaccinium* has slightly astringent and connective tissue supportive effects. *Calendula* may promote healing via enhancement of microcirculation below mucosal membranes, and *Centella* is another classic herb for ulcerated mucosal tissues.

Glycyrrhiza glabra
Vaccinium myrtillus leaves
Centella asiatica
Calendula officinalis

Combine in equal parts and steep 1 tablespoon per cup of hot water. Vitamins B and C and glutamine powder may be added to the formula or prepared separately in water. For best effect, sip in small quantities throughout the day.

Tincture for Oral Ulcers and Stomatitis

Teas may be superior to tinctures for stomatitis, as the alcohol in tinctures can irritate the mouth. However, tinctures may be more convenient to prepare and easier to pack when away from home or traveling. *Echinacea* has stabilizing effects on fibroblasts and may inhibit the breakdown of hyaluronic acid, making it appropriate for autoimmune-related oral ulcers and microvascular inflammation. When oral ulcers are related to herpes infections, *Commiphora*, *Hypericum*, *Glycyrrhiza*, *Calendula*, or *Melissa* are antiviral agents to consider, and are further addressed in the Topical Herbal Resin for Cold Sores and Antiviral Tincture for Cold Sores.

Glycyrrhiza glabra solid extract	15 ml
Echinacea angustifolia	10 ml
Centella asiatica	10 ml
Foeniculum vulgare	10 ml
Symphytum officinale	10 ml
Mentha piperita essential oil	5 ml

Place 1 teaspoon of the combined tincture in a small cup of hot water and let stand for 5 minutes to allow some of the alcohol to evaporate. Swish around mouth for 1 minute, then swallow, repeating 3 or more times daily.

Topical Herbal Resin for Cold Sores

Melissa essential oil can promote the healing of herpetic ulcers when applied topically and is included along with other essential oils in this formula. Propolis has broad antimicrobial effects, and its sticky nature helps this formula to dry as a shellac-like resin over cold sores. The amino acids lysine and glutamine, as well as vitamin C and bioflavonoids, may also deter herpes outbreaks and may be used complementarily.

Glycyrrhiza solid extract	1 ounce (30 ml)
Propolis glycerine extract	½ ounce (15 ml)
Aloe vera gel	¼ ounce (7.5 ml)
Hypericum perforatum oil	¼ ounce (7.5 ml)
Commiphora myrrha essential oil	10 drops
Melissa officinalis essential oil	10 drops

Combine ingredients in a 2-ounce (60 ml) glass jar or plastic salve container and homogenize. Use a cotton swab to apply to lesions as often as possible. *Calendula* salve may be applied to the external lips between uses of the topical "resin."

Antiviral Tincture for Cold Sores

This formula features systemic antiviral and immuno-modulating agents. While the Topical Herbal Resin for Cold Sores is best for topical use in acute herpetic flares, this formula's systemic effects might help to reduce chronic infections. For those with stress-related herpetic flares, amend the formula so it contains nervines (see chapter 3, Volume 4).

Hypericum perforatum	15 ml
Phytolacca americana	15 ml
Glycyrrhiza glabra	15 ml
Lomatium dissectum	15 ml

Take 1 teaspoon of the combined tincture 2 times a day to reduce frequent outbreaks. Take 1 teaspoon hourly for acute eruptions, reducing as ulcers scab over.

Tea for Gingivitis of Vascular Disease

When gingivitis is concomitant with heart disease, or when lab work suggests vascular inflammation, formulae may include high-flavonoid herbs and other vascular anti-inflammatories, such as *Vaccinium*. Alternatives to *Vaccinium* include *Ginkgo*, *Crataegus*, *Angelica*, and *Salvia miltiorrhiza*. Here, *Vaccinium* is combined with one connective tissue strengthening herb, *Centella*, and one herb specific for mucosal ulcers, *Glycyrrhiza*. Prepare as a tea to address gingivitis while systemically improving heart and blood vessel health. Coenzyme Q10 is a complementary supplement for both heart and periodontal disease.

Vaccinium myrtillus
Glycyrrhiza glabra
Centella asiatica

Combine in equal parts. Steep 1 tablespoon per cup of hot water for 10 minutes, then strain and drink freely.

Specific Causes of Gingivitis

Gingivitis of malnutrition. Low-grade scurvy causes friable tissues that bleed readily, and patients may present with red to bruised-looking gums. Blood-filled lesions and swollen gums with loose teeth are pathognomonic of scurvy. Pellagra and other B vitamin deficiencies also cause oral inflammation with easy bleeding, and patients often present with a bright red inflamed tongue, a sensation of burning in the mouth, and often red cracked lips. Inadequate protein intake is rare in the United States, but also impairs connective tissue integrity and contributes to gingivitis. Optimizing nutrition will often improve gum health in several months or less.

Diabetic gingivitis. The diabetic state is associated with both altered oral microflora and increased systemic inflammation. Treatment should address blood sugar and metabolism. One therapeutic approach might be to use a variety of dietary and herbal metabolic remedies systemically (as described in chapter 3, Volume 3), while also using one of the herbal mouthwash formulas for stomatitis found in this chapter. Because many people with diabetes are prone to chronic fungal and other infections, *Berberis*, *Commiphora*, and oregano oil may be useful in mouthwashes.

Gingivitis of pregnancy. Hormones appear to have a proliferative effect on the gums because gingivitis often occurs during pregnancy and in some women, premenstrually. Gingivitis of pregnancy may relate both to the nutritional demands placed on a woman's body and to the suspected proliferative effects of hormones.

Gingivitis of vascular disease. Gingivitis and heart disease often occur in tandem,[149] and the mechanisms that contribute to large blood vessel inflammation also contribute to microcirculatory and connective tissue damage in the gums.[150] Research suggests that those with gingivitis should undergo diagnostic testing for cardiovascular disease.[151] When heart disease is identified, supplementation with coenzyme Q10 may improve basic metabolism in heart muscle cells and the gingivae. Vascular tonics such as *Crataegus*, *Ginkgo*, and *Angelica* may also be considered in systemic support formulas in such cases.

Stimulating Tincture for Dry Mouth

Salivary gland insufficiency may result from atrophy due to old age, or in association with autoimmune diseases such as Sjögren's syndrome. Sialagogue herbs with warming and stimulating effects include *Zanthoxylum*, *Allium*, and *Armoracia*. These may be appropriate in stagnant or atonic conditions, but inappropriate in those with allergic or sensitive oral mucosa, or during acute inflammatory situations. Those with systemic dryness and hot, dry constitutions should be encouraged to drink ample water and drink demulcent, cooling teas on a regular basis, particularly with meals.

Zanthoxylum clavis-herculis	30 ml
Allium sativum	15 ml
Armoracia rusticana	15 ml

Take 10 to 30 drops of the combined tincture, holding in the mouth for 1 minute before swallowing. Take 3 or more times a day to stimulate circulation and salivation. This formula may be used prior to meals to enhance saliva flow and support salivary enzyme production.

Sialagogue Paste for Dry Mouth

A sialagogue is an agent that promotes salivation, and *Pilocarpus* is a powerful sialagogue used only in small amounts because it also promotes perspiration, lacrimation, and intestinal secretions. These secretory properties make *Pilocarpus* a little tricky to use for xerostomia. *Iris* is a milder sialagogue affecting primarily the gastrointestinal system. Aim to use small amounts of the tincture combined with oil or emulsifying demulcent herbs, directly on the tongue, frequently. Demulcents such as *Aloe vera*, *Symphytum*, *Althaea*, or *Glycyrrhiza* solid extracts make a good base; however, *Glycyrrhiza* and *Althaea* provide a particularly nutritive, trophorestorative base for these potentially harsh herbs.

Oral Allergic Reactivity

Allergic phenomena affect the lips and peri-oral skin more commonly than the oral cavity itself but may occasionally trigger stomatitis. Allergic individuals may develop oral lesions following the ingestion of allergenic foods and drinks. Any food may trigger a peculiar idiosyncratic reaction in an individual, but nuts, citrus, and food chemicals such as benzoates, sulfites, and artificial colors are especially common offenders. Chronic irritation following the ingestion of substantial quantities of tobacco, coffee, and alcohol may trigger stomatitis, especially in those with nutritional deficits. Heavy metal toxicity is also associated with oral lesions and mercury toxicity is especially noted to promote excessive salivation. An allergen-free diet for several weeks is advised for acute cases, and further lab testing can be done if there are no rapid improvements.

Glycyrrhiza is used in the form of a solid extract to help the medicine cling to the oral mucosa, increasing its local effects and helping to limit its systemic effects.

Glycyrrhiza glabra solid extract	20 ml
Althaea officinalis	10 ml
Sesamum indicum oil	10 ml
Iris versicolor	10 ml
Pilocarpus jaborandi	10 ml

Whisk together in a small, wide-mouth glass or plastic vial. Take ⅛ to ¼ teaspoon every few hours, and particularly prior to meals in cases of dysphagia. If possible, hold the formula in the mouth for several minutes before swallowing.

Tea and Mouthwash for Allergic Stomatitis

The following herbs are noted to help soothe allergic discomfort while treating the underlying hyperreactivity. These herbs are soothing, anti-inflammatory, and reasonable-tasting, making for an excellent mouthwash and tea. Antioxidants, quercetin, and elimination of allergens from the diet may be more effective as long-term therapies.

Matricaria chamomilla	2 ounces (60 g)
Calendula officinalis	2 ounces (60 g)
Glycyrrhiza glabra	2 ounces (60 g)
Foeniculum vulgare	2 ounces (60 g)

Steep 1 tablespoon of the combined herbs per cup of hot water for 10 minutes. Use as a mouthwash, a gargle, or a simple tea, as often as possible.

Oral Oil with Aloe for Xerostomia

This dry-mouth remedy is similar in concept to the Sialagogue Paste for Dry Mouth, but omits the licorice for those who dislike it, or for when it is contraindicated. This formula relies on *Aloe* gel and sea buckthorn oil to help the medicine cling to the mouth.

Aloe vera gel	2 ounces (60 g)
Sea buckthorn (*Hippophae rhamnoides*) oil	1 ounce (30 g)
Pilocarpus jaborandi tincture	1 ounce (30 g)
Mentha piperita essential oil	60 drops

Combine in a 4-ounce (120 ml) bottle and shake vigorously. Use 10 to 30 drops on the tongue every hour, or as needed for symptomatic relief and to aid in chewing and eating for those with dysphagia due to Sjögren's syndrome.

Tincture for Oral Leukoplakia

Viscum is included here as an anticancer herb, *Glycyrrhiza* and *Curcuma* have general anti-inflammatory and immunomodulating effects, and *Centella* may support connective tissue and promote healing. *Sanguinaria* is specific for oral lesions and is used in this formula in a small amount as a synergist and counterirritant.

Glycyrrhiza glabra	15 ml
Curcuma longa	15 ml
Centella asiatica	13 ml
Viscum album	10 ml
Sanguinaria canadensis	7 ml

Place 1 dropperful of the combined tincture in a shot glass of water and use as an oral rinse to swish around the mouth for several minutes before swallowing. Repeat 3 or 4 times daily.

Leukoplakia and Natural Medicine

Oral cancers are among the most commonly occurring cancers, and chronic exposure to oral irritants and nutritional deficits may predispose. Chronic alcohol and tobacco exposure may deplete tissue nutrients and allow oxidative damage to chromosomes. Research suggests that inadequate nutrition depletes mucosal glutathione, B vitamins, and other nutrients, impairing cellular nuclei stability, and impairing T-cell and immune response. Infectious agents such as herpes or the papilloma viruses may establish themselves more easily in such situations and may act on genes contributing to cancer epigenesis. Leukoplakia may precede the emergence of oral neoplasms, and one study on oral leukoplakia showed cigarette smoking to be the most significant risk factor, with diabetes being a lesser contributor.[152] Deficient antioxidants and B vitamins may be additional contributors.[153] Beta carotene may help resolve oral leukoplakia,[154] and vitamin C and tomato consumption may reduce the occurrence of oral precancerous lesions.[155] Chronic exposure to mucosal irritants such as alcohol,[156] chewing betel nuts,[157] and regular cooking over a woodstove[158] may increase the risk of oral cancers.

Botanical agents may both help prevent and treat oral lesions. *Camellia sinensis* may reduce the size of oral leukoplakia lesions clinically,[159] and has also been shown to decrease cellular markers indicative of cancer progression.[160] *Centella* may repair ulcers associated with alcohol or aspirin use, circulatory impairment, and gastric inflammation.[161] *Centella* may increase gastric mucous secretion and enhance glycoprotein integrity, helping protect oral tissues from stomach acid, irritants, and nutrient-depleting substances.[162] *Bupleurum falcatum*[163] and *Curcuma longa*[164] are both traditional remedies for chronic inflammation and reducing aberrant cellular proliferation and may be considered in formulas for leukoplakia.

Formulas for Oral Erythema Multiforme

The autoimmune disease erythema multiforme affects multiple organ systems including the joints, skin, respiratory passages, genitals, eyes, and mouth. Oral and other lesions can be extensive and acutely painful, and joint pain and arthritis may also be prominent. The oral lesions are typically raw and tend to bleed and may occur on the lips and diffusely throughout the oral cavity, slowly forming tender crusts and scabs. Simple upper respiratory symptoms and fever may precede the emergence of the lesions.

Because the oral lesions are so painful, the patient will often require special foods, from soups to smoothies, to purees, and creamed foods in order to obtain adequate nutrition, and may sometimes even require IV fluids to prevent dehydration. Herbal teas, hot or cold or iced as the person prefers, may be pain relieving and speed healing. Drinking with a straw or freezing herbal teas into ice chips may be ways to consume medicines with the least pain. Corticosteroid mouthwashes are sometimes used to suppress the lesions, and anesthetic rinses are often prescribed for pain relief. *Calendula* salve may be applied to lips frequently to speed healing.

Mouth Rinse for Erythema Multiforme

Erythema multiforme can involve painful oral lesions, and this formula aims to provide anodyne effects. The formula uses a *Calendula* succus as opposed to a tincture to avoid further stinging or irritating the mouth.

Glycyrrhiza glabra solid extract	1/3 ounce (10 g)
Calendula officinalis succus	1/3 ounce (10 g)
Aloe vera gel	1/3 ounce (10 g)
Mentha piperita essential oil	10 drops

Combine the ingredients and take 1 teaspoon directly, or dilute with water to prepare as a mouthwash, swallowing after rinsing. Repeat hourly until the pain subsides, reducing frequency as symptoms improve.

Herbal Tea for Erythema Multiforme

This formula complements the Mouth Rinse for Erythema Multiforme; both may be used separately, or the mouth rinse can be placed in this tea. The herbs are tasty and provide anti-inflammatory and antiulcerative effects for the oral mucosa. For significant pain, some patients find comfort in sipping this soothing tea constantly throughout the day.

Glycyrrhiza glabra

Centella asiatica
Calendula officinalis
Mentha piperita

Combine in equal parts. Steep 1 tablespoon per cup of hot water for 10 minutes. Strain and drink freely. This tea can also be frozen in ice cube trays, then pounded into slivers that may be placed in a bowl and eaten with a spoon for pain relief. Popsicle forms are also available for freezing fruit juices and herbal teas.

Formulas for Tonsillitis and Pharyngitis

Herbal formulas for tonsillitis and pharyngitis should usually include antimicrobial agents, anodyne ingredients to help allay pain, and in the case of enlarged tonsils, possibly lymphatic herbs. When acute viral symptoms of muscle aches, fever, or lung or influenza symptoms accompany such as cough and respiratory mucous, *Lomatium dissectum*, *Sambucus nigra*, *Eupatorium perfoliatum*, or *Glycyrrhiza glabra* may be foundation herbs. *Commiphora myrrha* is a specific for throat and oral infections, and *Phytolacca americana* is specific for enlarged tonsils.

Saline Gargle for Pharyngitis

Pharyngitis may be due to viral infections—including common rhinoviruses and sometimes mononucleosis and other chronic "slow" viruses—or may be bacterial, as with strep throat. In allergic people, sore throats may result from allergic inflammation. Sore throats may also result from acute irritation, such as exposure to smoke or airborne chemical irritants. When pharyngitis involves the tonsils, it can be referred to as tonsillitis, though the treatments are virtually the same for both. The term *tonsillitis* is used over pharyngitis when the tonsils are particularly enlarged, prominent, pocked, or covered with exudate, and exhibiting pus accumulation in the crypts and folds. Saline gargles have general antimicrobial effects and can be an all-purpose initial therapy for sore throats. A natural high-mineral salt is preferable for this gargle.

Salt, such as pink Himalayan salt

Dissolve 1 teaspoon of salt in 3 cups of water over low heat. Gargle ½ cup of the liquid at a time for at least 1 minute before spitting out. Repeat as often as hourly at the onset of pharyngitis. Complement with one of the antimicrobial formulas that follow.

Tincture for Pharyngitis with URI

At the onset of simple colds and rhinovirus infections, a formula such as this may be a good place to start when lung congestion, cough, or nasal congestion accompanies a sore throat. *Sambucus* or *Eupatorium* are other choices for those with accompanying chest

Hydrotherapy for Sore Throats

Cold applications to the throat can stimulate circulation to the trachea, pharynx, and larynx, both providing a pain-relieving effect and helping the immune system speed up resolving the infection. One method is to wrap an icy cold damp dish towel around the throat, cover it with a dry towel, and top that with a wool scarf. Such applications are referred to as a "heating" compress: The body will heat up the cold fabric, enhancing circulation and immune response in the process. The compress is removed once it has reached body temperature. Take care that the rest of the body is warm and comfortable when implementing a heating compress, and do not apply one to a chilled person.

cold symptoms, or for when flu viruses spread through a community.

Lomatium dissectum
Commiphora myrrha
Allium sativum
Phytolacca americana
Mentha piperita

Combine in equal parts and take ½ to 1 teaspoon hourly, reducing over several days' time as symptoms improve.

Throat Spray for Acute Pharyngitis

This pain-relieving formula can help relieve the discomfort of all types of sore throats, including that of strep throat. *Glycyrrhiza* is both antiviral and anti-inflammatory, and the use of a solid extract helps the formula cling to the inner throat. *Capsicum* exerts a warming sensation by triggering the release of substance P from nerve endings, overriding pain sensations. However, the "hot" flavor may

be unpalatable for children and may be omitted. The trick is to thin the solid extract down with other medicinal tinctures so that it will flow through a spray nozzle.

Glycyrrhiza glabra solid extract	¼ ounce (7.5 ml)
Commiphora myrrha tincture	¼ ounce (7.5 ml)
Phytolacca americana tincture	¼ ounce (7.5 ml)
Berberis aquifolium tincture	⅛ ounce (3.75 ml)
Capsicum frutescens tincture	⅛ ounce (3.75 ml)
Mentha piperita essential oil	30 drops

Combine in a 1-ounce (30 ml) bottle with a spray applicator, shake well, and spray into the back of the throat as often as every 15 to 30 minutes, reducing frequency as pain and infection abate.

"HEMP" Tincture for Tonsillitis

H stands for *Hydrastis*, E for *Echinacea*, M for *Myrrha* (now renamed as *Commiphora*), and P for *Phytolacca*. This formula is effective for strep throat, as well as for

---

Hydrastis for Mucous Membranes

Hydrastis excels at astringing mucous membranes and enhancing mucosal immune function, and is specific for purulence and damp, swollen mucous membranes. *Hydrastis* is an excellent antimicrobial astringent remedy for mucosal congestion and catarrhal states and is specific for a flabby swollen tongue with indentations of the teeth on the sides. *Hydrastis* is also specific for a bitter taste, or a hot peppery taste in the mouth, and for stomatitis and ulcerative lesions in the mouth and on the tongue. Use *Hydrastis* in tinctures, gargles, and mouthwashes for infectious stomatitis, oral ulcers, and dysbiosis. Use *Hydrastis* as an eyewash ingredient specifically for thick purulent mucous accumulations. *Hydrastis* contains well-studied alkaloids including berberine, hydrastine, and palmatine. Berberine-containing powders have been used in eye drops for infectious conjunctivitis including trachoma, a disease characterized by keratoconjunctivitis. Berberine inhibits lens aldolase reductase, which is elevated in diabetic retinopathy and cataracts. Use sunglasses or avoid strong sunlight when using berberine-containing eyewashes because increased photosensitivity may occur.[165] Discontinue *Hydrastis* eyewashes as soon as infections are improved.

Hydrastis canadensis, goldenseal

General Therapies for Throat Pain

The following herbs, nutrients, throat sprays, and other ideas may help to allay throat pain.

Ice chips. Ice chips can provide relief when swallowing is painful and improve hydration when patients are avoiding drinking.

Throat sprays. The placement of herbal formulas in a tincture bottle with a spray nozzle can be useful to spray medicines directly on the tonsils. Such formulas will stick to the tonsils better if a glycerite or solid extract is included in the recipe.

Cayenne. *Capsicum* tincture or the inclusion of cayenne in lozenges can numb throat pain.

Demulcents. Demulcents coat the throat to soothe a dry scratchy sensation. Simple herbal demulcents can be pain relieving when needed, and may be included in tinctures, or used separately as throat lozenges or as a tea. *Ulmus*, slippery elm, is available in commercial throat lozenges.

Zinc and vitamin C lozenges. These medicinal lozenges may speed resolution of acute tonsillitis and complement herbal formulas.

Essential oils. All essential oils are antimicrobial and many also help to open the airways when cough or bronchitis accompanies sore throat. Excellent options include citrus, mints, and eucalyptus oils. Essential oils may also be prepared into steams and inhalant therapies for laryngitis and used as anodynes.

Gargles. Saline gargles can offer antimicrobial action and immediately astringe and decongest the mucosal membranes of the pharynx and tonsils. Herbal teas such as *Thymus*, *Mentha*, or *Eucalyptus* may also be used, and a few drops of essential oils may allay pain as well as treat pathogens and help break up biofilms.

any infectious sore throat. The HEMP formula is not tasty, so glycerite forms of some of the herbs may be used, where available, to improve the flavor for young children and others.

Hydrastis canadensis
Echinacea angustifolia
Commiphora myrrha
Phytolacca americana

Combine in equal parts. Take 1 teaspoon hourly, reducing frequency over several days' time as symptoms improve.

Tincture for Chronic Sore Throat

For patients with chronic sore throats due to chronic fatigue syndrome or postmononucleosis, immunomodulators such as *Astragalus*, *Panax*, or *Eleutherococcus* may be included in formulas.

Astragalus membranaceus	20 ml
Eleutherococcus senticosus	20 ml
Baptisia tinctoria	10 ml
Phytolacca americana	10 ml

Take 1 dropperful of the tincture 3 or 4 times daily for 2 or 3 months, reducing as symptoms improve. Discontinue once chronic sore throat is resolved.

Tincture for Laryngitis with Dampness

Sambucus, *Lomatium*, and *Stillingia* serve as an antiviral base in this formula for infectious laryngitis. *Thymus* and *Allium* make this formula warming and drying to address mucous congestion in the throat. The use of one or more glycerites in the formula will improve surface contact in the throat and enhance efficacy.

Sambucus nigra
Lomatium dissectum
Stillingia sylvatica
Thymus vulgaris
Allium sativum

Combine in equal parts and take ½ teaspoon hourly, reducing as symptoms improve.

Tincture for Laryngitis with Dryness

This formula is similar to the Tincture for Laryngitis with Dampness, but it is made more specific for tightness with the antispasmodic *Mentha*. *Glycyrrhiza*—a soothing, demulcent, antiviral anti-inflammatory—may improve a dry sensation. For severe tightness and laryngospasm, *Drosera* would make a good addition. The use of one or more glycerites in the formula will improve surface contact in the throat and enhance efficacy.

Lomatium dissectum
Stillingia sylvatica
Glycyrrhiza glabra
Mentha piperita

Combine in equal parts and take ½ teaspoon hourly, reducing as symptoms improve.

Tincture for Vocal Strain

Collinsonia is specific for vocal cord swelling due to vocal strain. This formula omits the antivirals and includes the demulcent *Glycyrrhiza*, as well as the aromatic herbs *Mentha* and *Eucalyptus* to open and relax the throat.

Glycyrrhiza glabra solid extract	1 ounce (30 g)
Collinsonia canadensis	½ ounce (15 g)
Stillingia sylvatica	½ ounce (15 g)
Mentha piperita	10–20 drops
Eucalyptus globulus	10–20 drops

Combine in equal parts and take ½ teaspoon hourly, reducing as symptoms improve.

Demulcent Tea for Scratchy Throats

This tea features the demulcents *Verbascum*, *Althaea*, and *Tussilago*, and the antiviral herbs *Sambucus* and *Glycyrrhiza*.

Verbascum thapsus leaves	1 ounce (30 g)
Althaea officinalis root	1 ounce (30 g)
Glycyrrhiza glabra shredded	1 ounce (30 g)
Tussilago farfara leaves	1 ounce (30 g)
Mentha piperita leaves	1 ounce (30 g)
Sambucus nigra flowers	1 ounce (30 g)

Combine in equal parts. Steep 1 tablespoon per 1 cup (240 ml) of hot water for 10 to 15 minutes. Strain and drink freely.

Inhalant Steam for Laryngitis

Essential oils can provide immediate pain relief, relax coughs and spasms, and help open the airways.

Mentha piperita essential oil	10 drops
Eucalyptus globulus essential oil	10 drops

Place the essential oils in freshly boiled water, and breathe in the vapors for 10 minutes, inhaling through the mouth. Repeat 2 or more times each day, reducing as symptoms improve.

Herbs for Specific Types of Throat Pain

The following are specific indications from folklore of herbs for specific qualities of throat pain and laryngitis.

CONDITION	RECOMMENDED HERBS
Cough: gagging, croupy, with throat spasms; also tight sensations and spastic coughing	*Drosera rotundifolia*
Cough: with thick, sticky mucus; hacking; and hemming	*Allium sativum* *Armoracia rusticana*
Laryngitis: with hoarseness with full, heavy sensation; vascular and fluid congestion in throat	*Collinsonia canadensis* *Stillingia sylvatica*
Pharynx: sore, swollen, livid, or bluish	*Arnica montana*
Tonsils: chronically enlarged, ulcerated, and fetid, with purplish or bluish discoloration	*Baptisia tinctoria*
Tonsils: enlarged with lesions, exudates, ulcers, and foul breath or secretions	*Echinacea* spp.
Tonsils: painful, enlarged	*Commiphora myrrha* *Phytolacca americana*
Tonsils: ulcerated during viral infections	*Glycyrrhiza glabra* *Lomatium dissectum* *Sambucus nigra*
Tonsils: boggy, enlarged, with excess mucus and phlegm	*Salvia officinalis*
Throat: stiff and dry, accompanied by burning and dysphagia	*Guaiacum officinale*
Throat: infected, with cough and mucus	*Hydrastis canadensis* *Thymus vulgaris*

Specific Indications:
Herbs for Eye, Ear, Nose, Mouth, and Throat Conditions

The following descriptions offer symptom pictures, patterns, and presentations that may be indications for using these specific *materia medica* options. The plants are selected for their specificity for infections, allergic reactivity, autoimmune issues, and other conditions that afflict the eyes, ears, nose, oral cavity, and throat.

Achillea millefolium • Yarrow

The young leaves and flowering tops of yarrow have many medicinal virtues including that they are antimicrobial,[166] alterative, diaphoretic, vasodilating, hypotensive, and anti-inflammatory. Consider *Achillea* in formulas for oral complaints associated with intestinal inflammation, IBS, and liver congestion. *Achillea* is also helpful for eye symptoms related to upper respiratory allergies. Yarrow has a relaxing effect on tracheal smooth muscle via effects on β-adrenoreceptors contributing relaxing effects on the airways.[167] Although bitter, yarrow may be included in mouthwashes to treat oral infection and inflammation and chemotherapy-induced oral mucositis.[168]

Aconitum napellus • Aconite

Caution: The roots of aconite, also known as monkshood, are highly toxic and available as tincture to skilled clinicians only. Aconite is also baked or processed in China and used in many traditional formulas. Do not confuse TCM formulas that contain processed *Aconitum carmichaelii* (fu zhi) with the use of such formulas using unprocessed herbal tinctures or raw material. *Aconitum* tincture is used by herbalists as a single diluted drop, or a homeopathic preparation, for a sensation of numbness and tingling in the tongue, for a coated tongue, and for inflamed gums with significant pain or a sensation of burning. *Aconitum* is considered a very hot herb and is used to bring heat and stimulation to cold, deficiency symptoms. *Aconitum* contains neurotoxic alkaloids that affect ion channels that can have a numbing effect for eye, ear, or oral pain, but is also the mechanism of severe toxicity with overdose.[169]

Allium sativum • Garlic

Garlic bulbs contain sulfur-rich antimicrobial compounds, and garlic oils, pills, tinctures, and medicinal foods are used widely in herbal medicine. Use garlic as a supportive ingredient in formulas for eye infections. It is also useful for thinning thick mucous congestion. *Allium* is specific for excessive saliva with a sweet taste or quality, and for the sensation of a hair caught on the tongue or in the throat. Garlic is shown to protect retinal cells from lead toxicity[170] and from diabetic retinopathy.[171] The oral consumption of garlic or use of garlic containing medicines may help protect against cariogenic bacteria[172] and various garlic products may be useful in the treatment of thrush, periodontitis, and other oral infections.

Aloe vera • Aloe

The succulent leaves of various *Aloe* species can be used to extract gels and liquids that can allay pain and speed healing of mouth ulcers. The plant's demulcent effects are helpful in treating burning sensations in the mouth or eyes. *Aloe* gels improve the viscosity of herbal tinctures, helping them to coat the mouth and gums, and yield soothing mouth pastes and gels. Use *Aloe* topically and internally for lips that are cracked and dry. Folkloric herbals report that *Aloe* is specific for a bitter or sour taste in the mouth. *Aloe* gels and liquids may also be used directly in the eyes for ulcerative lesions, and for pain relief following irritants, burns, and eye trauma.

Angelica sinensis • Dong Quai

The roots of *Angelica* are one of the foundational "blood movers" of TCM as discussed in chapter 2, Volume 2. *Angelica* is well researched and shown to have many vascular anti-inflammatory effects and may be a key herb in formulas for diabetic retinopathy and other diseases of the eye.[173] Also consider *Angelica* for allergic stomatitis, rhinitis, and sinusitis.

Apis mellifica • Honeybee

Homeopathic preparations and mother tinctures of *Apis* have long been used to treat eye symptoms and allergic reactions. Use *Apis* mother tincture or homeopathic for red, puffy eyelid swelling, with stinging and burning sensations, injected conjunctival vessels, and photophobia. *Apis* may also be useful in formulas for keratitis and styes. Bee venom therapy, referred to as apitherapy, has been used since ancient times and may be employed for arthritis and chronic pain, as well as neurologic diseases and cancer.[174] Bee products such as honey, royal jelly, and propolis may be included in oral formulas for gingivitis, periodontal disease, and oral lesions such as aphthous stomatitis.[175]

Armoracia rusticana • Horseradish

Horseradish roots contain pungent sulfur-rich compounds with lacrimatory activity, making horseradish one of our best herbs for thinning upper respiratory mucus and expectorating thick stagnant congestion from the sinuses. The pungent compounds, including isothiocyanate, are metabolized into glucosinolates in the body.[176] Glucosinolates may help to activate hormone and toxin detoxification pathways in the body and may help treat and prevent cancer.[177] The disulfide compounds also have broad antimicrobial effects,[178] making horseradish medicines and foods useful in the treatment of sinusitis.

Arnica montana • Leopard's Bane

Arnica leaves and flowers are used to prepare tinctures, oils, creams, salves, and homeopathic medications all emphasized folklorically for trauma. Use *Arnica* salves for eyelid trauma and small amounts of *Arnica* tinctures diluted in formulas or homeopathic preparations both topically and orally for eye trauma, double vision following eye trauma, retinal hemorrhage following head trauma, and sore, bruised, aching pain in the tissues of the eyes. Use *Arnica* as a homeopathic or include small amounts of diluted tincture in mouthwashes to reduce pain and to speed healing after dental extractions. *Arnica* is also specific for a sore, swollen, livid, or bluish pharynx.

Artemisia absinthium • Wormwood

Wormwood is most commonly known as a bitter worming agent. It is also a known central nervous system toxin due to the presence of absinthol in the leaves and flowering tops. *Artemisia* species are traditional medicines typically used in small quantities only. Folkloric writings detail *Artemisia* as being specific for a large, swollen, protruding tongue and the tendency to bite the tongue accidentally. Use wormwood for insufficient digestive secretions, parasites, and atrophic glossitis. Limit therapy to short-term use and use only in small doses due to potential central nervous system toxicity with large and continuous usage.

Astragalus membranaceus • Milk Vetch

The roots of leguminous *Astragalus* are widely used in TCM for immunomodulation, allergies, and chronic infections. Include *Astragalus* as an immunomodulator for chronic infections, as well as EENT allergic conditions. Tinctures and encapsulations are readily available, and dry herbs can be used in teas. Use *Astragalus* powder in smoothies and medicinal foods. See also the *Astragalus membranaceus* entry in chapter 2 on page 86.

Atropa belladonna • Deadly Nightshade

The young leaves of belladonna have long been used as a pain medicine and have been prepared into specific medicines and homeopathics with an affinity for the eyes, head, and cranial nerves. Use belladonna in very small amounts as a specific for dry mouth, throbbing pains in the teeth and gums, boils on gums, and acute dental abscesses. *Atropa* is also specific for a "strawberry tongue" with pain and swelling. Use belladonna as a synergist in formulas for acute infections associated with fever and inflammation. Belladonna is potentially toxic and hallucinogenic and is considered a prescription-only botanical.

Azadirachta indica • Neem

The leaves and twigs of neem have been traditionally chewed to both treat and prevent dental and oral disease, and oil extracts are presently used in toothpastes, dental rinses, and custom formulations prepared by herbalists to treat infectious and inflammatory disease of the mouth. Neem oil may be used in very small doses for thrush, hairy tongue, and oral dysbiosis. Neem oil has been included in oil-pulling blends, a traditional Ayurvedic folk remedy for oral health in which high-quality oils are swished in the mouth like a mouthwash, then expectorated, often using sesame oil as a base. *Azadirachta* has anticariogenic activity and contains the antimicrobial compounds nimbidin, azadirachtin, and nimbinin.[179] Neem-containing mouthwashes can reduce plaque, and some products are commercially available. There are many options for using neem oil in custom-made rinses, oils, and oral health products.

Baptisia tinctoria • Wild Indigo

Baptisia roots and leaves are used to prepare medicines that are highly specific for oral ulcers, and pain and inflammation in the oral cavity, pharynx, and tonsils, especially when chronic and associated with fetid breath. *Baptisia* is indicated for sore throats with pain when swallowing and for gagging when swallowing solid food. *Baptisia* is also specific for a burning sensation of the tongue, a yellow-brown coating on the tongue with red edges, and a bitter taste in the mouth.

Berberis aquifolium • Oregon Grape

Berberis is an alterative herb especially indicated for liver congestion and slow digestion, intestinal dysbiosis, skin eruptions due to poor liver and digestive health, coated tongue, and dyspepsia. *Berberis* is also an excellent antimicrobial agent, especially useful for infected mucous membranes. *Berberis* may be prepared into eyewashes for thick purulent mucous accumulation in conjunctivitis. *Berberis* is also indicated as an alterative ingredient for chronic eye complaints associated with digestive or liver disorders. *Berberis* is a source of berberine that may be used to prepare eye drops, mouthwashes, throat gargles, sinus sprays, and neti pot rinses and ear lavages to treat fungal and other infections of the eyes, ears, nose, and throat. *B. aquifolium* is also known as *Mahonia aquifolium*.

Calendula officinalis • Pot Marigold

Dry *Calendula* petals are useful as an eyewash ingredient for traumatic, inflammatory, and infectious conditions of the conjunctiva. Use *Calendula* as a mucous membrane tonic in cases of dry eyes or conjunctival lesions and as an all-purpose mucous membrane rinse. Use *Calendula* as an infusion or diluted succus for pain, infection, and inflammation, and to soothe irritated mucosal membranes of the eyes, ears, nose, or throat. *Calendula* may enhance microcirculation below mucous membranes and support connective tissue structure and healing of mucosal lesions. *Calendula* has antimicrobial, anti-inflammatory, and vulnerary effects for oral and ocular lesions.

Camellia sinensis • Green Tea

Much of the anti-inflammatory effect of green tea is credited to the catechin polyphenolic compounds. Green tea catechins in *Camellia* leaves deter formation of *Streptococcus mutans* biofilms[180] and have potent antioxidant activity that can help reduce oral inflammation involved with the pathophysiology of dental caries and periodontal disease.[181] Green tea is a natural source of fluoride to include in mouth rinses and daily teas for general health. Oral gels prepared from green tea have been shown to improve periodontal health in human subjects with chronic gingivitis.[182]

Capsicum annuum, C. frutescens • Cayenne

The ripe fruits of cayenne pepper are used for their fiery resin, which contains capsaicin, and to prepare tincture, pain creams, oils, and other products for medicinal use. One drop of cayenne oil or tincture may be applied to oral herpes lesions, or added in small amounts to formulas for stomatitis, especially when there is a fetid odor in the mouth. *Capsicum* may be used as a synergist in cold constitutions and for the elderly with digestive atrophy, as well as for atonic dyspepsia with much flatulence. Small amounts of *Capsicum* in sore throat tinctures may offer temporary pain relief. See chapter 2 of Volume 4 for research on capsaicin's analgesic effects.

Centella asiatica • Gotu Kola

The leaves of this creeping tropical wetland plant have been used as a traditional food and medicine in many cultures. Use *Centella* teas, powders, or tinctures as a mucous membrane tonic in cases of chronic conjunctival derangements, and for eye trauma and visual complaints following head trauma. *Centella* is also indicated for styes and chronic ulcers or eyelid lesions and is also useful for complementary support for macular degeneration and atrophic conditions of the eye. *Centella* is folklorically emphasized for tissue ulceration, fibrosis, and malnutrition. *Centella* may act as a nutritive trophorestorative herb in formulas for herpetic, autoimmune, and aphthous ulcerations of the oral mucosa.

Chelidonium majus • Celandine

The young stems, leaves, and flowers of this poppy family plant are used medicinally, and some herbalists use the roots as well. *Chelidonium* can be used as an antimicrobial wash for profuse discharges and tearing in the eyes, for jaundiced eyes, and as an alterative ingredient in formulas for eye complaints associated with liver, gallbladder, or digestive disorders. Folkloric herbals claim *Chelidonium* to be specific for orbital neuralgia. *Chelidonium* is also specifically indicated for a large flabby tongue with indentations of teeth on lateral margins, and a yellowish coating. There has been little scientific or molecular research on *Chelidonium* to date.

Chionanthus virginicus • Fringe Tree

The root bark of this flowering shrub is used medicinally as a traditional biliary tonic with an ability to improve vascular and glandular congestion. *Chionanthus* is specific for a broad flabby tongue, and a dry sensation in the mouth that is not improved by drinking water. Decoctions may be superior to tincture for improving congestion in tissues.

Cinnamomum verum • Cinnamon

The bark of several *Cinnamomum* tree species is used to prepare important culinary spices and flavoring extracts, as well a wide variety of medicines. Cinnamon is useful as an oral antimicrobial and breath freshener in mouthwashes, tinctures, or essential oils. Cinnamon may also be used as an oral medicine as an adjuvant ingredient when vascular inflammation and retinal disease is related to underlying diabetes and metabolic dysfunction.

Collinsonia canadensis • Stoneroot

The tuberous roots of *Collinsonia* are a folkloric vascular tonic, specific for hoarseness with full, heavy sensation; laryngitis; and vascular congestion of the larynx following vocal stress. Consider *Collinsonia* in throat formulas for symptoms that accompany esophageal varices and other situations of vascular congestion. Some early American physicians prepared *Collinsonia* into syrups for treating chronic laryngitis and pharyngitis, as well as for some cases of otitis media in which increased secretions of the throat and glands lead to poor Eustachian tube drainage. *Collinsonia* has not yet been studied and is uncommon to find in commercial preparations other than hemorrhoid or varicose vein products.

Commiphora myrrha • Myrrh

The resinous gum from several *Commiphora* tree species are used to prepare antimicrobial, anti-inflammatory, expectorating, and other medicines. Myrrh is specific for oral and dental infections as a mouthwash, to apply to tooth sockets after dental extractions, and to include in rinses for treating gingivitis. Myrrh is also specific for swollen tonsils and is an all-purpose antimicrobial for throat infections. *Commiphora mukul* and *C. myrrha* may be used in similar ways. *C. myrrha* is also known as *C. molmol*.

Convallaria majalis • Lily of the Valley

All parts of this woodland ground cover plant have been used primarily as a remedy for congestive heart failure, but *Convallaria* is also specific for stasis of fluid. Folkloric writings report that *Convallaria* is also specifically indicated for a tongue that feels sore and scalded; a broad thick, heavy, dirty looking coating on the tongue; and a metallic taste in the mouth. *Convallaria* is appropriate in long-term formulas for those with both heart disease and chronic gingivitis and is specific for a sore throat that is worse inspiring and for grinding of the teeth. **Caution:** This plant contains cardiac glycosides and should be used only by experienced herbalists.

Coptis trifolia • Goldthread

The bright yellow, thin, cord-like roots of *Coptis* may be used to prepare antimicrobial washes for conjunctivitis, purulent discharges, and chronic infections in mucous membranes such as fungal infections related to intestinal dysbiosis, sluggish digestion, or diabetes-related infections. The plant contains isoquinoline alkaloids of the berberine group broadly studied for antimicrobial, metabolic, and hepatotonic effects. *Coptis* and other berberine-containing plants have been used in folkloric eye drop formulas to treat conjunctivitis.[183] *Coptis* and isolated berberine are also shown to deter *Streptococcus mutans*, one of the main microbes associated with dental plaque and gingivitis.[184] Mouthwashes containing *Coptis* have been shown to help treat chemotherapy-induced oral mucositis;[185] hand, foot, and mouth disease;[186] and gastritis-related halitosis.[187]

Crataegus oxyacantha • Hawthorn

Hawthorn berries, as well as the leaves and flowers, can be used as an ingredient in formulas to improve blood vessel health and integrity, as a vascular anti-inflammatory, for diabetic retinopathy, and to improve circulation in the eye for cases of cataracts, macular degeneration, and papilledema. Use hawthorn as an antioxidant and anti-inflammatory ingredient in formulas for chronic eye complaints, and as a supportive ingredient in formulas for chronic, degenerative, or infective eye complaints.

Curcuma longa • Turmeric

Curcuma may be used as a base herb in formulas for infectious and inflammatory oral lesions due to its broad antioxidant, anti-inflammatory, antimicrobial, alterative, and liver-supportive actions. *Curcuma* may protect connective tissue in cases of autoimmune disorders of the mouth. Curcumin, a principal curcuminoid phenolic compound in turmeric roots, is shown to exert an antiangiogenic effect in corneal disease as well as anti-inflammatory effects in allergic and dry eye disorders, and may provide beneficial effects in uveitis, pterygium, and other ocular disorders.[188] Curcumin's powerful antioxidant effect may also offer possible neuroprotective effects in glaucoma.[189] *Curcuma* may be beneficial when used topically in the eye, such as suspended in lipid nanoparticles.

Echinacea purpurea • Purple Coneflower

The fleshy taproots of *Echinacea* species were used by Great Plains native peoples to treat infections, and

Echinacea has been the subject of much research for immunomodulating activity. *Echinacea* is specific for septic conditions with emaciation and debility, and for oral canker sores, receding gums, and gums that bleed readily. *Echinacea* is also indicated for cheilosis, cracked swollen lips, a sore swollen tongue with a white coating, and excess saliva. *Echinacea* is also specific for inflamed tonsils that are purple or blue rather than red, for oral and tonsillar ulcers, and for grayish mucus and exudates in tonsillar crypts. Use *Echinacea* as an antimicrobial ingredient in formulas for infectious complaints of the eyes, including influenza with eye symptoms, herpes-induced ophthalmia, and infectious conjunctivitis.

Eleutherococcus senticosus • Siberian Ginseng

The roots of Siberian ginseng are a traditional adaptogen used to improve longevity, reduce stress reactions, and improve immune and metabolic disorders. Use it as a synergist in formulas for chronic digestive irritability associated with nervous affectations. Neuroprotective effects of *Eleutherococcus* may help reduce oxidative stress to neural cells of the retina[190] in various neurodegenerative diseases. This herb contains compounds called eleutherosides or acanthosides, which have broad anti-inflammatory, protective, and immunomodulating effects similar to the ginsenosides in *Panax ginseng*. *E. senticosus* is also known as *Acanthopanax senticosus*.

Ephedra sinica • Ma Huang

The young twigs of *Ephedra sinensis* and *E. sinica* have been used in TCM, and *E. nevadensis* (Mormon tea) has been used as a traditional beverage in the United States. *Ephedra* may be included in tea, when available, in formulas for eye complaints related to allergies and for hay fever. Due to a history of being abused, and even being processed into amphetamine-like substances, commercial products containing *Ephedra* are no longer available, and even the dry herb for use in teas is hard to come by without collecting the plant for one's self.

Equisetum arvense • Horsetail

The entire plant has been used in traditional teas, pastes, powders, tinctures, baths, and other medicinal preparations. Horsetail is a biosilicifier,[191] meaning it is known to bioaccumulate silica as well as other minerals, contributing to its traditional use as an agent capable of improving mineral density of teeth, nails, and bones. Consider *Equisetum* for oral ulcerations and as a trophorestorative, nutritive herb to help rebuild connective tissues and mucous membranes. Use thoroughly macerated teas to provide the most minerals in cases of malnutrition and malabsorption. Animal studies have shown various *Equisetum* species to have an antihistamine effect,[192] supporting traditional use in hay fever formulas to improve eye and respiratory symptoms.

Eupatorium perfoliatum • Boneset

The young flowering tops of boneset have been a traditional remedy against influenza and viral infections associated with deep musculoskeletal aching. Use *Eupatorium* for throat and oral symptoms in cases of colds and flu, for infectious viral epidemics, and for exhausted and "burnt out" presentations such as chronic fatigue patients who frequently suffer from low-grade, persistent sore throats. *Eupatorium* is specific for cheilosis, yellowish coating on the tongue, marked thirst, and a bitter taste in the mouth.

Euphrasia officinalis • Eyebright

Use *Euphrasia* for conjunctival inflammation, especially with profuse discharges; eye symptoms with sneezing, coughing, and respiratory mucus; and stinging and itching sensations in the eyes. *Euphrasia* is also helpful for symptoms of pain, itching, and irritation when due to a virus, contact irritant, or allergic response. *Euphrasia* is specifically for itching, tearing, and watering eyes, accompanied by bouts of sneezing. Leaf extracts have been shown to have numerous anti-inflammatory mechanisms including inhibition of lipopolysaccharide and NF-κB activation,[193] and a reduction in pro-inflammatory cytokine expression by human corneal cells, helping to explain traditional usage for allergy-induced inflammation of the eyes.[194] See also the *Euphrasia officinalis* entry in chapter 3, Volume 2.

Foeniculum vulgare • Fennel

The entire fennel plant may be used medicinally; fennel seeds are the plant part most commonly used to prepare tinctures and teas. Fennel is an excellent source of minerals and can be used simply to improve flavor of any mouthwash or paste. Fennel tea also serves as gentle eyewashes in cases of conjunctivitis and irritation. The young seeds and leaves can be chewed for mouth ulcers and gum diseases and to improve halitosis.[195] Fennel essential oil may exert an anticariogenic effect by inhibiting *Streptococcus mutans* biofilm formation and related plaque formation.[196]

Gelsemium sempervirens • Jessamine

The roots of *Gelsemium* have been used to prepare medicinal and homeopathic remedies. Use *Gelsemium* in small drop doses for neuralgic pains and burning pains of the eyes, throat, tongue, and ears. **Caution:** All parts of *Gelsemium* are highly toxic, and this herb is to be used only by skilled practitioners. *Gelsemium* is discussed in more detail in chapter 2, Volume 4.

Ginkgo biloba • Maidenhair Tree

Ginkgo leaves are used to prepare a variety of herbal medicines. Less commonly, the fruits and seeds have been used in TCM. Use *Ginkgo* leaf medications as an anti-inflammatory and antioxidant for chronic allergic conditions of the eyes. It is also a vascular protectant.[197] Include *Ginkgo* in formulas to enhance circulation to the eyes in cases of macular degeneration, diabetic retinopathy, cataracts, spots, and floaters. *Ginkgo* may help to slow the progression of or to treat glaucoma[198] and optimize ocular blood flow.[199] *Ginkgo* is specific for failing eyesight associated with poor circulation and for ringing in the ears, and *Ginkgo* is shown to have protective effect against noise-induced tinnitus.[200] Use *Ginkgo* in long-term systemic formulas for chronic gum disease concomitant with heart disease. For more information, see "*Ginkgo* for Dizziness, Vertigo, and Tinnitus" on page 165.

Glycyrrhiza glabra • Licorice

The sweet-tasting roots of *Glycyrrhiza* are widely used in herbal medicine. *Glycyrrhiza* is very useful in topical and local preparations for stomatitis, gingivitis, dry mouth, oral ulcers, aphthous ulcers, and allergic inflammation. Use licorice preparations as supportive antiviral ingredients in formulas to treat herpetic ophthalmia and eye symptoms associated with influenza. Use licorice in throat and mouthwash formulas for viral infections, herpes, and aphthous ulcers. *Glycyrrhiza* is one of the best-known sources of naturally occurring fluoride. Solid extracts of licorice can be used to make oral pastes that will stick in the mouth longer than teas or tinctures.

Gymnema sylvestre • Gurmar

Both leaves and roots of *Gymnema* are used in herbal medicine with long-standing traditions in Ayurvedic and Asian medicine. Consider *Gymnema* as a supportive measure in long-term systemic formulas where diabetes, especially type 1, is associated with microvascular disease, chronic gingivitis, and oral fungal infections.[201]

Gymnema may be contraindicated in those with rapid hypoglycemic reactivity.

Hamamelis virginiana • Witch Hazel

The bark from *Hamamelis* branches has been used as a remedy for vascular congestion and is also indicated for a burning sensation on the tongue, blisters on the sides of the tongue, and thirst. Use *Hamamelis* as an astringent ingredient in mouthwash formulas and eyewash formulas and in herbal teas for oral inflammation, flabby tongue, and thick slimy saliva. *Hamamelis* is helpful for infection or purulence due to irritation. *Hamamelis* is specific for bloodshot eyes with soreness, and for eye trauma with intraocular hemorrhage. Witch hazel extracts may be used in eye and nasal rinses for upper respiratory symptoms of the flu.[202]

Hippophae rhamnoides • Sea Buckthorn

The ripe fruits of *Hippophae* are edible and the small seeds may be used to extract a bright orange oil for oral and topical use. Sea buckthorn oil may be a base emollient ingredient for topical use on cheilosis and labial lesions, as well as in oral pastes crafted to allay pain or moisten atrophic, dry mucosal tissues. The oil may be taken internally as a source of essential fatty acids, but is more expensive than other options, such as fish oil or flaxseed oil. Sea buckthorn proanthocyanidins are shown to have a protective effect against retinal degeneration[203] and help resist damage to retinal ganglia in animal models of chronic hypertension and oxidative stress.[204] Sea buckthorn is also shown to protect the oral mucosa from chemotherapy-induced oral mucositis, reducing inflammation, ulceration, and pro-inflammatory cytokine levels.[205]

Humulus lupulus • Hops

The mature strobiles are dried or fresh tinctured and traditionally used as a calmative nervine, hormone tonic, and bitter digestant. Use *Humulus* as a complementary ingredient in formulas for oral irritation due to underlying gastric reflux disease, or digestive ailments in alcoholics, especially when accompanied by stress and insomnia.

Hypericum perforatum • St. Johnswort

The young flower buds of *Hypericum* contain a bright red pigment rich in flavonoids such as hypericin and hyperforin widely studied for photosensitizing, anti-inflammatory, and mood-altering effects. Use *Hypericum* for eye trauma and eye infection, such as in formulas for acute viral conjunctivitis and herpes-induced ophthalmia. *Hypericum* can act as a neural anti-inflammatory in formulas for retinopathy, retinitis, and ocular nerve disorders. *Hypericum* is also emphasized folklorically for traumatic injury to the tissues (especially trauma to the spine and highly innervated areas), for head trauma, for traumatic injuries to the eyes, and for vascular trauma resulting in extensive bruising and hematomas. Use *Hypericum* as an antiviral and nerve anti-inflammatory in cases of oral herpes, also known as herpes labialis or common cold sores. *Hypericum* is also useful topically following dental extractions and oral surgery to reduce swelling, speed healing, and allay pain. *Hypericum* is specifically indicated for sensations of numbness, tingling, and burning. *Hypericum* is also a useful connective tissue tonic for microvascular fragility seen in chronic gingivitis and erythema multiforme.

Iris versicolor • Iris

Iris roots may be used to increase saliva flow in cases of dry mouth, particularly when liver and digestive deficiency accompanies. As *Iris* is a warming stimulating herb, small doses of only a few drops, combined with other herbs, are all that is needed to gently stimulate the glands. *Iris* may be used as a secretory stimulant in cases of dry eyes, for headaches with blurred vision, and for eye pain.

Juglans cinera • Walnut

The inner hull of *Juglans* nuts are mainly a remedy for digestive and liver insufficiency with jaundice, headaches, and tissue congestion. *Juglans* is also indicated when there is a dry, acrid feeling in the throat and mouth. *Juglans* is specific for tonsillar soreness that is experienced as a sensation coming from the external neck and throat, rather than from the inner throat.

Lobelia inflata • Indian Tobacco

The young flower buds have been prepared into emetics and medicines, dried into smoking mixtures for asthma, and used topically for dyspnea and chest pain. Diluted and drop dosages of *Lobelia* may reduce excessive salivation, while larger dosages of tincture can promote saliva and digestive secretions. *Lobelia* is specific for a metallic taste in the mouth, a burning sensation, and tenacious mucus in the throat. Use small drop dosages for excessive salivation and nausea due to morning sickness, and for rapid onset of nausea and salivation after smelling tobacco or other offensive odor.

Lomatium dissectum • Biscuitroot

The roots of *Lomatium* have been used traditionally for respiratory issues and by some Native American peoples to promote a strong voice for orating and singing. *Lomatium* may be included as a supportive ingredient in formulas for viral respiratory infections that cause sore throat and laryngitis. Use moderate dosages combined with other herbs in formulas because large and repetitive dosages or concentrated products are sometimes observed to induce a pruritic maculopapular skin rash, although this is uncommon.

Lycopodium spp. • Club Moss

Lycopodium is not commonly available on the marketplace except in homeopathic form. The entire moss is used to prepare remedies primarily used for conditions of weakness, atony, emaciation, and senility, but *Lycopodium* is also specific for tooth pain with decay, tooth pain that is better with warm applications, dryness of the mouth without thirst, fissures or blisters on the tongue, and halitosis. *Lycopodium* is also indicated for ulcers of the tonsils, deposits in the back of the throat including diphtheria membrane formation, and laryngitis associated with ulceration or destruction of the larynx and vocal cords.

Matricaria chamomilla • Chamomile

Young chamomile flowers can be dried for use in tea, or fresh tinctured. Include *Matricaria* in formulas for teething in infants and toddlers and in mouthwashes for toothaches that are worse from hot food and liquids. A sesquiterpene compound in chamomile, α-bisabolol, is widely used in cosmetics, burn creams, and wound healing and is shown to have a healing and antinociceptive effect on corneal lesions.[206] Do not use *Matricaria* infusions in the eyes or as nasal lavage for those with known pollen allergies. *Matricaria* mouth rinses may help to prevent and treat oral mucositis related to chemotherapy; the constituent chamazulene inhibits leukotriene synthesis, thereby helping to limit the anti-inflammatory activity.[207] *Matricaria recutita* is also used medicinally and shown to treat oral mucositis due to chemotherapy or radiation,[208] and promote healing of oral ulcers in animal models of diabetes.[209]

Medicago sativa • Alfalfa

Medicago is a rich source of calcium, potassium, magnesium, phosphorus, and other nutrients.[210] The young leaves, which are also used as a livestock feed, can be consumed as a tea or extracted in vinegar. Use *Medicago* as a nutritive mineral and nutritional tonic in cases of malnutrition, osteopenia of the jaw, and weak, cracked teeth.

Melissa officinalis • Lemon Balm

The young aromatic leaves of lemon balm are used to prepare teas and tinctures with nervine, hormone-regulating, and anti-inflammatory properties. Use *Melissa* as a supportive ingredient in formulas for viral infections and respiratory infections symptoms. *Melissa* has antiviral effects against herpes, and the essential oil is useful topically on herpetic lesions. While lemon balm essential oil is expensive due to the difficulty of distillation, simple aqueous extracts are also shown to exert an antiviral effect against herpes simplex. Much of the herb's antiviral activity is credited to caffeic, rosmarinic, and ferulic acids.[211]

Mentha piperita • Mint

Mint leaves are dried for use in herbal teas, as well as used in large quantity to distill essential oil, prized as a source of menthol and widely used in commercial products from toothpaste to pain-relieving ointments. The essential oil or tincture is very helpful to allay oral and dental pain, to provide a cooling effect for burning sensations, and to help disinfect oral lesions and infections. *Mentha* is also specific for a dry, sticking sensation of the throat and to clear mucus from the throat. *Mentha* tea is pleasant tasting and useful in formulas for oral infections due to its broad antimicrobial effects. A drop of mint essential oil can be placed directly on the tongue to treat oral infections. Mint oil can help to break apart *Candida* biofilms for those with thrush.[212]

Momordica charantia • Bitter Melon

The fruits of this cucurbit, although bitter, are often eaten as a vegetable and are also a traditional medicine for improving digesting and blood sugar regulation.[213] Bitter melon is a natural source of organic fluoride, and is useful in gingivitis and tooth decay formulas related to diabetes.

Morella cerifera • Bayberry

The leaves and young twigs have been used traditionally to create astringent mouth rinses and teas and shown to have broad anti-inflammatory effects.[214] Use *Morella* to astringe spongy, swollen, bleeding gums and to clear thick, ropy, tenacious mucus in the throat and on the tongue. *Myrica* is specific for abundant mucus and a tight feeling in the throat that compels one to swallow frequently. Use *Morella* in formulas for chronic sore throats associated with nausea, loss of appetite, and long-lasting stomach discomfort after eating, a bitter taste in the mouth, and halitosis. *Morella* is indicated for oral congestion and dysbiosis related to underlying liver disease and hepatic congestion in which a person may crave or be better with acids. *Morella cerifera* is also known as *Myrica cerifera*. See also the *Morella cerifera* entry in chapter 3, Volume 2.

Nepeta cataria • Catnip

The young aerial parts of catnip are used as a carminative tonic in formulas where EENT symptoms are due to underlying URIs, especially in children where digestive upset and flatulence accompany. *Nepeta* is specific for irritable children with simple colds and upper respiratory symptoms, and its antiviral activity has been documented.[215]

Origanum vulgare • Oregano

Oregano leaves are used as a culinary spice, fresh tinctured, or used in large quantity to distill essential oil. Oregano essential oil is a powerful antimicrobial and a few drops may be included in mouthwashes, tinctures, gargles, and mouth pastes for infectious stomatitis and gingivitis, as well as for a hairy tongue, chronic infections, and for cases where poor oral health is due to underlying digestive difficulty. Oregano oil capsules and other products are also used to treat intestinal parasites and dysbiosis that can contribute to oral dysbiosis and symptoms. Clinical trials have shown oregano extracts to deter chronic *Candida* infections of the mouth in denture wearers.[216] Carvacrol is a leading component of oregano essential oil and is shown to have broad antimicrobial activity. Carvacrol injures bacterial cell walls and makes them more permeable, leading to instability and depolarization of the interior cytoplasmic membrane and impairing the ability of bacteria to adhere to tissues and form biofilms.[217]

Panax ginseng • Ginseng

Ginseng roots are one of the most esteemed longevity herbs in all of TCM. *Panax* is useful for diabetic patients with eye complaints and for failing eyesight in the elderly, especially when associated with fatigue and weakness. *Panax* will not greatly improve eyesight but may slow the demise of vision and help maintain strength. *Panax* is shown to improve dry eye in patients using ocular medicines for dry eyes.[218] Due to its neuro-protective effect, *Panax* ginsenosides have been prepared into eye drops,[219] and constituents are shown to bind lysophosphatidic acid receptors and speed healing of corneal injury.[220] *Panax* ginsenosides, which are a group of steroidal saponins, may also slow the progression of macular degeneration, cataracts, and glaucoma.[221]

Phytolacca americana • Poke

The fleshy roots of *Phytolacca* are mainly used as a remedy for tissue and glandular congestion with a tendency to infections. *Phytolacca* is indicated for a geographic tongue, blisters and lesions in the mouth with bleeding, and indentations on the sides of the tongue. *Phytolacca* is one of the best herbal remedies for tonsillitis, quinsy and tonsillar abscess, tonsillar pain and swelling, and enlargement of cervical lymph nodes. *Phytolacca* is useful for pain in the throat that radiates to the ear and for throat pain that is worse when swallowing. *Phytolacca* is also indicated for eye symptoms associated with infections, lymphatic congestion, and enlarged tonsils, and it is useful as an ingredient in formulas for orbital cellulitis and for chronic eyelid disorders with crusting, discharges, and styes.

Pilocarpus jaborandi • Jaborandi

Pilocarpus is a profound cholinergic agent that will promote salivary flow and lacrimation when used orally or in eye drops. This herb has a long history of use dating from the late 1800s as a topical glaucoma therapy.[222] The use of *Pilocarpus* is limited because it also promotes perspiration and digestive secretions. It is best used locally for dry eyes and mouth, such as in formulas for Sjögren's patients. Pilocarpine, the primary alkaloid in *Pilocarpus*, is used to treat glaucoma, dry mouth, and dry eyes due to muscarinic agonism.[223] Pilocarpine can also help optimize intraocular pressure in glaucoma patients, increasing Schlemm's canal outflow as much as 15 percent to 25 percent via effects on ciliary muscles.[224] See also "*Pilocarpus* for Dry Eyes" on page 151.

Pimpinella anisum • Anise

The seeds of anise may be used in teas as well as to prepare tinctures or essential oils. *Pimpinella* is useful to improve the flavor of other medicines, such as when *Commiphora myrrha* and *Hydrastis* are used for oral and throat ailments, especially in children. *Pimpinella* offers its own antimicrobial and anti-inflammatory effects as well as in cases of stomatitis or oral lesions. Anise seed teas and essential oils can be used in formulas for cold sores; *Pimpinella* has been shown to have activity against the herpes simplex virus.

Piper methysticum • Kava Kava

Kava has a numbing effect on the oral mucosa and may be considered as an anodyne ingredient in formulas for severe stomatitis, erythema multiforme, and throat pain. Due to concerns of hepatotoxicity, limit the use of kava to short term only. It is contraindicated for those who have liver disease or are using other hepatotoxic substances.

Propolis

This substance is made by bees from high-resin plant secretions, such as poplar bud resin, and has many antimicrobial properties. Due to its sticky nature, propolis can help herbal preparations cling to the throat or mouth. Propolis can also be used to prepare sticky medicines to use topically. Allow such preparations to dry on the lips or external ear canals, acting as an herbal bandage.

Punica granatum • Pomegranate

Tree bark and fruit rinds of *Punica* have astringent and anti-inflammatory effects. *Punica* is useful in mouthwashes for oral infections and in teas for traveler's diarrhea, dysbiosis, and intestinal parasites. *Punica* fruits and pomegranate juice are high in flavonoids credited with vasoprotective effects, which may offer antioxidant and stabilizing activity on the retina and vasculature of the eye.

Rheum palmatum • Chinese Rhubarb

The roots of this Asian species of rhubarb have been prepared into syrups for babies with teething difficulties, for children who develop digestive upset during dentition, for children who drool and become irritable because of teething, and for crying, whining children who develop sour-smelling sweat. *Rheum* is especially indicated when oral pain and teething are accompanied by colicky digestive pains centered about the umbilicus, and when bowel movements are passed with much straining, as well as for those who quickly feel fullness after eating only a small amount. Traditional syrups blended *Rheum* with peppermint essential oil and sodium bicarbonate.

Rubus idaeus • Raspberry

Rubus leaves may be used as an astringent ingredient in mouthwash formulas when vascular issues and friable tissues are present. *Rubus* is nutritive and can be included in tea blends to support healing, dentition, and mineral density. *Rubus* leaves also contain a variety of carotenoids and other flavonoids with anti-inflammatory and antioxidant effects.[225]

Salix alba • White Willow

A variety of *Salix* species have been mentioned in folkloric herbalism of ancient Egypt, Greece, and Rome for anodyne and antipyretic properties, known in modern times to be due to the oxidation of salicin into salicylic acid in the body following ingestion. *Salix* bark preparations may be used as an oral astringent and anti-inflammatory in mouthwash and gargle formulas, and systemically when fever, muscle pain, and hiccups accompany infectious illnesses and diarrhea.

Salvia miltiorrhiza • Red Sage

The reddish roots of *S. miltiorrhiza* are a traditional circulatory supportive herb, and their use is now supported by a large volume of modern research. Also known as dan shen in TCM, this species of *Salvia* has many beneficial effects on the vasculature and blood cells, reducing inflammation and enhancing organ perfusion. *Salvia miltiorrhiza* may act as a synergist in formulas where gingivitis is associated with heart disease, vascular inflammation, or diabetes.

Salvia officinalis • Sage

Sage leaves have numerous anti-inflammatory, antimicrobial, astringent, and cognitive-enhancing actions. Use *Salvia* as a drying ingredient in eyewash formulas for eye complaints with profuse discharges, excessive salivation, and perspiration. *Salvia* is also astringent for boggy enlarged tonsils, excessive mucus, and tonsillar and oral phlegm. Rosmarinic acid, carnosol, and ursolic acid are among the anti-inflammatory compounds contributing to antinociceptive effect in cases of oral pain with pharyngitis.[226] Sage also has significant antimicrobial effects against common pathogens, fungal infections, and dental microbes.

Sambucus nigra • Elderberry

Elderberries are widely used to prepare antiviral syrups and other medicine, and elder flowers are also used to prepare antiviral tea formulas for eye complaints associated with the flu and upper respiratory viruses. Use *Sambucus* berry preparations for their flavonoids as a vascular tonic for diabetics, to improve circulation, and to protect the vascular system from oxidative damage.

Sanguinaria canadensis • Bloodroot

Sanguinaria roots are used to prepare tincture and may be included in oral mouth rinses for poor dental and gingival health related to extensive plaque formation, and as a synergist in formulas where digestive symptoms are associated with burning sensations, hot flashes, and vasomotor symptoms. *Sanguinaria* tincture can have an irritating and even harmful effect on the tissues and thus is used in small, diluted amounts. Such diluted preparations can be used to stimulate secretions in cases of dry mouth, for tonsillitis, and stomatitis, and is specifically indicated when a sensation of dryness and burning in the mouth accompanies. **Caution:** Due to its potential causticity, *Sanguinaria* is considered prescription only.

Schisandra chinensis • Magnolia Vine

Schisandra fruits are the primary medicinal part, used historically to treat diseases of the gastrointestinal tract, respiratory failure, and cardiovascular diseases, as well as those who suffer from chronic fatigue and weakness, excessive sweating, and insomnia.[227] *Schisandra* is a liver supportive and blood moving herb to include in formulas where intestinal dysbiosis, alcoholism, and poor digestion contribute to oral lesions.

Silybum marianum • Milk Thistle

Use milk thistle seed preparations as supportive ingredients in formulas for chronic eye complaints that occur concomitantly with liver congestion and digestive organ inflammation or difficulty. The use of milk thistle is indicated to protect the eyes from pharmaceuticals known to have ocular toxicity.

Solidago canadensis • Goldenrod

The young flowering tops of goldenrod are a traditional remedy for allergy and hay fever. *Solidago* has an anti-inflammatory effect in the upper respiratory passages and will astringe catarrh in the throat. *Solidago* is a traditional remedy for allergic symptoms and genitourinary weakness and irritation.

Spilanthes acmella • Pinkroot

Spilanthes acmella contains mouth-numbing isobutylamides, including an alkaloid known as spilanthol that can be used to treat tongue paralysis, sore throat, and gum infections. The flowerheads and roots are used to treat toothache and can be chewed for stomatitis. They have been traditionally prepared into dental pastes for topical application. *Spilanthes* may also be a sialagogue for dry mouth syndromes.[228] Approximately 30 *Spilanthes* patents are registered in the United States mainly for pain relief in cases of dental pain, gum infections, and periodontitis.[229]

Syzygium aromaticum • Clove

The highly aromatic buds of clove are dense and woody and used traditionally as a culinary spice and medicine. Clove buds are commonly processed into essential oils used in oral pain medications, mouthwashes, and dental products. Use diluted clove essential oil topically to alleviate dental and gum pain, and clove tincture as an antimicrobial and anodyne for toothache and oral pain.[230] Place several drops of clove oil on a cold wet washrag for teething babies to gnaw on. Clove is also known as *Eugenia caryophyllata*.

Tabebuia impetiginosa • Pau D'Arco

The bark of this South American tree has been traditionally used as an anti-inflammatory and anti-microbial agent and modern research has shown anticancer effects. Use *Tabebuia* as an antifungal agent to include in formulas for oral thrush and infections and dysbiotic disorders of the oral and intestinal mucosa. *Tabebuia* tea can be used both locally and systemically when oral infections are related to intestinal dysbiosis, as is common in diabetes, and when ulcerative lesions are present. The most-studied compounds are lapachol and β-lapachone, which have antifungal, antimicrobial, antiparasitic, antioxidant, antiviral, anti-inflammatory, and antiulcerogenic properties.[231]

Tanacetum parthenium • Feverfew

The young leaves of feverfew are a long-standing Western herb used for inflammatory and allergic disorders. *Tanacetum* may be included in formulas for pain in the head, ears, eyes, teeth, or nose. Use *Tanacetum* orally to reduce allergic phenomena involving the eyes and to treat hay fever symptoms and migraines accompanied by eye symptoms. *Tanacetum* may also be used in formulas for unusual EENT complaints that are suspected to be due to underlying allergic phenomena. More than 30 different sesquiterpene lactones have been identified in *Tanacetum*; parthenolide is one of the most widely reported anti-inflammatory agents contributing to

Tanacetum's antiallergy effects. Parthenolide may desensitize the trigeminal and other nerves via effects on TRP channels.[232]

Taraxacum officinale • Dandelion

Although *Taraxacum* is mainly a liver supportive and alterative herb, medicines prepared from *Taraxacum* roots are also indicated for a mapped tongue with a white coat. Consider *Taraxacum* for raw, sore lesions on the tongue especially when associated with loss of appetite or when associated with liver or digestive disturbances. *Taraxacum* may also be used as a supportive ingredient in formulas where chronic eye complaints occur concomitantly with liver congestion and digestive organ inflammation or difficulty.

Thuja occidentalis • Northern White Cedar

Young *Thuja* leaves are a Native American medicine indicated for gum disease, receding gums, and dental decay, and tinctures or teas can be used to prepare mouthwash and other oral preparations. Folkloric herbals mention *Thuja* for inflammation of the tongue with pain, blisters on the tongue sides, and varicose veins under the tongue. Modern research suggests that *Thuja* may be useful for throat or tongue lesions due to viral infections, including HPV-related warty lesions.

Thymus vulgaris • Thyme

The diminutive leaves of thyme may be used in teas, prepared into tinctures, or distilled into powerful essential oils for medicinal use. Use thyme for its strong antimicrobial and astringent effects in the mouth and throat, or in mouthwashes to deter oral pathogens that contribute to plaque and caries. Thyme tea and tincture are warming and stimulating and may be used for cough and throat symptoms associated with URIs.

Ulmus rubra • Slippery Elm

The inner bark of the many species of elm trees yield a mucilaginous material historically used for pain, ulceration, and inflammation of oral and digestive mucous membranes. Slippery elm lozenges are a classic folkloric demulcent medicine to allay oral and throat pain. Because *Ulmus* species are high in minerals and have mucilaginous components, regular consumption of *Ulmus* may have trophorestorative effects on the oral mucosa. Elm bark constituents may improve cytokine profiles in mucous membranes and optimize white blood cell and mucous cell homeostasis.[233]

Herbs for Eye, Ear, Nose, Mouth, and Throat Conditions

Urtica dioica • Nettle

Stinging nettle is one of the most nourishing of all known species of plants. The leaves offer a rich source of bio-available minerals, vitamins, and a substantial amount of protein.[234] Use *Urtica* leaves as a general mineral tonic in formulas for malnutrition and tooth loss and for osteopenia in the elderly.

Vaccinium myrtillus • Blueberry

Vaccinium fruits and leaves as well as prepared solid extracts and tinctures act as a vascular tonic with antioxidant and anti-inflammatory properties on blood vessels, for diabetic retinopathy, and to improve circulation to the eyes in cases of macular degeneration, cataracts, and failing eyesight of the elderly. Berry anthocyanins are credited with much of the antioxidant and vascular protective effects.[235] Use *Vaccinium* leaf infusions in mouthwashes to protect the blood vessels in cases of hyperglycemia and hyperlipidemia and for local astringent effects in oral infections. *Vaccinium* is specific for diabetic microvascular damage. Include the fruits or fruit-based medicines in the diet to support and protect blood vessels.

Veratrum album • White Hellebore

Caution: *Veratrum* is a potentially toxic botanical and is mainly used homeopathically, and in drop dosages, for states of weakness and collapse. It is also specifically indicated for a pale tongue, sensation of coolness, a salty taste in the mouth, and toothaches with a sensation of extreme heaviness in the teeth.

Zanthoxylum clava-herculis • Southern Prickly Ash

The spiny bark of *Zanthoxylum* is used as a warming, stimulating remedy to bring heat to the stomach, increasing function. *Zanthoxylum* increases circulation and secretions in cases of digestive debility and insufficiency and is best in those with cold constitutions, weakness, lethargy, and poor circulation. *Zanthoxylum* has a carminative and antispasmodic action and is a mild appetite stimulant in cases of dyspepsia. *Zanthoxylum* may improve neuralgic pains in the mouth and lower jaw and may also improve pharyngitis with a sensation of dryness of the mouth.

Zingiber officinale • Ginger

Zingiber roots are widely used as a culinary spice and a medicine for their anti-inflammatory, antimicrobial, and digestion-enhancing effects. Ginger is a useful anti-inflammatory in cases of alcohol- or irritant-induced stomatitis, and when oral problems are a reflection of bowel atony. Ginger helps to optimize the digestive microbiota, gastric emptying, and mucosa-protective enzyme systems.[236] Energetically, *Zingiber* is warming and stimulating and may help improve microcirculation in the oral cavity when hyperglycemia and hyperlipidemia contribute to gingival diseases.

Herbs for Eye, Ear, Nose, Mouth, and Throat Conditions

— ACKNOWLEDGMENTS —

I would like to thank and acknowledge my colleagues who kindly shared their expertise and time to review the chapter contents, helping to make the text as up to date and strong as possible. I wish to thank Dr. Paul Anderson of the Anderson Medical Specialty Associates in Seattle, Washington, for his outstanding and generous leadership in representing the soul of natural medicine during the FDA hearings on over-the-counter herbal products, assuredly a frustrating and thankless task. I also wish to thank Dr. Brice Thompson, postdoctoral research fellow in the Department of Pharmaceutics at the University of Washington School of Pharmacy for his time and expertise. His suggestions and rigorous review offered many important improvements to chapter 2 of this volume. And I would like to thank Dr. Mary Bove, retired midwife, renowned physician, and beloved herb teacher and academic professor who, once again, blessed me with her valuable wisdom. She will forever be held in my utmost esteem for her integrity, her medical competence, her many impressive homesteading skills, and most of all her friendship. She really knows how to hold a sister up! Gratitude is also due to Ann Armbrecht for reading the intro, and to Roy Upton, executive editor of the *American Herbal Pharmacopoeia*, the most rigorous and authoritative series of herbal monographs in the English language. Roy's dedication to the plants, his bringing together multidisciplinary plant people, and his attention to scientific, botanical, historical, and legal issues in herbal medicine are impressive and make Roy a key figure in the field of herbal medicine. I would also like to thank my doctoral student Alexandra Loch-Mally for helping to tidy up all the references, no easy task in a research review of this size.

Thank you to my editor at Chelsea Green, Fern Marshall Bradley, for her professional prowess and for making this much more polished than it would have been without her guidance. Her attention to every detail was impressive, and my respect and admiration grew as we progressed through the work on these volumes. I wish to also thank Margo Baldwin, the publisher at Chelsea Green, for making this text possible at all, as well as Sean Maher and Pati Stone and the entire production team for helping to perfect the many small details that producing a reference text of this nature entails.

And I extend warm, fond appreciation to my kind and supportive sweetie, Warren Martin, for his patience, political updates, and willingness to pitch in with the day-to-day chores of running a household and a business. And thanks to Isaiah Alexander Stansbury for putting one foot in front of the other and walking a clean and clear path, bringing us hope and integrity of the spoken word as a model of healthy communication for all. And much love and affection to the vigorous and vivacious Sierra Ileta Stansbury Halberg and to Mike Halberg for taking such good care of her. You are an impressive team, and I am proud of you both. I can't wait to see what you create next! And this being the final volume, I wish to thank my sweet mother, Judy Deiken Flater Newell, for her strength, for being a domestic goddess, and for always being my biggest fan.

I would like to thank and honor all my teachers and the herb community for being my beloved tribe, filled with powerful, skillful women and gentle, reverent men, and for all the times that the healing plants have brought us together in beauty, to share our gifts and wisdom with one another. If I have never told you to your face (and to many of you I have), I love you!

And last, but not least, to Pachamama herself for her mystery, abundance, and amazing gifts. May we all Walk in Beauty and return her gifts tenfold. Blessed Be!

— SCIENTIFIC NAMES —
TO COMMON NAMES

The following list includes all the herbs, medicinal fungi, and homeopathic preparations mentioned in the text of this book.

Scientific Name	Common Name
Acacia catechu	catechu, khadira
Acanthopanax senticosus (also known as *Eleutherococcus senticosus*)	Siberian ginseng, eleuthero, ciwujia
Achillea millefolium	yarrow
Aconitum carmichaelii	fu zhi
Aconitum napellus	aconite, wolfsbane
Actaea cimicifuga (also known as *Cimicifuga foetida*), *A. heracleifolia*	shengma
Actaea racemosa (also known as *Cimicifuga racemosa*)	black cohosh, macrotys
Adhatoda vasica (also known as *Justicia adhatoda*)	Malabar nut
Aesculus hippocastanum	horse chestnut
Agaricus bisporus, *Agaricus* spp.	white button mushroom
Agave spp.	agave
Agrimonia eupatoria, *A. pilosa* (also known as *A. japonica*)	agrimony
Ajuga forrestii	ajuga
Albizia lebbeck	siris, lebbeck tree
Alcea rosea	hollyhock
Alisma plantago-aquatica (also known as *A. orientale*)	Asian water plantain
Allium cepa	onion
Allium sativum	garlic
Aloe barbadensis, *A. vera*, *A. ferox*, *A. arborescens*	aloe
Althaea officinalis	marshmallow
Amanita phalloides	death cap mushroom
Ammi visnaga, *A. majus*	khella, bishop's weed
Ananas comosus	pineapple
Andrographis paniculata	king of bitters, andrographis
Anemarrhena asphodeloides	anemarrhena, zhi mu
Angelica archangelica	garden angelica, garden archangel
Angelica dahurica	bai zhi
Angelica sinensis	angelica, dong quai
Apis mellifica	honey bee
Apium graveolens	celery
Apocynum cannabinum	dogbane
Aquilaria agallocha	agarwood, agaru
Arctium lappa	burdock, gobo root
Argania spinosa	argan tree
Armoracia rusticana (also known as *Cochlearia armoracia*)	horseradish
Arnica montana	arnica, leopard's bane
Artemisia absinthium	wormwood, sweet wormwood, qing hao su
Artemisia annua	sweet Annie, sweet wormwood, qinghao
Artemisia indica (also known as *A. asiatica*)	Indian wormwood
Asclepias tuberosa	pleurisy root, butterfly weed, milkweed
Asparagus cochinchinensis	Chinese asparagus
Astragalus membranaceus	milk vetch, huang qi
Atractylodes lancea, *A. macrocephala*	atractylodes, bai zhu
Atropa belladonna	belladonna, deadly nightshade
Avena sativa	oats

Azadirachta indica (also known as *Melia azadirachta*)	neem, bean tree, China tree	*Cedrus deodara*	Himalayan cedar, deodar cedar
Bacopa monnieri	brahmi	*Celastrus aculeatus*	staff vine, staff tree
Baptisia tinctoria	wild indigo, yellow wild indigo	*Centella asiatica*	gotu kola, pennywort, brahmi
Berberis aquifolium (also known as *Mahonia aquifolium*)	Oregon grape, mahonia	*Chelidonium majus*	celandine
		Chionanthus virginicus	fringe tree
Berberis nervosa (also known as *Mahonia nervosa*)	low Oregon grape, Oregon grape	*Cimicifuga racemosa* (also known as *Actaea racemosa*)	black cohosh
Berberis vulgaris	mahonia, barberry	*Cineraria maritima*	dusty miller, silver ragwort
Beta vulgaris	beet		
Bidens parviflora	beggarticks, Spanish needles	*Cinnamomum* spp.	cinnamon
		Cinnamomum camphora	camphor
Borago officinalis	borage	*Cinnamomum cassia*	cassia cinnamon, Chinese cinnamon
Boswellia sacra (also known as *B. carteri*)	frankincense		
		Cinnamomum seiboldii	Japanese cassia
Boswellia frereana	Coptic frankincense, dhidin	*Cinnamomum verum* (also known as *C. zeylanicum*)	Ceylon cinnamon, cinnamon
Boswellia papyrifera	Sudanese frankincense	*Citrus aurantium*	bitter orange
Boswellia serrata	frankincense, Indian frankincense	*Citrus* species	citrus
		Citrus reticulata	mandarin
Bupleurum chinense	chai hu, Chinese thoroughwax	*Claviceps purpurea*	ergot fungus
Bupleurum falcatum	Chinese thoroughwax, saiko	*Clerodendrum serratum*	glorybower, bagflower, bharangi
Calendula officinalis	pot marigold, calendula	*Cnicus benedictus*	blessed thistle
Camellia sinensis	green tea	*Cnidium monnieri*	snow parsley, osthole
Cannabis sativa	marijuana, hemp	*Cnidium officinale* (also known as *Ligusticum officinale*)	snow parsley
Cantharis vesicatoria	Spanish fly		
Capsella bursa-pastoris	Shepherd's purse	*Cochlearia armoracia* (also known as *Armoracia rusticana*)	horseradish
Capsicum annuum (also known as *C. frutescens*)	cayenne		
		Cocos nucifera	coconut
Capsicum spp.	hot peppers	*Colchicum autumnale*	autumn crocus
Cassia alata (also known as *Senna alata*)	candlebush, Christmas candle	*Coleus forskohlii* (also known as *Plectranthus forskohlii*)	coleus, Indian coleus
Cassia obtusifolia (also known as *Senna obtusifolia*)	coffeeweed, java bean	*Collinsonia canadensis*	stoneroot
		Commiphora mukul	guggul
Cassia tora (also known as *Senna tora*)	sickle senna, foetid senna	*Commiphora myrrha, C. molmol, C. wightii*	myrrh
Caulophyllum thalictroides	blue cohosh	*Conium maculatum*	poison hemlock
Ceanothus americanus	red root, New Jersey tea	*Convallaria majalis*	lily of the valley

Coptis chinensis	goldthread, duan e huang lian
Coptis trifolia	goldthread
Cordyceps militaris	caterpillar fungus
Cordyceps sinensis	caterpillar fungus, winter worm
Cornus mas	Cornelian cherry, European cornel
Cornus officinalis	Japanese cornel, shan zhu yu
Corydalis impatiens; C. spp.	corydalis
Corydalis yanhusuo	corydalis, turkey corn, yanhusuo
Crataegus monogyna, Crataegus oxyacantha, C. laevigata, C. monogyna	hawthorn
Crataegus spp.	hawthorn
Crinum glaucum	swamp lily
Cryptocarya aganthophylla (also known as *Ravensara aromatica*)	clove nutmeg
Curcuma aromatica	wild turmeric
Curcuma longa	turmeric
Cymbopogon citratus	lemongrass
Cynara scolymus	artichoke
Cyperus rotundus	nutgrass, tigernut
Cytisus laburnum (also known as *Laburnum anagyroides*)	laburnum, golden chain
Datura stramonium	jimson weed, devil's snare
Daucus carota	wild carrot, common carrot
Desmodium adscendens	desmodium
Dioscorea alata, D. belophylla, D. bulbifera, D. esculenta, D. hispida, D. oppositifolia, D. villosa, D. zingiberensis	wild yam
Drosera rotundifolia	sundew
Drymaria cordata	tropical chickweed
Dryobalanops sumatrensis	camphor tree, Borneo camphor
Echinacea angustifolia	coneflower
Echinacea pallida	pale coneflower
Echinacea purpurea	coneflower, purple coneflower
Eleutherococcus senticosus (also known as *Acanthopanax senticosus*)	Siberian ginseng, eleuthero, ciwujia
Emblica officinalis (also known as *Phyllanthus emblica*)	amla, amalaki, Indian gooseberry
Ephedra nevadensis	Mormon tea
Ephedra sinica	ma huang, ephedra
Ephedra spp.	ma huang, Mormon tea
Epimedium brevicornu (also known as *E. rotundatum*)	horny goatweed, yin yang huo, bishop's hat
Equisetum arvense, E. hyemale	horsetail, scouring rush
Equisetum giganteum	giant horsetail
Erigeron canadensis	Canada fleabane, Canadian horseweed
Erodium stephanianum	Stephan's storkbill, lao guan cao
Eschscholzia californica	California poppy
Eucalyptus globulus	eucalyptus
Eucommia ulmoides	Chinese rubber tree, hardy rubber tree
Eugenia aromatica, E. caryophyllata (also known as *Syzygium aromaticum*)	cloves
Eupatorium perfoliatum, E. chinense	boneset
Euphrasia officinalis	eyebright
Euphrasia stricta	eyebright
Filipendula ulmaria	meadowsweet
Foeniculum vulgare	fennel
Fucus vesiculosus	kelp, bladderwrack
Galium aparine	cleavers
Galphimia glauca	calderona amarilla, flor estrella
Ganoderma lucidum	reishi, reishi mushroom
Garcinia mangostana	Mangosteen
Gardenia jasminoides	cape jasmine, gardenia, cape jessamine
Gardenia latifolia	gardenia, Indian boxwood

Gastrodia elata	tian ma
Gaultheria fragrantissima, G. nummularioides, G. procumbens, G. subcorymbosa	wintergreen
Gelsemium sempervirens	false jasmine, yellow jessamine, yellow jasmine
Geranium maculatum, G. robertianum	wild geranium
Ginkgo biloba	ginkgo, maidenhair tree
Glycine max	soybean, soy
Glycyrrhiza glabra	licorice
Glycyrrhiza inflata, G. uralensis	Chinese licorice, Gan Cao
Grifola frondosa	maitake mushroom, hen of the woods
Grindelia squarrosa	gumweed
Guaiacum officinale	guaiacwood
Gymnema sylvestre	gurmar, sugar destroyer, cow plant
Hamamelis virginiana	witch hazel
Harpagophytum procumbens	devil's claw
Hedeoma pulegioides	pennyroyal
Hedyotis diffusa (synonym of *Oldenlandia diffusa*)	Snake needle grass, baihua sheshecao
Helichrysum italicum	immortelle, curry plant
Hibiscus sabdariffa	roselle, flor de Jamaica, hibiscus
Hippophae rhamnoides	sea buckthorn
Humulus lupulus	hops, common hop
Huperzia serrata	Chinese club moss, qian ceng ta
Hydrangea macrophylla	hydrangea
Hydrastis canadensis	goldenseal
Hypericum perforatum	St. Johnswort, St. John's wort
Hypericum scabrum	hofarighon, alafe chai
Inula cappa	sheep's ear
Inula helenium	elecampane
Inula racemosa	inula
Iris versicolor, I. tenax	wild iris, blue flag, vegetable mercury
Isatis tinctoria (also known as *I. indigotica*)	woad, dyer's woad, ban lan gen
Jacobaea maritima (also known as *Cineraria maritima*)	dusty miller, silver ragwort
Jatropha curcas	physic nut, Barbados nut
Juglans cinera, J. nigra	walnut
Juniperus communis, J. excelsa, J. phoenicea	juniper
Justicia adhatoda (also known as *Adhatoda vasica*)	Malabar nut
Laburnum anagyroides (*also known as Cytisus laburnum*)	laburnum, golden chain
Lavandula angustifolia (also known as *L. officinalis*)	lavender
Lentinula edodes	shiitake mushroom
Leonurus japonicus (also known as *L. heterophyllus*)	Chinese motherwort, yi mu co
Lepidium meyenii (also known as *L. peruvianum*)	maca
Ligusticum officinale (also known as *Cnidium officinale*)	snow parsley
Ligusticum striatum (also known as *L. chuanxiong* and *L. wallichii*)	Szechuan lovage, chuanxiong, ligusticum
Lindera aggregata	evergreen lindera, Japanese evergreen spicebush, wu yao
Linum usitatissimum	flax
Lobelia inflata	pokeweed, Indian tobacco, wild tobacco
Lomatium dissectum	biscuitroot
Lonicera japonica, L. caprifolium	honeysuckle, ren dong teng
Lycium barbarum, L. chinense	goji berry, wolfberry
Lycopodium complanatum	groundcedar
Lycopodium spp.	ground pine, creeping cedar, club moss
Magnolia officinalis	magnolia, houpu magnolia, magnolia-bark
Mahonia aquifolium (also known as *Berberis aquifolium*)	Oregon grape, mahonia
Mahonia nervosa (also known as *Berberis nervosa*)	low Oregon grape, Oregon grape

Malus domestica	apple
Matricaria chamomilla (also known as *M. recutita*)	chamomile, German chamomile
Medicago sativa	alfalfa, lucerne
Melaleuca alternifolia	tea tree
Melia azadirachta (also known as *Azadirachta indica*)	China tree, ku lian pi
Melissa officinalis	lemon balm
Mentha canadensis (also known as *Mentha haplocalyx*)	Canada mint, American wild mint
Mentha piperita	peppermint
Mentha spicata	spearmint
Momordica charantia	bitter melon
Morella cerifera (also known as *Myrica cerifera*)	southern wax myrtle, bayberry
Myrica californica	Pacific wax myrtle
Myrica cerifera (also known as *Morella cerifera*)	southern wax myrtle, bayberry
Nelumbo nucifera	lotus, Indian lotus
Nepeta cataria	catnip, catnep, catmint
Nicotiana tabacum	tobacco, cultivated tobacco
Nigella sativa	black cumin, black seed, love-in-a-mist
Ocimum tenuiflorum (also known as *O. sanctum*)	holy basil, tulsi
Oenothera biennis	evening primrose
Oldenlandia diffusa (synonym of *Hedyotis diffusa*)	Snake needle grass, baihua sheshecao
Opuntia ficus-indica	prickly pear, prickly pear cactus
Origanum vulgare, *O. marjorana*, *O. onites*	oregano
Paeonia lactiflora	white peony, bai shao yao, red peony
Paeonia spp.	peony
Paeonia × *suffruticosa*, *P. anomala*	moutan, peony, tree peony
Palmaria palmata	dulse
Panax ginseng	ginseng, ren shen
Panax notoginseng	san qi, notoginseng

Panax pseudoginseng	Pseudoginseng, Nepal ginseng, Hamalayan ginseng
Passiflora incarnata	passionflower
Perilla frutescens	shiso, perilla, Korean perilla
Persicaria tinctoria (also known as *Polygonum tinctorium*)	Chinese indigo
Petasites hybridus, *P. japonicus*	butterbur
Petroselinum crispum (also known as *P. sativum*)	parsley
Phellodendron amurense	Amur cork tree
Phellodendron chinense	Amur cork tree, huang bai
Phellodendron spp.	cork tree
Phoradendron chrysocladon (also known as *Viscum flavens*)	mistletoe
Phyllanthus amarus	chanca piedra, bahupatra, hurricane weed
Phyllanthus emblica (also known as *Emblica officinalis*)	amla, amalaki, Indian gooseberry
Phytolacca americana (also known as *P. decandra*)	pokeweed, pokeroot, poke
Picrorhiza kurroa	kutki
Pilocarpus alatus, *P. jaborandi*	jaborandi
Pimpinella anisum	anise, aniseed, anise seed
Pinellia ternata	crow dipper, ban xia
Pinus pinaster	maritime pine
Pinus sylvestris	Scots pine
Piper longum	pippali, long pepper
Piper methysticum	kava kava, kava
Piper nigrum	black pepper
Piscidia piscipula (also known as *P. erythrina*)	Jamaican dogwood
Platycodon grandiflorus	Balloon flower
Plectranthus forskohlii (also known as *Coleus forskohlii*)	coleus, Indian coleus
Pleurotus spp.	oyster mushrooms
Polygonum multiflorum (also known as *Reynoutria multiflora*)	fo ti, he shou wu, Chinese knotweed

Polygonum tinctorium (also known as *Persicaria tinctoria*)	Chinese indigo
Populus species	aspen
Poria cocos	hoelen mushroom, fu ling
Propolis	resin gathered from various plants by bees
Prunella vulgaris	common self-heal, heal-all
Prunus armeniaca	ansu apricot, Siberian apricot, Tibetan apricot
Prunus mume	Chinese plum, Japanese plum, Japanese apricot
Pterocarpus marsupium	kino tree
Pueraria candollei var. *mirifica*	kudzu, white kwao krua
Pueraria montana var. *chinensis* (also known as *P. thomsonii*), *Pueraria* spp.	kudzu
Pueraria montana var. *lobata*	kudzu, gegen, Japanese arrowroot
Pueraria tuberosa	kudzu, Indian kudzu
Punica granatum	pomegranate
Quercus alba	white oak
Quercus brantii (also known as *Q. persica*)	Brant's oak, Persian oak
Quercus robur	common oak
Quercus rubra	red oak
Ravensara aromatica (also known as *Cryptocarya aganthophylla*)	clove nutmeg
Rehmannia glutinosa	di huang, Chinese foxglove
Reynoutria multiflora (also known as *Polygonum multiflorum*)	fo ti, he shou wu, Chinese knotweed
Rheum officinale, *R. palmatum*, *R. emodi*	Chinese rhubarb, turkey rhubarb
Rhodiola rosea	arctic rose, rhodiola, golden root
Ribes nigrum	black currant
Ribes spp.	currant
Ricinus communis	castor, castor bean
Rosa canina	dog rose
Rosmarinus officinalis	rosemary
Rubus idaeus	raspberry, red raspberry
Rumex crispus	yellow dock, yellowdock, curly dock
Rumex spp.	dock
Ruta graveolens	rue
Salix alba	white willow
Salix spp.	willow
Salvia miltiorrhiza	red sage, dan shen
Salvia officinalis	sage
Sambucus canadensis, *S. nigra*	elderberry, elder, black elder
Sanguinaria canadensis	bloodroot, puccoon
Schisandra chinensis	magnolia vine, wu wei zi, five flavor fruit
Scrophularia nodosa	figwort
Scutellaria baicalensis	scute, huang qin, baical
Scutellaria lateriflora	skullcap, mad dog
Senna alata (also known as *Cassia alata*)	candlebush, Christmas candle
Senna obtusifolia (also known as *Cassia obtusifolia*)	coffeeweed, java bean
Senna tora (also known as *Cassia tora*)	sickle senna, foetid senna
Serenoa repens	saw palmetto
Sesamum indicum	black sesame seed, sesame seed
Silybum marianum	milk thistle
Simmondsia chinensis	jojoba
Smilax ornata	sarsaparilla, Honduran sarsaparilla, Jamaican sarsaparilla
Solanum xanthocarpum	yellow-fruit nightshade, yellow-berried nightshade, Thai green eggplant
Solidago odora, *S. canadensis*, *S. virgaurea*	goldenrod
Sophora alopecuroides	kudouzi
Sophora flavescens	shrubby sophora
Spigelia species	pinkroot

Spilanthes acmella	paracress
Stellaria media	chickweed
Stemona sessilifolia	bai bu
Stephania tetranda	han fang ji
Stevia rebaudiana	sweet leaf, stevia
Stillingia sylvatica	queen's root
Symphytum officinale	comfrey
Syzygium aromaticum (also known as *Eugenia caryophyllata*)	cloves
Tabebuia impetiginosa	pau d'arco, taheebo
Tamarindus indica	tamarind
Tanacetum parthenium	feverfew
Taraxacum officinale	dandelion
Tephrosia purpurea	wild indigo
Terminalia chebula	myrobalan, haritaki, black myrobalan
Thea sinensis (also known as *Camellia sinensis*)	green tea
Thuja occidentalis	northern white cedar
Thuja plicata	western red cedar, cedar
Thymus vulgaris	thyme
Tinospora cordifolia	guduchi, amrita
Tribulus terrestris	puncture vine
Trifolium pratense	red clover
Trigonella foenum-graecum	fenugreek
Tripterygium wilfordii	thunder god vine, lei gong teng
Tussilago farfara	coltsfoot
Tylophora asthmatica, T. indica	Indian ipecac
Ulmus fulva, U. rubra	slippery elm
Uncaria rhynchophylla, U. tomentosa	uña de gato, cat's claw
Urtica dioica, U. urens	nettle, stinging nettle

Vaccinium myrtillus	bilberry, blueberry
Valeriana officinalis	valerian
Valeriana sitchensis	Sitka valerian
Veratrum album	white hellebore, white false hellebore
Veratrum viride	false hellebore, corn lily, corn-lily
Verbascum thapsus	mullein
Verbena hastata, Verbena officinalis, Verbena spp.	vervain
Viburnum opulus	crampbark, guelder rose, snowball bush
Viburnum prunifolium	blackhaw
Vinca minor	lesser periwinkle
Viscum album	mistletoe
Viscum flavens (also known as *Phoradendron chrysocladon*)	mistletoe
Vitellaria paradoxa	shea butter tree
Vitex negundo	Chinese chaste tree
Vitis vinifera	common grape vine
Withania somnifera	ashwagandha, Indian ginseng
Yucca filamentosa	Adam's needle and thread
Yucca schidigera	Mojave yucca
Zanthoxylum americanum	northern prickly ash, common prickly ash, toothache tree
Zanthoxylum bungeanum	Chinese prickly ash, chuan hua jiao, Sichuan pepper
Zanthoxylum clava-herculis	southern prickly ash, Hercules' club, toothache tree
Zingiber officinale	ginger
Ziziphus jujuba	Chinese date, jujube

The following list includes all the herbs, medicinal fungi, and homeopathic preparations mentioned in the text of this book.

aconite	*Aconitum napellus*
Adam's needle and thread	*Yucca filamentosa*
agaru	*Aquilaria agallocha*
agarwood	*Aquilaria agallocha*
agave	*Agave* spp.
agrimony	*Agrimonia eupatoria, A. pilosa* (also known as *A. japonica*)
ajuga	*Ajuga forrestii*
alafe chai	*Hypericum scabrum*
alfalfa	*Medicago sativa*
aloe	*Aloe barbadensis, A. vera, A. ferox, A. arborescens*
amalaki	*Phyllanthus emblica* (also known as *Emblica officinalis*)
American wild mint	*Mentha canadensis* (also known as *Mentha haplocalyx*)
amla	*Phyllanthus emblica* (also known as *Emblica officinalis*)
amrita	*Tinospora cordifolia*
Amur cork tree	*Phellodendron amurense, P. chinense*
andrographis	*Andrographis paniculata*
anemarrhena	*Anemarrhena asphodeloides*
angelica	*Angelica sinensis*
anise	*Pimpinella anisum*
anise seed	*Pimpinella anisum*
aniseed	*Pimpinella anisum*
ansu apricot	*Prunus armeniaca*
apple	*Malus domestica*
arctic rose	*Rhodiola rosea*
argan tree	*Argania spinosa*
arnica	*Arnica montana*
artichoke	*Cynara scolymus*
ashwagandha	*Withania somnifera*
Asian water plantain	*Alisma plantago-aquatica* (also known as *A. orientale*)
aspen	*Populus* species
atractylodes	*Atractylodes lancea, A. macrocephala*
autumn crocus	*Colchicum autumnale*
bagflower	*Clerodendrum serratum*
bahupatra	*Phyllanthus amarus*
bai bu	*Stemona sessilifolia*
bai shao yao	*Paeonia lactiflora*
bai zhi	*Angelica dahurica*
bai zhu	*Atractylodes lancea, A. macrocephala*
baical	*Scutellaria baicalensis*
baihua sheshecao	*Oldenlandia diffusa* (synonym of *Hedyotis diffusa*)
balloon flower	*Platycodon grandiflorus*
ban lan gen	*Isatis tinctoria* (also known as *I. indigotica*)
ban xia	*Pinellia ternata*
Barbados nut	*Jatropha curcas*
barberry	*Berberis vulgaris*
bayberry	*Morella cerifera* (also known as *Myrica cerifera*)
bean tree	*Azadirachta indica* (also known as *Melia azadirachta*)
beet	*Beta vulgaris*
beggarticks	*Bidens parviflora*
belladonna	*Atropa belladonna*

bharangi	*Clerodendrum serratum*
bilberry	*Vaccinium myrtillus*
biscuitroot	*Lomatium dissectum*
bishop's hat	*Epimedium brevicornu* (also known as *E. rotundatum*)
bishop's weed	*Ammi visnaga, A. majus*
bitter melon	*Momordica charantia*
bitter orange	*Citrus aurantium*
black cohosh	*Actaea racemosa* (also known as *Cimicifuga racemosa*)
black cumin	*Nigella sativa*
black currant	*Ribes nigrum*
black elder	*Sambucus canadensis, S. nigra*
black myrobalan	*Terminalia chebula*
black pepper	*Piper nigrum*
black seed	*Nigella sativa*
black sesame seed	*Sesamum indicum*
blackhaw	*Viburnum prunifolium*
bladderwrack	*Fucus vesiculosus*
blessed thistle	*Cnicus benedictus*
bloodroot	*Sanguinaria canadensis*
blue cohosh	*Caulophyllum thalictroides*
blue flag	*Iris versicolor, I. tenax*
blueberry	*Vaccinium myrtillus*
boneset	*Eupatorium perfoliatum, E. chinense*
borage	*Borago officinalis*
Borneo camphor	*Dryobalanops sumatrensis*
brahmi	*Bacopa monnieri, Centella asiatica*
Brant's oak	*Quercus brantii* (also known as *Q. persica*)
burdock	*Arctium lappa*
butterbur	*Petasites hybridus, P. japonicus*
butterfly weed	*Asclepias tuberosa*
calderona amarilla	*Galphimia glauca*
calendula	*Calendula officinalis*
California poppy	*Eschscholzia californica*
camphor	*Cinnamomum camphora*
camphor tree	*Dryobalanops sumatrensis*
Canada fleabane	*Erigeron canadensis*
Canada mint	*Mentha canadensis* (also known as *Mentha haplocalyx*)
Canadian horseweed	*Erigeron canadensis*
candlebush	*Senna alata* (also known as *Cassia alata*)
cape jasmine	*Gardenia jasminoides*
cape jessamine	*Gardenia jasminoides*
cassia cinnamon	*Cinnamomum cassia*
castor	*Ricinus communis*
castor bean	*Ricinus communis*
cat's claw	*Uncaria rhynchophylla, U. tomentosa*
catechu	*Acacia catechu*
caterpillar fungus	*Cordyceps militaris, Cordyceps sinensis*
catmint	*Nepeta cataria*
catnep	*Nepeta cataria*
catnip	*Nepeta cataria*
cayenne	*Capsicum annuum* (also known as *C. frutescens*)
cedar	*Thuja plicata*
celandine	*Chelidonium majus*
celery	*Apium graveolens*
Ceylon cinnamon	*Cinnamomum verum* (also known as *C. zeylanicum*)
chai hu	*Bupleurum chinense*
chamomile	*Matricaria chamomilla* (also known as *M. recutita*)
chanca piedra	*Phyllanthus amarus*
chickweed	*Stellaria media*
China tree	*Azadirachta indica* (also known as *Melia azadirachta*)
Chinese asparagus	*Asparagus cochinchinensis*
Chinese chaste tree	*Vitex negundo*
Chinese cinnamon	*Cinnamomum cassia*
Chinese club moss	*Huperzia serrata*
Chinese date	*Ziziphus jujuba*
Chinese foxglove	*Rehmannia glutinosa*
Chinese indigo	*Persicaria tinctoria* (also known as *Polygonum tinctorium*)

Chinese knotweed	*Reynoutria multiflora* (also known as *Polygonum multiflorum*)
Chinese licorice	*Glycyrrhiza inflata, G. uralensis*
Chinese motherwort	*Leonurus japonicus* (also known as *Leonurus heterophyllus*)
Chinese plum	*Prunus mume*
Chinese prickly ash	*Zanthoxylum bungeanum*
Chinese rhubarb	*Rheum officinale, R. palmatum, R. emodi*
Chinese rubber tree	*Eucommia ulmoides*
Chinese thoroughwax	*Bupleurum chinense, B.m falcatum*
Christmas candle	*Senna alata* (also known as *Cassia alata*)
chuan hua jiao	*Zanthoxylum bungeanum*
chuanxiong	*Ligusticum striatum* (also known as *L. chuanxiong* and *L. wallichii*)
cinnamon	*Cinnamomum verum* (also known as *C. zeylanicum*), *Cinnamomum* spp.
citrus	*Citrus species*
ciwujia	*Eleutherococcus senticosus* (also known as *Acanthopanax senticosus*)
cleavers	*Galium aparine*
clove nutmeg	*Cryptocarya aganthophylla* (also known as *Ravensara aromatica*)
cloves	*Syzygium aromaticum* (also known as *Eugenia caryophyllata*)
club moss	*Lycopodium* spp.
coconut	*Cocos nucifera*
coffeeweed	*Senna obtusifolia* (also known as *Cassia obtusifolia*)
coleus	*Coleus forskohlii* (also known as *Plectranthus forskohlii*)
coltsfoot	*Tussilago farfara*
comfrey	*Symphytum officinale*
common carrot	*Daucus carota*
common grape vine	*Vitis vinifera*
common hop	*Humulus lupulus*
common oak	*Quercus robur*

common prickly ash	*Zanthoxylum americanum*
common self-heal	*Prunella vulgaris*
coneflower	*E. angustifolia, Echinacea purpurea*
Coptic frankincense	*Boswellia frereana*
cork tree	*Phellodendron* spp.
corn lily	*Veratrum viride*
corn-lily	*Veratrum viride*
Cornelian cherry	*Cornus mas*
corydalis	*Corydalis impatiens, C. yanhusuo, Corydalis* spp.
cow plant	*Gymnema sylvestre*
crampbark	*Viburnum opulus*
creeping cedar	*Lycopodium* spp.
crow dipper	*Pinellia ternata*
cultivated tobacco	*Nicotiana tabacum*
curly dock	*Rumex crispus*
currant	*Ribes* spp.
curry plant	*Helichrysum italicum*
dan shen	*Salvia miltiorrhiza*
dandelion	*Taraxacum officinale*
deadly nightshade	*Atropa belladonna*
death cap mushroom	*Amanita phalloides*
deodar cedar	*Cedrus deodara*
desmodium	*Desmodium adscendens*
devil's snare	*Datura stramonium*
devil's claw	*Harpagophytum procumbens*
dhidin	*Boswellia frereana*
di huang	*Rehmannia glutinosa*
dock	*Rumex* spp.
dog rose	*Rosa canina*
dogbane	*Apocynum cannabinum*
dong quai	*Angelica sinensis*
duan e huang lian	*Coptis chinensis*
dulse	*Palmaria palmata*
dusty miller	*Jacobaea maritima* (also known as *Cineraria maritima*)
dyer's woad	*Isatis tinctoria* (also known as *I. indigotica*)
elder	*Sambucus canadensis, S. nigra*

elderberry	*Sambucus canadensis, S. nigra*
elecampane	*Inula helenium*
eleuthero	*Eleutherococcus senticosus* (also known as *Acanthopanax senticosus*)
ephedra	*Ephedra sinica*
ergot fungus	*Claviceps purpurea*
eucalyptus	*Eucalyptus globulus*
European cornel	*Cornus mas*
evening primrose	*Oenothera biennis*
evergreen lindera	*Lindera aggregata*
eyebright	*Euphrasia officinalis, Euphrasia stricta*
false hellebore	*Veratrum viride*
false jasmine	*Gelsemium sempervirens*
fennel	*Foeniculum vulgare*
fenugreek	*Trigonella foenum-graecum*
feverfew	*Tanacetum parthenium*
figwort	*Scrophularia nodosa*
five flavor fruit	*Schisandra chinensis*
flax	*Linum usitatissimum*
flor de Jamaica	*Hibiscus sabdariffa*
flor estrella	*Galphimia glauca*
fo ti	*Reynoutria multiflora* (also known as *Polygonum multiflorum*)
foetid senna	*Senna tora* (also known as *Cassia tora*)
frankincense	*Boswellia sacra* (also known as *B. carteri*), *B. serrata*
fringe tree	*Chionanthus virginicus*
fu ling	*Poria cocos*
fu zhi	*Aconitum carmichaelii*
Gan Cao	*Glycyrrhiza vralensis*
garden angelica	*Angelica archangelica*
garden archangel	*Angelica archangelica*
gardenia	*Gardenia jasminoides, G. latifolia*
garlic	*Allium sativum*
gegen	*Pueraria montana* var. *lobata*
German chamomile	*Matricaria chamomilla* (also known as *M. recutita*)
giant horsetail	*Equisetum giganteum*
ginger	*Zingiber officinale*
ginkgo	*Ginkgo biloba*
ginseng	*Panax ginseng*
glorybower	*Clerodendrum serratum*
gobo root	*Arctium lappa*
goji berry	*Lycium barbarum, L. chinense*
golden chain	*Laburnum anagyroides* (also known as *Cytisus laburnum*)
golden root	*Rhodiola rosea*
goldenrod	*Solidago odora, S. canadensis, S. virgaurea*
goldenseal	*Hydrastis canadensis*
goldthread	*Coptis chinensis, C. trifolia*
gotu kola	*Centella asiatica*
green tea	*Camellia sinensis*
ground pine	*Lycopodium* spp.
groundcedar	*Lycopodium complanatum*
guaiacwood	*Guaiacum officinale*
guduchi	*Tinospora cordifolia*
guelder rose	*Viburnum opulus*
guggul	*Commiphora mukul*
gumweed	*Grindelia squarrosa*
gurmar	*Gymnema sylvestre*
Hamalayan ginseng	*Panax pseudoginseng*
han fang ji	*Stephania tetranda*
hardy rubber tree	*Eucommia ulmoides*
haritaki	*Terminalia chebula*
hawthorn	*Crataegus monogyna, Crataegus oxyacantha, C. laevigata, C. monogyna, Crataegus* spp.
he shou wu	*Reynoutria multiflora* (also known as *Polygonum multiflorum*)
heal-all	*Prunella vulgaris*
hemp	*Cannabis sativa*
hen of the woods	*Grifola frondosa*
Hercules' club	*Zanthoxylum clava-herculis*
hibiscus	*Hibiscus sabdariffa*
Himalayan cedar	*Cedrus deodara*
hoelen mushroom	*Poria cocos*
hofarighon	*Hypericum scabrum*

hollyhock	*Alcea rosea*
holy basil	*Ocimum tenuiflorum* (also known as *O. sanctum*)
Honduran sarsaparilla	*Smilax ornata*
honey bee	*Apis mellifica*
honeysuckle	*Lonicera japonica, L. caprifolium*
hops	*Humulus lupulus*
horny goatweed	*Epimedium brevicornu* (also known as *E. rotundatum*)
horse chestnut	*Aesculus hippocastanum*
horseradish	*Armoracia rusticana* (also known as *Cochlearia armoracia*)
horsetail	*Equisetum arvense, E. hyemale*
hot peppers	*Capsicum* spp.
houpu magnolia	*Magnolia officinalis*
huang bai	*Phellodendron chinense*
huang qi	*Astragalus membranaceus*
huang qin	*Scutellaria baicalensis*
hurricane weed	*Phyllanthus amarus*
hydrangea	*Hydrangea macrophylla*
immortelle	*Helichrysum italicum*
Indian boxwood	*Gardenia latifolia*
Indian coleus	*Coleus forskohlii* (also known as *Plectranthus forskohlii*)
Indian frankincense	*Boswellia serrata*
Indian ginseng	*Withania somnifera*
Indian gooseberry	*Phyllanthus emblica* (also known as *Emblica officinalis*)
Indian ipecac	*Tylophora asthmatica, T. indica*
Indian kudzu	*Pueraria tuberosa*
Indian lotus	*Nelumbo nucifera*
Indian tobacco	*Lobelia inflata*
Indian wormwood	*Artemisia indica* (also known as *A. asiatica*)
inula	*Inula racemosa*
jaborandi	*Pilocarpus alatus, P. jaborandi*
Jamaican dogwood	*Piscidia piscipula* (also known as *P. erythrina*)
Jamaican sarsaparilla	*Smilax ornata*

Japanese apricot	*Prunus mume*
Japanese arrowroot	*Pueraria montana* var. *lobata*
Japanese cassia	*Cinnamomum seiboldii*
Japanese cornel	*Cornus officinalis*
Japanese evergreen spicebush	*Lindera aggregata*
Japanese plum	*Prunus mume*
java bean	*Senna obtusifolia* (also known as *Cassia obtusifolia*)
jimson weed	*Datura stramonium*
jojoba	*Simmondsia chinensis*
jujube	*Ziziphus jujuba*
juniper	*Juniperus communis, J. excelsa, J. phoenicea*
kava	*Piper methysticum*
kava kava	*Piper methysticum*
kelp	*Fucus vesiculosus*
khadira	*Acacia catechu*
khella	*Ammi visnaga, A. majus*
king of bitters	*Andrographis paniculata*
kino tree	*Pterocarpus marsupium*
Korean perilla	*Perilla frutescens*
ku lian pi	*Melia azadirachta* (also known as *Azadirachta indica*)
kudouzi	*Sophora alopecuroides*
kudzu	*Pueraria candollei* var. *mirifica, P. montana* var. *chinensis* (also known as *P. thomsonii*), *P. montana* var. *lobata, P. tuberosa, Pueraria* spp.
kutki	*Picrorhiza kurroa*
laburnum	*Laburnum anagyroides* (also known as *Cytisus laburnum*)
lao guan cao	*Erodium stephanianum*
lavender	*Lavandula angustifolia* (also known as *L. officinalis*)
lebbeck tree	*Albizia lebbeck*
lei gong teng	*Tripterygium wilfordii*
lemon balm	*Melissa officinalis*
lemongrass	*Cymbopogon citratus*

leopard's bane	*Arnica montana*
lesser periwinkle	*Vinca minor*
licorice	*Glycyrrhiza glabra*
ligusticum	*Ligusticum striatum* (also known as *L. chuanxiong* and *L. wallichii*)
lily of the valley	*Convallaria majalis*
long pepper	*Piper longum*
lotus	*Nelumbo nucifera*
love-in-a-mist	*Nigella sativa*
low Oregon grape	*Berberis nervosa* (also known as *Mahonia nervosa*)
lucerne	*Medicago sativa*
ma huang	*Ephedra* spp.
maca	*Lepidium meyenii* (also known as *L. peruvianum*)
macrotys	*Actaea racemosa* (also known as *Cimicifuga racemosa*)
mad dog	*Scutellaria lateriflora*
magnolia	*Magnolia officinalis*
magnolia vine	*Schisandra chinensis*
magnolia-bark	*Magnolia officinalis*
mahonia	*Berberis aquifolium* (also known as *Mahonia aquifolium*), *B. vulgaris*
maidenhair tree	*Ginkgo biloba*
maitake mushroom	*Grifola frondosa*
Malabar nut	*Justicia adhatoda* (also known as *Adhatoda vasica*)
mandarin	*Citrus reticulata*
Mangosteen	*Garcinia mangostana*
marijuana	*Cannabis sativa*
maritime pine	*Pinus pinaster*
marshmallow	*Althaea officinalis*
meadowsweet	*Filipendula ulmaria*
milk thistle	*Silybum marianum*
milk vetch	*Astragalus membranaceus*
milkweed	*Asclepias tuberosa*
mistletoe	*Phoradendron chrysocladon* (also known as *Viscum flavens*), *V. album*
Mojave yucca	*Yucca schidigera*
Mormon tea	*Ephedra nevadensis*, *Ephedra* spp.

moutan	*Paeonia* × *suffruticosa*, *P. anomala*
mullein	*Verbascum thapsus*
myrobalan	*Terminalia chebula*
myrrh	*Commiphora myrrha*, *C. molmol*, *C. wightii*
neem	*Azadirachta indica* (also known as *Melia azadirachta*)
Nepal ginseng	*Panax pseudoginseng*
nettle	*Urtica dioica*, *U. urens*
New Jersey tea	*Ceanothus americanus*
northern prickly ash	*Zanthoxylum americanum*
northern white cedar	*Thuja occidentalis*
notoginseng	*Panax notoginseng*
nutgrass	*Cyperus rotundus*
oats	*Avena sativa*
onion	*Allium cepa*
oregano	*Origanum vulgare*, *O. marjorana*, *O. onites*
Oregon grape	*Berberis aquifolium* (also known as *Mahonia aquifolium*), *B. nervosa* (also known as *Mahonia nervosa*)
osthole	*Cnidium monnieri*
oyster mushrooms	*Pleurotus* spp.
Pacific wax myrtle	*Myrica californica*
pale coneflower	*Echinacea pallida*
paracress	*Spilanthes acmella*
parsley	*Petroselinum crispum* (also known as *P. sativum*)
passionflower	*Passiflora incarnata*
pau d'arco	*Tabebuia impetiginosa*
pennyroyal	*Hedeoma pulegioides*
pennywort	*Centella asiatica*
peony	*Paeonia* × *suffruticosa*, *P. anomala*, *Paeonia* spp.
peppermint	*Mentha piperita*
perilla	*Perilla frutescens*
Persian oak	*Quercus brantii* (also known as *Q. persica*)
physic nut	*Jatropha curcas*
pineapple	*Ananas comosus*

pinkroot	*Spigelia* species
pippali	*Piper longum*
pleurisy root	*Asclepias tuberosa*
poison hemlock	*Conium maculatum*
poke	*Phytolacca americana* (also known as *P. decandra*)
pokeroot	*Phytolacca americana* (also known as *P. decandra*)
pokeweed	*Lobelia inflata, Phytolacca americana* (also known as *P. decandra*)
pomegranate	*Punica granatum*
pot marigold	*Calendula officinalis*
prickly pear	*Opuntia ficus-indica*
prickly pear cactus	*Opuntia ficus-indica*
Pseudoginseng	*Panax pseudoginseng*
puccoon	*Sanguinaria canadensis*
puncture vine	*Tribulus terrestris*
purple coneflower	*Echinacea purpurea*
qian ceng ta	*Huperzia serrata*
qing hao su	*Artemisia absinthium*
qinghao	*Artemisia annua*
queen's root	*Stillingia sylvatica*
raspberry	*Rubus idaeus*
red clover	*Trifolium pratense*
red oak	*Quercus rubra*
red peony	*Paeonia lactiflora*
red raspberry	*Rubus idaeus*
red root	*Ceanothus americanus*
red sage	*Salvia miltiorrhiza*
reishi	*Ganoderma lucidum*
reishi mushroom	*Ganoderma lucidum*
ren dong teng	*Lonicera japonica, L. caprifolium*
ren shen	*Panax ginseng*
resin gathered from various plants by bees	Propolis
rhodiola	*Rhodiola rosea*
roselle	*Hibiscus sabdariffa*
rosemary	*Rosmarinus officinalis*

rue	*Ruta graveolens*
sage	*Salvia officinalis*
saiko	*Bupleurum falcatum*
san qi	*Panax notoginseng*
sarsaparilla	*Smilax ornata*
saw palmetto	*Serenoa repens*
Scots pine	*Pinus sylvestris*
scouring rush	*Equisetum arvense, E. hyemale*
scute	*Scutellaria baicalensis*
sea buckthorn	*Hippophae rhamnoides*
sesame seed	*Sesamum indicum*
shan zhu yu	*Cornus officinalis*
shea butter tree	*Vitellaria paradoxa*
sheep's ear	*Inula cappa*
shengma	*Actaea cimicifuga* (also known as *Cimicifuga foetida*), *A. heracleifolia*
Shepherd's purse	*Capsella bursa-pastoris*
shiitake mushroom	*Lentinula edodes*
shiso	*Perilla frutescens*
shrubby sophora	*Sophora flavescens*
Siberian apricot	*Prunus armeniaca*
Siberian ginseng	*Eleutherococcus senticosus* (also known as *Acanthopanax senticosus*)
Sichuan pepper	*Zanthoxylum bungeanum*
sickle senna	*Senna tora* (also known as *Cassia tora*)
silver ragwort	*Cineraria maritima, Jacobaea maritima* (also known as *Cineraria maritima*)
siris	*Albizia lebbeck*
Sitka valerian	*Valeriana sitchensis*
skullcap	*Scutellaria lateriflora*
slippery elm	*Ulmus fulva, U. rubra*
Snake needle grass	*Oldenlandia diffusa* (synonym of *Hedyotis diffusa*)
snow parsley	*Cnidium monnieri, Ligusticum officinale* (also known as *Cnidium officinale*)
snowball bush	*Viburnum opulus*

southern prickly ash	*Zanthoxylum clava-herculis*
southern wax myrtle	*Morella cerifera* (also known as *Myrica cerifera*)
soy	*Glycine max*
soybean	*Glycine max*
Spanish fly	*Cantharis vesicatoria*
Spanish needles	*Bidens parviflora*
spearmint	*Mentha spicata*
St. John's wort	*Hypericum perforatum*
St. Johnswort	*Hypericum perforatum*
staff tree	*Celastrus aculeatus*
staff vine	*Celastrus aculeatus*
Stephan's storkbill	*Erodium stephanianum*
stevia	*Stevia rebaudiana*
stinging nettle	*Urtica dioica, U. urens*
stoneroot	*Collinsonia canadensis*
Sudanese frankincense	*Boswellia papyrifera*
sugar destroyer	*Gymnema sylvestre*
sundew	*Drosera rotundifolia*
swamp lily	*Crinum glaucum*
sweet Annie	*Artemisia annua*
sweet leaf	*Stevia rebaudiana*
sweet wormwood	*Artemisia absinthium, A. annua*
Szechuan lovage	*Ligusticum striatum* (also known as *L. chuanxiong* and *L. wallichii*)
taheebo	*Tabebuia impetiginosa*
tamarind	*Tamarindus indica*
tea tree	*Melaleuca alternifolia*
Thai green eggplant	*Solanum xanthocarpum*
thunder god vine	*Tripterygium wilfordii*
thyme	*Thymus vulgaris*
tian ma	*Gastrodia elata*
Tibetan apricot	*Prunus armeniaca*
tigernut	*Cyperus rotundus*
tobacco	*Nicotiana tabacum*
toothache tree	*Zanthoxylum Americanum, Z. clava-herculis*
tree peony	*Paeonia × suffruticosa, P. anomala*
tropical chickweed	*Drymaria cordata*
tulsi	*Ocimum tenuiflorum* (also known as *O. sanctum*)
turkey corn	*Corydalis yanhusuo*
turkey rhubarb	*Rheum officinale, R. palmatum, R. emodi*
turmeric	*Curcuma longa*
uña de gato	*Uncaria rhynchophylla, U. tomentosa*
valerian	*Valeriana officinalis*
vegetable mercury	*Iris versicolor, I. tenax*
vervain	*Verbena hastata, Verbena officinalis, Verbena* spp.
walnut	*Juglans cinera, J. nigra*
western red cedar	*Thuja plicata*
white button mushroom	*Agaricus bisporus, Agaricus* spp.
white false hellebore	*Veratrum album*
white hellebore	*Veratrum album*
white kwao krua	*Pueraria candollei* var. *mirifica*
white oak	*Quercus alba*
white peony	*Paeonia lactiflora*
white willow	*Salix alba*
wild carrot	*Daucus carota*
wild geranium	*Geranium maculatum, G. robertianum*
wild indigo	*Baptisia tinctoria, Tephrosia purpurea*
wild iris	*Iris versicolor, I. tenax*
wild tobacco	*Lobelia inflata*
wild turmeric	*Curcuma aromatica*
wild yam	*Dioscorea alata, D. belophylla, D. bulbifera, D. esculenta, D. hispida, D. oppositifolia, D. villosa, D. zingiberensis*
willow	*Salix* spp.
winter worm	*Cordyceps sinensis*
wintergreen	*Gaultheria fragrantissima, G. nummularioides, G. procumbens, G. subcorymbosa*
witch hazel	*Hamamelis virginiana*

woad	*Isatis tinctoria* (also known as *I. indigotica*)
wolfberry	*Lycium barbarum, L. chinense*
wolfsbane	*Aconitum napellus*
wormwood	*Artemisia absinthium*
wu wei zi	*Schisandra chinensis*
wu yao	*Lindera aggregata*
yanhusuo	*Corydalis yanhusuo*
yarrow	*Achillea millefolium*
yellow dock	*Rumex crispus*
yellow jasmine	*Gelsemium sempervirens*

yellow jessamine	*Gelsemium sempervirens*
yellow wild indigo	*Baptisia tinctoria*
yellow-berried nightshade	*Solanum xanthocarpum*
yellow-fruit nightshade	*Solanum xanthocarpum*
yellowdock	*Rumex crispus*
yi mu co	*Leonurus japonicus* (also known as *L. heterophyllus*)
yin yang huo	*Epimedium brevicornu* (also known as *E. rotundatum*)
zhi mu	*Anemarrhena asphodeloides*

— GLOSSARY —
OF THERAPEUTIC TERMS

Abortifacient. An agent capable of promoting the expulsion of a developing fetus.

Absorbent. A drug that promotes the absorption of medicinal compounds.

Acidifier. An agent imparting acidity to body fluids, especially blood and urine.

Acute. A condition that has a new onset, comes on suddenly, and is relatively short-lasting in its entire duration.

Adaptogen. An agent that helps one to "adapt," a term stemming from research on herbs that improve resilience, immunity, and stress tolerance. The word has come to refer to herbs such as *Panax ginseng, Eleutherococcus, Withania,* and *Glycyrrhiza* that optimize HPA axis feedback loops and adrenal function.

Aerial parts. The parts of a plant that grow above ground.

Alkalinizer. An agent that increases the alkalinity of bodily fluids, especially the blood and urine.

Allopathic. A term applying to conventional, modern Western medicine. *Allo* refers to "opposite," and in this case, means to oppose pathology. For example: In cases of fever, an antipyretic is used; to treat inflammation, an anti-inflammatory is used; and to treat an infection, antimicrobials are used.

Alterative. An agent that favorably "alters" an individual's health. Alteratives stimulate digestive and absorptive functions while enhancing elimination of wastes. Alteratives are also traditionally said to "purify" the blood and optimize metabolic functions.

Amphoteric. An agent capable of both increasing and decreasing activity, such as estrogenic activity, in the body; a single compound that can both promote and reduce hormonal or other activity depending on the physiologic situation.

Analgesic. An agent that is pain-relieving.

Anaphrodisiac. An agent that diminishes sexual drive or function.

Anesthetic. An agent that diminishes pain and tactile sensations temporarily.

Anhydrotic. An agent that diminishes excessive sweating.

Anodyne. An agent that is pain-relieving.

Antacid. An agent that diminishes stomach acid.

Antagonist. An agent that opposes the action of some other medicine, usually a poison or toxic alkaloid.

Anthelmintic. An agent used to combat intestinal worms.

Antidote. A remedy to counteract the action of poisons or other strong actions.

Antiemetic. An agent that allays nausea and vomiting.

Antigalactogogue. An agent that diminishes lactation.

Antihemorrhagic. An agent that helps control excessive bleeding.

Anti-inflammatory. An agent that reduces inflammatory processes by a variety of mechanisms, reducing oxidative stress and protecting tissues from stress and damage.

Antilithic. An agent used to reduce the formation of stones and calculi in the body.

Antioxidant. An agent capable of accepting electrons or highly reactive molecules that could damage body membranes if left free in circulation.

Antiperiodic. An agent used to combat the periodic fevers of malaria.

Antiphlogistic. An agent used to reduce fever and inflammation.

Antiproliferative. An agent capable of reducing hormone-driven proliferation in hormone-sensitive tissues or of reducing other excessive stimulation of tissues (such as the endometrium or prostate) that can lead to enlargement or an increased risk of cancer.

Antipyretic. An agent used to reduce fever.

Antiscorbutic. An agent used to provide vitamin C and prevent or treat scurvy.

Antiseptic. An agent having antimicrobial capacity for the prevention of sepsis.

Antisialagogue. An agent capable of reducing salivation.

Antispasmodic. An agent capable of reducing painful spasms in muscles and hollow organs.

Antitussive. An agent used to diminish coughing.

Anxiolytic. An agent that reduces anxiety.

Aperient. A gentle nonirritating laxative.

Aphrodisiac. An agent used to stimulate the libido.

Aromatherapy. The use of essential oils for the purpose of inhaling their strong odors.

Aromatic. An agent with a strong fragrance to be inhaled or absorbed through the skin.

Astringent. An agent that dries, condenses, and shrinks inflamed or suppurative tissues.

Bitter. An agent that has a bitter flavor and is used to stimulate gastrointestinal tone and secretions. Bitters prepare mucosa for food, stimulate appetite, and enhance digestion.

Blood mover. A term from Traditional Chinese Medicine (TCM) used to refer to agents capable of improving circulation and relieving blood stagnation and tissue congestion.

Cardiotonic. An agent that improves heart function.

Carminative. An agent that reduces gas, bloating, flatulence, and associated pain.

Cathartic. A strong, potentially harsh laxative.

Caustic. An agent having a corrosive action on tissues.

Chi tonic. A term from TCM in which chi refers to the body's vital energy. Chi tonics are herbs purported to increase and support vitality, longevity, stamina, fertility, and other aspects of the vital force. Chi deficiency manifests as low energy and stamina, a weak voice, and coldness, as well as exercise intolerance, general fatigue, shortness of breath, and dizziness.

Cholagogue. An agent that increases gallbladder tone and the flow of bile from the gallbladder.

Choleretic. An agent that increases the production of bile.

Chronic. A condition that develops slowly over time and becomes persistent and sometimes permanent.

Corrigent. An agent that balances a harsh or strong action of another agent, a corrective.

Counterirritant. An agent that irritates local tissues to enhance blood flow to the area. Counterirritants are used to induce temporary hyperemia in chronic conditions in an attempt to relieve pain, promote healing, and reduce inflammation.

Dacryagogue. An agent that promotes the flow of tears (lacrimation).

Dampness. A term used in TCM and other energetic descriptions of physiologic tissue states that refers to fluid stagnation when evidenced by a coated tongue, chronic phlegm in the mucous membranes, fluid stagnation, and an increased tendency to opportunistic infections.

Deficient. Referring to low energy, low vitality, and poor functioning tissues (herbal medicine). In TCM, the term *chi deficiency* is used when the entire body is in a weakened state. The term may also be used to indicate a poorly functioning organ or biochemical state, such as digestive deficiency, circulatory deficiency, or metabolic deficiency.

Demulcent. A cooling, soothing, mucilagenous substance used internally or topically to emolliate abraded, inflamed, or irritated mucosal tissues.

Depurant. Any agent aimed at purifying, such as a liver depurant, a renal depurant, a blood depurant, and so on. Depurants have a purifying effect by promoting the elimination of wastes from the body.

Diaphoretic. An agent capable of inducing perspiration and often a temporary fever.

Diuretic. An agent that stimulates the production and flow of urine.

Ecbolic. An agent that stimulates childbirth (parturition).

Emetic. An agent that causes vomiting (emesis).

Emmenagogue. An agent that promotes menstrual flow.

Emollient. An agent that soothes and softens the skin and mucosal tissues.

Errhine. An agent that irritates the nasal mucosa and promotes sneezing and secretions.

Escharotic. Any caustic substances applied topically to diseased tissues to kill the cells and promote sloughing away. The word *eschar* means to cast off.

Essential oil. See *volatile oil*.

Excess. Indicates a condition beyond normal range, such as too much heat, overstimulated bowel or muscle tone, or other situations of excess in various physiologic functions.

Excitant. An agent that causes excitation of nervous, circulatory, or motor functions; however, in Latin America, the term is more often used to refer to an aphrodisiac, or sexual excitant.

Exhilarant. An agent that causes excitation of psychic functions and promotes euphoria.

Exogenous estrogen. Exposure to estrogenic compounds that did not originate in the body, but influence physiology due to the ingestion of hormones in animal products or exposure to a variety of chemicals that bind to estrogen receptors in the body and have an estrogenic effect. Estrogens from outside the body.

Expectorant. An agent that promotes the flow of secretions from the respiratory tract.

Febrifuge. An agent used to bring down the temperature in cases of fever.

Fibrinolytic. An agent capable of breaking down fibrin, which may be deposited in vein and artery walls, as well as numerous tissues, in response to inflammatory processes.

GABAergic. GABA is the abbreviation for gamma amino butyric acid, one of the primary inhibitory neurotransmitters of the central nervous systems, exerting a calming and relaxing affect. To match the terms *serotonergic* and *dopaminergic*, GABAergic is used to refer to promotion of GABA neurotransmission and signaling.

Galactogogue. An agent that stimulates lactation.

Glycerite. A liquid extract prepared using glycerine to extract the active ingredients from medicinal plant materials.

Hematic, hematinic. An agent that improves the quality of the blood, especially in cases of anemia, but may be used in other situations.

Hemostatic. An agent that reduces blood flow and promotes clotting in cases of hemorrhage, trauma, and internal bleeding.

Hepatic. An agent that improves the function of the liver.

Hepatotonic. An agent with a trophorestorative effect on the liver.

Hydragogue. An agent that promotes watery secretions.

Inotropic. An agent that supports ion flow in electrically active cardiac muscle and improves the contractile force of the heart, slowing and strengthening the heartbeat and improving circulation.

Irritant. An agent applied locally for the purposes of intentionally causing local hyperemia. See also *counterirritant*.

Laxative. An agent that promotes a mild and painless evacuation of the bowels.

Lipotropic. Literally translates as "fat mover," and used to refer to various alterative and cholagogue herbs, as well as substances such as choline that promote bile flow and biliary function, and thereby improve liver function, and hepatic clearance of lipids, carbohydrates, hormones, toxins, and chemicals.

Lithotriptic. An agent aimed at dissolving calculi within the body.

Material dose. A term used by herbalists and alternative medicine practitioners to distinguish between a highly diluted or homeopathic preparation of a substance and the substance given in a more substantial or "material dose."

Miotic, myotic. An agent that causes the pupil to contract (miosis).

Mydriatic. An agent that promotes dilation of the pupil (mydriasis).

Nanomedicine. A term most commonly used to refer to nanomaterials and biological devices engineered to an extremely small size of roughly 1 to 100 nanometers. The term may also refer to molecular nanotechnology, either natural or synthetic, having medicinal applications.

Narcotic. A drug that promotes stupor or sleep and is used to relieve pain or diminish consciousness.

Nervine. An agent having a tonifying effect on the nervous system, usually only used in the context of herbal medicine, with some herbs being referred to as nervine herbs.

Nociceptive. Relating to the perception or sensation of pain.

Nutrient, nutritive. An agent that enhances assimilation, metabolism, and nutrition.

Oxytocic. An agent that promotes uterine contractions and hastens childbirth.

Parturifacient. An agent that facilitates childbirth when taken during labor.

Partus preparator. An agent taken in the last months of pregnancy to tone the uterus and optimize labor and delivery.

Phytoestrogen. A steroid-like molecule occurring naturally in a plant and capable of exerting hormonal or other activity in animals via ligand activity at estrogen receptors.

Phytosterol. A steroid-like molecule occurring naturally in a plant and capable of exerting hormonal or other activity in animals.

Purgative. A strong laxative that may be irritating and cause cramping.

Refrigerant. An agent capable of imparting a cooling sensation when applied topically.

Revulsive. An agent used to enhance the blood flow to a particular body part (hand, foot) in order to draw it away from a congested, engorged area (head, uterus).

Rubefacient. An agent that promotes reddening or hyperemia of the tissues.

Sedative. An agent that calms in cases of nervousness, insomnia, and mania, and may be stronger and less tonifying than a nervine.

SERM. A selective estrogen receptor modulator, exerting effects only at particular estrogen receptors due to binding specific estrogen receptor types and subtypes, and variably in specific tissues.

Sialagogue, salivant. An agent that increases the flow of saliva.

Simple. A term used to refer to a single herb, not mixed with other herbs or used in a formula, but rather used as "a simple."

Specific. An agent thought to be of specific value for a collection of symptoms.

Sternutatory. An agent that promotes sneezing when inhaled.

Styptic. A strong astringent agent capable of reducing bleeding when applied topically.

Succus. An herbal product containing freshly expressed plant juice preserved with 10 percent ethyl alcohol and employed in a manner similar to a tincture.

Sudorific. An agent capable of inducing perspiration and regarded as being stronger than a diaphoretic.

Synergist. An agent that duplicates, enhances, or pulls together the action of a group of medicinal substances.

Taenicide. An agent that kills or weakens tapeworms.

Tonic. An agent that has a positive effect on the function of an organ or tissue and suggests an ability to restore normal function be it excess or deficient, atonic or hypertonic, overstimulated or understimulated. Tonic supports the optimal physiologic state.

Toxicity. A deranged, inflammatory, or otherwise corrupted or polluted biochemical state in the body.

Vasoconstrictor. An agent that constricts the blood vessels.

Vasodepressant. An agent that slows the pulse rate and lowers the pressure.

Vasodilator. An agent used to dilate the vasculature, usually used in cases of hypertension.

Vermifuge. An agent that promotes the expulsion of intestinal worms.

Vesicant. An agent that promotes blistering or vesication of the skin.

Volatile oil. An essential oil. Aromatic plants are high in volatile oils, so named due to the fact that they are small, light molecules that readily volatilize into the air, contributing to the aromatic quality. Volatile oils are often distilled out of aromatic plants such as mint, thyme, citrus, and numerous others and sold in small bottles to use in aromatherapy, to make body products, and for other purposes.

Xenoestrogen. A compound with an estrogenic effect on the tissue that originated from outside the body, such as an animal hormone, or a chemical having an estrogenic effect in the tissues.

— NOTES —

Introduction

1. Marcia Angell, "Drug Companies & Doctors: A Story of Corruption," *New York Review of Books*, January 15, 2009, https://www.nybooks.com/articles/2009/01/15/drug-companies-doctorsa-story-of-corruption/.

2. Richard W. Smith, "In Search of an Optimal Peer Review System," *Journal of Participatory Medicine* 1, no. 1 (October 2009): e13, https://participatorymedicine.org/journal/opinion/2009/10/21/in-search-of-an-optimal-peer-review-system/.

3. Abigail Zuger, "A Drumbeat on Profit Takers," *New York Times*, March 19, 2012, https://www.nytimes.com/2012/03/20/science/a-drumbeat-on-profit-takers.html.

4. Harris Coulter, *Divided Legacy, Volume I: The Patterns Emerge Hippocrates to Paracelsus* (Berkeley, CA: North Atlantic Books, 1982).

5. "CPG Sec. 450.200 Drugs—General Provisions and Administrative Procedures for Recognition as Safe and Effective," US Food and Drug Administration, updated March 1995, https://www.fda.gov/regulatory-information/search-fda-guidance-documents/cpg-sec-450200-drugs-general-provisions-and-administrative-procedures-recognition-safe-and-effective.

6. "Bulk Drug Substances Used in Compounding Under Section 503A of the FD&C Act," US Food and Drug Administration, updated February 20, 2020, https://www.fda.gov/drugs/human-drug-compounding/bulk-drug-substances-used-compounding-under-section-503a-fdc-act#:~:text=FDA%20would%20consider%20taking%20action,of%20the%20identified%20safety%20risks.

7. "Bulk Drug Substances Nominated for Use in Compounding Under Section 503A of the Federal Food, Drug, and Cosmetic Act," US Food and Drug Administration, updated July 1, 2020, https://www.fda.gov/media/94155/download; "Bulk Drug Substances Used in Compounding Under Section 503A of the FD&C Act."

8. Paul Anderson, "Exclusive Interview with Dr. Paul Anderson: The FDA and the Fate of Compounded Medicines," *Journal of Restorative Medicine* 8, no. 1 (January 2019): 1–5, https://restorativemedicine.org/journal/fda-fate-compounded-medicines/.

9. "Bulk Drug Substances Nominated for Use in Compounding Under Section 503A of the Federal Food, Drug, and Cosmetic Act"; "Bulk Drug Substances Used in Compounding Under Section 503A of the FD&C Act."

10. Anderson, "Exclusive Interview with Dr. Paul Anderson."

11. Anderson, "Exclusive Interview with Dr. Paul Anderson."

12. Anderson, "Exclusive Interview with Dr. Paul Anderson."

Chapter 2: Creating Herbal Formulas for Allergic and Autoimmune Conditions

1. G. Wohlleben and K. J. Erb, "Atopic Disorders: A Vaccine around the Corner?" *Trends in Immunology* 22, no. 11 (2001): 518–26, https://doi.org/10.1016/S1471-4906(01)02055-5.

2. M. C. Di Prisco et al., "Association between Giardiasis and Allergy," *Annals of Allergy, Asthma and Immunology* 81, no. 3 (1998): 261–65, https://doi.org/10.1016/S1081-1206(10)62823-2.

3. Kirsi Karvala et al., "Occupational Rhinitis in Damp and Moldy Workplaces," *American Journal of Rhinology* 22, no. 5 (2008): 457–62, https://doi.org/10.2500/ajr.2008.22.3209.

4. E. Isolauri et al., "Modulation of the Maturing Gut Barrier and Microbiota: A Novel Target in Allergic Disease," *Current Pharmaceutical Design* 14, no. 14 (2008): 1368–75, https://doi.org/10.2174/138161208784480207.

5. M. Feuerecker et al., "A Corticoid-Sensitive Cytokine Release Assay for Monitoring Stress-Mediated Immune Modulation," *Clinical and Experimental Immunology* 172, no. 2 (2013): 290–99, https://doi.org/10.1111/cei.12049.

6. I. J. Elenkov and G. P. Chrousos, "Stress Hormones, Th1/Th2 Patterns, Pro/Anti-Inflammatory Cytokines and Susceptibility to Disease," *Trends in Endocrinology and Metabolism* 10, no. 9 (1999): 359–68, https://doi.org/10.1016/s1043-2760(99)00188-5.

7. Mübeccel Akdis et al., "T Helper (Th) 2 Predominance in Atopic Diseases Is Due to Preferential Apoptosis of Circulating Memory/Effector Th1 Cells," *FASEB Journal* 17, no. 9 (2003): 1026–35, https://doi.org/10.1096/fj.02-1070com.

8. Akdis et al., "T Helper (Th) 2 Predominance in Atopic Diseases."

9. U. Nurmatov et al., "Allergen Immunotherapy for IgE-Mediated Food Allergy: A Systematic Review and Meta-Analysis," *Allergy* 72, no. 8 (2017): 1133–47, https://doi.org/10.1111/all.13124.

10. S. M. Tariq et al., "The Prevalence of and Risk Factors for Atopy in Early Childhood: A Whole Population Birth Cohort Study," *Journal of Allergy and Clinical Immunology* 101, no. 5 (1998): 587–93, https://doi.org/10.1016/S0091-6749(98)70164-2.

11. Sadia Hayat Khan et al., "Respiratory Virus and Asthma: The Role of Immunoglobulin E," *Clinical Therapeutics* 30 (January 2008): 1017–24, https://doi.org/10.1016/j.clinthera.2008.06.002.

12. Rafea Shaaban et al., "Rhinitis and Onset of Asthma: A Longitudinal Population-Based Study," *Lancet* 372, no. 9643 (2008): 1049–57, https://doi.org/10.1016/S0140-6736 (08)61446-4.

13. G. W. K. Wong et al., "Symptoms of Asthma and Atopic Disorders in Preschool Children: Prevalence and Risk Factors," *Clinical and Experimental Allergy* 37, no. 2 (2007): 174–79, https://doi.org/10.1111/j.1365-2222.2007.02649.x.

14. K. Wickens et al., "The Association of Early Life Exposure to Antibiotics and the Development of Asthma, Eczema and Atopy in a Birth Cohort: Confounding or Causality?" *Clinical and Experimental Allergy* 38, no. 8 (2008): 1318–24, https://doi.org/10.1111/j.1365-2222.2008.03024.x.

15. B. M. Simpson et al., "NAC Manchester Asthma and Allergy Study (NACMAAS): Risk Factors for Asthma and Allergic Disorders in Adults," *Clinical and Experimental Allergy* 31, no. 3 (2001): 391–99, https://doi.org/10.1046/j.1365-2222 .2001.01050.x.

16. Erkki Savilahti, "Interaction of Early Infant Feeding, Heredity and Other Environmental Factors as Determinants in the Development of Allergy and Sensitization," *Nestlé Nutrition Workshop Series Pediatric Program* 62 (2008): 168–72, https://doi.org/10.1159/000146258.

17. Tariq et al., "The Prevalence of and Risk Factors for Atopy."

18. Andrea Giacometti et al., "Prevalence of Intestinal Parasites among Individuals with Allergic Skin Diseases," *Journal of Parasitology* 89, no. 3 (2003): 490–92, https://doi.org/10 .1645/0022-3395(2003)089[0490:POIPAI]2.0.CO;2.

19. Mathilde Versini et al., "Unraveling the Hygiene Hypothesis of Helminthes and Autoimmunity: Origins, Pathophysiology, and Clinical Applications," *BMC Medicine* 13, no. 81 (2015), https://doi.org/10.1186/s12916-015-0306-7.

20. O. A. Nyan et al., "Atopy, Intestinal Helminth Infection and Total Serum IgE in Rural and Urban Adult Gambian Communities," *Clinical and Experimental Allergy* 31, no. 11 (2001): 1672–78, https://doi.org/10.1046/j.1365-2222.2001.00987.x.

21. Anita H. J. van den Biggelaar et al., "Long-Term Treatment of Intestinal Helminths Increases Mite Skin-Test Reactivity in Gabonese Schoolchildren," *Journal of Infectious Diseases* 189, no. 5 (2004): 892–900, https://doi.org/10.1086/381767.

22. Hyun-Shiek Yeum et al., "*Fritillaria cirrhosa, Anemarrhena asphodeloides*, Lee-Mo-Tang and Cyclosporine A Inhibit Ovalbumin-Induced Eosinophil Accumulation and Th2-Mediated Bronchial Hyperresponsiveness in a Murine Model of Asthma," *Basic and Clinical Pharmacology and Toxicology* 100, no. 3 (2007): 205–13, https://doi.org/10.1111 /j.1742-7843.2007.00043.x.

23. Ji Young Kim et al., "DA-9601, *Artemisia asiatica* Herbal Extract, Ameliorates Airway Inflammation of Allergic Asthma in Mice," *Molecules and Cells* 22, no. 1 (2006): 104–12.

24. H. Shibata et al., "L-Ephedrine Is a Major Constituent of Mao-Bushi-Saishin-To, One of the Formulas of Chinese Medicine, Which Shows Immediate Inhibition after Oral Administration of Passive Cutaneous Anaphylaxis in Rats," *Inflammation Research* 49, no. 8 (2000): 398–403, https:// doi.org/10.1007/s000110050607.

25. T. Enomoto et al., "Clinical Effects of Apple Polyphenols on Persistent Allergic Rhinitis: A Randomized Double-Blind Placebo-Controlled Parallel Arm Study," *Journal of Investigational Allergology and Clinical Immunology* 16, no. 5 (2006): 283–89, http://www.jiaci.org/issues /vol16issue05/3.pdf.

26. Hien-Trung Trinh et al., "Evaluation of Antipruritic Effects of Red Ginseng and Its Ingredients in Mice," *Planta Medica* 74, no. 3 (2008): 210–14, https://doi.org/10.1055/s-2008-1034313.

27. N. A. Hayes and J. C. Foreman, "The Activity of Compounds Extracted from Feverfew on Histamine Release from Rat Mast Cells," *Journal of Pharmacy and Pharmacology* 39, no. 6 (1987): 466–70, https://doi.org/10.1111/j.2042-7158.1987 .tb03421.x.

28. Eliane Torri et al., "Anti-Inflammatory and Antinociceptive Properties of Blueberry Extract (*Vaccinium corymbosum*)," *Journal of Pharmacy and Pharmacology* 59, no. 4 (2007): 591–96, https://doi.org/10.1211/jpp.59.4.0015.

29. Zhengmin Liang et al., "Rosmarinic Acid Attenuates Airway Inflammation and Hyperresponsiveness in a Murine Model of Asthma," *Molecules* 21, no. 6 (2016): 769, https://doi.org /10.3390/molecules21060769.

30. Zheng-Hai Qu et al., "Inhibition Airway Remodeling and Transforming Growth Factor-β1/Smad Signaling Pathway by Astragalus Extract in Asthmatic Mice," *International Journal of Molecular Medicine* 29, no. 4 (2012): 564–68, https://doi.org/10.3892/ijmm.2011.868.

31. Andrea Giacometti et al., "Prevalence of Intestinal Parasites among Individuals with Allergic Skin Diseases," *Journal of Parasitology* 89, no. 3 (2003): 490–92, https://doi.org/10 .1645/0022-3395(2003)089[0490:POIPAI]2.0.CO;2.

32. Schandra Purnamawati et al., "The Rose of Moisturizers in Addressing Various Kinds of Dermatitis: A Review," *Clinical Medicine and Research* 15, nos. 3–4 (2017): 75–87, https:// doi.org/10.3121/cmr.2017.1363.

33. Gary B. Huffnagle, "The Microbiota and Allergies/Asthma," *PLoS Pathogens* 6, no. 5 (2010): e1000549, https://doi.org /10.1371/journal.ppat.1000549.

34. Scott H. Sicherer and Hugh A. Sampson, "Food Allergy: Recent Advances in Pathophysiology and Treatment," *Annual Review of Medicine* 60 (February 2009): 261–77, https://doi.org/10.1146/annurev.med.60.042407.205711.

35. Ichiro Nomura et al., "Non-IgE-Mediated Gastrointestinal Food Allergies: Distinct Differences in Clinical Phenotype Between Western Countries and Japan," *Current Allergy and Asthma Reports* 12, no. 4 (2012): 297–303, https://doi .org/10.1007/s11882-012-0272-5.

36. Anne-Marie Irani and Elias G. Akl, "Management and Prevention of Anaphylaxis," *F1000Research* 4 (2015): 1492, https://doi.org/10.12688/f1000research.7181.1.

37. Scott H. Sicherer and Donald Y. M. Leung, "Advances in Allergic Skin Disease, Anaphylaxis, and Hypersensitivity

Reactions to Foods, Drugs, and Insects," *Journal of Allergy and Clinical Immunology* 118, no. 1 (2006): 170–77, https://doi.org/10.1016/j.jaci.2006.04.018.

38. Wesley Burks, Mike Kulis, and Laurent Pons, "Food Allergies and Hypersensitivity: A Review of Pharmacotherapy and Therapeutic Strategies," *Expert Opinion on Pharmacotherapy* 9, no. 7 (2008): 1145–52, https://doi.org/10.1517/14656566.9.7.1145.

39. Jacob D. Kattan et al., "Pharmacological and Immunological Effects of Individual Herbs in the Food Allergy Herbal Formula-2 (FAHF-2) on Peanut Allergy," *Phytotherapy Research* 22, no. 5 (2008): 651–59, https://doi.org/10.1002/ptr.2357; C. Qu et al., "Induction of Tolerance after Establishment of Peanut Allergy by the Food Allergy Herbal Formula-2 Is Associated with Up-Regulation of Interferon-Gamma," *Clinical and Experimental Allergy* 37, no. 6 (2007): 846–55, https://doi.org/10.1111/j.1365-2222.2007.02718.x.

40. Sae-Hae Kim et al., "Suppression of Th2-Type Immune Response-Mediated Allergic Diarrhea Following Oral Administration of Traditional Korean Medicine: *Atractylodes macrocephala* Koidz," *Immunopharmacology and Immunotoxicology* 27, no. 2 (2005): 331–43, https://doi.org/10.1081/iph-200067950.

41. Weifeng Li et al., "Effect of Chelerythrine Against Endotoxic Shock in Mice and Its Modulation of Inflammatory Mediators in Peritoneal Macrophages through the Modulation of Mitogen-Activated Protein Kinase (MAPK) Pathway," *Inflammation* 35, no. 6 (2012): 1814–24, https://doi.org/10.1007/s10753-012-9502-1.

42. Nan Yang et al., "Berberine and Limonin Suppress IgE Production by Human B Cells and Peripheral Blood Mononuclear Cells from Food-Allergic Patients," *Annals of Allergy, Asthma and Immunology* 113, no. 5 (2014): 556–64. E4, https://doi.org/10.1016/j.anai.2014.07.021.

43. Changda Liu et al., "Anti-Inflammatory Effects of *Ganoderma lucidum* Triterpenoid in Human Crohn's Disease Associated with Downregulation of NF-κB Signaling," *Inflammatory Bowel Diseases* 21, no. 8 (2015): 1918–25, https://doi.org/10.1097/MIB.0000000000000439.

44. Takeo Yoshikawa and Kazuhiko Yanai, "Histamine Clearance through Polyspecific Transporters in the Brain," in *Histamine and Histamine Receptors in Health and Disease,* ed. Y. Hattori and R. Seifert, *Handbook of Experimental Pharmacology,* vol. 241 (Basel: Springer, 2017), 173–878, https://doi.org/10.1007/164_2016_13.

45. Hee Soon Shin et al., "Preventive Effects of Skullcap (*Scutellaria baicalensis*) Extract in a Mouse Model of Food Allergy," *Journal of Ethnopharmacology* 153, no, 3 (2014): 667–73, https://doi.org/10.1016/j.jep.2014.03.018.

46. Hee Soon Shin et al., "Skullcap (*Scutellaria baicalensis*) Extract and Its Active Compound, Wogonin, Inhibit Ovalbumin-Induced Th2-Mediated Response," *Molecules* 19, no. 2 (2014): 2536–45, https://doi.org/10.3390/molecules19022536.

47. Yue Hao, Xiangshu Piao, and Xianglan Piao, "Saikosaponin-d Inhibits β-Conglycinin Induced Activation of Rat Basophilic Leukemia-2H3 Cells," *International Immunopharmacology* 13, no. 3 (2012): 257–63, https://doi.org/10.1016/j.intimp.2012.04.021.

48. Yuanmin Yin et al., "The Immune Effects of Edible Fungus Polysasccharides Compounds in Mice," *Asia Pacific Journal of Clinical Nutrition* 16, suppl. 1 (2007): 258–60, http://apjcn.nhri.org.tw/server/APJCN/16%20Suppl%201//258.pdf.

49. Julie Wang and Xiu-Min Li, "Chinese Herbal Therapy for the Treatment of Food Allergy," *Current Allergy and Asthma Reports* 12, no. 4 (2012): 332–38, https://doi.org/10.1007/s11882-012-0265-4.

50. Julie Wang et al., "Safety, Tolerability, and Immunologic Effects of a Food Allergy Herbal Formula in Food Allergic Individuals: A Randomized, Double-Blinded, Placebo-Controlled, Dose Escalation, Phase 1 Study," *Annals of Allergy, Asthma and Immunology* 105, no. 1 (2010): 75–84, https://doi.org/10.1016/j.anai.2010.05.005.

51. Ruoling Guo, Max H. Pittler, and Edzard Ernst, "Herbal Medicines for the Treatment of Allergic Rhinitis: A Systematic Review," *Annals of Allergy, Asthma and Immunology* 99, no. 6 (2007): 483–95, https://doi.org/10.1016/S1081-1206(10)60375-4.

52. Guo, Pittler, and Ernst, "Herbal Medicines for the Treatment of Allergic Rhinitis."

53. Andreas Schapowal and Study Group, "Treating Intermittent Allergic Rhinitis: A Prospective, Randomized, Placebo and Antihistamine-Controlled Study of Butterbur Extract Ze 339," *Phytotherapy Research* 19, no. 6 (2005): 530–37, https://doi.org/10.1002/ptr.1705.

54. D. K. C. Lee et al., "A Placebo-Controlled Evaluation of Butterbur and Fexofenadine on Objective and Subjective Outcomes in Perennial Allergic Rhinitis," *Clinical and Experimental Allergy* 34, no. 4 (2004): 646–49, https://doi.org/10.1111/j.1365-2222.2004.1903.x.

55. Ulrich C. Danesch, "*Petasites hybridus* (Butterbur Root) Extract in the Treatment of Asthma—an Open Trial," *Alternative Medicine Review: A Journal of Clinical Therapeutics* 9, no. 1 (2004): 54–62, http://archive.foundationalmedicinereview.com/publications/9/1/54.pdf.

56. H. C. Diener, V. W. Rahlfs, and U. Danesch, "The First Placebo-Controlled Trial of a Special Butterbur Root Extract for the Prevention of Migraine: Reanalysis of Efficacy Criteria," *European Neurology* 51, no. 2 (2004): 89–97, https://doi.org/10.1159/000076535.

57. B. L. Fiebich et al., "*Petasites hybridus* Extracts In Vitro Inhibit COX-2 and PGE2 Release by Direct Interaction with the Enzyme and by Preventing P42/44 MAP Kinase Activation in Rat Primary Microglial Cells," *Planta Medica* 71, no. 1 (2005): 12–19, https://doi.org/10.1055/s-2005-837744.

58. Majid Sheykhzade et al., "S-Petasin and Butterbur Lactones Dilate Vessels through Blockage of Voltage Gated Calcium Channels and Block DNA Synthesis," *European Journal of*

Pharmacology 593, nos. 1–3 (2008): 79–86, https://doi.org
/10.1016/j.ejphar.2008.07.004.

59. Ling-Hung Lin et al., "Bronchodilatory Effects of
S-Isopetasin, an Antimuscarinic Sesquiterpene of *Petasites
formosanus*, on Obstructive Airway Hyperresonsiveness,"
European Journal of Pharmacology 584, nos. 2–3 (2008):
398–404, https://doi.org/10.1016/j.ejphar.2008.02.034.

60. D. K. C. Lee et al., "Butterbur, a Herbal Remedy, Confers
Complementary Anti-Inflammatory Activity in Asthmatic
Patients Receiving Inhaled Corticosteroids," *Clinical and
Experimental Allergy* 34, no. 1 (2004): 110–14, https://doi
.org/10.1111/j.1365-2222.2004.01838.x.

61. Fiebich et al., "*Petasites hybridus* Extracts."

62. "Petasites–Zeller," *Drugs in R&D* 4, no. 6 (2003): 378–79,
https://doi.org/10.2165/00126839-200304060-00011.

63. Ling-Ling Chang et al., "Effects of S-Petasin on Cyclic
AMP Production and Enzyme Activity of P450scc in Rat
Zona Fasciculata-Reticularis Cells," *European Journal of
Pharmacology* 489, nos. 1–2 (2004): 29–37, https://doi
.org/10.1016/j.ejphar.2004.02.029; Ling-Ling Chang et al.,
"Effects of S-Petasin on Corticosterone Release in Rats,"
Chinese Journal of Physiology 45, no. 4 (2002): 137–42,
https://pubmed.ncbi.nlm.nih.gov/12817704/.

64. Elena S. Resnick, Brett P. Bielory, and Leonard Bielory,
"Complementary Therapy in Allergic Rhinitis," *Current
Allergy and Asthma Reports* 8, no. 2 (2008): 118–25, https://
doi.org/10.1007/s11882-008-0021-y.

65. M. Stoss et al., "Prospective Cohort Trial of *Euphrasia* Single-
Dose Eye Drops in Conjunctivitis," *Journal of Alternative
and Complementary Medicine* 6, no. 6 (2000): 499–508,
https://doi.org/10.1089/acm.2000.6.499.

66. Desiderio Passali et al., "Nasal Muco-Ciliary Transport
Time Alteration: Efficacy of 18 B Glycyrrhetinic Acid,"
Multidisciplinary Respiratory Medicine 12 (November 2017):
article 29, https://doi.org/10.1186/s40248-017-0110-7.

67. Carmen Cantisani et al., "Unusual Food Allergy: Alioidea
Allergic Reactions Overview," *Recent Patents on
Inflammation and Allergy Drug Discovery* 8, no. 3 (2014):
178–84, https://doi.org/10.2174/1872213x08666141107170159.

68. Ravindra G. Mali and Avinash S. Dhake, "A Review on Herbal
Antiasthmatics," *Oriental Pharmacy and Experimental
Medicine* 11, no. 2 (2011): 77–90, https://doi.org/10.1007
/s13596-011-0019-1.

69. Maha E. Houssen et al., "Natural Anti-Inflammatory Products
and Leukotriene Inhibitors as Complementary Therapy for
Bronchial Asthma," *Clinical Biochemistry* 43, nos. 10–11 (2010):
887–90, https://doi.org/10.1016/j.clinbiochem.2010.04.061.

70. Hiwa M. Ahmed, "Ethnomedicinal, Phytochemical and
Pharmacological Investigations of *Perilla frutescens* (L.)
Britt," *Molecules* 24, no. 1 (2018): 102, https://doi.org
/10.3390/molecules24010102.

71. Ahmed, "Investigations of *Perilla frutescens* (L.) Britt," 102.

72. Deung Dae Park et al., "*Perilla frutescens* Extracts Protects
against Dextran Sulfate Sodium-Induced Murine Colitis:

NF-κB, STAT3, and Nrf2 as Putative Targets," *Frontiers in
Pharmacology* 8 (August 2017): 482, https://doi.org/10.3389
/fphar.2017.00482.

73. Sybille Buchwald-Werner et al., "Perilla Extract Improves
Gastrointestinal Discomfort in a Randomized Placebo
Controlled Double Blind Human Pilot Study," *BMC
Complementary and Alternative Medicine* 14 (May 2014):
article 173, https://doi.org/10.1186/1472-6882-14-173.

74. Gianluigi Marseglia et al., "A Polycentric, Randomized, Parallel-
Group, Study on Letral®, a Multicomponent Nutraceutical,
as Preventive Treatment in Children with Allergic
Rhinoconjunctivitis: Phase II," *Italian Journal of Pediatrics* 45,
no. 1 (2019): 84, https://doi.org/10.1186/s13052-019-0678-y.

75. Kristin Kelly-Pieper et al., "Safety and Tolerability of an
Antiasthma Herbal Formula (ASHMI™) in Adult Subjects
with Asthma: A Randomized, Double-Blind, Placebo-
Controlled, Dose-Escalation Phase I Study," *Journal of
Alternative and Complementary Medicine* 15, no. 7 (2009):
735–43, https://doi.org/10.1089/acm.2008.0543.

76. H. Kakegawa, H. Matsumoto, and T. Satoh, "Inhibitory
Effects of Some Natural Products on the Activation of
Hyaluronidase and Their Anti-Allergic Actions," *Chemical
and Pharmaceutical Bulletin* 40, no. 6 (1992): 1439–42,
https://doi.org/10.1248/cpb.40.1439.

77. Hung-Chou Chang et al., "A Nebulized Complex Traditional
Chinese Medicine Inhibits Histamine and IL-4 Production
by Ovalbumin in Guinea Pigs and Can Stabilize Mast Cells *In
Vitro*," *BMC Complementary and Alternative Medicine* 13 (July
2013): article 174, https://doi.org/10.1186/1472-6882-13-174.

78. Eric Dubuis et al., "Theophylline Inhibits the Cough Reflex
through a Novel Mechanism of Action," *Journal of Allergy
and Clinical Immunology* 133, no. 6 (2014): 1588–98, https://
doi.org/10.1016/j.jaci.2013.11.017.

79. M. Demirtürk et al., "The Importance of Mold Sensitivity
in Nonallergic Rhinitis Patients," *International Forum of
Allergy and Rhinology* 6, no. 7 (2016): 716–21, https://doi.org
/10.1002/alr.21731.

80. William J. Sheehan and Wanda Phipatanakul, "Indoor
Allergen Exposure and Asthma Outcomes," *Current
Opinion in Pediatrics* 28, no. 6 (2016): 772–77, https://doi.org
/10.1097/MOP.0000000000000421.

81. Ozlem Atan Sahin et al., "The Association of Residential Mold
Exposure and Adenotonsillar Hypertrophy in Children
Living in Damp Environments," *International Journal
of Pediatric Otorhinolaryngology* 88 (September 2016):
233–38, https://doi.org/10.1016/j.ijporl.2016.07.018.

82. Krzysztof Kołodziejczyk and Andrzej Bozek, "Clinical
Distinctness of Allergic Rhinitis in Patients with Allergy to
Molds," *BioMed Research International* 2016 (May 2016):
article 3171594, https://doi.org/10.1155/2016/3171594.

83. Ritchie C. Shoemaker, Dennis House, and James C.
Ryan, "Structural Brain Abnormalities in Patients with
Inflammatory Illness Acquired Following Exposure to
Water-Damaged Buildings: A Volumetric MRI Study

Using NeuroQuant®," *Neurotoxicology and Teratology* 45 (September–October 2014): 18–26, https://doi.org/10.1016/j.ntt.2014.06.004.

84. Robert K. Bush et al., "The Medical Effects of Mold Exposure," *Journal of Allergy and Clinical Immunology* 117, no. 2 (2006): 326–33, https://doi.org/10.1016/j.jaci.2005.12.001.

85. James M. Seltzer and Marion J. Fedoruk, "Health Effects of Mold in Children," *Pediatric Clinics of North America* 54, no. 2 (2007): 309–33, https://doi.org/10.1016/j.pcl.2007.02.001.

86. B. Yang et al., "Effect of Dietary Supplementation with Sea Buckthorn (*Hippophaë rhamnoides*) Seed and Pulp Oils on the Fatty Acid Composition of Skin Glycerophospholipids of Patients with Atopic Dermatitis," *Journal of Nutritional Biochemistry* 11, no. 6 (2000): 338–40, https://doi.org/10.1016/s0955-2863(00)00088-7.

87. E. Ianev et al., "The Effect of an Extract of Sea Buckthorn (*Hippophae rhamnoides* L.) on the Healing of Experimental Skin Wounds in Rats," *Khirurgiia* 48, no. 3 (1995): 30–33.

88. Asheesh Gupta et al., "Influence of Sea Buckthorn (*Hippophae rhamnoides* L.) Flavone on Dermal Wound Healing in Rats," *Molecular and Cellular Biochemistry* 290, nos. 1–2 (2006): 193–98, https://doi.org/10.1007/s11010-006-9187-6.

89. T. Beveridge et al., "Sea Buckthorn Products: Manufacture and Composition," *Journal of Agricultural and Food Chemistry* 47, no. 9 (1999): 3480–88, https://doi.org/10.1021/jf981331m.

90. Asheesh Gupta et al., "A Preclinical Study on the Effects of Seabuckthorn (*Hippophae rhamnoides* L.) Leaf Extract on Cutaneous Wound Healing in Albino Rats," *International Journal of Lower Extremity Wounds* 4, no. 2 (2005): 88–92, https://doi.org/10.1177/1534734605277401.

91. Jun Kunisawa et al., "Dietary ω3 Fatty Acid Exerts Anti-Allergic Effect through the Conversion to 17,18-Epoxyeicosatetraenoic Acid in the Gut," *Scientific Reports* 5 (June 2015): article 9750, https://doi.org/10.1038/srep09750.

92. Wei Yang et al., "Effects of Flaxseed Oil on Anti-Oxidative System and Membrane Deformation of Human Peripheral Blood Erythrocytes in High Glucose Level," *Lipids in Health and Disease* 11 (July 2012): article 88, https://doi.org/10.1186/1476-511X-11-88.

93. Uzdan Uz et al., "Effects of Thymoquinone and Montelukast on Sinonasal Ciliary Beat Frequency," *American Journal of Rhinology and Allergy* 28, no. 2 (2014): 122–25, https://doi.org/10.2500/ajra.2014.28.4010.

94. Magdalena Timoszuk, Katarzyna Bielawska, and Elżbieta Skrzydlewska, "Evening Primrose (*Oenothera biennis*) Biological Activity Dependent on Chemical Composition," *Antioxidants* 7, no. 8 (2018): 108, https://doi.org/10.3390/antiox7080108.

95. C. W. Cutler and R. Jotwani, "Dendritic Cells at the Oral Mucosal Interface," *Journal of Dental Research* 85, no. 8 (2006): 678–89, https://doi.org/10.1177/154405910608500801.

96. Marzia Caproni et al., "Celiac Disease and Dermatologic Manifestations: Many Skin Clue to Unfold Gluten-Sensitive Enteropathy," *Gastroenterology Research and Practice* 2012, (May 2012): article 952753, https://doi.org/10.1155/2012/952753.

97. Jan Faergemann, "Atopic Dermatitis and Fungi," *Clinical Microbiology and Reviews* 15, no. 4 (2002): 545–63, https://doi.org/10.1128/cmr.15.4.545-563.2002.

98. Claudia Pföhler et al., "Contact Allergic Gastritis: An Underdiagnosed Entity?" *BMJ Case Reports* 2012 (2012): bcr2012006916, https://doi.org/10.1136/bcr-2012-006916.

99. Paola Lucia Minciullo et al., "Unmet Diagnostic Needs in Contact Oral Mucosal Allergies," *Clinical and Molecular Allergy* 14, no. 1 (2016): 10, https://doi.org/10.1186/s12948-016-0047-y.

100. Minciullo et al., "Unmet Diagnostic Needs."

101. D. B. Gandhi Babu et al., "Low Level Laser Therapy to Reduce Recurrent Oral Ulcers in Behçet's Disease," *Case Reports in Dentistry* 2016 (July 2016): article 4283986, https://doi.org/10.1155/2016/4283986.

102. "Stevens-Johnson Syndrome/Toxic Epidermal Necrolysis," NIH: National Center for Advancing Translational Sciences and GARD: Genetic and Rare Diseases Information Center, last modified October 8, 2018, https://rarediseases.info.nih.gov/diseases/7700/stevens-johnson-syndrometoxic-epidermal-necrolysis.

103. J. Chen, X. Wu, and H. Li, "A Treatment Combined Chinese with Western Medicine for Microcirculation Diseases in Tongue," *Hua Xi Kou Qiang Yi Xue Za Zhi* 18, no. 2 (2000): 101–2, 108.

104. Aaron W. Michels and David A. Ostrov, "New Approaches for Predicting T Cell-Mediated Drug Reactions: A Role for Inducible and Potentially Preventable Autoimmunity," *Journal of Allergy and Clinical Immunology* 136, no. 2 (2015): 252–57, https://doi.org/10.1016/j.jaci.2015.06.024.

105. Wolfgang Uter et al., "Contact Allergy: A Review of Current Problems from a Clinical Perspective," *International Journal of Environmental Research and Public Health* 15, no. 6 (2018): 1108, https://doi.org/10.3390/ijerph15061108.

106. Tsugunobu et al., "Inhibitory Effects of the Methanol Extract of *Ganoderma lucidum* on Mosquito Allergy-Induced Itch-Associated Responses in Mice," *Journal of Pharmacological Sciences* 114, no. 3 (2010): 292–97, https://doi.org/10.1254/jphs.10180fp.

107. Rohit Kumar et al., "Anti-Inflammatory Effect of *Picrorhiza kurroa* in Experimental Models of Inflammation," *Planta Medica* 82, no. 16 (20160: 1403–9, https://doi.org/10.1055/s-0042-106304.

108. S. Y. Ryu et al., "Anti-Allergic and Anti-Inflammatory Triterpenes from the Herb of *Prunella vulgaris*," *Planta Medica* 66, no. 4 (2000): 358–60, https://doi.org/10.1055/s-2000-8531.

109. Brit Jeffrey Long, Alex Koyfman, and Michael Gottlieb, "Evaluation and Management of Angioedema in the Emergency Department," *Western Journal of Emergency Medicine* 20, no. 4 (2019): 587–600, https://doi.org/10.5811/westjem.2019.5.42650.

110. Hayes and Foreman, "The Activity of Compounds Extracted from Feverfew."

111. Daniel B. Lyle et al., "Screening Biomaterials for Stimulation of Nitric Oxide-Mediated Inflammation," *Journal of Biomedical Materials Research* 90, no. 1 (2009): 82–93, https://doi.org/10.1002/jbm.a.32060.

112. Jolanta Parada-Turska et al., "Parthenolide Inhibits Proliferation of Fibroblast-Like Synoviocytes *In Vitro*," *Inflammation* 31, no. 4 (2008): 281–85, https://doi.org/10.1007/s10753-008-9076-0.

113. Gang Xie, Igor A. Schepetkin, and Mark T. Quinn, "Immunomodulatory Acitivty of Acidic Polysaccharides Isolated from *Tanacetum vulgare* L," *International Immunopharmacology* 7, no. 13 (2007): 1639–50, https://doi.org/10.1016/j.intimp.2007.08.013.

114. Chin-Fu Chen and Albert Y. Leung, "Gene Response of Human Monocytic Cells for the Detection of Antimigraine Activity of Feverfew Extracts," *Canadian Journal of Physiology and Pharmacology* 85, no. 11 (2007): 1108–15, https://doi.org/10.1139/Y07-097.

115. Sonal Grover et al., "Role of Inflammation in Bladder Function and Interstitial Cystitis," *Therapeutic Advances in Urology* 3, no. 1 (2011): 19–33, https://doi.org/10.1177/1756287211398255.

116. Lori Birder and Karl-Erik Andersson, "Urothelial Signaling," *Physiological Reviews* 93, no. 2 (2013): 653–80, https://doi.org/10.1152/physrev.00030.2012.

117. Nicolas Montalbetti et al., "Urothelial Tight Junction Barrier Dysfunction Sensitizes Bladder Afferents," *eNeuro* 4, no. 3 (2017): ENEURO.0381–16.2017, https://doi.org/10.1523/ENEURO.0381-16.2017.

118. Suming Xu et al., "Transgenic Mice Expressing MCP-1 by the Urothelium Demonstrate Bladder Hypersensitivity, Pelvic Pain and Voiding Dysfunction: A Multidisciplinary Approach to the Study of Chronic Pelvic Pain Research Network Animal Model Study," *PLoS ONE* 11, no. 9 (2016): e0163829, https://doi.org/10.1371/journal.pone.0163829.

119. Geoffrey Burnstock, "Purinergic Signalling: Therapeutic Developments," *Frontiers in Pharmacology* 8 (September 2017): article 661, https://doi.org/10.3389/fphar.2017.00661.

120. Lori A. Birder et al., "Beyond Neurons: Involvement of Urothelial and Glial Cells in Bladder Function," *Neurourology and Urodynamics* 29, no. 1 (2010): 88–96, https://doi.org/10.1002/nau.20747.

121. Amin S. Herati et al., "Effects of Foods and Beverages on the Symptoms of Chronic Prostatitis/Chronic Pelvic Pain Syndrome," *Urology* 82, no. 6 (2013): 1373–80, https://doi.org/10.1016/j.urology.2013.07.015; Barbara Shorter et al., "Effects of Comestibles on Symptoms of Interstitial Cystitis," *Journal of Urology* 178, no. 1 (2007): 145–52, https://doi.org/10.1016/j.juro.2007.03.020.

122. E. S. Lukacz et al., "A Healthy Bladder: A Consensus Statement," *International Journal of Clinical Practice* 65, no. 10 (2011): 1026–36, https://doi.org/10.1111/j.1742-1241.2011.02763.x.

123. Mehmet İnci et al., "Toxic Effects of Formaldehyde on the Urinary System," *Turkish Journal of Urology* 39, no. 1 (2013): 48–52, https://doi.org/10.5152/tud.2013.010.

124. Hyun-Wook Lee et al., "E-Cigarette Smoke Damages DNA and Reduces Repair Activity in Mouse Lung, Heart, and Bladder as Well as in Human Lung and Bladder Cells," *Proceedings of the National Academy of Sciences of the United States of America* 115, no. 7 (2018): E1560–69, https://doi.org/10.1073/pnas.1718185115.

125. Cynthia Delago, Martin A Finkel, and Esther Deblinger, "Urogenital Symptoms in Premenarchal Girls: Parents' and Girls' Perceptions and Associations with Irritants," *Journal of Pediatric and Adolescent Gynecology* 25, no. 1 (2012): 67–73, https://doi.org/10.1016/j.jpag.2011.08.002.

126. Seung I. Jeong et al., "Alpha-Spinasterol Isolated from the Root of *Phytolacca americana* and Its Pharmacological Property of Diabetic Nephropathy," *Planta Medica* 70, no. 8 (2004): 736–39, https://doi.org/10.1055/s-2004-827204.

127. Youme Ko et al., "Efficacy and Safety of *Ojeok-san* in Korean Female Patients with Cold Hypersensitivity in the Hands and Feet: Study Protocol for a Randomized, Double-Blinded, Placebo-Controlled, Multicenter Pilot Study," *Trials* 19, no. 1 (2018): 662, https://doi.org/10.1186/s13063-018-3013-9.

128. C. P. Sung et al., "Effects of Extracts of *Angelica polymorpha* on Reaginic Antibody Production," *Journal of Natural Products* 45, no. 4 (1982): 406, https://doi.org/10.1021/np50022a006.

129. S. J. Sheu et al., "Analysis and Processing of Chinese Herbal Drugs; VI. The Study of *Angelicae radix*," *Planta Medica* 53, no. 4 (1987): 377–78, https://doi.org/10.1055/s-2006-962742.

130. J. Y. Tao et al., "Studies on the Antiasthmatic Action of Ligustilide of Dang-Gui, *Angelica sinensis* (Oliv.) Diels," *Yao Xue Xue Bao* 19, no. 8 (1984): 561–65.

131. T. Suzuki et al., "Calcium Anatagonist-Like Actions of Coumarins Isolated from 'Quan-Hu' on Anaphylactic Mediator Release from Mast Cell Induced by Concanavalin A," *Journal of Pharmacobio-Dynamics* 8, no. 4 (1985): 257–63, https://doi.org/10.1248/bpb1978.8.257.

132. Tae Chul Moon et al., "The Effects of Isoimperatorin Isolated from *Agelicae dahuricae* on Cyclooxygenase-2 and 5-Lipoxygenase in Mouse Bone Marrow-Derived Mast Cells," *Archives of Pharmacal Research* 31, no. 2 (2008): 210–15, https://doi.org/10.1007/s12272-001-1143-0.

133. "'Dong Quai' or 'Angelica sinensis,'" *Positive Health News*, no. 17 (Fall 1998): 15–16.

134. Z. P. He et al. "Treating Amenorrhea in Vital Energy-Deficient Patients with *Angelica sinensis–Astragalus membranaceus* Menstruation-Regulating Decoction," *Chung I Tsa Chih Ying Wen Pan* 6, no. 3 (1986): 187–90.

135. Hideaki Matsuda et al., "Anti-Allergic Effects of Cnidii Monnieri Fructus (Dried Fruits of *Cnidium monnieri*) and Its Major Component, Osthol," *Biological and Pharmaceutical Bulletin* 25, no. 6 (2002): 809–12, https://doi.org/10.1248/bpb.25.809.

136. R. A. Momin and M. G. Nair, "Antioxidant, Cyclooxygenase and Topoisomerase Inhibitory Compounds from *Apium graveolens* Linn. Seeds," *Phytomedicine* 9, no. 4 (2002): 312–18, https://doi.org/10.1078/0944-7113-00131.

137. F. N. Ko et al., "Inhibition of Platelet Thromboxane Formation and Phosphoinositides Breakdown by Osthole from *Angelica pubescens*," *Thrombosis and Haemostasis* 62, no. 3 (1989): 996–99.

138. F. N. Ko et al., "Vasorelaxation of Rat Thoracic Aorta Caused by Osthole Isolated from *Angelica pubescens*," *European Journal of Pharmacology* 219, no. 1 (1992): 29–34, https://doi.org/10.1016/0014-2999(92)90576-p.

139. Guo Hua Zheng et al., "*Aloe vera* for Prevention and Treatment of Infusion Phlebitis," *The Cochrane Database of Systematic Reviews* 2014, no. 6 (2014): CD009162, https://doi.org/10.1002/14651858.CD009162.pub2.

140. Pier Giorgio Neri et al., "Oral *Echinacea purpurea* Extract in Low-Grade, Steroid-Dependent, Autoimmune Idiopathic Uveitis: A Pilot Study," *Journal of Ocular Pharmacology and Therapeutics* 22, no. 6 (2006): 431–36, https://doi.org/10.1089/jop.2006.22.431.

141. Danielle Delorme and Sandra C. Miller, "Dietary Consumption of Echinacea by Mice Afflicted with Autoimmune (Type I) Diabetes: Effect of Consuming the Herb on Hemopoietic and Immune Cell Dynamics," *Autoimmunity* 38, no. 6 (2005): 453–61, https://doi.org/10.1080/08916930500221761.

142. Virgilijus Zitkevicius et al., "Influence of *Echinacea purpurea* (L.) Moench Extract on the Toxicity of Cadmium," *Annals of the New York Academy of Sciences* 1095 (2007): 585–92, https://doi.org/10.1196/annals.1397.063.

143. Margherita Massa et al., "Proinflammatory Responses to Self HLA Epitopes Are Triggered by Molecular Mimicry to Epstein-Barr Virus Proteins in Oligoarticular Juvenile Idiopathic Arthritis," *Arthritis and Rheumatism* 46, no. 10 (2002): 2721–29, https://doi.org/10.1002/art.10564.

144. Bin Wang et al., "Vaccinations and Risk of Systemic Lupus Erythematosus and Rheumatoid Arthritis: A Systematic Review and Meta-Analysis," *Autoimmunity Reviews* 16, no. 7 (2017): 756–65, https://doi.org/10.1016/j.autrev.2017.05.012.

145. Kathrin Thell et al., "Immunosuppressive Peptides and Their Therapeutic Applications," *Drug Discovery Today* 19, no. 5 (2014): 645–53, https://doi.org/10.1016/j.drudis.2013.12.002.

146. Shivaprasad H. Venkatesha et al., "Control of Autoimmune Inflammation by Celastrol, a Natural Triterpenoid," *Pathogens and Disease* 74, no. 6 (2016): ftw059, https://doi.org/10.1093/femspd/ftw059.

147. J. L. Lamaison, C. Petitjean-Freytet, and A. Carnat, "Medicinal Lamiaceae with Antioxidant Properties, a Potential Source of Rosmarinic Acid," *Pharmaceutica Acta Helvetiae* 66, no. 7 (1991): 185–88.

148. Hyun-A Oh et al., "Effect of *Perilla frutescens* var. *acuta* Kudo and Rosmarinic Acid on Allergic Inflammatory Reactions," *Experimental Biology and Medicine* 236, no. 1 (2011): 99–106, https://doi.org/10.1258/ebm.2010.010252.

149. Da Yeon Jung et al., "Prolonged Survival of Islet Allografts in Mice Treated with Rosmarinic Acid and Anti-CD154 Antibody," *Experimental and Molecular Medicine* 40, no. 1 (2008): 1–10, https://doi.org/10.3858/emm.2008.40.1.1.

150. Hnin Thanda Aung et al., "Rosmarinic Acid in *Argusia argentea* Inhibits Snake Venom-Induced Hemorrhage," *Journal of Natural Medicines* 64, no. 4 (2010): 482–86, https://doi.org/10.1007/s11418-010-0428-3.

151. Rodrigo Lucarini et al., "*In Vivo* Analgesic and Anti-Inflammatory Activities of *Rosmarinus officinalis* Aqueous Extracts, Rosmarinic Acid and Its Acetyl Ester Derivative," *Pharmaceutical Biology* 51, no. 9 (2013): 1087–90, https://doi.org/10.3109/13880209.2013.776613.

152. Chiaki Sanbongi et al., "Rosmarinic Acid Inhibits Lung Injury Induced by Diesel Exhaust Particles," *Free Radical Biology and Medicine* 34, no. 8 (2003): 1060–69, https://doi.org/10.1016/s0891-5849(03)00040-6.

153. Nan Huang et al., "Rosmarinic Acid in *Prunella vulgaris* Ethanol Extract Inhibits Lipopolysaccharide-Induced Prostaglandin E2 and Nitric Oxide in RAW 264.7 Mouse Macrophages," *Journal of Agricultural and Food Chemistry* 57, no. 22 (2009): 10579–89, https://doi.org/10.1021/jf9023728.

154. Yun-Gyoung Hur et al., "Rosmarinic Acid Induces Apoptosis of Activated T Cells from Rheumatoid Arthritis Patients via Mitochondrial Pathway," *Journal of Clinical Immunology* 27, no. 1 (2007): 36–45, https://doi.org/10.1007/s10875-006-9057-8.

155. Chiaki Sanbongi et al., "Rosmarinic Acid Inhibits Lung Injury Induced by Diesel Exhaust Particles," *Free Radical Biology and Medicine* 34, no. 8 (2003): 1060–69, https://doi.org/10.1016/s0891-5849(03)00040-6.

156. Eun-Ju Yang et al., "Barrier Protective Effects of Rosmarinic Acid on HMGB1-Induced Inflammatory Responses *In Vitro* and *In Vivo*," *Journal of Cellular Physiology* 228, no. 5 (2013): 975–82, https://doi.org/10.1002/jcp.24243.

157. Hnin Thanda Aung et al., "Biological and Pathological Studies of Rosmarinic Acid as an Inhibitor of Hemorrhagic *Trimeresurus flavoviridis* (Habu) Venom," *Toxins* 2, no. 10 (2010): 2478–89, https://doi.org/10.3390/toxins2102478.

158. R. Tundis et al., "Potential Role of Natural Compounds against Skin Aging," *Current Medicinal Chemistry* 22, no. 12 (2015): 1515–38, https://doi.org/10.2174/0929867322666150227151809.

159. Jeehee Youn et al., "Beneficial Effects of Rosmarinic Acid on Suppression of Collagen Induced Arthritis," *Journal of Rheumatology* 30, no. 6 (2003): 1203–7, https://www.jrheum.org/content/jrheum/30/6/1203.full.pdf.

160. Hur et al., "Rosmarinic Acid Induces Apoptosis of Activated T Cells."

161. Bo-Jun Xiong et al., "Analgesic Effects and Pharmacological Mechanisms of the *Gelsemium* Alkaloid Koumine on a Rat Model of Postoperative Pain," *Scientific Reports* 7, no. 1 (2017): 14269, https://doi.org/10.1038/s41598-017-14714-0.

162. Rongcai Yue et al., "Immunoregulatory Effect of Koumine on Nonalcoholic Fatty Liver Disease Rats," *Journal of*

Immunology Research 2019, (February 2019): article 8325102, https://doi.org/10.1155/2019/8325102.

163. Xiao-ke Zheng et al., "Fingerprint Research on Immunosuppressive Fraction of *Rehmanniae radix* by HPLC," *Zhong Ya Cai* 36, no. 12 (2013): 1933–36.

164. N. M. Nwinuka, B. W. Abbey, and E. O. Ayalogy, "Effects of Processing on Flatus Producing Oligosaccharides in Cowpea (*Vigna unguiculata*) and the Tropical African Yam Bean (*Sphenostylis stenocarpa*)," *Plant Foods for Human Nutrition* 51, no. 3 (1997): 209–18, https://doi.org/10.1023/a:1007945100867.

165. Ting Li, Xinshan Lu, and Xingbin Yang, "Stachyose-Enriched α-Galacto-Oligosaccharides Regulate Gut Microbiota and Relieve Constipation in Mice," *Journal of Agricultural and Food Chemistry* 61, no. 48 (2013): 11825–31, https://doi.org/10.1021/jf404160e.

166. Jiang-Yue Lie, "Catalpol Protect against Diabetic Vascular Endothelial Function by Inhibiting NADPH Oxidase," *Zhongguo Zhong Yao Za Zhi* 39, no. 15 (2014): 2936–41.

167. Cheuk-Lun Liu et al., "Bioassay-Guided Isolation of Anti-Inflammatory Components from the Root of *Rehmannia glutinosa* and Its Underlying Mechanism via Inhibition of iNOS Pathway," *Journal of Ethnopharmacology* 143, no. 3 (2012): 867–75, https://doi.org/10.1016/j.jep.2012.08.012.

168. Kai Fu et al., "Protective Effect of Catalpol on Lipopolysaccharide-Induced Acute Lung Injury in Mice," *International Immunopharmacology* 23, no. 2 (2014): 400–406, https://doi.org/10.1016/j.intimp.2014.07.011.

169. Fan-Ji Meng et al., "The Protective Effect of Picroside II against Hypoxia/Reoxygenation Injury in Neonatal Rat Cardiomyocytes," *Pharmaceutical Biology* 50, no. 10 (2012): 1226–32, https://doi.org/10.3109/13880209.2012.664555; Yan-Ru Liu et al., "Catalpol Provides Protective Effects against Cerebral Ischaemia/Reperfusion Injury in Gerbils," *Journal of Pharmacy and Pharmacology* 66, no. 9 (2014): 1265–70, https://doi.org/10.1111/jphp.12261.

170. Yoon-Young Sung et al., "Topical Application of *Rehmannia glutinosa* Extract Inhibits Mite Allergen-Induced Atopic Dermatitis in NC/Nga Mice," *Journal of Ethnopharmacology* 134, no. 1 (2011): 37–44, https://doi.org/10.1016/j.jep.2010.11.050.

171. Jun Zhou et al., "*Rehmannia glutinosa* (Gaertn.) DC. Polysaccharide Ameliorates Hyperglycemia, Hyperlipidemia and Vascular Inflammation in Streptozotocin-Induced Diabetic Mice," *Journal of Ethnopharmacology* 164 (2015): 229–38, https://doi.org/10.1016/j.jep.2015.02.026.

172. Kendrick Co Shih et al, "Systematic Review of Randomized Controlled Trials in the Treatment of Dry Eye Disease in Sjogren Syndrome," *Journal of Inflammation* 14 (November 2017): article 26, https://doi.org/10.1186/s12950-017-0174-3.

173. Hairong Zhong et al., "Antiaging Effects of Skin through Activating Nrf2 and Inhibiting NF-κB," *Evidence-Based Complementary and Alternative Medicine* 2019 (May 2019): article 5976749, https://doi.org/10.1155/2019/5976749.

174. Stephen D. Hsu et al., "Green Tea Polyphenols Reduce Autoimmune Symptoms in a Murine Model for Human Sjogren's Syndrome and Protect Human Salivary Acinar Cells from TNF-Alpha-Induced Cytotoxicity," *Autoimmunity* 40, no. 2 (2007): 138–47, https://doi.org/10.1080/08916930601167343.

175. Shih et al., "Treatment of Dry Eye Disease in Sjogren's Syndrome."

176. Shih et al., "Treatment of Dry Eye Disease in Sjogren's Syndrome."

177. D. J. Aframian et al., "Pilocarpine Treatment in a Mixed Cohort of Xerostomic Patients," *Oral Diseases* 13, no. 1 (2007): 88–92, https://doi.org/10.1111/j.1601-0825.2006.01252.x.

178. Magdalena Działo et al., "The Potential of Plant Phenolics in Prevention and Therapy of Skin Disorders," *International Journal of Molecular Sciences* 17, no. 2 (2016): 160, https://doi.org/10.3390/ijms17020160.

179. Golnaz Sarafian et al., "Topical Turmeric Microemulgel in the Management of Plaque Psoriasis: A Clinical Evaluation," *Iranian Journal of Pharmaceutical Research* 14, no. 3 (2015): 865–76, https://www.ncbi.nlm.nih.gov/pmc/articles/PMC4518115/pdf/ijpr-14-865.pdf.

180. Yang-Yang He et al., "Methyl Salicylate 2-O-β-D-Lactoside Alleviates the Pathological Progression of Pristane-Induced Systemic Lupus Erythematosus-Like Disease in Mice via Suppression of Inflammatory Response and Signal Transduction," *Drug Design, Development and Therapy* 2016, no. 10 (September 2016): 3183–96, https://doi.org/10.2147/DDDT.S114501.

181. B. Dasgeb et al., "Colchicine: An Ancient Drug with Novel Applications," *British Journal of Dermatology* 178, no. 2 (2018): 350–56, https://doi.org/10.1111/bjd.15896.

182. Ming-Chun Kuo, Shun-Jen Chang, and Ming-Chia Hsieh, "Colchicine Significantly Reduces Incident Cancer in Gout Male Patients: A 12-Year Cohort Study," *Medicine* 94, no. 50 (December 2015): e1570, https://doi.org/10.1097/MD.0000000000001570.

183. Dongjie Wu, Weiwei Lin, and Ka-Wang Wong, "Herbal Medicine (Gancao Xiexin Decoction) for Behcet Disease: A Systematic Review Protocol," *Medicine* 97, no. 37 (2018): e12324, https://doi.org/10.1097/MD.0000000000012324.

184. Yunkai Dai et al., "Efficacy and Safety of Modified Banxia Xiexin Decoction (Pinellia Decoction for Draining the Heart) for Gastroesophageal Reflux Disease in Adults: A Systematic Review and Meta-Analysis," *Evidence-Based Complementary and Alternative Medicine* 2017 (February 2017): article 9591319, https://doi.org/10.1155/2017/9591319.

185. Louise Oni and Sunil Sampath, "Childhood IgA Vasculitis (Henoch Schonlein Purpura)—Advances and Knowledge Gaps," *Frontiers in Pediatrics* 7 (June 2019): article 257, https://doi.org/10.3389/fped.2019.00257.

186. J. W. C. Dieker, J. van der Vlag, and J. H. M. Berden, "Triggers for Anti-Chromatin Autoantibody Production in SLE," *Lupus* 11, no. 12 (2002): 856–64, https://doi.org/10.1191/0961203302lu307rr.

187. Jiali Zhu et al., "*Naja naja atra* Venom Protects against Manifestations of Systemic Lupus Erythematosus in MRL/lpr Mice," *Evidence-Based Complementary and Alternative Medicine* 2014 (June 2014): article 969482, https://doi.org/10.1155/2014/969482.

188. Zhe Cai et al., "Anti-Inflammatory Activities of *Ganoderma lucidum* (*Lingzhi*) and San-Miao-San Supplements in MRL/lpr Mice for the Treatment of Systemic Lupus Erythematosus," *Chinese Medicine* 11 (April 2016): article 23, https://doi.org/10.1186/s13020-016-0093-x.

189. Ouyang Jin et al., "A Pilot Study of the Therapeutic Efficacy and Mechanism of Artesunate in the MRL/lpr Murine Model of Systemic Lupus Erythematosus," *Cellular and Molecular Immunology* 6, no. 6 (2009): 461–67, https://doi.org/10.1038/cmi.2009.58.

190. Hee-Kap Kang et al., "Apigenin, a Non-Mutagenic Dietary Flavonoid, Suppresses Lupus by Inhibiting Autoantigen Presentation for Expansion of Autoreactive Th1 and Th17 Cells," *Arthritis Research and Therapy* 11, no. 2 (2009): R59, https://doi.org/10.1186/ar2682.

191. Shuk-Man Ka et al., "Citral Alleviates an Accelerated and Severe Lupus Nephritis Model by Inhibiting the Activation Signal of NLRP3 Inflammasome and Enhancing Nrf2 Activation," *Arthritis Research and Therapy* 17 (November 2015): article 331, https://doi.org/10.1186/s13075-015-0844-6.

192. Xiaoli Nie et al., "Reno-Protective Effect and Mechanism Study of Huang Lian Jie Du Decoction on Lupus Nephritis MRL/lpr Mice," *BMC Complementary and Alternative Medicine* 16, no. 1 (2016): 448, https://doi.org/10.1186/s12906-016-1433-1.

193. Linda L.D. Zhong et al., "Chinese Herbal Medicine (Zi Shen Qing) for Mild-to-Moderate Systematic Lupus Erythematosus: A Pilot Prospective, Single-Blinded, Randomized Controlled Study," *Evidence-Based Complementary and Alternative Medicine* 2013 (May 2013): article 327245, https://doi.org/10.1155/2013/327245.

194. Sung-Kyun Kim et al., "Korean Herbal Medicine for Treating Henoch-Schonlein Purpura with Yin Deficiency: Five Case Reports," *Journal of Pharmacopuncture* 17, no. 4 (2014): 70–75, https://pubmed.ncbi.nlm.nih.gov/25780723/.

195. Mahboube Ganji-Arjenaki and Mahmoud Rafieian-Kopaei, "Phytotherapies in Inflammatory Bowel Disease," *Journal of Research in Medical Sciences* 24 (May 2019): 42, https://doi.org/10.4103/jrms.JRMS_590_17.

196. Aditya Reddy and Bernard Fried, "An Update on the Use of Helminths to Treat Crohn's and Other Autoimmune Diseases," *Parasitology Research* 104, no. 2 (2009): 217–21, https://doi.org/10.1007/s00436-008-1297-5.

197. Jia Liu et al., "Supercritical Fluid Extract of *Angelica sinensis* and *Zingiber officinale* Roscoe Ameliorates TNBS-Induced Colitis in Rats," *International Journal of Molecular Sciences* 20, no. 15 (August 2019): 3816, https://doi.org/10.3390/ijms20153816.

198. Richard Li et al., "An Old Herbal Medicine with a Potentially New Therapeutic Application in Inflammatory Bowel Disease," *International Journal of Clinical and Experimental Medicine* 4, no. 4 (2011): 309–19, https://www.ncbi.nlm.nih.gov/pmc/articles/PMC3228586/pdf/ijcem0004-0309.pdf.

199. Alan Gaby, "Multiple Sclerosis," *Global Advances in Health and Medicine* 2, no. 1 (2013): 50–56, https://doi.org/10.7453/gahmj.2013.2.1.009.

200. Sina Mojaverrostami et al., "A Review of Herbal Therapy in Multiple Sclerosis," *Advanced Pharmaceutical Bulletin* 8, no. 4 (2018): 575–90, https://doi.org/10.15171/apb.2018.066.

201. Jonathan B. Baell et al., "Khellinone Derivatives as Blockers of the Voltage-Gated Potassium Channel Kv1.3: Synthesis and Immunosuppressive Activity," *Journal of Medicinal Chemistry* 47, no. 9 (2004): 2326–36, https://doi.org/10.1021/jm030523s.

202. J. C. Bertoglio et al., "*Andrographis paniculata* Decreases Fatigue in Patients with Relapsing-Remitting Multiple Sclerosis: A 12-Month Double-Blind Placebo-Controlled Pilot Study," *BMC Neurology* 16 (May 2016): article 77, https://doi.org/10.1186/s12883-016-0595-2.

203. Ramu Venkatesan, Eunhee Ji, and Sun Yeou Kim, "Phytochemicals that Regulate Neurodegenerative Disease by Targeting Neurotrophins: A Comprehensive Review," *BioMed Research International* 2015 (May 2015): article 814068, https://doi.org/10.1155/2015/814068.

204. Hadi Ghaffari et al., "Rosmarinic Acid Mediated Neuroprotective Effects against H2O2-Induced Neuronal Damage in N2A Cells," *Life Sciences* 113, nos. 1–2 (2014): 7–13, https://doi.org/10.1016/j.lfs.2014.07.010; S. Fallarini et al., "Clovamide and Rosmarinic Acid Induce Neuroprotective Effects *In Vitro* Models of Neuronal Death," *British Journal of Pharmacology* 157, no. 6 (2009): 1072–84, https://doi.org/10.1111/j.1476-5381.2009.00213.x.

205. Tatiane Teixeira Oliveira et al., "Potential Therapeutic Effect of *Allium cepa* L. and Quercetin in a Murine Model of *Blomia tropicalis* Induced Asthma," *DARU Journal of Pharmaceutical Sciences* 23 (February 2015): article 18, https://doi.org/10.1186/s40199-015-0098-5.

206. Rodrigo Arreola et al., "Immunomodulation and Anti-Inflammatory Effects of Garlic Compounds" *Journal of Immunology Research* 2015 (April 2015): article 401630, https://doi.org/10.1155/2015/401630.

207. Heike Wulff and Boris S. Zhorov, "K+ Channel Modulators for the Treatment of Neurological Disorders and Autoimmune Diseases," *Chemical Reviews* 108, no. 5 (2008): 1744–73, https://doi.org/10.1021/cr078234p.

208. Jonathan B. Baell et al., "Khellinone Derivatives as Blockers of the Voltage-Gated Potassium Channel Kv1.3: Synthesis and Immunosuppressive Activity," *Journal of Medicinal Chemistry* 47, no. 9 (2004): 2326–36, https://doi.org/10.1021/jm030523s.

209. Wei Wang et al., "Immunomodulatory Activity of Andrographolide on Macrophage Activation and Specific Antibody Response," *Acta Pharmacological Sinica* 31, no. 2 (2010): 191–201, https://doi.org/10.1038/aps.2009.205.

210. Kathrin S. Michelsen et al., "HMPL-004 (*Andrographis paniculata* Extract) Prevents Development of Murine

Colitis by Inhibiting T-Cell Proliferation and TH1/TH17 Responses," *Inflammatory Bowel Disease* 19, no. 1 (2013): 151–64, https://www.ncbi.nlm.nih.gov/pmc/articles /PMC4465822/pdf/nihms694674.pdf.

211. Xin Nie et al., "Attenuation of Innate Immunity by Andrographolide Derivatives through NF-κB Signaling Pathway," *Scientific Reports* 7, no. 1 (2017): 4738, https://doi .org/10.1038/s41598-017-04673-x.

212. Julie Wang and Xiu-Min Li, "Chinese Herbal Therapy for the Treatment of Food Allergy," *Current Allergy and Asthma Reports* 12, no. 4 (2012): 332–38, https://doi.org/10.1007 /s11882-012-0265-4.

213. Yuwein Pan et al., "Synergistic Effect of Ferulic Acid and Z-Ligustilide, Major Components of *A. sinensis*, on Regulating Cold-Sensing Protein TRPM8 and TPRA1 *In Vitro*," *Evidence-Based Complementary and Alternative Medicine* 2016 (June 2016): article 3160247, https://doi.org/10.1155/2016/3160247.

214. Wanida Sukketsiri et al., "Effects of *Apium graveolens* Extract on the Oxidative Stress in the Liver of Adjuvant-Induced Arthritic Rats," *Preventive Nutrition and Food Science* 21, no. 2 (2016): 79–84, https://doi.org/10.3746/pnf.2016.21.2.79.

215. Hua-Hao Shen et al., "*Astragalus membranaceus* Prevents Airway Hyperreactivity in Mice Related to Th2 Response Inhibition," *Journal of Ethnopharmacology* 116, no. 2 (2008): 363–69, https://doi.org/10.1016/j.jep.2007.12.002.

216. Su-Jin Lee et al., "Oral Administration of *Astragalus membranaceus* Inhibits the Development of DNFB-Induced Dermatitis in NC/Nga Mice," *Biological and Pharmaceutical Bulletin* 30, no. 8 (2007): 1468–71, https://doi.org/10.1248/bpb.30.1468.

217. Maryline Criquet et al., "Safety and Efficacy of Personal Care Products Containing Colloidal Oatmeal," *Clinical, Cosmetic and Investigational Dermatology* 2012, no. 5 (2012): 183–93, https://doi.org/10.2147/CCID.S31375.

218. Rafie Hamidpour et al., "Frankincense (乳香 Rǔ Xiāng; Boswellia Species): From the Selection of Traditional Applications to the Novel Phytotherapy for the Prevention and Treatment of Serious Diseases," *Journal of Traditional and Complementary Medicine* 3, no. 4 (2013): 221–26, https://doi.org/10.4103/2225-4110.119723.

219. A. M. Malfait et al., "The Nonpsychoactive Cannabis Constituent Cannabidiol Is an Oral Anti-Arthritic Therapeutic in Murine Collagen-Induced Arthritis," *PNAS* 97, no. 17 (August 2000): 9561–66, https://doi.org/10.1073/pnas.160105897.

220. Brian Astry, "Celastrol, a Chinese Herbal Compound, Controls Autoimunne Inflammation by Altering the Balance of Pathogenic and Regulatory T Cells in the Target Organ," *Clinical Immunology* 157, no. 2 (2015): 228–38, https://doi.org/10.1016/j.clim.2015.01.011.

221. Astry, "Celastrol."

222. Siddaraju M. Nanjundaiah et al., "Celastrus and Its Bioactive Celastrol Protect against Bone Damage in Autoimmune Arthritis by Modulating Osteoimmune Cross-Talk," *Journal of Biological Chemistry* 287, no. 26 (2012): 22216–26, https:// doi.org/10.1074/jbc.M112.356816.

223. Li Tong and Kamal D. Moudgil, "*Celastrus aculeatus* Merr. Suppresses the Induction and Progression of Autoimmune Arthritis by Modulating Immune Response to Heat-Shock Protein 65," *Arthritis Research and Therapy* 9, no. 4 (2007): R70, https://doi.org/10.1186/ar2268.

224. Shivaprasad H. Venkatesha et al., "Celastrus-Derived Celastrol Suppresses Autoimmune Arthritis by Modulating Antigen-Induced Cellular and Humoral Effector Responses," *Journal of Biological Chemistry* 286, no. 17 (2011): 15138–46, https://doi.org/10.1074/jbc .M111.226365.

225. Himanshu Sharma, Prerna Chauhan, and Surender Singh, "Evaluation of the Anti-Arthritic Activity of *Cinnamomum cassia* Bark Extract in Experimental Models," *Integrative Medicine Research* 7, no. 4 (2018): 366–73, https://doi.org /10.1016/j.imr.2018.08.002.

226. Ho-Keun Kwon et al., "Cinnamon Extract Suppresses Experimental Colitis through Modulation of Antigen-Presenting Cells," *World Journal of Gastroenterology* 17, no. 8 (20110: 976–86, https://doi.org/10.3748/wjg.v17.i8.976.

227. Badal Rathi et al., "Ameliorative Effects of a Polyphenolic Fraction of *Cinnamomum zeylanicum* L. Bark in Animal Models of Inflammation and Arthritis," *Scientia Pharmaceutica* 81, no. 2 (2013): 567–89, https://doi.org /10.3797/scipharm.1301-16.

228. Won-Yong Kim et al., "A Herbal Formula, *Atofreellage*, Ameliorates Atopic Dermatitis-Like Skin Lesions in an NC/ Nga Mouse Model," *Molecules* 21, no. 1 (2015): E35, https:// doi.org/10.3390/molecules21010035.

229. Chen-Yuan Chiang et al., "Osthole Treatment Ameliorates Th2-Mediated Allergic Asthma and Exerts Immunomodulatory Effects on Dendritic Cell Maturation and Function," *Cellular and Molecular Immunology* 14, no. 11 (2017): 935–47, https://doi.org/10.1038/cmi.2017.71.

230. B. Dasgeb et al., "Colchicine: An Ancient Drug with Novel Applications," *British Journal of Dermatology* 178, no. 2 (2018): 350–56, https://doi.org/10.1111/bjd.15896.

231. Xiaohua Du et al., "Isoforskolin and Forskolin Attenuate Lipopolysaccharide-Induced Inflammation through TLR4/MyD88/NF-κB Cascades in Human Mononuclear Leukocytes," *Phytotherapy Research* 33, no. 3 (2019): 602–9, https://doi.org/10.1002/ptr.6248.

232. Chuan-Fa Hsieh et al., "Prescribed Renoprotective Chinese Herbal Medicines Were Associated with a Lower Risk of All-Cause and Disease-Specific Mortality among Patients with Chronic Kidney Disease: A Population-Based Follow-Up Study in Taiwan," *Evidence-Based Complementary and Alternative Medicine* 2017 (July 2017): article 5632195, https://doi.org/10.1155/2017/5632195.

233. Monika E. Czerwińska and Matthias F. Melzig, "*Cornus mas* and *Cornus officinalis*—Analogies and Differences of Two Medicinal Plants Traditionally Used," *Frontiers in Pharmacology* 9 (August 2018): article 894, https://doi.org /10.3389/fphar.2018.00894.

234. Yu Dong et al., "*Corni Fructus*: A Review of Chemical Constituents and Pharmacological Activities," *Chinese Medicine* 13 (June 2018): article 34, https://doi.org/10.1186/s13020-018-0191-z.

235. Chinnasamy Elango and Sivasithambaram Niranjali Devaraj, "Immunomodulatory Effect of Hawthorn Extract in an Experimental Stroke Model," *Journal of Neuroinflammation* 7 (December 2010): article 97, https://doi.org/10.1186/1742-2094-7-97.

236. Yusuke Takahashi et al., "Prophylactic and Therapeutic Effects of *Acanthopanax senticosus* Harms Extract on Murine Collagen-Induced Arthritis," *Phytotherapy Research* 28, no. 10 (2014): 1513–19, https://doi.org/10.1002/ptr.5157.

237. Nan Zhang et al., "A Herbal Composition of *Scutellaria baicalensis* and *Eleutherococcus senticosus* Shows Potent Anti-Inflammatory Effects in an *Ex Vivo* Human Mucosal Tissue Model," *Evidence-Based Complementary and Alternative Medicine* 2012 (January 2012): article 673145, https://doi.org/10.1155/2012/673145.

238. Shuang-Feng Lin, Xue-Hong Ke, and Gang Wei, "Investigate into of Effective Constituent Transference of Herba *Ephedrae* and Cortex *Magnoliae officinalis* in Preparation Course of Shujin Kechuan Capsule," *Zhongguo Zhong Yao Za Zhi* 31, no. 23 (2006): 1950–52; Qinglin Zha et al., "Xiaoqinglong Granules as Add-On Therapy for Asthma: Latent Class Analysis of Symptom Predictors of Response," *Evidence-Based Complementary and Alternative Medicine* 2013 (February 2013): article 759476, https://doi.org/10.1155/2013/759476.

239. Amanda de Araújo Lopes et al., "Eugenol as a Promising Molecule for the Treatment of Dermatitis: Antioxidant and Anti-inflammatory Activities and Its Nanoformulation," *Oxidative Medicine and Cellular Longevity* 2018, (December 2018): article 8194849, https://doi.org/10.1155/2018/8194849.

240. Jingjing Li et al., "Protective Effect of Fucoidan from *Fucus vesiculosus* on Liver Fibrosis via the TGF-β1/Smad Pathway-Mediated Inhibition of Extracellular Matrix and Autophagy," *Drug Design, Development and Therapy* 10 (2016): 619–30, https://doi.org/10.2147/DDDT.S98740.

241. Qi Ying Lean et al., "Fucoidan Extracts Ameliorate Acute Colitis," *PLoS One* 10, no. 6 (2015): e0128453, https://doi.org/10.1371/journal.pone.0128453.

242. Peng-Yun Wang, Xiao-Ling Zhu, and Zhi-Bin Lin, "Antitumor and Immunomodulatory Effects of Polysaccharides from Broken-Spore of *Ganoderma lucidum*," *Frontiers in Pharmacology* 3 (July 2012): article 135, https://doi.org/10.3389/fphar.2012.00135.

243. Thulasi G. Pillai et al., "Fungal Beta Glucan Protects Radiation Induced DNA Damage in Human Lymphocytes," *Annals of Translational Medicine* 2, no. 2 (2014): 13, https://doi.org/10.3978/j.issn.2305-5839.2014.02.02.

244. Jin-Yuarn Lin, Miaw-Ling Chen, and Bi-Fong Lin, "*Ganoderma tsugae In Vivo* Modulates Th1/Th2 and Macrophage Responses in an Allergic Murine Model," *Food and Chemical Toxicology* 44, no. 12 (2006): 2025–32, https://doi.org/10.1016/j.fct.2006.07.002.

245. Lin, Chen, and Lin, "*Ganoderma tsugae In Vivo*."

246. Hsien-Yeh Hsu et al., "Reishi Protein LZ-8 Induces FOXP3(+) Treg Expansion via a CD45-Dependent Signaling Pathway and Alleviates Acute Intestinal Inflammation in Mice," *Evidence-Based Complementary and Alternative Medicine* 2013 (June 2013): article 513542, https://doi.org/10.1155/2013/513542.

247. Chuan M. Yeh et al., "Extracellular Expression of a Functional Recombinant *Ganoderma lucidium* Immunomodulatory Protein by *Bacillus subtilis* and *Lactococcus lactis*," *Applied and Environmental Microbiology* 74, no. 4 (2008): 1039–49, https://doi.org/10.1128/AEM.01547-07.

248. Ling Xie et al., "The Effects of Freeze-Dried *Ganoderma lucidum* Mycelia on a Recurrent Oral Ulceration Rat Model," *BMC Complementary and Alternative Medicine* 17, no. 1 (2017): 511, https://doi.org/10.1186/s12906-017-2021-8.

249. Yan Zhou et al., "Inflammatory Modulation Effect of Glycopeptide from *Ganoderma capense* (Lloyd) Teng," *Mediators of Inflammation* 2014 (May 2014): article 691285, https://doi.org/10.1155/2014/691285.

250. Arulmani Manavalan et al., "*Gastrodia elata* Blume (tianma) Mobilizes Neuro-Protective Capacities," *International Journal of Biochemistry and Molecular Biology* 3, no. 2 (2012): 219–41, https://www.ncbi.nlm.nih.gov/pmc/articles/PMC3388733/pdf/ijbmb0003-0219.pdf.

251. Jay M. Hendricks et al., "18β-Glycyrrhetinic Acid Delivered Orally Induces Isolated Lymphoid Follicle Maturation at the Intestinal Mucosa and Attenuates Rotavirus Shedding," *PLoS ONE* 7, no. 11 (2012): e49491, https://doi.org/10.1371/journal.pone.0049491.

252. Lihua Hou et al., "A Water-Soluble Polysaccharide from *Grifola frondosa* Induced Macrophages Activation via TLR4-MyD88-IKKβ-NF-κB P65 pathways," *Oncotarget* 8, no. 49 (2017): 86604–14, https://doi.org/10.18632/oncotarget.21252.

253. Jong Suk Lee et al., "*Grifola frondosa* Water Extract Alleviates Intestinal Inflammation by Suppressing TNF-α Production and Its Signaling," *Experimental and Molecular Medicine* 42, no. 2 (2010): 143–54, https://doi.org/10.3858/emm.2010.42.2.016.

254. Hong Lin et al., "Maitake Beta-Glucan Promotes Recovery of Leukocytes and Myeloid Cell Function in Peripheral Blood from Paclitaxel Hematotoxicity," *Cancer Immunology, Immunotherapy* 59, no. 6 (2010): 885–97, https://doi.org/10.1007/s00262-009-0815-3.

255. Weiwei Zhang et al., "A Polysaccharide-Peptide with Mercury Clearance Activity from Dried Fruiting Bodies of Maitake Mushroom *Grifola frondosa*," *Scientific Reports* 8, no. 1 (2018): 17630, https://doi.org/10.1038/s41598-018-35945-9.

256. Sanhong Yu et al., "The Effects of Whole Mushrooms during Inflammation," *BMC Immunology* 10 (February 2009): article 12, https://doi.org/10.1186/1471-2172-10-12; Yu-Ri Seo et al., "Structural Elucidation and Immune-Enhancing

Effects of Novel Polysaccharide from *Grifola frondosa*," *BioMed Research International* 2019 (April 2019): article 7528609, https://doi.org/10.1155/2019/7528609.

257. Kanika Patel and Dinesh Kumar Patel, "Medicinal Important, Pharmacological Activities, and Analytical Aspects of Hispidulin: A Concise Report," *Journal of Traditional and Complementary Medicine* 7, no. 3 (2016): 360–66, https://doi.org/10.1016/j.jtcme.2016.11.003.

258. Dong Eun Kim et al., "Hispidulin Inhibits Mast Cell-Mediated Allergic Inflammation through Down-Regulation of Histamine Release and Inflammatory Cytokines," *Molecules* 24, no. 11 (2019): 2131, https://doi.org/10.3390/molecules24112131.

259. Linda L. Theisen et al., "Tannins from *Hamamelis virginiana* Bark Extract: Characterization and Improvement of the Antiviral Efficacy against Influenza A Virus and Human Papillomavirus," *PLoS ONE* 9, no. 1 (2014): e88062, https://doi.org/10.1371/journal.pone.0088062.

260. Renata Dawid-Pać, "Medicinal Plants Used in Treatment of Inflammatory Skin Diseases," *Postepy Dermatologii i Alergologii = Advances in Dermatology and Allergology* 30, no. 3 (2013): 170–77, https://doi.org/10.5114/pdia.2013.35620.

261. Ralph M. Trüeb, "North American Virginian Witch Hazel (*Hamamelis virginiana*): Based Scalp Care and Protection for Sensitive Scalp, Red Scalp, and Scalp Burn-Out," *International Journal of Trichology* 6, no. 3 (2014): 100–103, https://doi.org/10.4103/0974-7753.139079.

262. Saima Jadoon et al., "Anti-Aging Potential of Phytoextract Loaded-Pharmaceutical Creams for Human Skin Cell Longevity," *Oxidative Medicine and Cellular Longevity* 2015 (September 2015): article 709628, https://doi.org/10.1155/2015/709628.

263. K. Neukam et al., "Supplementation of Flaxseed Oil Diminishes Skin Sensitivity and Improves Skin Barrier Function and Condition," *Skin Pharmacology and Physiology* 24, no. 2 (2011): 67–74, https://doi.org/10.1159/000321442.

264. Stephanie P.B. Caligiuri et al., "Elevated Levels of Pro-Inflammatory Oxylipins in Older Subjects Are Normalized by Flaxseed Consumption," *Experimental Gerontology* 59 (November 2014): 51–57, https://doi.org/10.1016/j.exger.2014.04.005.

265. Tzu-Kai Lin, Lily Zhong, and Juan Luis Santiago, "Anti-Inflammatory and Skin Barrier Repair Effects of Topical Application of Some Plant Oils," *International Journal of Molecular Sciences* 19, no. 1 (2017): 70, https://doi.org/10.3390/ijms19010070.

266. C. F. Carson, K. A. Hammer, and T. V. Riley, "*Melaleuca alternifolia* (Tea Tree) Oil: A Review of Antimicrobial and Other Medicinal Properties," *Clinical Microbiology Reviews* 19, no. 1 (2006): 50–62, https://doi.org/10.1128/CMR.19.1.50-62.2006.

267. Wissal Dhifi et al., "Essential Oils' Chemical Characterization and Investigation of Some Biological Activities: A Critical Review," *Medicines* 3, no. 4 (2016): 25, https://doi.org/10.3390/medicines3040025.

268. Lotfy T. Elsaie et al., "Effectiveness of Topical Peppermint Oil on Symptomatic Treatment of Chronic Pruritus," *Clinical, Cosmetic and Investigational Dermatology* 9 (2016): 333–38, https://doi.org/10.2147/CCID.S116995.

269. Marjan Akhavan Amjadi, Faraz Mojab, and Seyedeh Bahareh Kamranpoura, "The Effect of Peppermint Oil on Symptomatic Treatment of Pruritus in Pregnant Women," *Iranian Journal of Pharmaceutical Research* 11, no. 4 (2012): 1073–77, https://www.ncbi.nlm.nih.gov/pmc/articles/PMC3813175/pdf/ijpr-11-1073.pdf.

270. Sonali S. Bharate and Sandip B. Bharate, "Modulation of Thermoreceptor TRPM8 by Cooling Compounds," *ACS Chemical Neuroscience* 3, no. 4 (2012): 248–67, https://doi.org/10.1021/cn300006u.

271. Farimah Beheshti, Majid Khazaei, and Mahmoud Hosseini, "Neuropharmacological Effects of *Nigella sativa*," *Avicenna Journal of Phytomedicine* 6, no. 1 (2016): 104–16, https://www.ncbi.nlm.nih.gov/pmc/articles/PMC4884225/pdf/AJP-6-124.pdf.

272. Munawar Alam Ansari et al., "*Nagilla sativa*: A Non-Conventional Herbal Option for the Management of Seasonal Allergic Rhinitis," *Pakistan Journal of Pharmacology* 23, no. 2 (2006): 31–35, http://probotanic.com/pdf_istrazivanja/crni_kumin/Crni%20kumin%20-%20nekonvencionalna%20prirodna%20opcija%20u%20tretmanu%20sezonskog%20alergijskog%20rinitisa.pdf.

273. Neveen A. Noor et al., "*Nigella sativa* Amliorates Inflammation and Demyelination in the Experimental Autoimmune Encephalomyelitis-Induced Wistar Rats," *International Journal of Clinical and Experimental Pathology* 8, no. 6 (2015): 6269–86, https://www.ncbi.nlm.nih.gov/pmc/articles/PMC4525838/pdf/ijcep0008-6269.pdf.

274. Seyedeh-Masomeh Derakhshandeh-Rishehri et al., "Role of Fatty Acids Intake in Generalized Vitiligo," *International Journal of Preventive Medicine* 10, no. 1 (2019): 52, https://doi.org/10.4103/ijpvm.IJPVM_47_17.

275. Dagmar Simon et al., "Gamma-Linolenic Acid Levels Correlate with Clinical Efficacy of Evening Primrose Oil in Patients with Atopic Dermatitis," *Advances in Therapy* 31, no. 2 (2014): 180–88, https://doi.org/10.1007/s12325-014-0093-0.

276. Ruoling Guo, Max H. Pittler, and Edzard Ernst, "Herbal Medicines for the Treatment of Allergic Rhinitis: A Systematic Review," *Annals of Allergy, Asthma and Immunology* 99, no. 6 (2007): 483–95, https://doi.org/10.1016/S1081-1206(10)60375-4.

277. De-Kui Zhang et al., "A *Picrorhiza kurroa* Derivative, Picroliv, Attenuates the Development of Dextran-Sulfate-Sodium-Induced Colitis in Mice," *Mediators of Inflammation* 2012 (October 2012): article 751629, https://doi.org/10.1155/2012/751629.

278. Arshad Hussain et al., "Protective Effects of *Picrorhiza kurroa* on Cyclophosphamide-Induced Immunosuppression in Mice," *Pharmacognosy Research* 5, no. 1 (2013): 30–35, https://doi.org/10.4103/0974-8490.105646.

279. Ching-Mao Chang et al., "The Core Pattern Analysis on Chinese Herbal Medicine for Sjögren's Syndrome: A Nationwide Population-Based Study," *Scientific Reports* 5 (April 2015): article 9541, https://doi.org/10.1038/srep09541.

280. Yoon-Young Sung et al., "Topical Application of *Rehmannia glutinosa* Extract Inhibits Mite Allergen-Induced Atopic Dermatitis in NC/Nga Mice," *Journal of Ethnopharmacology* 134, no. 1 (2011): 37–44, https://doi.org10.1016/j.jep.2010.11.050.

281. Qian Li et al., "Role of Catalpol in Ameliorating the Pathogenesis of Experimental Autoimmune Encephalomyelitis by Increasing the Level of Noradrenaline in the Locus Coeruleus," *Molecular Medicine Reports* 17, no. 3 (2018): 4163–72, https://doi.org/10.3892/mmr.2018.8378.

282. Jing-Yan Han et al., "Ameliorating Effects of Compounds Derived from *Salvia miltiorrhiza* Root Extract on Microcirculatory Disturbance and Target Organ Injury by Ischemia and Reperfusion," *Pharmacology and Therapeutics* 117, no. 2 (2008): 280–95, https://doi.org/10.1016/j.pharmthera.2007.09.008.

283. Hyun Lim et al., "Inhibition of Chronic Skin Inflammation by Topical Anti-Inflammatory Flavonoid Preparation, Ato Formula," *Archives of Pharmacal Research* 29, no. 6 (2006): 503–7, https://doi.org/10.1007/BF02969424.

284. M. Kumar Roy et al., "Baicalein, a Flavonoid Extracted from a Methanolic Extract of *Oroxylum indicum* Inhibits Proliferation of a Cancer Cell Line *In Vitro* via Induction of Apoptosis," *Die Pharmazie* 62, no. 2 (2007): 149–53.

285. Chia-Jung Hsieh et al., "Baicalein Inhibits IL-1Beta- and TNF-Alpha-Induced Inflammatory Cytokine Production from Human Mast Cells via Regulation of the NF-kappaB Pathway," *Clinical and Molecular Allergy* 5 (November 2007): article 5, https://doi.org/10.1186/1476-7961-5-5.

286. T. Nakajima et al., "Inhibitory Effect of Baicalein, a Flavonoid in Scutellaria Root, on Eotaxin Production by Human Dermal Fibroblasts," *Planta Medica* 67, no. 2 (2001): 132–35, https://doi.org/10.1055/s-2001-11532.

287. Bok-Soo Lee et al., "Wogonin Suppresses TARC Expression Induced by Mite Antigen via Heme Oxygenase 1 in Human Keratinocytes. Suppressive Effect of Wogonin on Mite Antigen-Induced TARC Expression," *Journal of Dermatological Science* 46, no. 1 (2007): 31–40, https://doi.org/10.1016/j.jdermsci.2007.01.001.

Chapter 3: Creating Herbal Formulas for Musculoskeletal Conditions

1. R. Lahesmaa-Rantala et al., "Intestinal Permeability in Patients with Yersinia Triggered Reactive Arthritis," *Annals of the Rheumatic Diseases* 50, no. 2 (1991): 91–94, https://doi.org/10.1136/ard.50.2.91; R. Serrander, K. E. Magnusson, and T. Sundqvist, "Acute Infections with *Giardia lamblia* and Rotavirus Decrease Intestinal Permeability to Low-Molecular Weight Polyethylene Glycols (PEG 400)," *Scandinavian Journal of Infectious Diseases* 16, no. 4 (1984): 339–44, https://doi.org/10.3109/00365548409073958; Le Shen and Jerrold R. Turner, "Role of Epithelial Cells in Initiation and Propagation of Intestinal Inflammation. Eliminating the Static: Tight Junction Dynamics Exposed," *American Journal of Physiology: Gastrointestinal and Liver Physiology* 290, no. 4 (2006): G577–82, https://doi.org/10.1152/ajpgi.00439.2005.

2. R. T. Jenkins et al., "Increased Intestinal Permeability in Patients with Rheumatoid Arthritis: A Side-Effect of Oral Nonsteroidal Anti-Inflammatory Drug Therapy?" *British Journal of Rheumatology* 26, no. 2 (1987): 103–7, https://doi.org/10.1093/rheumatology/26.2.103; I. Bjarnason et al., "Effect of Non-Steroidal Anti-Inflammatory Drugs on the Human Small Intestine," *Drugs* 32, suppl. 1 (1986): 35–41, https://doi.org/10.2165/00003495-198600321-00007.

3. M. P. Hazenberg et al., "Are Intestinal Bacteria Involved in the Etiology of Rheumatoid Arthritis? Review Article," *Acta Pathologica, Microbiologica, et Immunologica Scandinavica* 100, no. 1 (1992): 1–9, https://doi.org/10.1111/j.1699-0463.1992.tb00833.x; S. M. Dearlove et al., "The Effect of Non-Steroidal Anti-Inflammatory Drugs on Faecal Flora and Bacterial Antibody Levels in Rheumatoid Arthritis," *British Journal of Rheumatology* 31, no. 7 (1992): 443–47, https://doi.org/10.1093/rheumatology/31.7.443.

4. P. J. Rooney, R. T. Jenkins, and W. W. Buchanan, "A Short Review of the Relationship between Intestinal Permeability and Inflammatory Joint Disease," *Clinical and Experimental Rheumatology* 8, no. 1 (1990): 75–83.

5. L. Sköldstam, L. Larsson, and F. D. Lindström, "Effect of Fasting and Lactovegetarian Diet on Rheumatoid Arthritis," *Scandinavian Journal of Rheumatology* 8, no. 4 (1979): 249–55, https://doi.org/10.3109/03009747909114631.

6. Roberto D'Anchise, Michael Bulitta, and Bruno Giannetti, "Comfrey Extract Ointment in Comparison to Diclofenac Gel in the Treatment of Acute Unilateral Ankle Sprains (Distortions)," *Arzneimittelforschung* 57, no. 11 (2007): 712–16, https://doi.org/10.1055/s-0031-1296672; R. Koll et al., "Efficacy and Tolerance of a Comfrey Root Extract (Extr. Rad. Symphyti) in the Treatment of Ankle Distortions: Results of a Multicenter, Randomized, Placebo-Controlled, Double-Blind Study," *Phytomedicine* 11, no. 6 (2004): 470–77, https://doi.org/10.1016/j.phymed.2004.02.001.

7. B. Somashekar Shetty et al., "Effect of *Centella asiatica* L (Umbelliferae) on Normal and Dexamethasone-Suppressed Wound Healing in Wistar Albino Rats," *International Journal of Lower Extremity Wounds* 5, no. 3 (2006): 137–43, https://doi.org/10.1177/1534734606291313.

8. Bharat B. Aggarwal et al., "Curcumin: The Indian Solid Gold," *Advances in Experimental Medicine and Biology* 595 (2007): 1–75, https://doi.org/10.1007/978-0-387-46401-5_1.

9. Xiaoping Yang et al., "Curcumin Inhibits Platelet-Derived Growth Factor-Stimulated Vascular Smooth Muscle Cell Function and Injury-Induced Neointima Formation," *Arteriosclerosis, Thrombosis, and Vascular Biology* 26, no. 1 (2006): 85–90, https://doi.org/10.1161/01.ATV.0000191635.00744.b6.

10. S. C. Fu et al., "Total Flavones of *Hippophae rhamnoides* Promotes Early Restoration of Ultimate Stress of Healing Patellar Tendon in a Rat Model," *Medical Engineering and Physics* 27, no. 4 (2005): 313–21, https://doi.org/10.1016/j.medengphy.2004.12.011.

11. D. H. Garabrant and C. Dumas, "Epidemiology of Organic Solvents and Connective Tissue Disease," *Arthritis Research* 2, no. 1 (2000): 5–15, https://doi.org/10.1186/ar65.

12. Carolina Barragán-Martínez et al., "Organic Solvents as Risk Factor for Autoimmune Diseases: A Systematic Review and Meta-Analysis," *PLoS ONE* 7, no. 12 (2012): e51506, https://doi.org/10.1371/journal.pone.0051506.

13. Patrizia Fuschiotti, "Current Perspectives on the Immunopathogenesis of Systemic Sclerosis," *ImmunoTargets and Therapy* 2016, no. 5 (2016): 21–35, https://doi.org/10.2147/ITT.S82037; Birgitta Kütting, Wolfgang Uter, and Hans Drexler, "Is Occupational Exposure to Solvents Associated with an Increased Risk for Developing Systemic Scleroderma?" *Journal of Occupational Medicine and Toxicology* 1 (July 2006): article 15, https://doi.org/10.1186/1745-6673-1-15.

14. Glinda S. Cooper et al., "Occupational and Environmental Exposures and Risk of Systemic Lupus Erythematosus: Silica, Sunlight, Solvents," *Rheumatology* 49, no. 11 (2010): 2172–80, https://doi.org/10.1093/rheumatology/keq214.

15. Daniel Smyk et al., "Hair Dyes as a Risk for Autoimmunity: From Systemic Lupus Erythematosus to Primary Biliary Cirrhosis," *Autoimmunity Highlights* 4, no. 1 (2012): 1–9, https://doi.org/10.1007/s13317-011-0027-7.

16. K. Michael Pollard, Per Hultman, and Dwight H. Kono, "Toxicology of Autoimmune Diseases," *Chemical Research in Toxicology* 23, no. 3 (2010): 455–66, https://doi.org/10.1021/tx9003787.

17. K. Pytlakowska et al., "Multi-Element Analysis of Mineral and Trace Elements in Medicinal Herbs and Their Infusions," *Food Chemistry* 135, no. 2 (2012): 494–501, https://doi.org/10.1016/j.foodchem.2012.05.002.

18. Rajinder Singh, Subrata De, and Asma Belkheir, "*Avena sativa* (Oat), a Potential Neutraceutical and Therapeutic Agent: An Overview," *Critical Reviews in Food Science and Nutrition* 53, no. 2 (2013): 126–44, https://doi.org/10.1080/10408398.2010.526725; Masood Sadiq Butt et al., "Oat: Unique among the Cereals," *European Journal of Nutrition* 47, no. 2 (2008): 68–79, https://doi.org/10.1007/s00394-008-0698-7.

19. A. J. Afolayan and F. O. Jimoh, "Nutritional Quality of Some Wild Leafy Vegetables in South Africa," *International Journal of Food Sciences and Nutrition* 60, no. 5 (2009): 424–31, https://doi.org/10.1080/09637480701777928; Florence Jimoh et al., "Polyphenolic and Biological Activities of Leaves Extracts of *Argemone subfusiformis* (Papaveraceae) and *Urtica urens* (Urticaceae)," *Revista de Biologia Tropical* 58, no. 4 (2010): 1517–31, https://doi.org/10.15517/rbt.v58i4.5428.

20. Alpaslan Oztürk et al., "The Effects of Phytoestrogens on Fracture Healing: Experimental Research in New Zealand White Rabbits," *Ulusal Travma Ve Acil Cerrahi Dergisi* 14, no. 1 (2008): 21–27.

21. P. Bolle et al., "Estrogen-Like Effect of a *Cimicifuga racemosa* Extract Sub-Fraction as Assessed by *In vivo*, *Ex Vivo* and *In Vitro* Assays," *Journal of Steroid Biochemistry and Molecular Biology* 107, nos. 3–5 (2007): 262–69, https://doi.org/10.1016/j.jsbmb.2007.03.044.

22. Z. L. Wang et al., "Pharmacological Studies of the Large-Scaled Purified Genistein from Huaijiao (*Sophora japonica*—Leguminosae) on Anti-Osteoporosis," *Phytomedicine* 13, nos. 9–10 (2006): 718–23, https://doi.org/10.1016/j.phymed.2005.09.005.

23. Qing Yang et al., "Effect of *Angelica sinensis* on the Proliferation of Human Bone Cells," *Clinical Chimica Acta* 324, nos. 1–2 (2002): 89–97, https://doi.org/10.1016/s0009-8981(02)00210-3.

24. Abdullah bin Habeeballah bin Abdullah Juma, "The Effects of *Lepidium sativum* Seeds on Fracture-Induced Healing in Rabbits," *MedGenMed* 9, no. 2 (2007): 23, https://www.ncbi.nlm.nih.gov/pmc/articles/PMC1994840/.

25. Nazanin Arbabzadegan et al., "Effect of *Equisetum arvense* Extract on Bone Mineral Density in Wistar Rats via Digital Radiography," *Caspian Journal of Internal Medicine* 10, no. 2 (2019): 176–82, https://doi.org/10.22088/cjim.10.2.176.

26. Swati D. Kotwal and Smita R. Badole, "Anabolic Therapy with *Equisetum arvense* along with Bone Mineralising Nutrients in Ovariectomized Rat Model of Osteoporosis," *Indian Journal of Pharmacology* 48, no. 3 (2016): 312–15, https://doi.org/10.4103/0253-7613.182880.

27. Mirian Farinon et al., "Effect of Aqueous Extract of Giant Horsetail (*Equisetum giganteum* L.) in Antigen-Induced Arthritis," *Open Rheumatology Journal* 7 (2013): 129–33, https://doi.org/10.2174/1874312901307010129.

28. Taylor C. Peak et al., "Role of Collagenase *Clostridium histolyticum* in Peyronie's Disease," *Biologics: Targets and Therapy* 9 (2015): 107–16, https://doi.org/10.2147/BTT.S65619.

29. Joanna Narbutt, Agnieszka Hołdrowicz, and Aleksandra Lesiak, "Morphea—Selected Local Treatment Methods and Their Effectiveness," *Reumatologia* 55, no. 6 (2017): 305–13, https://doi.org/10.5114/reum.2017.72628.

30. W. A. Townley et al., "Dupuytren's Contracture Unfolded," *BMJ* 332, no. 7538 (2006): 397–400, https://doi.org/10.1136/bmj.332.7538.397.

31. A. Drovanti, A. A. Bignamini, and A. L. Rovati, "Therapeutic Activity of Oral Glucosamine Sulfate in Osteoarthrosis: A Placebo-Controlled Double-Blind Investigation," *Clinical Therapeutics* 3, no. 4 (1980): 260–72.

32. I. Setnikar, C. Giachetti, and G. Zanolo, "Absorption, Distribution and Excretion of Radioactivity after a Single Intravenous or Oral Administration of [14C] Glucosamine to the Rat," *Pharmatherapeutica* 3, no. 8 (1984): 538–50.

33. I. Setnikar, C. Giacchetti, and G. Zanolo, "Pharmacokinetics of Glucosamine in the Dog and in Man," *Arzneimittelforschung* 36, no. 4 (1986): 729–35.

34. E. D'Ambrosio et al., "Glucosamine Sulphate: A Controlled Clinical Investigation in Arthrosis," *Pharmatherapeutica* 2, no. 8 (1981): 504–8; I. Setnikar, M. A. Pacini, and L. Revel, "Antiarthritic Effects of Glucosamine Sulfate Studied in Animal Models," *Arzneimittelforschung* 41, no. 5 (1991): 542–45.

35. T. C. Welbourne, "Increased Plasma Bicarbonate and Growth Hormone after an Oral Glutamine Load," *American Journal of Clinical Nutrition* 61, no. 5 (1995): 1058–61.

36. Patrick I. Emelife, Russell E. Kling, and Ronit Wollstein, "Postoperative Management of Dupuytren's Disease with Topical Nitroglycerin," *Journal Canadien de Chirurgie Plastque* 20, no. 4 (2012): 249–50, https://doi.org/10.1177/229255031202000412.

37. Seung-Hee Ryu et al., "Protective Effect of α-Lipoic Acid against Radiation-Induced Fibrosis in Mice," *Oncotarget* 7, no. 13 (2016): 15554–65, https://doi.org/10.18632/oncotarget.6952.

38. Anna Carolina Miola et al., "Randomized Clinical Trial Testing the Efficacy and Safety of 0.5% Colchicine Cream versus Photodynamic Therapy with Methyl Aminolevulinate in the Treatment of Skin Field Cancerization: Study Protocol," *BMC Cancer* 18, no. 1 (2018): 340, https://doi.org/10.1186/s12885-018-4288-7.

39. Ara DerMarderosian and John A. Beutler, eds., "Kava," in *The Review of Natural Products*, 2nd ed. (St. Louis, MI, 2002).

40. G. Boonen and H. Häberlein, "Influence of Genuine Kavapyrone Enanntiomers on the GABA-A Binding Site," *Planta Medica* 64, no. 6 (1998): 504–6, https://doi.org/10.1055/s-2006-957502.

41. L. Davies et al., "Effects of Kava on Benzodiazepine and GABA Receptor Binding," *European Journal of Pharmacology* 183, no. 2 (July 1990): 558, https://doi.org/10.1016/0014-2999(90)93467-5.

42. R. Kretzschmar, H. J. Meyer, and J. H. Teschendorf, "Strychnine Antagonistic Potency of Pyrone Compounds of Kavaroot (*Piper methysticum* Forst.)," *Experientia* 26, no. 3 (1970): 283–84, https://doi.org/10.1007/BF01900097.

43. Rajendra Pavan et al., "Properties and Therapeutic Application of Bromelain: A Review," *Biotechnology Research International* 2012 (2012): article 976203, https://www.ncbi.nlm.nih.gov/pmc/articles/PMC3529416/pdf/BTRI2012-976203.pdf; Mario Roxas, "The Role of Enzyme Supplementation in Digestive Disorders," *Alternative Medicine Review* 13, no. 4 (2008): 307–14, http://archive.foundationalmedicinereview.com/publications/13/4/307.pdf; H. R. Maurer, "Bromelain: Biochemistry, Pharmacology and Medical Use," *Cellular and Molecular Life Sciences* 58, no. 9 (2001): 123–45, https://doi.org/10.1007/PL00000936; Laura P. Hale, "Proteolytic Activity and Immunogenicity of Oral Bromelain within the Gastrointestinal Tract of Mice," *International Immunopharmacology* 4, no. 2 (2004): 255–64, https://doi.org/10.1016/j.intimp.2003.12.010.

44. Sarah Brien et al., "Bromelain as a Treatment for Osteoarthritis: A Review of Clinical Studies," *Evidence-Based Complementary and Alternative Medicine* 1, no. 3 (2004): 251–57, https://doi.org/10.1093/ecam/neh035.

45. A. I. Aiyegbusi et al., "The Role of Aqueous Extract of Pineapple Fruit Parts on the Healing of Acute Crush Tendon Injury," *Nigerian Quarterly Journal of Hospital Medicine* 20, no. 4 (2010): 223–27; S. Gumina et al., "Arginine L-Alpha-Ketoglutarate, Methylsulfonylmethane, Hydrolyzed Type I Collagen and Bromelain in Rotator Cuff Tear Repair: A Prospective Randomized Study," *Current Medical Research and Opinion* 28, no. 11 (2012): 1767–74, https://doi.org/10.1185/03007995.2012.737772; A. I. Aiyegbusi et al., "Bromelain in the Early Phase of Healing in Acute Crush Achilles Tendon Injury," *Phytotherapy Research* 25, no. 1 (2011): 49–52, https://doi.org/10.1002/ptr.3199; A. I. Aiyegbusi et al., "A Comparative Study of the Effects of Bromelain and Fresh Pineapple Juice on the Early Phase of Healing in Acute Crush Achilles Tendon Injury," *Journal of Medicinal Food* 14, no. 4 (2011): 348–52, https://doi.org/10.1089/jmf.2010.0078.

46. Thierry Conrozier et al., "A Complex of Three Natural Anti-Inflammatory Agents Provides Relief of Osteoarthritis Pain," *Alternative Therapies in Health and Medicine* 20, suppl 1. (2014): 32–37.

47. A. F. Walker et al., "Bromelain Reduces Mild Acute Knee Pain and Improves Well-Being in a Dose-Dependent Fashion in an Open Study of Otherwise Healthy Adults," *Phytomedicine* 9, no. 8 (2002): 681–86, https://doi.org/10.1078/094471102321621269; G. H. Tilwe et al., "Efficacy and Tolerability of Oral Enzyme Therapy as Compared to Diclofenac in Active Osteoarthrosis of Knee Joint: An Open Randomized Controlled Clinical Trial," *Journal of the Association of Physicians of India* 49 (2001): 617–21; G. Klein and W. Kullich, "Reducing Pain by Oral Enzyme Therapy in Rheumatic Diseases," *Wiener Medizinische Wochenschrift* 149, nos. 21–22 (1999): 577–80.

48. Naseer M. Akhtar et al., "Oral Enzyme Combination versus Diclofenac in the Treatment of Osteoarthritis of the Knee—A Double-Blind Prospective Randomized Study," *Clinical Rheumatology* 23, no. 5 (2004): 410–15, https://doi.org/10.1007/s10067-004-0902-y; G. M. M. J. Kerkhoffs et al., "A Double Blind, Randomised, Parallel Group Study on the Efficacy and Safety of Treating Acute Lateral Ankle Sprain with Oral Hydrolytic Enzymes," *British Journal of Sports Medicine* 38, no. 4 (2004): 431–35, https://doi.org/10.1136/bjsm.2002.004150; G. Klein et al., "Efficacy and Tolerance of an Oral Enzyme Combination in Painful Osteoarthritis of the Hip. A Double-Blind, Randomised Study Comparing Oral Enzymes with Non-Steroidal Anti-Inflammatory Drugs," *Clinical and Experimental Rheumatology* 24, no. 1 (2006): 25–30.

49. Ahmet Sahbaz et al., "Bromelain: A Natural Proteolytic for Intra-Abdominal Adhesion Prevention," *International Journal of Surgery* 14 (February 2015): 7–11, https://doi.org/10.1016/j.ijsu.2014.12.024.

50. V. Kameníček, P. Holán, and P. Franěk, "Systemic Enzyme Therapy in the Treatment and Prevention of Post-Traumatic and Postoperative Swelling," *Acta Chirurgiae Orthopaedicae et Traumatologiae Cechoslovaca* 68, no. 1 (2001): 45–49.

51. Chit Moy Ley et al., "A Review of the Use of Bromelain in Cardiovascular Diseases," *Zhong Xi Yi Jie He Xue Bao* 9, no. 7 (2011): 702–10.

52. Sangita Dutta and Debasish Bhattacharyya, "Enzymatic, Antimicrobial and Toxicity Studies of the Aqueous Extract of *Ananas comosus* (Pineapple) Crown Leaf," *Journal of Ethnopharmacology* 150, no. 2 (2013): 451–57, https://doi.org/10.1016/j.jep.2013.08.024.

53. K. Johann, K. Eschmann, and P. Meiser, "No Clinical Evidence for an Enhanced Bleeding Tendency Due to Perioperative Treatment with Bromelain," *Sportverletzung Sportschaden* 25, no. 2 (2011): 108–13, https://doi.org/10.1055/s-0031-1273307.

54. Micha Abeles et al., "Update on Fibromyalgia Therapy," *American Journal of Medicine* 121, no. 7 (2008): 555–61, https://doi.org/10.1016/j.amjmed.2008.02.036.

55. Chi-Un Pae et al., "The Relationship between Fibromyalgia and Major Depressive Disorder: A Comprehensive Review," *Current Medical Research and Opinion* 24, no. 8 (2008): 2359–71, https://doi.org/10.1185/03007990802288338.

56. Ali Gur and Pelin Oktayoglu, "Central Nervous System Abnormalities in Fibromyalgia and Chronic Fatigue Syndrome: New Concepts in Treatment," *Current Pharmaceutical Design* 14, no. 13 (2008): 1274–94, https://doi.org/10.2174/138161208799316348.

57. Laurence A. Bradley, "Pathophysiologic Mechanisms of Fibromyalgia and Its Related Disorders," *Journal of Clinical Psychiatry* 69, suppl. 2 (2008): 6–13.

58. C. K. Payne and S. Browning, "Graded Potassium Chloride Testing in Interstitial Cystitis," *Journal of Urology* 155 (1996); C. L. Parsons, "Potassium Sensitivity Test," *Techniques in Urology* 2, no. 3 (1996): 171–73.

59. Zhiming Cao et al., "*Ginkgo biloba* Extract EGb 761 and Wisconsin Ginseng Delay Sarcopenia in *Caenorhabditis elegans*," *Journal of Gerontology. Series A, Biological Sciences and Medical Sciences* 62, no. 12 (2007): 1337–45, https://doi.org/10.1093/gerona/62.12.1337.

60. Mária Báthori et al., "Phytoecdysteroids and Anabolic-Androgenic Steroids—Structure and Effects on Humans," *Current Medicinal Chemistry* 15, no. 1 (2008): 75–91, https://doi.org/10.2174/092986708783330674.

61. Volker Fintelmann and Joerg Gruenwald, "Efficacy and Tolerability of a *Rhodiola rosea* Extract in Adults with Physical and Cognitive Deficiencies," *Advances in Therapy* 24, no. 4 (2007): 929–39, https://doi.org/10.1007/BF02849986; A. G. Arbuzov et al., "Antihypoxic, Cardioprotective, and Antifibrillation Effects of a Combined Adaptogenic Plant Preparation," *Bulletin of Experimental Biology and Medicine* 142, no. 2 (2006): 212–15, https://doi.org/10.1007/s10517-006-0330-x.

62. Mayumi Ikeuchi et al., "Effects of Fenugreek Seeds (*Trigonella foenum greaecum*) Extract on Endurance Capacity in Mice," *Journal of Nutritional Science and Vitaminology* 52, no. 4 (2006): 287–92, https://doi.org/10.3177/jnsv.52.287.

63. A. Schiavone et al., "Use of *Silybum marianum* Fruit Extract in Broiler Chicken Nutrition: Influence on Performance and Meat Quality," *Journal of Animal Physiology and Animal Nutrition* 91, nos. 5–6 (2007): 256–62, https://doi.org/10.1111/j.1439-0396.2007.00701.x.

64. Agnieszka Szopa, Radosław Ekiert, and Halina Ekiert, "Current Knowledge of *Schisandra chinensis* (Turcz.) Baill. (Chinese Magnolia Vine) as a Medicinal Plant Species: A Review on the Bioactive Components, Pharmacological Properties, Analytical and Biotechnological Studies," *Phytochemistry Reviews* 16, no. 2 (2017): 195–218, https://doi.org/10.1007/s11101-016-9470-4.

65. Eun-Jung Kim et al., "Inhibition of RANKL-Stimulated Osteoclast Differentiation by *Schisandra chinensis* through Down-Regulation of NFATc1 and c-Fos Expression," *BMC Complementary and Alternative Medicine* 18, no. 1 (2018): 270, https://doi.org/10.1186/s12906-018-2331-5.

66. Adriana Nowak et al., "Potential of *Schisandra chinensis* (Turcz.) Baill. in Human Health and Nutrition: A Review of Current Knowledge and Therapeutic Perspectives," *Nutrients* 11, no. 2 (2019): 333, https://doi.org/10.3390/nu11020333.

67. Maha Sellami et al., "Herbal Medicine for Sports: A Review," *Journal of the International Society of Sports Nutrition* 15 (March 2018): article 14, https://doi.org/10.1186/s12970-018-0218-y.

68. Yuhsuke Ohmi et al., "Sialylation Converts Arthritogenic IgG into Inhibitors of Collagen-Induced Arthritis," *Nature Communications* 7 (April 2016): article 11205, https://doi.org/10.1038/ncomms11205.

69. Chang Deok Kim et al., "Inhibition of Mast Cell-Dependent Allergy Reaction by Extract of Black Cohosh (*Cimicifuga racemosa*)," *Immunopharmacology and Immunotoxicology* 26, no. 2 (2004): 299–308, https://doi.org/10.1081/iph-120037728.

70. A. Kusano et al., "Effects of Fukinolic Acid and Cimicifugic Acids from Cimicifuga Species on Collagenolytic Activity," *Biological and Pharmaceutical Bulletin* 24, no. 10 (2001): 1198–201, https://doi.org/10.1248/bpb.24.1198.

71. Jin Qi et al., "Iridoid Glycosides from *Harpagophytum procumbens* D.C. (Devil's Claw)," *Phytochemistry* 67, no. 13 (2006): 1372–77, https://doi.org/10.1016/j.phytochem.2006.05.029; M. Pignet and A. Lecomte, "The Effects of *Harpagophytum* Capsules in Degenerative Rheumatology," *Medicine Actuelle* 12, no. 4 (1985): 65–76; Joel J. Gagnier et al., "Herbal Medicine for Low Back Pain: A Cochrane Review," *Spine* 32, no. 1 (2007): 82–92, https://doi.org/10.1097/01.brs.0000249525.70011.fe.

72. Shinya Uchida et al., "Antinociceptive Effects of St. John's Wort, *Harpagophytum procumbens* Extract and Grape Seed Proanthocyanidins Extract in Mice," *Biological and Pharmaceutical Bulletin* 31, no. 2 (2008): 240–45, https://doi.org/10.1248/bpb.31.240.

73. J. E. Chrubasik, B. D. Roufogalis, and S. Chrubasik, "Evidence of Effectiveness of Herbal Antiinflammatory Drugs in the Treatment of Painful Osteoarthritis and Chronic Low Back Pain," *Phytotherapy Research* 21, no. 7 (2007): 675–83, https://doi.org/10.1002/ptr.2142; Sarah Brien, George T. Lewith, and Gerry McGregor, "Devil's Claw (*Harpagophytum procumbens*) as a Treatment for Osteoarthritis: A Review of Efficacy and Safety," *Journal of Alternative and Complementary Medicine* 12, no. 10 (2006): 981–93, https://doi.org/10.1089/acm.2006.12.981.

74. H. P. T. Ammon, "Boswellic Acids in Chronic Inflammatory Diseases," *Planta Medica* 72, no. 12 (2006): 1100–16, https://doi.org/10.1055/s-2006-947227.

75. S. Singh et al., "Boswellic Acids: A Leukotriene Inhibitor Also Effective through Topical Application in Inflammatory Disorders," *Phytomedicine* 15, nos. 6–7 (2008): 400–7, https://doi.org/10.1016/j.phymed.2007.11.019.

76. Norihiro Banno et al., "Anti-Inflammatory Activities of the Triterpene Acids from the Resin of *Boswellia carteri,*" *Journal of Ethnopharmacology* 107, no. 2 (2006): 249–53, https://doi.org/10.1016/j.jep.2006.03.006; N. Kimmatkar et al., "Efficacy and Tolerability of *Boswellia serrata* Extract in Treatment of Osteoarthritis of Knee—A Randomized Double Blind Placebo Controlled Trial," *Phytomedicine* 10, no. 1 (2003): 3–7, https://doi.org/10.1078/094471103321648593.

77. John J. Bright, "Curcumin and Autoimmune Disease," *Advances in Experimental Medicine and Biology* 595 (2007): 425–51, https://doi.org/10.1007/978-0-387-46401-5_19; Hua Zhou et al., "Suppressive Effects of JCICM-6, the Extract of an Anti-Arthritic Herbal Formula, on the Experimental Inflammatory and Nociceptive Models in Rodents," *Biological and Pharmaceutical Bulletin* 29, no. 2 (2006): 253–60, https://doi.org/10.1248/bpb.29.253; R. R. Kulkarni et al., "Treatment of Osteoarthritis with a Herbomineral Formulation: A Double-Blind, Placebo-Controlled, Cross-Over Study," *Journal of Ethnopharmacology* 33, nos. 1–2 (1991): 91–95, https://doi.org/10.1016/0378-8741(91)90167-c.

78. Reinhard Grzanna, Lars Lindmark, and Carmelita G. Frondoza, "Ginger—An Herbal Medicinal Product with Broad Anti-Inflammatory Actions," *Journal of Medicinal Food* 8, no. 2 (2005): 125–32, https://doi.org/10.1089/jmf.2005.8.125.

79. K. C. Srivastava and T. Mustafa, "Ginger (*Zingiber officinale*) in Rheumatism and Musculoskeletal Disorders," *Medical Hypotheses* 39, no. 4 (1992): 342–48, https://doi.org/10.1016/0306-9877(92)90059-l.

80. Olivier Huck et al., "Identification of a Kavain Analog with Efficient Anti-inflammatory Effects," *Scientific Reports* 9, no. 1 (2019): 12940, https://doi.org/10.1038/s41598-019-49383-8.

81. C. Randall et al., "Randomized Controlled Trial of Nettle Sting for Treatment of Base-of-Thumb Pain," *Journal of the Royal Society of Medicine* 93, no. 6 (2000): 305–9, https://doi.org/10.1177/014107680009300607.

82. W. K. Kim et al., "Molt Performance and Bone Density of Cortical, Medullary, and Cancellous Bone in Laying Hens during Feed Restriction or Alfalfa-Based Feed Molt," *Poultry Science* 86, no. 9 (2007): 1821–30, https://doi.org/10.1093/ps/86.9.1821.

83. Seok-Jong Suh et al., "Stimulative Effects of *Ulmus davidiana* Planch (Ulmaceae) on Osteoblastic MC3T3-E1 Cells," *Journal of Ethnopharmacology* 109, no. 3 (2007): 480–85, https://doi.org/10.1016/j.jep.2006.08.030.

84. Shweta Kanna, Kumar Sagar Jaiswal, and Bhawna Gupta, "Managing Rheumatoid Arthritis with Dietary Interventions," *Frontiers in Nutrition* 4 (November 2017): article 52, https://doi.org/10.3389/fnut.2017.00052.

85. R. F. Weiss, *Herbal Medicine Second Edition, Revised and Expanded* (New York: Thieme, 2000), 259–61.

86. Stephen Hsu and Douglas Dickinson, "A New Approach to Managing Oral Manifestations of Sjogren's Syndrome and Skin Manifestations of Lupus," *Journal of Biochemistry and Molecular Biology* 39, no. 3 (2006): 229–39, https://doi.org/10.5483/bmbrep.2006.39.3.229.

87. P. Rohdewald, "A Review of the French Maritime Pine Bark Extract (Pycnogenol), a Herbal Medication with a Diverse Clinical Pharmacology," *International Journal of Clinical Pharmacology and Therapeutics* 40, no. 4 (2002): 158–68, https://doi.org/10.5414/cpp40158.

88. N. F. Childers and M. S. Margoles, "An Apparent Relation of Nightshades (*Solanaceae*) to Arthritis," *Journal of Neurological and Orthopedic Medical Surgery* 12 (1993): 227–31, http://noarthritis.com/research.htm#:~:text=Plants%20in%20the%20drug%20family,in%20arthritis%20in%20sensitive%20people.&text=Rigid%20omission%20of%20Solanaceae%2C%20with,in%20arthritis%20and%20general%20health.

89. Amar A. Vishal, Artatrana Mishra, and Siba P. Raychaudhuri, "A Double Blind, Randomized, Placebo Controlled Clinical Study Evaluates the Early Efficacy of Aflapin in Subjects with Osteoarthritis of the Knee," *International Journal of Medical Sciences* 8, no. 7 (2011): 615–22, https://doi.org/10.7150/ijms.8.615; Krishanu Sengupta et al., "Comparative Efficacy and Tolerability of 5-Loxin and Aflapin against Osteoarthritis of the Knee: A Double Blind, Randomized, Placebo Controlled Clinical Study," *International Journal of Medical Sciences* 7, no. 6 (2010): 366–77, https://doi.org/10.7150/ijms.7.366.

90. N. Kimmatkar et al., "Efficacy and Tolerability of *Boswellia serrata* Extract in Treatment of Osteoarthritis of Knee—A Randomized Double Blind Placebo Controlled Trial," *Phytomedicine* 10, no. 1 (2003): 3–7, https://doi.org/10.1078/094471103321648593.

91. Krishanu Sengupta et al., "A Double Blind, Randomized, Placebo Controlled Study of the Efficacy of 5-Loxin for Treatment of Osteoarthritis of the Knee," *Arthritis Research and Therapy* 10, no. 4 (2008): R85, https://doi.org/10.1186/ar2461.

92. Krishanu Sengupta et al., "Comparative Efficacy and Tolerability of 5-Loxin and Aflapin Against Osteoarthritis of the Knee: A Double Blind, Randomized, Placebo

Controlled Clinical Study," *International Journal of Medical Sciences* 7, no. 6 (2010): 366–77, https://doi.org/10.7150/ijms .7.366; Sengupta et al., "A Double Blind, Randomized, Placebo Controlled Study of the Efficacy of 5-Loxin for Treatment of Osteoarthritis of the Knee."

93. Mona Abdel-Tawab, Oliver Werz, and Manfred Schubert-Zsilavecz, "*Boswellia serrata*: An Overall Assessment of *In Vitro*, Preclinical, Pharmacokinetic and Clinical Data," *Clinical Pharmacokinetics* 50, no. 6 (2011): 349–69, https:// doi.org/10.2165/11586800-000000000-00000; Moritz Verhoff et al., "Tetra- and Pentacyclic Triterpene Acids from the Ancient Anti-Inflammatory Remedy Frankincense as Inhibitors of Microsomal Prostaglandin E(2) Synthase-1," *Journal of Natural Products* 77, no. 6 (2014): 1445–51, https:// doi.org/10.1021/np500198g; Miao Liu et al., "A Boswellic Acid-Containing Extract Attenuates Hepatic Granuloma C57BL/6 Mice Infected with *Schistosoma japonicum*," *Parasitology Research* 112, no. 3 (2013): 1105–11, https://doi .org/10.1007/s00436-012-3237-7; Liu et al., "A Boswellic Acid-Containing Extract Attenuates Hepatic Granuloma C57BL/6 Mice Infected with *Schistosoma japonicum*"; Sadiq Umar et al., "*Boswellia serrata* Extract Attenuates Inflammatory Mediators and Oxidative Stress in Collagen Induced Arthritis," *Phytomedicine* 21, no. 6 (2014): 847–56, https://doi.org/10.1016/j.phymed.2014.02.001; Arieh Moussaieff et al., "Protective Effects of Incensole Acetate on Cerebral Ischemic Injury," *Brain Research* 1443 (2012): 89–97, https://doi.org/10.1016/j.brainres.2012.01.001; Ruiqi Wang et al., "The Comparative Study of Acetyl-11-Ket-Beta-Boswellic Acid (AKBA) and Aspirin in the Prevention of Intestinal Adenomatous Polyposis in APC(Min/+) Mice," *Drug Discoveries and Therapeutics* 8, no. 1 (2014): 25–32, https://doi.org/10.5582/ddt.8.25; Vivek R. Yadav et al., "Boswellic Acid Inhibits Growth and Metastasis of Human Colorectal Cancer in Orthotopic Mouse Model by Downregulating Inflammatory, Proliferative, Invasive and Angiogenic Biomarkers," *International Journal of Cancer* 130, no. 9 (2012): 2176–84, https://doi.org/10.1002/ijc.26251.

94. M. Z. Siddiqui, "*Boswellia serrata*, a Potential Antiinflammatory Agent: An Overview," *Indian Journal of Pharmaceutical Sciences* 73, no. 3 (2011): 255–61, https:// www.ncbi.nlm.nih.gov/pmc/articles/PMC3309643/.

95. Krishanu Sengupta et al., "Cellular and Molecular Mechanisms of Anti-Inflammatory Effect of Aflapin: A Novel *Boswellia serrata* Extract," *Molecular and Cellular Biochemistry* 354, nos. 1–2 (2011): 189–97, https://doi.org /10.1007/s11010-011-0818-1.

96. R. Christensen et al., "Does the Hip Powder of *Rosa canina* (Rosehip) Reduce Pain in Osteoarthritis Patients?—A Meta-Analysis of Randomized Controlled Trials," *Osteoarthritis and Cartilage* 16, no. 9 (2008): 965–72, https://doi.org/10 .1016/j.joca.2008.03.001.

97. Mehdi Shakibaei et al., "Resveratrol Inhibits Il-1 Beta-Induced Stimulation of Caspase-3 and Cleavage of PARP in

Human Articular Chondrocytes *In Vitro*," *Annals of the New York Academy of Sciences* 1095 (January 2007): 554–63, https://doi.org/10.1196/annals.1397.060; Ali Mobasheri, "Intersection of Inflammation and Herbal Medicine in the Treatment of Osteoarthritis," *Current Rheumatology Reports* 14, no. 6 (2012): 604–16, https://doi.org/10.1007 /s11926-012-0288-9; Constanze Csaki et al., "Regulation of Inflammation Signalling by Resveratrol in Human Chondrocytes *In Vitro*," *Biochemical Pharmacology* 75, no. 3 (2008): 677–87, https://doi.org/10.1016/j.bcp.2007.09.014.

98. Yiru Wang et al., "A Multicenter, Randomized, Double-Blind, Placebo-Controlled Trial Evaluating the Efficacy and Safety of Huangqi Guizhi Wuwutang Granule in Patients with Rheumatoid Arthritis," *Medicine* 98, no. 11 (2019): e14888, https://doi.org/10.1097/MD.0000000000014888.

99. Wei Bao et al., "Curcumin Alleviates Ethanol-Induced Hepatocytes Oxidative Damage Involving Heme Oxygenase-1 Induction," *Journal of Ethnopharmacology* 128, no. 2 (2010): 549–53, https://doi.org/10.1016/j.jep .2010.01.029.

100. Bharat B. Aggarwal et al., "Curcumin-Free Turmeric Exhibits Anti-Inflammatory and Anticancer Activities: Identification of Novel Components of Turmeric," *Molecular Nutrition and Food Research* 57, no. 9 (2013): 1529–42, https://doi.org/10.1002/mnfr.201200838; Leelavinothan Pari, Daniel Tewas, and Juergen Eckel, "Role of Curcumin in Health and Disease," *Archives of Physiology and Biochemistry* 114, no. 2 (2008): 127–49, https://doi.org /10.1080/13813450802033958.

101. G. Shoba et al., "Influence of Piperine on the Pharmacokinetics of Curcumin in Animals and Human Volunteers," *Planta Medica* 64, no. 4 (1998): 353–56, https:// doi.org/10.1055/s-2006-957450.

102. K. Taty Anna et al., "Anti-Inflammatory Effect of *Curcuma longa* (Turmeric) on Collagen-Induced Arthritis: An Anatomico-Radiological Study," *La Clinica Terapeutica* 162, no. 3 (2011): 201–7, http://www.seu-roma.it/riviste/clinica _terapeutica/apps/autos.php?id=867; Janet L. Funk et al., "Turmeric Extracts Containing Curcuminoids Prevent Experimental Rheumatoid Arthritis," *Journal of Natural Products* 69, no. 3 (2006): 351–55, https://doi.org/10.1021 /np050327j; Nilson Nonose et al., "Oral Administration of Curcumin (*Curcuma longa*) Can Attenuate the Neutrophil Inflammatory Response in Zymosan-Induced Arthritis in Rats," *Acta Cirurgica Brasileira* 29, no. 11 (2014): 727–34, https://doi.org/10.1590/s0102-86502014001800006; Gamal Ramadan, Mohammed Ali Al-Kahtani, and Wael Mohamed El-Sayed, "Anti-Inflammatory and Anti-Oxidant Properties of *Curcuma longa* (Turmeric) versus *Zingiber officinale* (Ginger) Rhizomes in Rat Adjuvant-Induced Arthritis," *Inflammation* 34, no. 4 (2011): 291–301, https:// doi.org/10.1007/s10753-010-9278-0.

103. Cheol Park et al., "Curcumin Induces Apoptosis and Inhibits Prostaglandin E(2) Production in Synovial Fibroblasts of

Patients with Rheumatoid Arthritis," *International Journal of Molecular Medicine* 20, no. 3 (2007): 365–72, https://doi.org/10.3892/ijmm.20.3.365.

104. Janet L. Funk et al., "Anti-Arthritis Effects and Toxicity of the Essential Oils of Turmeric (*Curcuma longa* L.)," *Journal of Agricultural and Food Chemistry* 58, no. 2 (2010): 842–49, https://doi.org/10.1021/jf9027206.

105. Janet L. Funk et al., "Efficacy and Mechanism of Action of Turmeric Supplements in the Treatment of Experimental Arthritis," *Arthritis and Rheumatism* 54, no. 11 (2006): 3452–64, https://doi.org/10.1002/art.22180.

106. Abigail L. Clutterbuck et al., "Interleukin-1Beta-Induced Extracellular Matrix Degradation and Glycosaminoglycan Release Is Inhibited by Curcumin in an Explant Model of Cartilage Inflammation," *Annals of the New York Academy of Sciences* 1171 (August 2009): 428–35, https://doi.org/10.1111/j.1749-6632.2009.04687.x.

107. Ahmet Türkoğlu et al., "Effect of Intraperitoneal Curcumin Instillation on Postoperative Peritoneal Adhesions," *Medical Principles and Practice* 24, no. 2 (2015): 153–58, https://doi.org/10.1159/000369020.

108. Biji T. Kurien, Anil D'Souza, and R. Hal Scofield, "Heat-Solubilized Curry Spice Curcumin Inhibits Antibody-Antigen Interaction in *In Vitro* Studies: A Possible Therapy to Alleviate Autoimmune Diseases," *Molecular Nutrition and Food Research* 54, no. 8 (2010): 1202–9, https://doi.org/10.1002/mnfr.200900106; Dong Kyun Kim et al., "Curcumin Inhibits Cellular Condensation and Alters Microfilament Organization during Chondrogenic Differentiation of Limb Bud Mesenchymal Cells," *Experimental and Molecular Medicine* 41, no. 9 (2009): 656–64, https://doi.org/10.3858/emm.2009.41.9.072.

109. Gamal Ramadan, Mohammed Ali Al-Kahtani, and Wael Mohamed El-Sayed, "Anti-Inflammatory and Anti-Oxidant Properties of *Curcuma longa* (Turmeric) versus *Zingiber officinale* (Ginger) Rhizomes in Rat Adjuvant-Induced Arthritis," *Inflammation* 34, no. 4 (2011): 291–301, https://doi.org/10.1007/s10753-010-9278-0.

110. Nyoman Kertia et al., "Ability of Curcuminoid Compared to Diclofenac Sodium in Reducing the Secretion of Cycloxygenase-2 Enzyme by Synovial Fluid's Monocytes of Patients with Osteoarthritis," *Acta Medica Indonesiana* 44, no. 2 (2012): 105–13; Piya Pinsornsak and Sunyark Niempoog, "The Efficacy of *Curcuma longa* L. Extract as an Adjuvant Therapy in Primary Knee Osteoarthritis: A Randomized Controlled Trial," *Chotmaihet Thangphaet* 95, suppl. 1 (2012): S51–58.

111. Vilai Kuptniratsaikul et al., "Efficacy and Safety of *Curcuma domestica* Extracts Compared with Ibuprofen in Patients with Knee Osteoarthritis: A Multicenter Study," *Clinical Interventions in Aging* 2014, no. 9 (2014): 451–58, https://doi.org/10.2147/CIA.S58535; Vilai Kuptniratsaikul et al., "Efficacy and Safety of *Curcuma domestica* Extracts in Patients with Knee Osteoarthritis," *Journal of Alternative*

and Complementary Medicine 15, no. 8 (2009): 891–97, https://doi.org/10.1089/acm.2008.0186.

112. Eswar Krishnan, "Gout and the Risk for Incident Heart Failure and Systolic Dysfunction," *BMJ Open* 2, no. 1 (2012): e000282, https://doi.org/10.1136/bmjopen-2011-000282.

113. A. Elisabeth Hak et al., "Menopause, Postmenopausal Hormone Use and Risk of Incident Gout," *Annals of the Rheumatic Diseases* 69, no. 7 (2010): 1305–9, https://doi.org/10.1136/ard.2009.109884.

114. Joseph Jamnik et al., "Fructose Intake and Risk of Gout and Hyperuricemia: A Systematic Review and Meta-Analysis of Prospective Cohort Studies," *BMJ Open* 6, no. 10 (2016): e013191, https://doi.org/10.1136/bmjopen-2016-013191.

115. Hyon K. Choi and Gary Curhan, "Coffee Consumption and Risk of Incident Gout in Women: The Nurses' Health Study," *American Journal of Clinical Nutrition* 92, no. 4 (2010): 922–27, https://doi.org/10.3945/ajcn.2010.29565.

116. Hyon K. Choi, Xiang Gao, and Gary Curhan, "Vitamin C Intake and the Risk of Gout in Men: A Prospective Study," *Archives of Internal Medicine* 169, no. 5 (2009): 502–7, https://doi.org/10.1001/archinternmed.2008.606.

117. Yuqing Zhang et al., "Purine-Rich Foods Intake and Recurrent Gout Attacks," *Annals of the Rheumatic Diseases* 71, no. 9 (2012): 1448–53, https://doi.org/10.1136/annrheumdis-2011-201215.

118. Christopher M. Burns and Robert L. Wortmann, "Latest Evidence on Gout Management: What the Clinician Needs to Know," *Therapeutic Advances in Chronic Disease* 3, no. 6 (2012): 271–86, https://doi.org/10.1177/2040622312462056.

119. Sung Mun Jung et al., "Reduction of Urate Crystal-Induced Inflammation by Root Extracts from Traditional Oriental Medicinal Plants: Elevation of Prostaglandin D2 Levels," *Arthritis Research and Therapy* 9, no. 4 (2007): R64, https://doi.org/10.1186/ar2222.

120. Mahaboobkhan Rasool and Palaninathan Varalakshmi, "Suppressive Effect of *Withania somnifera* Root Powder on Experimental Gouty Arthritis: An *In Vivo* and *In Vitro* Study," *Chemico-Biological Interactions* 164, no. 3 (2006): 174–80, https://doi.org/10.1016/j.cbi.2006.09.011.

121. Vaidehi N. Sarvaiya et al., "Evaluation of Antigout Activity of *Phyllanthus emblica* Fruit Extracts on Potassium Oxonate-Induced Gout Rat Model," *Veterinary World* 8, no. 10 (2015): 1230–36, https://doi.org/10.14202/vetworld.2015.1230-1236.

122. Manjusha Choudhary et al., "Medicinal Plants with Potent Anti-Arthritic Activity," *Journal of Intercultural Ethnopharmacology* 4, no. 2 (2015): 147–79, https://doi.org/10.5455/jice.20150313021918.

123. Jerzy Falandysz and Jan Borovička, "Macro and Trace Mineral Constituents and Radionuclides in Mushrooms: Health Benefits and Risks," *Applied Microbiology and Biotechnology* 97, no. 2 (2013): 477–501, https://doi.org/10.1007/s00253-012-4552-8.

124. Małgorzata Drewnowska and Jerzy Falandysz, "Investigation on Mineral Composition and Accumulation by Popular

Edible Mushroom Common Chanterelle (*Cantharellus cibarius*)," *Ecotoxicology and Environmental Safety* 113 (March 2015): 9–17, https://doi.org/10.1016/j.ecoenv.2014.11.028.

125. F. Rey-Crespo, M. López-Alonso, and M. Miranda, "The Use of Seaweed from the Galician Coast as a Mineral Supplement in Organic Dairy Cattle," *Animal* 8, no. 4 (204): 580–86, https://doi.org/10.1017/S1751731113002474.

126. R. Giardino et al., "Effects of Synthetic Salmon Calcitonin and Alendronate on Bone Quality in Ovariectomized Rats," *Minerva Medica* 88, no. 11 (1997): 469–77.

127. Youngzhong Zhang et al., "Effect of Ethanol Extract of *Lepidium meyenii* Walp. on Osteoporosis in Ovariectomized Rat," *Journal of Ethnopharmacology* 105, nos. 1–2 (2006): 274–79, https://doi.org/10.1016/j.jep.2005.12.013.

128. Sok Kuan Wong, Kok-Yong Chin, and Soelaiman Ima-Nirwana, "The Osteoprotective Effects of Kaempferol: The Evidence from *In Vivo* and *In Vitro* Studies," *Drug Design, Development and Therapy* 2019, no. 13 (2019): 3497–514, https://doi.org/10.2147/DDDT.S227738.

129. Fang Cao et al., "Combined Treatment with an Anticoagulant and a Vasodilator Prevents Steroid-Associated Osteonecrosis of Rabbit Femoral Heads by Improving Hypercoagulability," *BioMed Research International* 2017 (October 2017): article 1624074, https://doi.org/10.1155/2017/1624074.

130. Joseph Pizzorno, "The Kidney Dysfunction Epidemic, Part 2: Intervention," *Integrative Medicine* 15, no. 1 (2016): 8–12.

131. Hassanali Vatanparast and Philip D. Chilibeck, "Does the Effect of Soy Phytoestrogens on Bone in Postmenopausal Women Depend on the Equol-Producing Phenotype?" *Nutrition Reviews* 65, no. 6 (2007): 294–99, https://doi.org/10.1301/nr.2007.jun.294-299.

132. Silvina Levis et al., "Design and Baseline Characteristics of the Soy Phytoestrogens as Replacement Estrogen (SPARE) Study—A Clinical Trial of the Effects of Soy Isoflavones in Menopausal Women," *Contemporary Clinical Trials* 31, no. 4 (2010): 293–302, https://doi.org/10.1016/j.cct.2010.03.007.

133. Jian Xin Li et al., "The Effect of Traditional Medicines on Bone Resorption Induced by Parathyroid Hormone (PTH) in Tissue Culture: A Detailed Study on *Cimicifugae* Rhizoma," *Wakan Iyakugaku Zasshi* 13 (1996): 50–58.

134. Jian Xin Li et al., "Anti-Osteoporotic Activity of Traditional Medicines: Active Constituents of *Cimicifugae* Rhizoma," *Wakan Iyakugaku Zasshi* 12 (1995): 316–17.

135. Xiao-Qin Wang, Xin-Rong Zou, and Yuan Clare Zhang, "From 'Kidneys Govern Bones' to Chronic Kidney Disease, Diabetes Mellitus, and Metabolic Bone Disorder: A Crosstalk between Traditional Chinese Medicine and Modern Science," *Evidence-Based Complementary and Alternative Medicine* 2016 (September 2016): article 4370263, https://doi.org/10.1155/2016/4370263.

136. Rosalind S. Gibson et al., "A Review of Phytate, Iron, Zinc, and Calcium Concentrations in Plant-Based Complementary Foods Used in Low-Income Countries and Implications for Bioavailability," *Food and Nutrition Bulletin* 31, suppl. 2 (2010): S134–46, https://doi.org/10.1177/15648265100312S206.

137. Anne Lise Tang et al., "Phytase Activity from *Lactobacillus* spp. in Calcium-Fortified Soymilk," *Journal of Food Science* 75, no. 6 (2010): M373–76, https://doi.org/10.1111/j.1750-3841.2010.01663.x.

138. Celia J. Prynne et al., "Dietary Fibre and Phytate—A Balancing Act: Results from Three Time Points in a British Birth Cohort," *British Journal of Nutrition* 103, no. 2 (2010): 274–80, https://doi.org/10.1017/S0007114509991644.

139. E. Knudsen, B. Sandström, and P. Solgaard, "Zinc, Copper and Magnesium Absorption from a Fibre-Rich Diet," *Journal of Trace Elements in Medicine and Biology* 10, no. 2 (1996): 68–76, https://doi.org/10.1016/S0946-672X(96)80014-9.

140. A. S. Sandberg et al., "The Effect of Wheat Bran on the Absorption of Minerals in the Small Intestine," *British Journal of Nutrition* 48, no. 2 (1982): 185–91, https://doi.org/10.1079/bjn19820103.

141. Ulrich Schlemmer et al., "Phytate in Foods and Significance for Humans: Food Sources, Intake, Processing, Bioavailability, Protective Role and Analysis," *Molecular Nutrition and Food Research* 53, suppl. 2 (2009): S330–75, https://doi.org/10.1002/mnfr.200900099.

142. Ok-Hee Kim et al., "β-Propeller Phytase Hydrolyzes Insoluble Ca(2+)-Phytate Salts and Completely Abrogates the Ability of Phytate to Chelate Metal Ions," *Biochemistry* 49, no. 47 (2010): 10216–27, https://doi.org/10.1021/bi1010249.

143. Richard Hurrell and Ines Egli, "Iron Bioavailability and Dietary Reference Values," *American Journal of Clinical Nutrition* 91, no. 5 (2010): 1461S–7S, https://doi.org/10.3945/ajcn.2010.28674F.

144. T. A. Woyengo et al., "Histomorphology and Small Intestinal Sodium-Dependent Glucose Transporter 1 Gene Expression in Piglets Fed Phytic Acid and Phytase-Supplemented Diets," *Journal of Animal Science* 89, no. 8 (2011): 2485–90, https://doi.org/10.2527/jas.2010-3204.

145. Luis Raul Tovar, Manuel Olivos, and Ma Eugenia Gutierrez, "Pulque, an Alcoholic Drink from Rural Mexico, Contains Phytase. Its *In Vitro* Effects on Corn Tortilla," *Plant Foods for Human Nutrition* 63, no. 4 (2008): 189–94, https://doi.org/10.1007/s11130-008-0089-5.

146. R. R. Recker, "Calcium Absorption and Achlorhydria," *New England Journal of Medicine* 313, no. 2 (1985): 70–73, https://doi.org/10.1056/NEJM198507113130202.

147. Ali A. El Gamal et al., "Beetroot (*Beta vulgaris* L.) Extract Ameliorates Gentamicin-Induced Nephrotoxicity Associated Oxidative Stress, Inflammation, and Apoptosis in Rodent Model," *Mediators of Inflammation* 2014 (October 2014): article 983952, https://doi.org/10.1155/2014/983952; Jinhee Cho et al., "Beetroot (*Beta vulgaris*) Rescues Mice from γ-Ray Irradiation by Accelerating Hematopoiesis and Curtailing Immunosuppression," *Pharmaceutical Biology* 55, no. 1 (2017): 306–19, https://doi.org/10.1080/13880209.2016.1237976.

148. Isabella Villa et al., "Betaine Promotes Cell Differentiation of Human Osteoblasts in Primary Culture," *Journal of Translational Medicine* 15, no. 1 (2017): 132, https://doi.org /10.1186/s12967-017-1233-5.

149. H. Spencer et al., "Fluoride Therapy in Metabolic Bone Disease," *Israel Journal of Medical Sciences* 20, no. 5 (1984): 373–80.

150. Rose G. Long et al., "Design Requirements for Annulus Fibrosus Repair: Review of Forces, Displacements, and Material Properties of the Intervertebral Disk and a Summary of Candidate Hydrogels for Repair," *Journal of Biomechanical Engineering* 138, no. 2 (2016): 021007, https://doi.org/10.1115/1.4032353.

151. Fabrice Külling et al., "Implantation of Juvenile Human Chondrocytes Demonstrates No Adverse Effect on Spinal Nerve Tissue in Rats," *European Spine Journal* 25, no. 9 (2016): 2958–66, https://doi.org/10.1007/s00586-016-4558-5.

152. David Oehme et al., "Reconstitution of Degenerated Ovine Lumbar Discs by STRO-3-Positive Allogeneic Mesenchymal Precursor Cells Combined with Pentosan Polysulfate," *Journal of Neurosurger: Spine* 24, no. 5 (2016): 715–26, https://doi.org/10.3171/2015.8.SPINE141097.

153. Aldemar Andres Hegewald et al., "Regenerative Treatment Strategies in Spinal Surgery," *Frontiers in Bioscience* 13 (January 2008): 1507–26, https://doi.org/10.2741/2777.

154. Rose G. Long et al., "Design Requirements for Annulus Fibrosus Repair: Review of Forces, Displacements, and Material Properties of the Intervertebral Disk and a Summary of Candidate Hydrogels for Repair," *Journal of Biomechanical Engineering* 138, no. 2 (2016): 021007, https://doi.org/10.1115/1.4032353.

155. Wim J. van Blitterswijk, Jos C. M. van de Nes, and Paul I. J. M. Wuisman, "Glucosamine and Chondroitin Sulfate Supplementation to Treat Symptomatic Disc Degeneration: Biochemical Rationale and Case Report," *BMC Complementary and Alternative Medicine* 3 (June 2003): article 2, https://doi.org/10.1186/1472-6882-3-2; V. Palepu, M. Kodigudla, and V. K. Goel, "Biomechanics of Disc Degeneration," *Advances in Orthopedics* 2012 (June 2012): article 726210, https://doi.org/10.1155/2012/726210.

156. Crista Raak et al., "*Hypericum perforatum* to Improve Post-Operative Pain Outcome after Monosegmental Spinal Microdiscectomy (HYPOS): A Study Protocol for a Randomised, Double-Blind, Placebo-Controlled Trial," *Trials* 19, no. 1 (2018): 253, https://doi.org/10.1186/s13063-018-2631-6.

157. Fuat Uslusoy, Mustafa Nazıroğlu, and Bilal Çiğ, "Inhibition of the TRPM2 and TRPV1 Channels through *Hypericum perforatum* in Sciatic Nerve Injury-Induced Rats Demonstrates Their Key Role in Apoptosis and Mitochondrial Oxidative Stress of Sciatic Nerve and Dorsal Root Ganglion," *Frontiers in Physiology* 8 (May 2017): article 335, https://doi.org/10.3389/fphys.2017.00335.

158. Chang Liu et al., "Extraction, Purification, Structural Characteristics, Biological Activities and Pharmacological

Applications of Acemannan, a Polysaccharide from *Aloe vera*: A Review," *Molecules* 24, no. 8 (2019): 1554, https:// doi.org/10.3390/molecules24081554.

159. Sarah Brien et al., "Bromelain as a Treatment for Osteoarthritis: A Review of Clinical Studies," *Evidence-Based Complementary and Alternative Medicine* 1, no. 3 (2004): 251–57, https://doi.org/10.1093/ecam/neh035.

160. Mitchell Low et al., "An *In Vitro* Study of Anti-Inflammatory Activity of Standardised *Andrographis paniculata* Extracts and Pure Andrographolide," *BMC Complementary and Alternative Medicine* 15 (February 2015): article 18, https:// doi.org/10.1186/s12906-015-0525-7.

161. Yinxian Web et al., "*Angelica sinensis* Polysaccharides Stimulated UDP-Sugar Synthase Genes through Promoting Gene Expression of IGF-1 and IGF1R in Chondrocytes: Promoting Anti-Osteoarthritic Activity," *PLoS ONE* 9, no. 9 (2014): e107024, https://doi.org/10.1371/journal.pone.0107024.

162. Dong Wook Lim and Yun Tai Kim, "Anti-Osteoporotic Effects of *Angelica sinensis* (Oliv.) Diels Extract on Ovariectomized Rats and Its Oral Toxicity in Rats," *Nutrients* 6, no. 10 (2014): 4362–72, https://doi.org/10.3390/nu6104362.

163. F. Yang et al., "Ligustilide, a Major Bioactive Component of *Angelica sinensis*, Promotes Bone Formation via the GPR30/EGFR Pathway," *Scientific Reports* 9, no. 1 (2019): 6991, https://doi.org/10.1038/s41598-019-43518-7.

164. Young-Kyu Kim et al., "Inhibitory Effect and Mechanism of *Arctium lappa* Extract on NLRP3 Inflammasome Activation," *Evidence-Based Complementary and Alternative Medicine* 2018 (January 2018): article 6346734, https://doi.org/10.1155/2018/6346734.

165. Ali Totonchi and Bahman Guyuron, "A Randomized, Controlled Comparison Between Arnica and Steroids in the Management of Postrhinoplasty Ecchymosis and Edema," *Plastic and Reconstructive Surgery* 120, no. 1 (2007): 271–74, https://doi.org/10.1097/01.prs.0000264397.80585.bd.

166. Honghai Zhao et al., "Berberine Ameliorates Cartilage Degeneration in Interleukin-1β-Stimulated Rat Chondrocytes and in a Rat Model of Osteoarthritis via Akt Signalling," *Journal of Cellular and Molecular Medicine* 18, no. 2 (2014): 283–92, https://doi.org/10.1111/jcmm.12186.

167. Nahid Akhtar and Tariq M Haqqi, "Current Nutraceuticals in the Management of Osteoarthritis: A Review," *Therapeutic Advances in Musculoskeletal Disease* 4, no. 3 (2012): 181–207, https://doi.org/10.1177/1759720X11436238.

168. Yeonju Lee et al., "Anti-Inflammatory Effect of Triterpene Saponins Isolated from Blue Cohosh (*Caulophyllum thalictroides*)," *Evidence-Based Complementary and Alternative Medicine* 2012 (September 2012): article 798192, https:// doi.org/10.1155/2012/798192.

169. Sandipan Datta et al., "Toxins in Botanical Dietary Supplements: Blue Cohosh Components Disrupt Cellular Respiration and Mitochondrial Membrane Potential," *Journal of Natural Products* 77, no. 1 (2014): 111–17, https:// doi.org/10.1021/np400758t.

170. Mohamed Fizur Nagoor Meeran et al., "Pharmacological Properties, Molecular Mechanisms, and Pharmaceutical Development of Asiatic Acid: A Pentacyclic Triterpenoid of Therapeutic Promise," *Frontiers in Pharmacology* 9 (September 2018): article 892, https://doi.org/10.3389/fphar.2018.00892.

171. Badal Rathi et al., "Ameliorative Effects of a Polyphenolic Fraction of *Cinnamomum zeylanicum* L. Bark in Animal Models of Inflammation and Arthritis," *Scientia Pharmaceutica* 81, no. 2 (2013): 567–89, https://doi.org/10.3797/scipharm.1301-16.

172. Jianbo He et al., "Therapeutic Anabolic and Anticatabolic Benefits of Natural Chinese Medicines for the Treatment of Osteoporosis," *Frontiers in Pharmacology* 10 (November 2019): article 1344, https://doi.org/10.3389/fphar.2019.01344.

173. Yu Sun et al., "Reduction of Intraarticular Adhesion by Topical Application of Colchicine Following Knee Surgery in Rabbits," *Scientific Reports* 4 (September 2014): article 6405, https://doi.org/10.1038/srep06405.

174. R. Kowshik Aravilli, S. Laveen Vikram, and V. Kohila, "Phytochemicals as Potential Antidotes for Targeting NF-κB in Rheumatoid Arthritis," *3 Biotech* 7, no. 4 (2017): article 253, https://doi.org/10.1007/s13205-017-0888-1.

175. Reecha Madaan and S. Kumar, "Screening of Alkaloidal Fraction of *Conium maculatum* L. Aerial Parts for Analgesic and Antiinflammatory Activity," *Indian Journal of Pharmaceutical Sciences* 74, no. 5 (2012): 457–60, https://doi.org/10.4103/0250-474X.108423.

176. Daniel A. Todd et al., "Ethanolic *Echinacea purpurea* Extracts Contain a Mixture of Cytokine-Suppressive and Cytokine-Inducing Compounds, Including Some that Originate from Endophytic Bacteria," *PLoS ONE* 10, no. 5 (2015): e0124276, https://doi.org/10.1371/journal.pone.0124276.

177. Milian Fedurco et al., "Modulatory Effects of *Eschscholzia californica* Alkaloids on Recombinant GABAA Receptors," *Biochemistry Research International* 2015 (October 2015): article 617620, https://doi.org/10.1155/2015/617620.

178. Piotr Michel et al., "Metabolite Profiling of Eastern Teaberry (*Gaultheria procumbens* L.) Lipophilic Leaf Extracts with Hyaluronidase and Lipoxygenase Inhibitory Activity," *Molecules* 22, no. 3 (2017): 412, https://doi.org/10.3390/molecules22030412.

179. Jing-Yang Zhang and Yong-Xiang Wang, "Gelsemium Analgesia and the Spinal Glycine Receptor/Allopregnanolone Pathway," *Fitoterapia* 100 (2015): 35–43, https://doi.org/10.1016/j.fitote.2014.11.002.

180. Zhe Wu et al., "Effects of the Extract of *Ginkgo biloba* on the Differentiation of Bone Marrow Mesenchymal Stem Cells *In Vitro*," *American Journal of Translational Research* 8, no. 7 (2016): 3032–40, https://www.ncbi.nlm.nih.gov/pmc/articles/PMC4969439/pdf/ajtr0008-3032.pdf; Ao Wong et al., "*Ginkgo biloba* L. Extract Reduces H2O2-Induced Bone Marrow Mesenchymal Stem Cells Cytotoxicity by Regulating Mitogen-Activated Protein Kinase (MAPK) Signaling Pathways and Oxidative Stress," *Medical Science Monitor* 24 (May 2018): 3159–67, https://doi.org/10.12659/MSM.910718.

181. I. Damlar et al., "Effects of *Hypericum perforatum* on the Healing of Xenografts: A Histomorphometric Study in Rabbits," *British Journal of Oral and Maxillofacial Surgery* 55, no. 4 (2017): 383–87, https://doi.org/10.1016/j.bjoms.2016.12.003.

182. E. Sherry, H. Boeck, and P. H. Warnke, "Percutaneous Treatment of Chronic MRSA Osteomyelitis with a Novel Plant-Derived Antiseptic," *BMC Surgery* 1 (May 2001): article 1, https://doi.org/10.1186/1471-2482-1-1.

183. Hyeong-Geug Kim et al., "Antifatigue Effects of *Panax ginseng* C.A. Meyer: A Randomised, Double-Blind, Placebo-Controlled Trial," *PLoS ONE* 8, no. 4 (2013): e61271, https://doi.org/10.1371/journal.pone.0061271.

184. Peter Jörg Rohdewald, "Review on Sustained Relief of Osteoarthritis Symptoms with a Proprietary Extract from Pine Bark, Pycnogenol," *Journal of Medicinal Food* 21, no. 1 (2018): 1–4, https://doi.org/10.1089/jmf.2017.0015.

185. Olivier Huck et al., "Reduction of Articular and Systemic Inflammation by Kava-241 in a *Porphyromonas gingivalis*-Induced Arthritis Murine Model," *Infection and Immunity* 86, no. 9 (2018): e00356–18, https://doi.org/10.1128/IAI.00356-18.

186. Mehdi Shakibaei et al., "Botanical Extracts from Rosehip (*Rosa canina*), Willow Bark (*Salix alba*), and Nettle Leaf (*Urtica dioica*) Suppress IL-1β-Induced NF-κB Activation in Canine Articular Chondrocytes," *Evidence-Based Complementary and Alternative Medicine* 2012 (March 2012): article 509383, https://doi.org/10.1155/2012/509383.

187. Nazir M. Khan et al., "A Wogonin-Rich-Fraction of *Scutellaria baicalensis* Root Extract Exerts Chondroprotective Effects by Suppressing IL-1β-Induced Activation of AP-1 in Human OA Chondrocytes," *Scientific Reports* 7 (March 2017): article 43789, https://doi.org/10.1038/srep43789.

188. Bahram H. Arjmandi et al., "A Combination of *Scutellaria baicalensis* and *Acacia catechu* Extracts for Short-Term Symptomatic Relief of Joint Discomfort Associated with Osteoarthritis of the Knee," *Journal of Medicinal Food* 17, no. 6 (2014): 707–13, https://doi.org/10.1089/jmf.2013.0010; Mesfin Yimam et al., "Cartilage Protection and Analgesic Activity of a Botanical Composition Comprised of *Morus alba*, *Scutellaria baicalensis*, and *Acacia catechu*," *Evidence-Based Complementary and Alternative Medicine* 2017 (August 2017): article 7059068, https://doi.org/10.1155/2017/7059068.

189. Sarah L. Morgan et al., "The Safety of Flavocoxid, a Medical Food, in the Dietary Management of Knee Osteoarthritis," *Journal of Medicinal Food* 12, no. 5 (2009): 1143–48, https://doi.org/10.1089/jmf.2008.0244.

190. Jayaraj Paulraj, Raghavan Govindarajan, and Pushpangadan Palpu, "The Genus *Spilanthes* Ethnopharmacology, Phytochemistry, and Pharmacological Properties: A Review," *Advances in Pharmaceutical Sciences* 2013 (December 2013): article 510298, https://doi.org/10.1155/2013/510298.

191. S. Kawakita et al., "Effect of an Isoflavones-Containing Red Clover Preparation and Alkaline Supplementation on Bone Metabolism in Ovariectomized Rats," *Clinical Interventions in Aging* 2009, no. 4 (February 2009): 91–100, https://doi.org/10.2147/cia.s4164; Urszula Cegieła et al., "Effects of Extracts from *Trifolium medium* L. and *Trifolium pratense* L. on Development of Estrogen Deficiency-Induced Osteoporosis in Rats," *Evidence-Based Complementary and Alternative Medicine* 2012 (November 2012): article 921684, https://doi.org/10.1155/2012/921684.

192. Dorota Kregiel, Ewelina Pawlikowska, and Hubert Antolak, "*Urtica* spp.: Ordinary Plants with Extraordinary Properties," *Molecules* 23, no. 7 (2018): 1664, https://doi.org/10.3390/molecules23071664.

193. S. Klingelhoefer et al., "Antirheumatic Effect of IDS 23, a Stinging Nettle Leaf Extract, on *In Vitro* Expression of T Helper Cytokines," *Journal of Rheumatology* 26, no. 12 (1999): 2517–22.

194. Vladana Domazetovic et al., "Blueberry Juice Protects Osteocytes and Bone Precursor Cells against Oxidative Stress Partly through SIRT1," *FEBS Open Bio* 9, no. 6 (2019): 1082–96, https://doi.org/10.1002/2211-5463.12634.

195. R. Bingham et al., "Yucca Plant Saponin in the Management of Arthritis," *Journal of Applied Nutrition* 27 (1975): 45–51; Beata Olas et al., "Comparative Anti-Platelet and Antioxidant Properties of Polyphenol-Rich Extracts from: Berries of *Aronia melanocarpa*, Seeds of Grape and Bark of *Yucca schidigera In Vitro*," *Platelets* 19, no. 1 (2008): 70–77, https://doi.org/10.1080/09537100701708506; Ying Zhang et al., "Steroidal Saponins from the Stem of *Yucca elephantipes*," *Phytochemistry* 69, no. 1 (2008): 264–70, https://doi.org/10.1016/j.phytochem.2007.06.015; Yu-Lan Jin et al., "A New Steroidal Saponin, Yuccalan, from the Leaves of *Yucca smalliana*," *Archives of Pharmacal Research* 30, no. 5 (2007): 543–46, https://doi.org/10.1007/BF02977645.

196. Lu Qu et al., "Spirostane-Type Saponins Obtained from *Yucca schidigera*," *Molecules* 23, no. 1 (2018): 167, https://doi.org/10.3390/molecules23010167.

197. P. R. Cheeke, S. Piacente, and W. Oleszek, "Anti-Inflammatory and Anti-Arthritic Effects of *Yucca schidigera*: A Review," *Journal of Inflammation* 3 (March 2006): article 6, https://doi.org/10.1186/1476-9255-3-6.

198. K. C. Srivastava and T. Mustafa, "Ginger (*Zingiber officinale*) and Rheumatic Disorders," *Medical Hypotheses* 29, no. 1 (1989): 25–28, https://doi.org/10.1016/0306-9877(89)90162-x.

199. Dorin Dragos et al., "Phytomedicine in Joint Disorders," *Nutrients* 9, no. 1 (2017): 70, https://doi.org/10.3390/nu9010070.

200. Søren Ribel-Madsen et al., "A Synoviocyte Model for Osteoarthritis and Rheumatoid Arthritis: Response to Ibuprofen, Betamethasone, and Ginger Extract-A Cross-Sectional *In Vitro* Study," *Arthritis* 2012 (2012): article 505842, https://www.ncbi.nlm.nih.gov/pmc/articles/PMC3546442/; Abdullah Al-Nahain, Rownak Jahan, and Mohammed Rahmatullah, "*Zingiber officinale*: A Potential Plant against Rheumatoid Arthritis," *Arthritis* 2014 (May 2014): article 159089, https://www.hindawi.com/journals/arthritis/2014/159089/.

Chapter 4: Creating Herbal Formulas for Eye, Ear, Nose, Mouth, and Throat Conditions

1. Sergio Fanella, Alex Singer, and Joanne Embree, "Presentation and Management of Pediatric Orbital Cellulitis," *The Canadian Journal of Infectious Diseases and Medical Microbiology* 22, no. 3 (2011): 97–100, https://doi.org/10.1155/2011/626809.

2. Bipasha Mukherjee, Nirav Dilip Raichura, and Md. Shahid Alam, "Fungal Infections of the Orbit," *Indian Journal of Ophthalmology* 64, no. 5 (2016): 337–45, https://doi.org/10.4103/0301-4738.185588.

3. Rupesh Chawla, James D. Kellner, and William F. Astle, "Acute Infectious Conjunctivitis in Childhood," *Pediatrics and Child Health* 6, no. 6 (2001): 329–35, https://doi.org/10.1093/pch/6.6.329.

4. Chawla, Kellner, and Astle, "Acute Infectious Conjunctivitis in Childhood."

5. Sadigheh Ghaemi et al., "Evaluation of Preventive Effects of Colostrum against Neonatal Conjunctivitis: A Randomized Clinical Trial," *Journal of Education and Health Promotion* 2014, no. 3 (2014): 63.

6. Amir A. Azari and Neal P. Barney, "Conjunctivitis: A Systematic Review of Diagnosis and Treatment," *Journal of the American Medical Association* 310, no. 16 (2013): 1721–29, https://doi.org/10.1001/jama.2013.280318.

7. M. Stoss et al., "Prospective Cohort Trial of *Euphrasia* Single-Dose Eye Drops in Conjunctivitis," *The Journal of Alternative and Complementary Medicine: Research on Paradigm, Practice, and Policy* 6, no. 6 (2000): 499–508, https://doi.org/10.1089/acm.2000.6.499.

8. Theoharis C. Theoharides and Leonard Bielory, "Mast Cells and Mast Cell Mediators as Targets of Dietary Supplements," *Annals of Allergy, Asthma and Immunology* 93, no. 2 Suppl 1 (2004): S24–34, https://doi.org/10.1016/s1081-1206(10)61484-6; Y.-D. Min et al., "Quercetin Inhibits Expression of Inflammatory Cytokines Through Attenuation of NF-kappaB and p38 MAPK in HMC-1 Human Mast Cell Line," *Inflammation Research* 56, no. 5 (2007): 210–15, https://doi.org/10.1007/s00011-007-6172-9.

9. Po-Chuen Chan, Qingsu Xia, and Peter P. Fu, "*Ginkgo biloba* Leave Extract: Biological, Medicinal, and Toxicological Effects," *Journal of Environmental Science and Health* 25, no. 3 (2007): 211–44, https://doi.org/10.1080/10590500701569414.

10. Maurizio Rolando and Cristiana Valente, "Establishing the Tolerability and Performance of Tamarind Seed Polysaccharide (TSP) in Treating Dry Eye Syndrome: Results of a Clinical Study," *BioMed Central Ophthalmology* 7, no. 5 (2007), https://doi.org/10.1186/1471-2415-7-5.

11. Aihua Liu and Jian Ji, "Omega-3 Essential Fatty Acids for Dry Eye Syndrome: A Meta-Analysis of Randomized Controlled Studies," *Medical Science Monitor: International Medical Journal of Experimental and Clinical Research* 2014, no. 20 (2014): 1583–89, https://doi.org/10.12659/MSM.891364.

12. H. Selek et al., "Evaluation of Retinoic Acid Ophthalmic Emulsion in Dry Eye," *European Journal of Ophthalmology* 10, no. 2 (2000): 121–7, https://doi.org/10.1177 /112067210001000205.

13. Sam Young Yoon et al., "Low Serum 25-Hydroxyvitamin D Levels Are Associated with Dry Eye Syndrome," *PLoS One* 11, no. 1 (2016): e0147847, https://doi.org/10.1371 /journal.pone.0147847; Seok Hyun Bae et al., "Vitamin D Supplementation for Patients with Dry Eye Syndrome Refractory to Conventional Treatment," *Scientific Reports* 2016, no. 6 (2016): 33083, https://doi.org/10.1038/srep33083.

14. Patrick M. Azcarate et al., "Androgen Deficiency and Dry Eye Syndrome in the Aging Male," *Investigative Ophthalmology and Visual Science* 55, no. 8 (2014): 5046-53, https://doi.org /10.1167/iovs.14-14689; Juan Ding and David A. Sullivan, "Aging and Dry Eye Disease," *Experimental Gerontology* 47, no. 7 (2012): 483–90, https://doi.org/10.1016/j.exger.2012.03.020.

15. Maya Salomon-Ben Zeev, Darby Douglas Miller, and Robert Latkany, "Diagnoses of Dry Eye Disease and Emerging Technologies," *Clinical Ophthalmology* 2014; no. 8 (2014): 581–90, https://doi.org/10.2147/OPTH.S45444.

16. Qing-Hua Peng et al., "Effects of Extract of *Buddleja officinalis* Eye Drops on Androgen Receptors of Lacrimal Gland Cells of Castrated Rats with Dry Eye," *International Journal of Ophthalmology* 3, no. 1 (2010): 43–48, https://doi.org /10.3980/j.issn.2222-3959.2010.01.10.

17. J. Schmidgall, E. Schnetz, and A. Hensel, "Evidence for Bioadhesive Effects of Polysaccharides and Polysaccharide-Containing Herbs in an *Ex Vivo* Bioadhesion Assay on Buccal Membranes," *Planta Medica* 66, no. 1 (2000): 48–53, https://doi.org/10.1055/s-2000-11118.

18. Christopher D. Coldren et al., "Gene Expression Changes in the Human Fibroblast Induced by *Centella asiatica* Triterpenoids," *Planta Medica* 69, no. 8 (2003): 725–32, https://doi.org/10.1055/s-2003-42791.

19. Asim V. Farooq and Deepak Shukla, "Herpes Simplex Epithelial and Stromal Keratitis: An Epidemiologic Update," *Survey of Ophthalmology* 57, no. 5 (2012): 448–62, https://doi.org/10.1016/j.survophthal.2012.01.005.

20. Jeffrey I. Cohen, "Licking Latency with Licorice," *The Journal of Clinical Investigation* 115, no. 3 (2005): 591–93, https://doi.org/10.1172/JCI24507; Jung Chung Lin, "Mechanism of Action of Glycyrrhizic Acid in Inhibition of Epstein-Barr Virus Replication *In Vitro*," *Antiviral Research* 59, no. 1 (2003): 41–47, https://doi.org/10.1016/s0166 -3542(03)00030-5; Y. P. Chang, W. X. Bi, and G. Z. Yang, "Studies on the Anti-Virus Effect of *Glycyrrhiza uralensis* Fish. Polysaccharide," *Zhongguo Zhong yao za zhi = Zhongguo zhongyao zazhi = China Journal of Chinese*

Materia Medica 14, no. 4 (1989): 236–38 and 255–56; R. Pompei et al., "Glycyrrhizic Acid Inhibits Virus Growth and Inactivates Virus Particles," *Nature* 281, no. 5733 (1979): 689–90, https://doi.org/10.1038/281689a0.

21. M.S. Zheng, "An Experimental Study of the Anti-HSV-II Action of 500 Herbal Drugs," *Journal of Traditional Chinese Medicine = Chung i tsa chih ying wen pan* 9, no. 2 (1989): 113–16; Daniela Fritz et al., "Herpes Virus Inhibitory Substances from *Hypericum connatum* Lam., A Plant Used in Southern Brazil to Treat Oral Lesions," *Journal of Ethnopharmacology* 113, no. 3 (2007): 517–20, https://doi.org/10.1016/j.jep.2007.07.013.

22. Silke Nolkemper et al., "Antiviral Effect of Aqueous Extracts from Species of the Lamiaceae Family against Herpes Simplex Virus Type 1 and Type 2 *In Vitro*," *Planta Medica* 72, no. 15 (2006): 1378–82, https://doi.org/10.1055/s-2006 -951719; A. Allahverdiyev et al., "Antiviral Activity of the Volatile Oils of *Melissa officinalis* L. against Herpes Simplex Virus Type-2," *Phytomedicine* 11, nos. 7-8 (2004): 657–61, https://doi.org/10.1016/j.phymed.2003.07.014; R. Koytchev, R. G. Alken, and S. Dundarov, "Balm Mint Extract (Lo-701) for Topical Treatment of Recurring Herpes Labialis," *Phytomedicine* 6, no. 4 (1999): 225–30, https://doi. org/10.1016/S0944-7113(99)80013-0.

23. James B. Hudson, "Applications of the Phytomedicine *Echinacea purpurea* (Purple Coneflower) in Infectious Diseases," *Journal of Biomedicine and Biotechnology* 2012 (2012): 769896, https://doi.org/10.1155/2012/769896.

24. Ali Salehi et al., "A Double Blind Clinical Trial on the Efficacy of Honey Drop in Vernal Keratoconjunctivitis," *Evidence-Based Complementary and Alternative Medicine: eCAM* 2014 (2014): 287540, https://doi.org/10.1155/2014/287540.

25. Sandamali A. Ekanayaka et al., "Glycyrrhizin Reduces HMGB1 and Bacterial Load in *Pseudomonas aeruginosa* Keratitis," *Investigative Ophthalmology and Visual Science* 57, no. 13 (2016): 5799–5809, https://doi.org/10.1167/iovs.16-20103.

26. S. H. Lin et al., "Fungal Corneal Ulcers of Onion Harvesters in Southern Taiwan," *Occupational and Environmental Medicine* 56, no. 6 (1999): 423–25, https://doi.org/10.1136/oem.56.6.423.

27. Mirko R. Jankov II et al., "Corneal Collagen Cross-Linking," *Middle East African Journal of Ophthalmology* 17, no. 1 (2010): 21–27; Mitra Zamani, Mahmoodreza Panahi-Bazaz, and Mona Assadi, "Corneal Collagen Cross-Linking for Treatment of Non-Healing Corneal Ulcers," *Journal of Ophthalmic and Vision Research* 10, no. 1 (2015): 16–20, https://doi.org/10.4103/2008-322X.156087; Adel Alhayek and Pei-Rong Lu, "Corneal Collagen Crosslinking in Keratoconus and Other Eye Disease," *International Journal of Ophthalmology* 8, no. 2 (2015): 407–18, https://doi.org /10.3980/j.issn.2222-3959.2015.02.35.

28. Junji Narioka and Yuichi Ohashi, "Changes in Lumen Width of Nasolacrimal Drainage System after Adrenergic and Cholinergic Stimulation," *American Journal of Ophthalmology* 141, no. 4 (2006): 689–98, https://doi.org /10.1016/j.ajo.2005.11.040.

29. Cheng-Han Wu et al., "Pilocarpine Hydrochloride for the Treatment of Xerostomia in Patients with Sjögren's Syndrome in Taiwan—A Double-Blind, Placebo-Controlled Trial," *Journal of the Formosan Medical Association* [*Taiwan Yi Zhi*] 105, no. 10 (2006): 796–803, https://doi.org/10.1016/S0929-6646(09)60266-7.

30. P. Aragona et al., "Conjunctival Epithelium Improvement after Systemic Pilocarpine in Patients with Sjögren's Syndrome," *The British Journal of Ophthalmology* 90, no. 2 (2006): 166–170, https://doi.org/10.1136/bjo.2005.078865.

31. G. Thiagarajan et al., "Antioxidant Properties of Green and Black Tea, and Their Potential Ability to Retard the Progression of Eye Lens Cataract," *Experimental Eye Research* 73, no. 3 (2001): 393–401, https://doi.org/10.1006/exer.2001.1049.

32. M. V. Albal, A. G. Chandorkar, and P. M. Bulakh, "Evaluation of Catalin, Succus *Cineraria maritima* and Catobell in Goat Lens Cultures," *Indian Journal of Ophthalmology* 29, no. 3 (1981): 147–49.

33. Suresh K, Gupta et al., "*Ocimum sanctum* Modulates Selenite-Induced Cataractogenic Changes and Prevents Rat Lens Opacification," *Current Eye Research* 30, no. 7 (2005): 583–91, https://doi.org/10.1080/02713680590968132.

34. A. Zh Fursova et al., "Dietary Supplementation with Bilberry Extract Prevents Macular Degeneration and Cataracts in Senesce-Accelerated OXYS Rats," *Uspekhi Gerontologii = Advances in Gerontology* 2005, no. 16 (2005): 76–79.

35. M. S. Moghaddam et al., "Effect of Diabecon on Sugar-Induced Lens Opacity in Organ Culture: Mechanism of Action," *Journal of Ethnopharmacology* 97, no. 2 (2005): 397–403, https://doi.org/10.1016/j.jep.2004.11.032.

36. V. Vats et al., "Anti-Cataract Activity of *Pterocarpus marsupium* Bark and *Trigonella foenum-graecum* Seeds Extract in Alloxan Diabetic Rats," *Journal of Ethnopharmacology* 93, nos. 2-3 (2004): 289–94, https://doi.org/10.1016/j.jep.2004.03.032.

37. Kristina Lindsley, Jason J. Nichols, and Kay Dickersin, "Interventions for Acute Internal Hordeolum," *The Cochrane Database of Systematic Reviews* 2010: no. 9 (2010): CD007742, https://doi.org/10.1002/14651858.CD007742.pub2.

38. J. P. McCulley, "Blepharoconjunctivitis," *International Ophthalmology Clinics* 24, no. 2 (1984): 65–77, https://doi.org/10.1097/00004397-198424020-00009.

39. Roman Paduch et al., "Assessment of Eyebright (*Euphrasia officinalis* L.) Extract Activity in Relation to Human Corneal Cells Using *In Vitro* Tests," *Balkan Medical Journal* 31, no. 1 (2014): 29–36, https://doi.org/10.5152/balkanmedj.2014.8377.

40. Jingbo Liu, Hosam Sheha, and Scheffer C.G. Tseng, "Pathogenic Role of *Demodex* Mites in Blepharitis," *Current Opinion in Allergy and Clinical Immunology* 10, no. 5 (2010): 505–10, https://doi.org/10.1097/ACI.0b013e32833df9f4.

41. Hyun Koo et al., "Ocular Surface Discomfort and *Demodex*: Effect of Tea Tree Oil Eyelid Scrub in *Demodex* Blepharitis," *Journal of Korean Medical Science* 27, no. 12 (2012): 1574–79, https://doi.org/10.3346/jkms.2012.27.12.1574; Y.-Y. Gao et al., "*In Vitro* and *In Vivo* Killing of Ocular *Demodex* by Tea Tree Oil," *The British Journal of Ophthalmology* 89, no. 11 (2005): 1468–73, https://doi.org/10.1136/bjo.2005.072363.

42. Alberto Izzotti et al., "Open Angle Glaucoma: Epidemiology, Pathogenesis and Prevention [Article in Italian]," *Recenti Progressi in Medicina = Recent Advances in Medicine* 97, no. 1 (2006): 37–45.

43. A. S. Verkman, "More than Just Water Channels: Unexpected Cellular Roles of Aquaporins," *Journal of Cell Science* 118, pt. 15 (2005): 3225–32, https://doi.org/10.1242/jcs.02519.

44. Syril Dorairaj, Robert Ritch, and Jeffrey M. Liebmann, "Visual Improvement in a Patient Taking *Ginkgo biloba* Extract: A Case Study," *Explore (New York, N.Y.)* 3, no. 4 (2007): 391–95, http://doi.org/10.1016/j.explore.2007.04.011; R. Ritch, "Potential Rose for *Ginkgo biloba* Extract in the Treatment of Glaucoma," *Medical Hypotheses* 54, no. 2 (2000): 221–35, https://doi.org/10.1054/mehy.1999.0025; Kazuyuki Hirooka et al., "The *Ginkgo biloba* Extract (EGb 761) Provides a Neuroprotective Effect on Retinal Ganglion Cells in a Rat Model of Chronic Glaucoma," *Current Eye Research* 28, no. 3 (2004): 153–57, https://doi.org/10.1076/ceyr.28.3.153.26246.

45. H. S. Chung et al., "*Ginkgo biloba* Extract Increases Ocular Blood Flow Velocity," *Journal of the Ocular Pharmacology and Therapeutics* 15, no. 3 (1999): 233–40, https://doi.org/10.1089/jop.1999.15.233.

46. Hannah Bartlett and Frank Eperjesi, "An Ideal Ocular Nutritional Supplement?" *Ophthalmic and Physiological Optics* 24, no. 4 (2004): 339–49, https://doi.org/10.1111/j.1475-1313.2004.00218.x; Luciano Quaranta et al., "Effect of *Ginkgo biloba* Extract on Preexisting Visual Field Damage in Normal Tension Glaucoma," *Ophthalmology* 10, no. 2 (2003): 359–62, https://doi.org/10.1016/S0161-6420(02)01745-1.

47. M. D. Zhu and F. Y. Cai, "The Effect of Inj. *Salviae miltiorrhizae* Co. on the Retrograde Axoplasmic Transport in the Optic Nerve of Rabbits with Chronic IOP Elevation [Article in Chinese]," *Zhonghua yan ke za zhi = Chinese Journal of Opthalmology* 27, no. 3 (1991): 174–78.

48. Kevin K. M. Yue et al., "Danshen Prevents the Occurrence of Oxidative Stress in the Eye and Aorta of Diabetic Rats Without Affecting the Hyperglycemic State," *Journal of Ethnopharmacology* 106, no. 1 (2006): 136–41, https://doi.org/10.1016/j.jep.2005.12.026.

49. Shamkant B. Badgujar, Vainav V. Patel, and Atmaram H. Bandivdekar, "*Foeniculum vulgare* Mill: A Review of its Botany, Phytochemistry, Pharmacology, Contemporary Application, and Toxicology," *BioMed Research International* 2014 (2014): 842674, https://doi.org/10.1155/2014/842674.

50. Dehong Gao et al., "An Aqueous Extract of Radix Astragali, *Angelica sinensis*, and *Panax notoginseng* Is Effective in Preventing Diabetic Retinopathy," *Evidence-Based Complementary and Alternative Medicine: eCAM* 2013 (2013): 578165, https://doi.org/10.1155/2013/578165.

51. Chuangxin Huang et al., "Herbal Compound Naoshuantong Capsules Attenuates Retinal Injury in Ischemia/Reperfusion Rat Model by Inhibiting Apoptosis,"

International Journal of Clinical and Experimental Medicine 8, no. 8 (2015): 12252–63.

52. Lin Cheng et al., "Systematic Review and Meta-Analysis of 16 Randomized Clinical Trials of Radix Astragali and Its Prescriptions for Diabetic Retinopathy," *Evidence-Based Complementary and Alternative Medicine: eCAM* 2013 (2013): 762783, https://doi.org/10.1155/2013/762783.

53. Suk-Yee Li et al., "*Lycium barbarum* Polysaccharides Reduce Neuronal Damage, Blood-Retinal Barrier Disruption and Oxidative Stress in Retinal Ischema/Reperfusion Injury," *PLoS One* 6, no. 1 (2011): e16380, https://doi.org/10.1371/journal.pone.0016380; Meihua He et al., "Activation of the Nrf2/HO-1 Antioxidant Pathway Contributes to the Protective Effects of *Lycium barbarum* Polysaccharides in the Rodent Retina after Ischemia-Reperfusion-Induced Damage," *PLoS One* 9, no. 1 (2014): e84800, https://doi.org/10.1371/journal.pone.0084800; M. K. Song, B. D. Roufogalis, and T. H. W. Huang, "Reversal of the Caspase-Dependent Apoptotic Cytotoxicity Pathway by Taurine from *Lycium barbarum* (Goji Berry) in Human Retinal Pigment Epithelial Cells: Potential Benefit in Diabetic Retinopathy," *Evidence-Based Complementary and Alternative Medicine: eCAM* 2012 (2012): 323784, https://doi.org/10.1155/2012/323784.

54. Paul E. Milbury et al., "Bilberry (*Vaccinium myrtillus*) Anthocyanins Modulate Heme Oxygenase-1 and Glutathione S-Transferase-Pi Expression in ARPE-19 Cells," *Investigative Ophthalmology and Visual Science* 48, no. 5 (2007): 2343–49, https://doi.org/10.1167/iovs.06-0452; Han Kook Chung et al., "Efficacy of Troxerutin on Streptozotocin-Induced Rat Model in the Early Stage of Diabetic Retinopathy," *Arzneimittel-Forschung = Drug Research* 55, no. 10 (2005): 573–80, https://doi.org/10.1055/s-0031-1296907.

55. Peter James Francis and Michael L. Klein, "Update on the Role of Genetics in the Onset of Age-Related Macular Degeneration," *Clinical Ophthalmology (Auckland, N.Z.)* 5 (2011): 1127–33, https://doi.org/10.2147/OPTH.S11627.

56. Lai Wei et al., "Hypomethylation of IL17RC Promoter Associatese with Age-Related Macular Degeneration," *Cell Reports* 2, no. 5 (2012): 1151–58, https://doi.org/10.1016/j.celrep.2012.10.013.

57. Karen A. Weikel, Chung-Jung Chiu, and Allen Taylor, "Nutritional Modulation of Age-Related Macular Degeneration," *Molecular Aspects of Medicine* 33, no. 4 (2012): 318–75, https://doi.org/10.1016/j.mam.2012.03.005; Jingsheng Tuo et al., "A High Omega-3 Fatty Acid Diet Reduces Retinal Lesions in a Murine Model of Macular Degeneration," *The American Journal of Pathology* 175, no. 2 (2009): 799–807, https://doi.org/10.2353/ajpath.2009.090089.

58. J. A. Mares et al., "Healthy Lifestyles Related to Subsequent Prevalence of Age-Related Macular Degeneration," *Archives of Ophthalmology (Chicago, Ill.:1960)* 129, no. 4 (2011): 470–80, https://doi.org/10.1001/archophthalmol.2010.314.

59. P. Fies and Angelika Dienel, "Ginkgo Extract in Impaired Vision—Treatment with Special Extract EGb 761 of Impaired Vision Due to Dry Senile Macular Degeneration [Article in German]," *Wiener Medizinische Wochenschrift = Vienna Medical Weekly* 152, nos. 5-6 (2002): 423–26, https://doi.org/10.1046/j.1563-258x.2002.02066.x.

60. Fursova et al., "Dietary Supplementation with Bilberry Extract Prevents Macular Degeneration and Cataracts in Senesce-Accelerated OXYS Rats."

61. Luba Robman et al., "Dietary Lutein, Zeaxanthin, and Fats and the Progression of Age-Related Macular Degeneration," *Canadian Journal of Ophthalmology* 42, no. 5 (2007): 720–26, https://doi.org/10.3129/i07-116.

62. Shusheng Wang and Khrishen Cunnusamy, "Traditional Chinese Medicine (TCM) for the Treatment of Age-Related Macular Degeneration—Evaluation of WO2012079419," *Expert Opinion on Therapeutic Patents* 23, no. 2 (2013): 269–72, https://doi.org/10.1517/13543776.2013.751972.

63. José Faibes Lubianca Neto, Lucas Hemb, and Daniela Brunelli e Silva, "Systematic Literature Review of Modifiable Risk Factors for Recurrent Acute Otitis Media in Childhood," *Jornal de Pediatria = Journal of Pediatrics* 82, no. 2 (2006): 87–96, https://doi.org/10.2223/JPED.1453; Laura L. Jones et al., "Parental Smoking and the Risk of Middle Ear Disease in Children: A Systematic Review and Meta-Analysis," *Archives of Pediatrics and Adolescent Medicine* 166, no. 1 (2012): 18–27, https://doi.org/10.1001/archpediatrics.2011.158.

64. B. Duncan et al., "Exclusive Breast-Feeding for at Least 4 Months Protects against Otitis Media," *Pediatrics* 91, no. 5 (1993): 867–72.

65. Charles Ps Hui, Canadian Paediatric Society, and Infectious Diseases and Immunization Committee, "Acute Otitis Externa," *Pediatrics and Child Health* 18, no. 2 (2013): 96–101, https://doi.org/10.1093/pch/18.2.96.

66. N. Lilic, M. T. Mowjood, and M. H. W. Wong, "A Rare and Sinister Variant of a Common Ailment: Fungal Malignant Otitis Externa," *Journal of Surgical Case Reports* 2012, no. 9 (2012): 4, https://doi.org/10.1093/jscr/2012.9.4.

67. R. Gürkov et al., "What Is Ménière's Disease? A Contemporary Re-Evaluation of Endolymphatic Hydrops," *Journal of Neurology* 263, Suppl 1 (2016): S71–81, https://doi.org/10.1007/s00415-015-7930-1.

68. Giuseppe Chiarella, C. Petrolo, and E. Cassandro, "The Genetics of Ménière's Disease," *The Application of Clinical Genetics* 8 (2015): 9–17, https://doi.org/10.2147/TACG.S59024.

69. Andrew F. Long et al., "Exploring the Evidence Base for Acupuncture in the Treatment of Ménière's Syndrome—A Systematic Review" *Evidence-Based Complementary and Alternative Medicine: eCAM* 2011 (2011): 429102, https://doi.org/10.1093/ecam/nep047; Jiajun He et al.,

"Acupuncture Points Stimulation for Meniere's Disease/Syndrome: A Promising Therapeutic Approach," *Evidence-Based and Complementary Medicine: eCAM* 2016 (2016): 6404197, https://doi.org/10.1155/2016/6404197.

70. Andrew A. McCall et al., "Drug Delivery for Treatment of Inner Ear Disease: Current State of Knowledge," *Ear and Hearing* 31, no. 2 (2010): 156–65, https://doi.org/10.1097/AUD.0b013e3181c351f2.

71. Coran M. H. Watanabe et al., "The *In Vivo* Neuromodulatory Effects of the Herbal Medicine *Ginkgo biloba*," *Proceedings of the National Academy of Sciences of the United States of America* 98, no. 12 (2001): 6577–80, https://doi.org/10.1073/pnas.111126298.

72. Larysa Sokolova, Robert Hoerr, and Tamara Mischchenko, "Treatment of Vertigo: A Randomized, Double-Blind Trial Comparing Efficacy and Safety of *Ginkgo biloba* Extract EGb 761 and Betahistine," *International Journal of Otolaryngology* 2014 (2014): 682439, https://doi.org/10.1155/2014/682439.

73. Andreas G. Franke et al., "The Use of *Ginkgo biloba* in Healthy Elderly," *Age (Dordrecht, Netherlands)* 36, no. 1 (2014); 435–44, https://doi.org/10.1007/s11357-013-9550-y.

74. Sittah Czeche et al., "Dosage Strength Is Associated with Medication Persistence with *Ginkgo biloba* Drug Products: A Cohort Study of Ambulatory Drug Claims Data in Germany," *BMC Complementary and Alternative Medicine* 13 (2013): 278, https://doi.org/10.1186/1472-6882-13-278.

75. Larysa Sokolova, Robert Hoerr, and Tamara Mischchenko, "Treatment of Vertigo: A Randomized, Double-Blind Trial Comparing Efficacy and Safety of *Ginkgo biloba* Extract EGb 761 and Betahistine," *International Journal of Otolaryngology* 2014 (2014): 682439, https://doi.org/10.1155/2014/682439.

76. A. Kumar, R. M. Raizada, and V. N. Chaturvedi, "Role of *Ginkgo biloba* Extract in Acquired Sensorineural Hearing Loss," *Indian Journal of Otolaryngology and Head and Neck Surgery* 52, no. 3 (2000): 212–19.

77. M. A. Burschka et al., "Effect of Treatment with *Ginkgo biloba* Extract EGb 761 (Oral) on Unilateral Idiopathic Sudden Hearing Loss in a Prospective Randomized Double-Blind Study of 106 Outpatients," *European Archives of Oto-Rhino-Laryngology* 258, no. 5 (2001): 213–19, https://doi.org/10.1007/s004050100343.

78. C. H. Reisser and H. Weidauer, "*Ginkgo biloba* Extract EGb 761 or Petoxifylline for the Treatment of Sudden Deafness: A Randomized, Reference-Controlled, Double-Blind Study," *Acta Oto-Laryngologica* 121, no. 5 (2001): 579–84.

79. Berthold Langguth, Richard Salvi, and Ana Belén Elgoyhen, "Emerging Pharmacotherapy of Tinnitus," *Expert Opinion on Emerging Drugs* 14, no. 4 (2009): 687–702, https://doi.org/10.1517/14728210903206975.

80. Yan-Yan Chen et al., "Puerarin and Betahistine Treatment of Vertebrobasilar Ischemia Vertigo: A Meta-Analysis of Randomized Controlled Trials," *Experimental and Therapeutic Medicine* 11, no. 3 (2016): 1051–58, https://doi.org/10.3892/etm.2016.3004.

81. Bing Li et al., "Evaluating the Effects of Danhong Injection in Treatment of Acute Ischemic Stroke: Study Protocol for a Multicenter Randomized Controlled Trial," *Trials* 16 (2015): 561, https://doi.org/10.1186/s13063-015-1076-4.

82. J. F. Golding and J. R. Scott, "Comparison of the Effects of a Selective Muscarinic Receptor Antagonist and Hyoscine (Scopolamine) on Motion Sickness, Skin Conductance and Heart Rate," *British Journal of Clinical Pharmacology* 43, no. 6 (1997): 633–37, https://doi.org/10.1046/j.1365-2125.1997.00606.x.

83. Feng-Qian Li et al., "A Novel Spray Dried Nanoparticles-in-Microparticles System for Formulating Scopolamine Hydrobromide into Orally Disintegrating Tablets," *International Journal of Nanomedicine* 6 (2011): 897–904, https://doi.org/10.2147/IJN.S17900.

84. Alexander Choukèr et al., "Motion Sickness, Stress and the Endocannabinoid System," *PLoS One* 5, no. 5 (2010): e10752, https://doi.org/10.1371/journal.pone.0010752.

85. Keith A. Sharkey, Nissar A. Darmani, and Linda A. Parker, "Regulation of Nausea and Vomiting by Cannabinoids and the Endocannabinoid System," *European Journal of Pharmacology* 722 (2014): 134–46, https://doi.org/10.1016/j.ejphar.2013.09.068; Marta Duran et al., "Preliminary Efficacy and Safety of an Oromucosal Standardized Cannabis Extract in Chemotherapy-Induced Nausea and Vomiting," *British Journal of Clinical Pharmacology* 70, no. 5 (2010): 656–63, https://doi.org/10.1111/j.1365-2125.2010.03743.x.

86. Linda A. Parker, Erin M. Rock, and Cheryl L. Limebeer, "Regulation of Nausea and Vomiting by Cannabinoids," *British Journal of Pharmacology* 163, no. 7 (2011): 1411–22, https://doi.org/10.1111/j.1476-5381.2010.01176.x.

87. E. M. Rock et al., "Cannabidiol, a Non-Psychotropic Component of Cannabis, Attenuates Vomiting and Nausea-Like Behaviour via Indirect Agonism of 5-HT$_1$A Somatodendritic Autoreceptors in the Dorsal Raphe Nucleus," *British Journal of Pharamcology* 165, no. 8 (2012): 2620–34, https://doi.org/10.1111/j.1476-5381.2011.01621.x.

88. Chen et al., "Puerarin and Betahistine Treatment of Vertebrobasilar Ischemia Vertigo."

89. Behnaz Farahanikia et al., "Phytochemical Investigation of *Vinca minor* Cultivated in Iran," *Iranian Journal of Pharmaceutical Research* 10, no. 4 (2011): 777–85.

90. Lu Wang et al., "Phytochemical and Pharmacological Review of Da Chuanxiong Formula: A Famous Herb Pair Composed of Chuanxiong Rhizoma and Gastrodiae Rhizoma for Headache," *Evidence-Based Complementary and Alternative Medicine: eCAM* 2013 (2013): 425369, https://doi.org/10.1155/2013/425369.

91. Pragnadyuti Mandal et al., "The Efficacy of Ginger Added to Ondansetron for Preventing Postoperative Nausea and Vomiting in Ambulatory Surgery," *Pharmacognosy Research* 6, no. 1 (2014): 52–57, https://doi.org/10.4103/0974-8490.122918.

92. Wolfgang Marx et al., "Can Ginger Ameliorate Chemotherapy-Induced Nausea? Protocol of a Randomized Double Blind, Placebo-Controlled Trial," *BMC Complementary and Alternative Medicine* 14 (2014): 134, https://doi.org/10.1186/1472-6882-14-134; Julie L. Ryan et al., "Ginger (*Zingiber officinale*) Reduces Acute Chemotherapy-Induced Nausea: A URCC CCOP Study of 576 Patients," *Supportive Care in Cancer* 20, no. 7 (2012): 1479–89, https://doi.org/10.1007/s00520-011-1236-3.

93. Fanak Fahimi et al., "Evaluating the Effect of *Zingiber officinale* on Nausea and Vomiting in Patients Receiving Cisplatin Based Regimens," *Iranian Journal of Pharmaceutical Research* 10, no. 2 (2011): 379–84.

94. Iñaki Lete and José Allué, "The Effectiveness of Ginger in the Prevention of Nausea and Vomiting during Pregnancy and Chemotherapy," *Integrative Medicine Insights* 11 (2016): 11–17, https://doi.org/10.4137/IMI.S36273.

95. Abolfazl Aslani, Alireza Ghannadi, and Farnaz Rostami, "Design, Formulation, and Evaluation of Ginger Medicated Chewing Gum," *Advanced Biomedical Research* 5 (2016): 130, https://doi.org/10.4103/2277-9175.187011.

96. D. Passali et al., "Phytoneering: A New Way of Therapy for Rhinosinusitis," *Acta Otorhinolaryngologica Italica = Italian Society of Otorhinolaryngology* 35, no. 1 (2015): 1–8; Koosha Ghazi Moghadam et al., "Phytomedicine in Otorhinolaryngology and Pulmonology: Clinical Trials with Herbal Remedies," *Pharmaceuticals (Basel, Switzerland)* 5, no. 8 (2012): 853–74, https://doi.org/10.3390/ph5080853.

97. Nan Zhang et al., "A Herbal Composition of *Scutellaria baicalensis* and *Eleutherococcus senticosus* Shows Potent Anti-Inflammatory Effects in an *Ex Vivo* Human Mucosal Tissue Model," *Evidence-Based Complementary and Alternative Medicine: eCAM* 2012 (2012): 673145, https://doi.org/10.1155/2012/673145.

98. Neil Foden et al., "A Guide to the Management of Acute Rhinosinusitis in Primary Care Management Strategy Based on Best Evidence and Recent European Guidelines," *The British Journal of General Practice* 63, no. 616 (2013): 611–13, https://doi.org/10.3399/bjgp13X674620.

99. Supriya D. Hayer et al., "Effectiveness of Nasal Irrigation for Chronic Rhinosinusitis and Fatigue in Patients with Gulf War Illness: Protocol for a Randomized Controlled Trial," *Contemporary Clinical Trials* 41 (2015): 219–26, https://doi.org/10.1016/j.cct.2015.01.008; Rachel B. Cain and Devyani Lal, "Update on the Management of Chronic Rhinosinusitis," *Infection and Drug Resistance* 6 (2013): 1–14, https://doi.org/10.2147/IDR.S26134; David Rabago and Aleksandra Zgierska, "Saline Nasal Irrigation for Upper Respiratory Conditions," *American Family Physician* 80, no. 10 (2009): 1117–19.

100. Barira Islam, Shahper N. Khan, and Asad U. Khan, "Dental Caries: From Infection to Prevention," *Medical Science Monitor* 13, no. 11 (2007): RA196–203.

101. K. C. Godowski, "Antimicrobial Action of Sanguinarine," *The Journal of Clinical Dentistry* 1, no. 4 (1989): 96–101; D. S. Harper et al., "Effect of 6 Months Use of a Dentrifice and Oral Rinse Containing Sanguinaria Extract and Zinc Chloride Upon the Microflora of the Dental Plaque and Oral Soft Tissues," Journal of Periodontology 61, no. 6 (1990): 359–63, https://doi.org/10.1902/jop.1990.61.6.359.

102. K. Ikeno, T. Ikeno, and C. Miyazawa, "Effects of Propolis on Dental Caries in Rats," *Caries Research* 25, no. 5 (1991): 347–51, https://doi.org/10.1159/000261390.

103. X. C. Li, L. Cai, and C. D. Wu, "Antimicrobial Compounds from *Ceanothus americanus* against Oral Pathogens," *Phytochemistry* 46, no. 1 (1997): 97–102, https://doi.org/10.1016/s0031-9422(97)00222-7.

104. S. Shapiro and B. Guggenheim, "The Action of Thymol on Oral Bacteria," *Oral Microbiology and Immunology* 10, no. 4 (1995): 241–46, https://doi.org/10.1111/j.1399-302x.1995.tb00149.x.

105. R. Wong and B. Rabie, "Effect of Puerarin on Bone Formation," *Osteoarthritis and Cartilage* 15, no. 8 (2007): 894–99, https://doi.org/10.1016/j.joca.2007.02.009.

106. Enas Ahmed Elgendy, Shereen Abdel-Moula Ali, and Doaa Hussien Zineldeen, "Effect of Local Application of Tea Tree (*Melaleuca alternifolia*) Oil Gel on Long Pentraxin Level Used as Adjunctive Treatment of Chronic Periodontitis: A Randomized Controlled Clinical Study," *Journal of Indian Society of Periodontology* 17, no. 4 (2013): 444–48, https://doi.org/10.4103/0972-124X.118314.

107. Anirban Chatterjee et al., "Green Tea: A Boon for Periodontal and General Health," *Journal of Indian Society of Periodontology* 16, no. 2 (2012): 161–67, https://doi.org/10.4103/0972-124X.99256.

108. Abdulsalam Alshammari et al., "Kava-241 Reduced Periodontal Destruction in a Collagen Antibody Primed *Porphyromonas gingivalis* Model of Periodontitis," *Journal of Clinical Periodontology* 44, no. 11 (2017): 1123–32, https://doi.org/10.1111/jcpe.12784.

109. Sun Young Yin, Hyoung Jin Kim, and Hong-Jin Kim, "Protective Effect of Dietary Xylitol on Influenza A Virus Infection," *PLoS One* 9, no. 1 (2014): e84633, https://doi.org/10.1371/journal.pone.0084633.

110. B. E. Fustafsson et al., "The Vipeholm Dental Caries Study: The Effect of Different Levels of Carbohydrate Intake on Caries Activity in 436 Individuals Observed for Five Years," *Acta Odontologica Scandinavica* 11, nos. 3-4 (1954): 232–64, https://doi.org/10.3109/00016355308993925; D. Kandelman and G. Gagnon, "A 24-Month Clinical Study of the Incidence and Progression of Dental Caries in Relation to Consumption of Chewing Gum Containing Xylitol in School Preventive Programs," *Journal of Dental Research* 69, no. 11 (1990): 1771–75, https://doi.org/10.1177/00220345900690111201.

111. "Fluoride," USDA–Dr. Duke's Phytochemical and Ethnobotanical Databases, accessed January 2015, https://phytochem.nal.usda.gov/phytochem/chemicals/show/8512?qlookup=Momordica+charantia+fluoride&offset=0&max=20&et=C.

112. Michael P. Whyte et al., "Skeletal Fluorosis and Instant Tea," *The American Journal of Medicine* 118, no. 1 (2005): 78–82, https://doi.org/10.1016/j.amjmed.2004.07.046; Annekathrin Behrendt, Volker Oberste, and Willi Eckhard Wetzel, "Fluoride Concentration and pH of Iced Tea Products," *Caries Research* 36, no. 6 (2002): 405–10, https://doi.org/10.1159/000066532.

113. J. Moran, M. Addy, and S. Roberts, "A Comparison of Natural Product, Triclosan and Chlorhexidine Mouthrinses on 4-Day Plaque Regrowth," *Journal of Clinical Periodontology* 19, no. 8 (1992): 578–82, https://doi.org/10.1111/j.1600-051x.1992.tb00686.x; H. Tenenbaum, M. Dahan, and M. Soell, "Effectiveness of a Sanguinarine Regimen after Scaling and Root Planing," *Journal of Periodontology* 70, no. 3 (1999): 307–11, https://doi.org/10.1902/jop.1999.70.3.307; J. J. Hannah, J. D. Johnson, and M. M. Kuftinec, "Long-Term Clinical Evaluation of Toothpaste and Oral Rinse Containing Sanguinaria Extract in Controlling Plaque, Gingival Inflammation, and Sulcular Bleeding during Orthodontic Treatment," *American Journal of Orthodontics and Dentofacial Orthopedics* 96, no. 3 (1989): 199–207, https://doi.org/10.1016/0889-5406(89)90456-3; M. E. Mallatt et al., "Clinical Effect of a Sanguinaria Dentifrice on Plaque and Gingivitis in Adults," *Journal of Periodontology* 60, no. 2 (1989): 91–95, https://doi.org/10.1902/jop.1989.60.2.91.

114. Irene Paterniti et al., "Effects of *Hypericum perforatum* in a Rodent Model of Periodontitis," *BMC Complementary and Alternative Medicine* 10 (2010): 73, https://doi.org/10.1186/1472-6882-10-73.

115. Ahmet Altan et al., "The Effect of *Hypericum perforatum* on Wound Healing of Oral Mucosa in Diabetic Rats," *European Oral Research* 52, no. 3 (2018): 143–49, https://doi.org/10.26650/eor.2018.505.

116. S. Okabe et al., "Effects of Acetylsalicylic Acid (ASA), ASA Plus L-Glutamine and L-Glutamine on Healing of Chronic Gastric Ulcer in the Rat," *Digestion* 14, no. 1 (1976): 85–88, https://doi.org/10.1159/000197802.

117. K. Thongprasom, P. Youngnak, and V. Aneksuk, "Folate and Vitamin B12 Levels in Patients with Oral Lichen Planus, Stomatitis or Glossitis," *The Southeast Asian Journal of Tropical Medicine and Public Health* 32, no. 3 (2001): 643–47.

118. P. J. Drinka et al., "Nutritional Correlates of Atrophic Glossitis: Possible Role of Vitamin E in Papillary Atrophy," *Journal of the American College of Nutrition* 12, no. 1 (1993): 14–20, https://doi.org/10.1080/07315724.1993.10718276.

119. Robert B. Saper and Rebecca Rash, "Zinc: An Essential Micronutrient," *American Family Physician* 79, no. 9 (2009): 768–72.

120. T. Bøhmer and M. Mowé, "Tongue Atrophy—A Marker of Malnutrition [Article in Norwegian]," *Tidsskrift for den Norske laegeforening = Journal of the Norwegian Medical Association* 120, no. 8 (2000): 900–903.

121. T. Bøhmer and M. Mowé, "The Association between Atrophic Glossitis and Protein-Calorie Malnutrition in Old Age," *Age and Ageing* 29, no. 1 (2000): 47–50, https://doi.org/10.1093/ageing/29.1.47.

122. Grindwit Sastravaha et al., "Adjunctive Periodontal Treatment with *Centella asiatica* and *Punica granatum* Extracts in Supportive Periodontal Therapy," *Journal of the International Academy of Periodontology* 7, no. 3 (2005): 70–79.

123. Supanee Rassameemasmaung et al., "Effects of Herbal Mouthwash Containing the Pericarp Extract of *Garcinia mangostana* L on Halitosis, Plaque and Papillary Bleeding Index," *Journal of the International Academy of Periodontology* 9, no. 1 (2007): 19–25.

124. Moran et al., "A Comparison of Natural Product, Triclosan and Chlorhexidine Mouthrinses on 4-Day Plaque Regrowth"; H. Tenenbaum, M. Dahan, and M. Soell, "Effectiveness of a Sanguinarine Regimen after Scaling and Root Planing," *Journal of Periodontology* 70, no. 3 (1999): 307–11, https://doi.org/10.1902/jop.1999.70.3.307; R. A. Kopczyk et al., "Clinical and Microbiological Effects of a Sanguinaria-Containing Mouthrinse and Dentifrice with and without Fluoride during 6 Months of Use," *Journal of Periodontology* 62, no. 10 (1991): 617–22, https://doi.org/10.1902/jop.1991.62.10.617; M. E. Mallatt et al., "Clinical Effect of a Sanguinaria Dentifrice on Plaque and Gingivitis in Adults," *Journal of Periodontology* 60, no. 2 (1989): 91–95, https://doi.org/10.1902/jop.1989.60.2.91.

125. J. J. Hannah, J. D. Johnson, and M. M. Kuftinec, "Long-Term Clinical Evaluation of Toothpaste and Oral Rinse Containing Sanguinaria Extract in Controlling Plaque, Gingival Inflammation, and Sulcular Bleeding during Orthodontic Treatment," *American Journal of Orthodontics and Dentofacial Orthopedics* 96, no. 3 (1989): 199–207, https://doi.org/10.1016/0889-5406(89)90456-3.

126. K. C. Godowski, "Antimicrobial Action of Sanguinarine," *The Journal of Clinical Dentistry* 1, no. 4 (1989): 96–101.

127. D. Scott Harper et al, "Effect of 6 Months Use of a Dentifrice and Oral Rinse Containing Sanguinaria Extract and Zinc Chloride Upon the Microflora of the Dental Plaque and Oral Soft Tissues," *Journal of Peridontology* 61, no. 6 (June 1990): 359–63, https://doi.org/10.1902/jop.1990.61.6.359.

128. Andrew Croaker et al., "*Sanguinaria canadensis*: Traditional Medicine, Phytochemical Composition, Biological Activities and Current Uses," *International Journal of Molecular Sciences* 17, no. 9 (August 2016): 1414, https://doi.org/10.3390/ijms17091414.

129. L. R. Eversole, G. M. Eversole, and J. Kopcik, "Sanguinaria-Associated Oral Leukoplakia: Comparison with Other Benign and Dysplastic Leukoplakic Lesions," *Oral Surgery, Oral Medicine, Oral Pathology, Oral Radiology, and Endodontics* 89, no. 4 (2000): 455–65, https://doi.org/10.1016/s1079-2104(00)70125-9; Ana Karina Mascarenhas, Carl M. Allen, and Melvin L. Moeschberger, "The Association Between Viadent Use and Oral Leukoplakia—Results of a Matched Case-Control Study," *Journal of Public Health Dentistry* 62, no. 3 (2002): 158–62, https://doi.org/10.1111/j.1752-7325.2002.tb03437.x.

130. I. C. Munro et al., "Viadent Usage and Oral Leukoplakia: A Spurious Association," *Regulatory Toxicology and Pharmacology* 30, no. 3 (1999): 182–96, https://doi.org/10.1006/rtph.1999.1339.

131. Tatsuo Suzutani et al., "Anti-Herpesvirus Activity of an Extract of *Ribes nigrum* L," *Phytotherapy Research* 17, no. 6 (2003): 609–13, https://doi.org/10.1002/ptr.1207.

132. L. Zhang, Y. Liu, and Z. Yu, "Study of the Anti-Herpes Simplex Virus Activity of a Suppository or Ointment Form of *Astragalus membranaceus* Combined with Interferon Alpha 2b in Human Diploid Cell Culture [Article in Chinese]," *Zhonghua shi yah he lin chuang bing du xue za zhi = Zhonghua shiyan he linchuang bingduxue zazhi = Chinese Journal of Experimental and Clinical Virology* 12, no. 3 (1998): 269–71.

133. Cristina Fiore et al., "Antiviral Effects of *Glycyrrhiza* Species," *Phytotherapy Research* 22, no. 2 (2008): 141–48, https://doi.org/10.1002/ptr.2295; R. Pompei et al., "Glycyrrhizic Acid Inhibits Virus Growth and Inactivates Virus Particles," *Nature* 281, no. 5733 (1979): 689–90, https://doi.org/10.1038/281689a0; Jung Chung Lin, "Mechanism of Action of Glycyrrhizic Acid in Inhibition of Epstein-Barr Virus Replication *In Vitro*," *Antiviral Research* 59, no. 1 (2003): 41–47, https://doi.org/10.1016/s0166-3542(03)00030-5; T. Sekizawa, K. Yanagi, and Y. Itoyama, "Glycyrrhizin Increases Survival of Mice with Herpes Simplex Encephalitis," *Acta Virologica* 45, no. 1 (2001): 51–54.

134. Ch Dubois-Gosnet, "The Oral Mucosa and Delayed Hypersensitivity [Article in French]," *Allergie et Immunologie = Allergy and Immunology* 34, no. 1 (2002): 19–21.

135. Alexander M. Maley and Jack L. Arbiser, "Gentian Violet: A 19th Century Drug Re-Emerges in the 21st Century," *Experimental Dermatology* 22, no. 12 (2013): 775–80, https://doi.org/10.1111/exd.12257.

136. Pranab K. Mukherjee et al., "Topical Gentian Violet Compared to Nystatin Oral Suspension for the Treatment of Oropharyngeal Candidiasis in HIV-1 Infected Participants," *AIDS (London, England)* 31, no. 1 (2017): 81–88, https://doi.org/10.1097/QAD.0000000000001286.

137. Cibele Nasri-Heir et al., "Burning Mouth Syndrome: Current Concepts," *Journal of Indian Prosthodontic Society* 15, no. 4 (2015): 300–307, https://doi.org/10.4103/0972-4052.171823; Zuzanna Ślebioda and Elżbieta Szponar, "Burning Mouth Syndrome—A Common Dental Problem in Perimenopausal Women," Przeglad Menopauzalny = Menopause Review 13, no. 3 (2014): 198–202, https://doi.org/10.5114/pm.2014.43825.

138. K. A. Kamala et al., "Burning Mouth Syndrome," *Indian Journal of Palliative Care* 22, no. 1 (2016): 74–79, https://doi.org/10.4103/0973-1075.173942.

139. Isabel Adler et al, "*Helicobacter pylori* and Oral Pathology: Relationship with the Gastric Infection," *World Journal of Gastroenterology* 20, no. 29 (2014): 9922–35, https://doi.org/10.3748/wjg.v20.i29.9922.

140. Grigoriy E. Gurvits and Amy Tan, "Burning Mouth Syndrome," *World Journal of Gastroenterology* 19, no. 5 (2013): 665–72, https://doi.org/10.3748/wjg.v19.i5.665.

141. Parveen Dahiya et al., "Burning Mouth Syndrome and Menopause," *International Journal of Preventive Medicine* 4, no. 1 (2013): 15–20.

142. Vanessa Maria de Campos Rasteiro et al., "Essential Oil of *Melaleuca alternifolia* for the Treatment of Oral Candidiasis in an Immunosuppressed Mouse Model," *BMC Complementary and Alternative Medicine* 14 (2014): 489, https://doi.org/10.1186/1472-6882-14-489.

143. Walicyranison Plinio Silva-Rocha et al., "Effect of the Crude Extract of *Eugenia uniflora* in Morphogenesis and Secretion of Hydrolytic Enzymes in *Candida albicans* from the Oral Cavity of Kidney Transplant Recipients," *BMC Complementary and Alternative Medicine* 15, no. 6 (2015), https://doi.org/10.1186/s12906-015-0522-x.

144. P. Dolara et al., "Local Anesthetic, Antibacterial and Antifungal Properties of Sesquiterpenes from Myrrh," *Planta Medica* 66, no. 4 (2000): 356–58, https://doi.org/10.1055/s-2000-8532.

145. Dolara et al., "Local Anesthetic, Antibacterial and Antifungal Properties of Sesquiterpenes from Myrrh."

146. Jeffrey A Burgess et al., "Review of Over-the-Counter Treatments for Aphthous Ulceration and Results from Use of a Dissolving Oral Patch Containing Glycyrrhiza Complex Herbal Extract," *The Journal of Contemporary Dental Practice* 9, no. 3 (2008): 88–98.

147. Lúcia Helena Denardi Roveroni-Favaretto, Karina Bortolin Lodi, and Janete Dias Almeida, "Topical *Calendula officinalis* L. Successfully Treated Exfoliative Cheilitis: A Case Report," *Cases Journal* 2 (2009): 9077, https://doi.org/10.1186/1757-1626-2-9077.

148. Kui Young Park et al., "The Effect of Evening Primrose Oil for the Prevention of Xerotic Cheilitis in Acne Patients Being Treated with Isotretinoin: A Pilot Study," *Annals of Dermatology* 26, no. 6 (2014): 706–12, https://doi.org/10.5021/ad.2014.26.6.706.

149. Jukka H. Meurman et al., "Dental Infections and Serum Inflammatory Markers in Patients with and without Severe Heart Disease," *Oral Surgery, Oral Medicine, Oral Pathology, Oral Radiology, and Oral Endontics* 96, no. 6 (2003): 695–700, https://doi.org/10.1016/j.tripleo.2003.08.017; P. V. Ylöstalo et al., "Gingivitis, Dental Caries and Tooth Loss: Risk Factors for Cardiovascular Diseases or Indicators of Elevated Health Risks," *Journal of Clinical Periodontology* 33, no. 2 (2006): 92–101, https://doi.org/10.1111/j.1600-051X.2005.00875.x; G. J. Seymour et al., "Relationship between Periodontal Infections and Systemic Disease," *Clinical Microbiology and Infection* 13, suppl. 4 (2007): 3–10, https://doi.org/10.1111/j.1469-0691.2007.01798.x.

150. Maurizio Battino et al., "Antioxidant Status (CoQ10 and Vit. E Levels) and Immunohistochemical Analysis of Soft Tissues in Periodontal Diseases," *BioFactors (Oxford,*

England) 25, nos. 1–4 (2005): 213–17, https://doi.org/10.1002/biof.5520250126.

151. Sok-Ja Janket et al., "Asymptotic Dental Score and Prevalent Coronary Heart Disease," *Circulation* 109, no. 9 (2004): 1095–1100, https://doi.org/10.1161/01.CIR.0000118497.44961.1E.

152. Thomas Dietrich, Peter A. Reichart, and Christian Scheifele, "Clinical Risk Factors of Oral Leukoplakia in a Representative Sample of the US Population," *Oral Oncology* 40, no. 2 (2004): 158–63, https://doi.org/10.1016/s1368-8375(03)00145-3.

153. F. Cianfriglia et al., "Retinol Dietary Intake and Oral Leukoplakia Development," *Journal of Experimental & Clinical Cancer Research* 17, no. 3 (1998): 331–36.

154. H. S. Garewal et al., "Beta-Carotene Produces Sustained Remissions in Patients with Oral Leukoplakia: Results of a Multicenter Prospective Trial," *Archives of Otolaryngology—Head and Neck Surgery* 125, no. 12 (1999): 1305–310; Giovanni Lodi et al., "Systematic Review of Randomized Trials for the Treatment of Oral Leukoplakia," Journal of Dental Education 66, no. 8 (2002): 896–902.

155. P. C. Gupta et al., "Dietary Factors in Oral Leukoplakia and Submucous Fibrosis in a Population-Based Case Control Study in Gujarat, India," *Oral Diseases* 4, no. 3 (1998): 200–206, https://doi.org/10.1111/j.1601-0825.1998.tb00279.x.

156. C. O. Enwonwu and V. I. Meeks, "Bionutrition and Oral Cancer in Humans," *Critical Review in Oral Biology and Medicine* 6, no. 1 (1995): 5–17, https://doi.org/10.1177/10454411950060010401.

157. Amy Ming-Fang Yet, Shao-Ching Chen, and Tony Hsiu-His Chen, "Dose-Response Relationships of Oral Habits Associated with the Risk of Oral Pre-Malignant Lesions among Men Who Chew Betel Quid," *Oral Oncology* 43, no. 7 (2007): 634–38, https://doi.org/10.1016/j.oraloncology.2006.05.001.

158. D. M. Winn, "Diet and Nutrition in the Etiology of Oral Cancer," *The American Journal of Clinical Nutrition* 61, no. 2 (1995): 437S–45S, https://doi.org/10.1093/ajcn/61.2.437S.

159. N. Li et al., "The Chemopreventive Effects of Tea on Human Oral Precancerous Mucosa Lesions," *Proceedings of the Society for Experimental Biology and Medicine* 220, no. 4 (1999): 218–24, https://doi.org/10.1046/j.1525-1373.1999.d01-37.x.

160. Ajanta Halder et al., "Black Tea (*Camellia sinensis*) as a Chemopreventive Agent in Oral Precancerous Lesions," *Journal of Environmental Pathology, Toxicology and Oncology* 24, no. 2 (2005): 141–44, https://doi.org/10.1615/jenvpathtoxoncol.v24.i2.70; Yung-Chuan Ho et al., "Epigallocatechin-3-Gallate Inhibits the Invasion of Human Oral Cancer Cells and Decreases the Productions of Matrix Metalloproteinases and Urokinase-Plasminogen Activator," *Journal of Oral Pathology and Medicine* 36, no. 10 (2007): 588–93, https://doi.org/10.1111/j.1600-0714.2007.00588.x; Periasamy Srinivasan et al., "Chemopreventive and Therapeutic Modulation of Green Tea Polyphenols on Drug Metabolizing Enzymes in 4-Nitroquinoline 1-Oxide Induced Oral Cancer," *Chemico-Biological Interactions* 172, no. 3 (2008): 224–34, https://doi.org/10.1016/j.cbi.2008.01.010.

161. C. L. Cheng and M. W. Koo, "Effects of *Centella asiatica* on Ethanol Induced Gastric Mucosal Lesions in Rats," *Life Sciences* 67, no. 21 (2000): 2647–53, https://doi.org/10.1016/s0024-3205(00)00848-1; Chuen Lung Cheng et al., "The Healing Effects of *Centella* Extract and Asiaticoside on Acetic Acid Induced Gastric Ulcers in Rats," *Life Sciences* 74, no. 18 (2004): 2237–49, https://doi.org/10.1016/j.lfs.2003.09.055.

162. K. Sairam, C. V. Rao, and R. K. Goel, "Effect of *Centella asiatica* Linn on Physical and Chemical Factors Induced Gastric Ulceration and Secretion in Rats," *Indian Journal of Experimental Biology* 39, no. 2 (2001): 137–42.

163. X. B. Sun, T. Matsumoto, and H. Yamada, "Effects of a Polysaccharide Fraction from the Roots of *Bupleurum falcatum* L. on Experimental Gastric Ulcer Models in Rats and Mice," *The Journal of Pharmacy and Pharmacology* 43, no. 10 (1991): 699–704, https://doi.org/10.1111/j.2042-7158.1991.tb03461.x.

164. E. M. Bruzell, E. Morisbak, and H. H. Tønnesen, "Studies on Curcumin and Curcuminoids. XXIX. Photoinduced Cytotoxicity of Curcumin in Selected Aqueous Preparations," *Photochemical and Photobiological Sciences* 4, no. 7 (2005): 523–30, https://doi.org/10.1039/b503397g.

165. Colin F. Chignell et al., "Photochemistry and Photocytotoxicity of Alkaloids from *Goldenseal* (*Hydrastis canadensis* L.) 3: Effect on Human Lens and Retinal Pigment Epithelial Cells," *Photochemistry and Photobiology* 83, no. 4 (July/August 2007): 938–43, https://doi.org/10.1111/j.1751-1097.2007.00086.x.

166. Chaker El-Kalamouni et al., "Antioxidant and Antimicrobial Activities of the Essential Oil of *Achillea millefolium* L. Grown in France," *Medicines (Basel, Switzerland)* 4, no. 2 (2017): 30, https://doi.org/10.3390/medicines4020030.

167. H. Koushyar et al., "The Effect of Hydroethanol Extract of *Achillea millefolium* on β-Adrenoreceptors of Guinea Pig Tracheal Smooth Muscle," *Indian Journal of Pharmaceutical Sciences* 75, no. 4 (2013): 400–405.

168. Sedigheh Miranzadeh et al., "A New Mouthwash for Chemotherapy Induced Stomatitis," *Nursing and Midwifery Studies* 3, no. 3 (2014): e20249.

169. Hironori Tsuchiya, "Anesthetic Agents of Plant Origin: A Review of Phytochemicals with Anesthetic Activity," *Molecules (Basel, Switzerland)* 22, no. 8 (2017): 1369, https://doi.org/10.3390/molecules22081369.

170. Elnaz Khordad, Alireza Fazel, and Alireza Ebrahimzadeh Bideskan, "The Effect of Ascorbic Acid and Garlic Administration on Lead-Induced Apoptosis in Rat Offspring's Eye Retina," *Iranian Biomedical Journal* 17, no. 4 (2013): 203–13, https://doi.org/10.6091/ibj.1229.2013.

171. Ajda Ota and Nataša P. Ulrih, "An Overview of Herbal Products and Secondary Metabolites Used for Management

of Type Two Diabetes," *Frontiers in Pharmacology* 8 (2017): 436, https://doi.org/10.3389/fphar.2017.00436.

172. Minal Madhukar Kshirsagar et al., "Antibacterial Activity of Garlic Extract on Cariogenic Bacteria: An *In Vitro* Study," *Ayu* 39, no. 3 (2018): 165–68, https://doi.org/10.4103/ayu .AYU_193_16.

173. Lisha He et al., "Administration of Traditional Chinese Blood Circulation Activating Drugs for Microvascular Complications in Patients with Type 2 Diabetes Mellitus," *Journal of Diabetes Research* 2016 (2016): 1081657, https:// doi.org/10.1155/2016/1081657.

174. Rim Wehbe et al., "Bee Venom: Overview of Main Compounds and Bioactivities for Therapeutic Interests," *Molecules (Basel, Switzerland)* 24, no. 16 (2019): 2997, https://doi.org/10.3390/molecules24162997.

175. Mona G. Arafa et al., "Propolis-Based Niosomes as Oromuco-Adhesive Films: A Randomized Clinical Trial of a Therapeutic Drug Delivery Platform for the Treatment of Oral Recurrent Aphthous Ulcers," *Scientific Reports* 8, no. 1 (2018): 18056, https://doi.org/10.1038/s41598-018-37157-7.

176. Regina Bertóti et al., "Variability of Bioactive Glucosoinolates, Isothiocyanates and Enzyme Patterns in Horseradish Hairy Root Cultures Initiated from Different Organs," *Molecules (Basel, Switzerland)* 24, no. 15 (2019): 2828, https://doi.org/10.3390/molecules24152828.

177. Günther Bonifert et al., "Recombinant Horseradish Peroxidase Variants for Targeted Cancer Treatment," *Cancer Medicine* 5, no. 6 (20160: 1194–203, https://doi.org /10.1002/cam4.668.

178. Peyman Mikaili et al., "Therapeutic Uses and Pharmacological Properties of Garlic, Shallot, and Their Biologically Active Compounds," *Iranian Journal of Basic Medical Sciences* 16, no. 10 (2013): 1031–48.

179. T. Lakshmi et al., "*Azadirachta indica*: A Herbal Panacea in Dentistry—An Update," *Pharmacognosy Reviews* 9, no. 17 (2015): 41–44, https://doi.org/10.4103/0973-7847.156337.

180. Lanlan Bai et al., "Antimicrobial Activity of Tea Catechin against Canine Oral Bacterial and the Functional Mechanisms," *The Journal of Veterinary Medical Science* 78, no. 9 (2016): 1439–45, https://doi.org/10.1292/jvms.16-0198.

181. Chaitali U. Hambire, "Comparing the Antiplaque Efficacy of 0.5% *Camellia sinensis* Extract, 0.05% Sodium Fluoride, and 0.2% Chlorhexidine Gluconate Mouthwash in Children," *Journal of International Society of Preventive & Community Dentistry* 5, no. 3 (2015): 218–26, https://doi.org/10.4103 /2231-0762.158016.

182. Hannaneh Safiaghdam et al., "Medicinal Plants for Gingivitis: A Review of Clinical Trials," *Iranian Journal of Basic Medical Sciences* 21, no. 10 (2018): 978–91, https:// doi.org/10.22038/IJBMS.2018.31997.7690.

183. Latif Abdul et al., "Anti-Inflammatory and Antihistaminic Study of a Unani Eye Drop Formulation," *Ophthalmology and Eye Diseases* 2 (2010): 17–22, https://doi.org/10.4137 /oed.s3612.

184. Arkadiusz Dziedzic, Robert D. Wojtyczka, and Robert Kubina, "Inhibition of Oral Streptococci Growth Induced by the Complementary Action of Berberine Chloride and Antibacterial Compounds," *Molecules (Basel, Switzerland)* 20, no. 8 (2015): 13705–24, https://doi.org/10.3390 /molecules200813705.

185. Kanako Miyano et al., "The Japanese Herbal Medicine Hangeshashinto Enhances Oral Keratinocyte Migration to Facilitate Healing of Chemotherapy-Induced Oral Ulcerative Mucositis," *Scientific Reports* 10, no. 1 (2020): 625, https://doi.org/10.1038/s41598-019-57192-2.

186. Xiuhui Li et al., "Comparison between Chinese Herbal Medicines and Conventional Therapy in the Treatment of Severe Hand, Foot, and Mouth Disease: A Randomized Controlled Trial," *Evidence-Based Complementary and Alternative Medicine: eCAM* 2014 (2014): 140764, https:// doi.org/10.1155/2014/140764.

187. Xinyu Wu et al., "Whether Chinese Medicine Have Effect on Halitosis: A Systematic Review and Meta-Analysis," *Evidence-Based Complementary and Alternative Medicine: eCAM* 2018 (2018): 4347378, https://doi.org/10.1155 /2018/4347378.

188. Xiu-Fen Liu et al., "Curcumin, a Potential Therapeutic Candidate for Anterior Segment Eye Diseases: A Review" *Frontiers in Pharmacology* 8 (2017): 66, https://doi.org /10.3389/fphar.2017.00066; Dorota M. Radomska-Leśniewska et al., "Therapeutic Potential of Curcumin in Eye Diseases," *Central-European Journal of Immunology* 44, no. 2 (2019): 181–89, https://doi.org/10.5114/ceji.2019.87070.

189. Benhamin M. Davis et al., "Topical Curcumin Nanocarriers Are Neuroprotective in Eye Disease," *Scientific Reports* 8, no. 1 (2018): 11066, https://doi.org/10.1038/s41598-018-29393-8.

190. Alexander Panossian et al., "Adaptogens Stimulate Neuropeptide Y and Hsp72 Expression and Release in Neuroglia Cells," *Frontiers in Neuroscience* 6 (2012): 6, https://doi.org/10.3389/fnins.2012.00006.

191. Chinnoi Law and Christopher Exley, "New Insight into Silica Deposition in Horsetail (*Equisetum arvense*)," *BMC Plant Biology* 11, 112 (2011), https://doi.org/10.1186/1471 -2229-11-112.

192. Shahrukh Ali et al., "A Pharmacological Evidence for the Presence of Antihistaminic and Anticholinergic Activities in *Equisetum debile* Roxb," *Indian Journal of Pharmacology* 49, no. 1 (2017): 98–101.

193. Ying Liu et al., "Green Synthesis of Gold Nanoparticles Using *Euphrasia officinalis* Leaf Extract to Inhibit Lipopolysaccharide-Induced Inflammation through NF-κB and JAK/STAT Pathways in RAW 264.7 Macrophages," *International Journal of Nanomedicine* 14 (2019): 2945–59, https://doi.org/10.2147/IJN.S199781.

194. Roman Paduch et al., "Assessment of Eyebright (*Euphrasia officinalis* L.) Extract Activity in Relation to Human Corneal Cells Using *In Vitro* Tests," *Balkan Medical Journal* 31, no. 1 (2014): 29–36, https://doi.org/10.5152/balkanmedj.2014.8377.

195. Shamkant B. Badgujar, Vainav V. Patel, and Atmaram H. Bandivdekar, "*Foeniculum vulgare* Mill: A Review of Its Botany, Phytochemistry, Pharmacology, Contemporary Application, and Toxicology," *BioMed Research International* 2014 (2014): 842674, https://doi.org/10.1155/2014/842674.

196. Kornsit Wiwattanarattanabut, Suwan Choonharuangdejm, and Theerathavaj Srithavaj, "*In Vitro* Anticariogenic Plaque Effects of Essential Oils Extracted from Culinary Herbs," *Journal of Clinical and Diagnostic Research* 11, no. 9 (2017): DC30–35, https://doi.org/10.7860/JCDR/2017/28327.10668.

197. Sun, M. et al, "Efficacy and Safety of Ginkgo Biloba Pills for Coronary Heart Disease with Impaired Glucose Regulation: Study Protocol for a Series of *N*-of-1 Randomized, Double-Blind, Placebo-Controlled Trials," *Evidence-Based Complementary and Alternative Medicine* 2018 (May 2018): 7571629, https://doi.org/10.1155/2018/7571629.

198. A. K. Cybulska-Heinrich, M. Mozaffarieh, and J. Flammer, "*Ginkgo biloba*: An Adjuvant Therapy for Progressive Normal and High Tension Glaucoma," *Molecular Vision* 18 (2012): 390–402.

199. Jong Woon Park et al., "Short-Term Effects of *Ginkgo biloba* Extract on Peripapillary Retinal Blood Flow in Normal Tension Glaucoma," *Korean Journal of Ophthalmology* 25, no. 5 (2011): 323–28, https://doi.org/10.3341/kjo.2011.25.5.323.

200. Konstantin Tziridis et al., "Protective Effects of *Ginkgo biloba* Extract EGb 761 against Noise Trauma-Induced Hearing Loss and Tinnitus Development," *Neural Plasticity* 2014 (2014): 427298, https://doi.org/10.1155/2014/427298.

201. Govindsamy Vediyappan et al., "Gymnemic Acids Inhibit Hyphal Growth and Virulence in *Candida albicans*," *PLoS One* 8, no. 9 (2013): e74189, https://doi.org/10.1371/journal.pone.0074189.

202. Linda L. Theisen et al., "Tannins from *Hamamelis virginiana* Bark Extract: Characterization and Improvement of the Antiviral Efficacy against Influenza A Virus and Human Papillomavirus," *PLoS One* 9, no. 1 (2014): e88062, https://doi.org/10.1371/journal.pone.0088062.

203. Yong Wang et al., "Protective Effect of Proanthocyanidins from Sea Buckthorn (*Hippophae rhamnoides* L.) Seed against Visible Light-Induced Retinal Degeneration *In Vivo*," *Nutrients* 8, no. 5 (2016): 245, https://doi.org/10.3390/nu8050245.

204. Konstantinos Bouras et al., "Effects of Dietary Supplementation with Sea Buckthorn (*Hippophae rhamnoides* L.) Seed Oil on an Experimental Model of Hypertensive Retinopathy in Wistar Rats," *Biomedicine Hub* 2, no. 1 (2017): 1–12, https://doi.org/10.1159/000456704.

205. Ozan Kuduban et al., "The Effect of *Hippophae rhamnoides* Extract on Oral Mucositis Induced in Rats with Methotrexate," *Journal of Applied Oral Science* 24, no. 5 (2016): 423–30, https://doi.org/10.1590/1678-775720160139.

206. Gisele Façanha Diógenes Teixeira, Flávio Nogueira da Costa, and Adriana Rolim Camposa, "Corneal Antinociceptive Effect of (-)-α-Bisabolol," *Pharmaceutical Biology* 55, no. 1 (2017): 1089–92, https://doi.org/10.1080/13880209.2017.1285944.

207. Azar Aghamohamamdi and Seyed Jalal Hosseinimehr, "Natural Products for Management of Oral Mucositis Induced by Radiotherapy and Chemotherapy," *Integrative Cancer Therapies* 15, no. 1 (2016): 60–68, https://doi.org/10.1177/1534735415596570.

208. Vânia Thais Silva Gomes et al., "Effects of *Matricaria recutita* (L.) in the Treatment of Oral Mucositis," *The Scientific World Journal* 2018 (2018): 4392184, https://doi.org/10.1155/2018/4392184.

209. Bruna Vasconcelos Oliveira et al., "TNF-Alpha Expression, Evaluation of Collagen, and TUNEL of *Matricaria recutita* L. Extract and Triamcinolone on Oral Ulcer in Diabetic Rats," *Journal of Applied Oral Science* 24, no. 3 (2016): 278–90, https://doi.org/10.1590/1678-775720150481.

210. Congjun Jia et al., "Identification of Genetic Loci Associated with Crude Protein and Mineral Concentrations in Alfalfa (*Medicago sativa*) Using Association Mapping," *BMC Plant Biology* 17, no. 1 (2017): 97, https://doi.org/10.1186/s12870-017-1047-x.

211. Sepide Miraj, Rafieian-Kopaei, and Sara Kiani, "*Melissa officinalis* L.: A Review Study with an Antioxidant Prospective," *Journal of Evidence-Based Complementary and Alternative Medicine* 22, no. 3 (2017): 385–94, https://doi.org/10.1177/2156587216663433.

212. Chahrazed Benzaid et al., "The Effects of *Mentha x piperita* Essential Oil of *C. albicans* Growth, Transition, Biofilm Formation, and the Expression of Secreted Aspartyl Proteinases Genes," *Antibiotics (Basel, Switzerland)* 8, no. 1 (2019): 10, https://doi.org/10.3390/antibiotics8010010.

213. Shuo Jia et al., "Recent Advances in *Momordica charantia*: Functional Components and Biological Activities," *International Journal of Molecular Sciences* 18, no. 12 (2017): 2555, https://doi.org/10.3390/ijms18122555.

214. Bruno J. C. Silva et al., "Recent Breakthroughs in the Antioxidant and Anti-Inflammatory Effects of Morella and Myrica Species," *International Journal of Molecular Sciences* 16, no. 8 (2015): 17160–80, https://doi.org/10.3390/ijms160817160.

215. Raimundas Lelešius et al., "*In Vitro* Antiviral Activity of Fifteen Plant Extracts against Avian Infectious Bronchitis Virus," *BMC Veterinary Research* 15, no. 1 (2019): 178, https://doi.org/10.1186/s12917-019-1925-6.

216. Vinaya Bhat et al., "Characterization of Herbal Antifungal Agent, *Origanum vulgare* against Oral *Candida spp.* Isolated from Patients with Candida-Associated Denture Stomatitis: An *In vitro* Study," *Contemporary Clinical Dentistry* 9, Suppl 1. (2018): S3–10, https://doi.org/10.4103/ccd.ccd_537_17.

217. Adrian Man et al., "Antimicrobial Activity of Six Essential Oils Against a Group of Human Pathogens: A Comparative Study," *Pathogens (Basel, Switzerland)* 8, no. 1 (2019): 15, https://doi.org/10.3390/pathogens8010015.

218. Hyoung Won Bae et al., "Effect of Korean Red Ginseng Supplementation on Dry Eye Syndrome in Glaucoma—A Randomized, Double-Blind, Placebo-Controlled Study," *Journal of Ginseng Research* 39, no. 1 (2015): 7–13, https://doi.org/10.1016/j.jgr.2014.07.002.

219. Mengshuang Li et al., "Novel Ultra-Small Micelles Based on Ginsenoside Rb1: A Potential Nanoplatform for Ocular Drug Delivery," *Drug Delivery* 26, no. 1 (2019): 481–89, https://doi.org/10.1080/10717544.2019.1600077.

220. Hyeon-Joong Kim et al., "Gintoin, an Exogenous Ginseng-Derived LPA Receptor Ligand, Promotes Corneal Wound Healing," *Journal of Veterinary Science* 18, no. 3 (2017): 387–97, https://doi.org/10.4142/jvs.2017.18.3.387.

221. Tuan-Phat Huynh, Shivani N. Mann, and Nawajes A. Mandal, "Botanical Compounds: Effects on Major Eye Diseases," *Evidence-Based Complementary and Alternative Medicine: eCAM* 2013 (2013): 549174, https://doi.org/10.1155/2013/549174.

222. Leonard K. Seibold et al., "The Diurnal and Nocturnal Effects of Pilocarpine on Intraocular Pressure in Patients Receiving Prostaglandin Analog Monotherapy," *Journal of Ocular Pharmacology and Therapeutics* 34, no. 8 (2018): 590–95, https://doi.org/10.1089/jop.2018.0050.

223. Alexey N. Pronin, Qiang Wang, and Vladlen Z. Slepak, "Teaching an Old Drug New Tricks: Agonism, Antagonism, and Biased Signaling of Pilocarpine through M3 Muscarinic Acetylcholine Receptor," *Molecular Pharmacology* 92, no. 5 (2017): 601–12, https://doi.org/10.1124/mol.117.109678.

224. Guorong Li et al., "Pilocarpine-Induced Dilation of Schlemm's Canal and Prevention of Lumen Collapse at Elevated Intraocular Pressures in Living Mice Visualized by OCT," *Investigative Ophthalmology and Visual Science* 55, no. 6 (2014): 3737–46, https://doi.org/10.1167/iovs.13-13700.

225. Alicja Ponder and Ewelina Hallmann, "Phenolics and Carotenoid Contents in the Leaves of Different Organic and Conventional Raspberry (*Rubus idaeus* L.) Cultivars and Their *In Vitro* Activity," *Antioxidants (Basel, Switzerland)* 8, no. 10 (2019): 458, https://doi.org/10.3390/antiox8100458.

226. Ahmad Ghorbani and Mahdi Esmaeilizadeh, "Pharmacological Properties of *Salvia officinalis* and Its Components," *Journal of Traditional and Complementary Medicine* 7, no. 4 (2017): 433–40, https://doi.org/10.1016/j.jtcme.2016.12.014.

227. Agnieszka Szopa, Radosław Ekiert, and Halina Ekiert, "Current Knowledge of *Schisandra chinensis* (Turcz.) Baill. (Chinese Magnolia Vine) as a Medicinal Plant Species: A Review on the Bioactive Components, Pharmacological Properties, Analytical and Biotechnological Studies," *Phytochemistry Reviews* 16, no. 2 (2017): 195–218, https://doi.org/10.1007/s11101-016-9470-4.

228. Suchita Dubey et al., "Phytochemistry, Pharmacology and Toxicology of *Spilanthes acmella*: A Review," *Advances in Pharmacological Sciences* 2013 (2013): 423750, https://doi.org/10.1155/2013/423750.

229. Jayaraj Paulraj, Raghavan Govindarajan, and Pushpangadan Palpu, "The Genus Spilanthes Ethnopharmacology, Phytochemistry, and Pharmacological Properties: A Review," *Advances in Pharmacological Sciences* 2013 (2013): 510298, https://doi.org/10.1155/2013/510298.

230. Yousef A. Taher et al., "Experimental Evaluation of Anti-Inflammatory, Antinociceptive and Antipyretic Activities of Clove Oil in Mice," *The Libyan Journal of Medicine* 10 (2015): 28685, https://doi.org/10.3402/ljm.v10.28685.

231. Neila C. de Sousa et al., "Modulatory Effects of *Tabebuia impetiginosa* (Lamiales, Bignoniaceae) on Doxorubicin-Induced Somatic Mutation and Recombination in *Drosophila melanogaster*," *Genetics and Molecular Biology* 32, no. 2 (2009): 382–88, https://doi.org/10.1590/S1415-47572009005000042.

232. Serena Materazzi et al., "Parthenolide Inhibits Nociception and Neurogenic Vasodilation in the Trigeminovascular System by Targeting the TRPA1 Channel," *Pain* 154, no. 12 (2013): 2750–58, https://doi.org/10.1016/j.pain.2013.08.002.

233. Han-Sung Lee et al., "*Ulmus davidiana* var. *japonica* Nakai Upregulates Eosinophils and Suppresses Th1 and Th17 Cells in the Small Intestine," *PLoS One* 8, no. 10 (2013): e76716, https://doi.org/10.1371/journal.pone.0076716.

234. Sabrina Esposito et al., "Therapeutic Perspectives of Molecules from *Urtica dioica* Extracts for Cancer Treatment," *Molecules (Basel, Switzerland)* 24, no. 15 (2019): 2753, https://doi.org/10.3390/molecules24152753.

235. Yong Wang et al., "Retinoprotective Effects of Bilberry Anthocyanins via Antioxidant, Anti-Inflammatory, and Anti-Apoptotic Mechanisms in a Visible Light-Induced Retinal Degeneration Model in Pigmented Rabbits," *Molecules (Basel, Switzerland)* 20, no. 12 (2015): 22395–410, https://doi.org/10.3390/molecules201219785.

236. Mehrnaz Nikkhah Bodagh, Iradj Maleki, and Azita Hekmatdoost, "Ginger in Gastrointestinal Disorders: A Systematic Review of Clinical Trials," *Food Science and Nutrition* 7, no. 1 (2018): 96–108, https://doi.org/10.1002/fsn3.807.

— INDEX —

Page numbers in *italics* refer to figures and illustrations. Page numbers followed by *t* refer to tables.

— ABOUT THE AUTHOR —

Shelly Fry of Battle Ground

Dr. Jill Stansbury is a naturopathic physician with over 30 years of clinical experience. She served as the chair of the Botanical Medicine Department of the National University of Natural Medicine in Portland, Oregon, for more than 20 years. She remains on the faculty, teaching herbal medicine and medicinal plant chemistry and leading ethnobotany field courses in the Amazon. Dr. Stansbury presents numerous original research papers each year and writes for health magazines and professional journals. She serves on scientific advisory boards for several medical organizations. She is the author of *Herbal Formularies for Health Professionals*, Volumes 1, 2, 3, and 4, and is the author of *Herbs for Health and Healing* and coauthor of *The PCOS Health and Nutrition Guide*. Dr. Stansbury lives in Battle Ground, Washington, and is the medical director of Battle Ground Healing Arts. She also runs the Healing Arts Apothecary, offering the best quality medicines from around the world, featuring many of her own custom tea formulas, blends, powders, and medicinal foods.

www.healingartsapothecary.org

HERBAL FORMULARIES FOR HEALTH PROFESSIONALS

This comprehensive five-volume set by Dr. Jill Stansbury serves as a practical and necessary reference manual for herbalists, physicians, nurses, and allied health professionals everywhere. This set is organized by body system, and each volume includes hundreds of formulas to treat common health conditions, as well as formulas that address specific energetic or symptomatic presentations.

VOLUME 1

DIGESTION AND ELIMINATION

INCLUDING THE GASTROINTESTINAL SYSTEM, LIVER AND GALLBLADDER, URINARY SYSTEM, AND THE SKIN

9781603587075
Hardcover

VOLUME 2

CIRCULATION AND RESPIRATION

INCLUDING THE CARDIOVASCULAR, PERIPHERAL VASCULAR, PULMONARY, AND RESPIRATORY SYSTEMS

9781603587983
Hardcover

VOLUME 3

ENDOCRINOLOGY

INCLUDING THE ADRENAL AND THYROID SYSTEMS, METABOLIC ENDOCRINOLOGY, AND THE REPRODUCTIVE SYSTEMS

9781603588553
Hardcover

VOLUME 4

NEUROLOGY, PSYCHIATRY, AND PAIN MANAGEMENT

INCLUDING COGNITIVE AND NEUROLOGIC CONDITIONS AND EMOTIONAL CONDITIONS

9781603588560
Hardcover

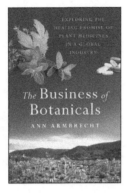
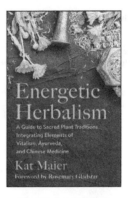
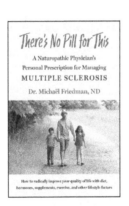